Mycoheterotrophy

Vincent S.F.T. Merckx
Editor

Mycoheterotrophy

The Biology of Plants Living on Fungi

Springer

Editor
Vincent S.F.T. Merckx
Naturalis Biodiversity Center
Leiden University
Leiden, The Netherlands

ISBN 978-1-4614-5208-9 ISBN 978-1-4614-5209-6 (eBook)
DOI 10.1007/978-1-4614-5209-6
Springer New York Heidelberg Dordrecht London

Library of Congress Control Number: 2012951288

Printed on acid-free paper

Springer is part of Springer Science+Business Media (www.springer.com)

Preface

Over 450 million years ago plants started to colonize land and "greened" the continents. The appearance of land plants is a key event in the history of life and shaped the evolution of all terrestrial ecosystems on our planet. Remarkably, an alliance with fungi was probably essential for the plants' conquest of land. Today, this alliance between plants and soil fungi, known as the "mycorrhizal symbiosis," remains of vital importance: the vast majority of plants are dependent on mycorrhizal fungi for their uptake of minerals and water from the soil. Mycorrhizal fungi are not drawn into this interaction by philanthropy: they obtain essential photosynthetically fixed carbohydrates from their plant partners in reward for their efforts. However, over the course of evolution several plant lineages have found ways to subvert this *quid pro quo* interaction, and are able to obtain water, minerals, *and* carbohydrates from fungi. Some plants are able to exploit fungi to such an extent that they lost the need for photosynthesis. The ability of a plant to live on fungal carbon is known as mycoheterotrophy. This intriguing process has fascinated botanists for centuries, yet many aspects of mycoheterotrophy have remained elusive for a long time. And despite recent advances in our understanding of the process mycoheterotrophy and its protagonists, this volume also illustrates that there is still much to learn.

The idea for a book providing an overview of the biology of mycoheterotrophs sprouts from my passion for these plants and the lack of recent volumes offering general insights into their fascinating ecology, diversity, and evolution. Rather than assembling a volume consisting of research papers, I aimed to provide the reader with a thematic overview of different aspects of mycoheterotrophy. However, a multiauthor book like this will never be fully coherent, and inevitably there is a certain overlap between the contents of the chapters. I am also aware that there are gaps, and thus some topics did not receive the coverage they deserve. Despite these shortcomings, I hope that this book proves a valuable tool for everyone who is interested in the process of mycoheterotrophy, and offers strong stimuli for further research on these intriguing plants.

This book owes its existence to Sean Graham. Sean approached me after a talk I gave at the 2008 Monocots meeting in Copenhagen. I had concluded my talk with the suggestion that the time was ripe for a book on mycoheterotrophy, and Sean enthusiastically linked me with Springer. The project was further outlined in a meeting in Leuven at which many of the contributing authors were present. I was able to organize this meeting with the support provided by Erik Smets and Suzy Huysmans. Without their involvement, the book would never have been put together. Editing this book has been a definite challenge and I could not have completed it without the help and dedication of all the authors, who also displayed considerable patience as the book slowly came together. I would like to thank them for their excellent work. I also deeply acknowledge the help of the many colleagues who provided direct or indirect support in the development of manuscript, including Tom Bruns, Anthony Amend, Steven Janssens, Benny Lemaire, Martin Brazeau, Rudi Smith, Joep Moonen, Mark Wapstra, and Sainge Moses. I also thank the people at Springer for their continuous support and advice. Over the years, my research on mycoheterotrophic plants has been possible with financial support from the Belgian and Dutch National Science Foundations (FWO Vlaanderen and NWO), the agency for Innovation by Science and Technology (IWT Vlaanderen), and the Belgian American Educational Foundation (BAEF), for which I am grateful. Finally, I also take this opportunity to thank my family, colleagues, and friends who have encouraged, motivated, and inspired me.

Leiden, The Netherlands Vincent S.F.T. Merckx

Contents

Contributors

Iris K.U. Adam Laboratory of Isotope Biogeochemistry, Bayreuth Center of Ecology and Environmental Research (BayCEER), University of Bayreuth, Bayreuth, Germany

Craig F. Barrett Department of Plant Biology, Cornell University, Ithaca, NY, USA; Department of Biological Sciences, California State University, Los Angeles, CA, USA

Gemma E. Beatty School of Biological Sciences, Queen's University Belfast, Belfast, UK

Martin I. Bidartondo Imperial College London and Royal Botanic Gardens, Kew, UK

Maarten J.M. Christenhusz Botanical Garden and Herbarium, Finnish Museum of Natural History, Helsinki University, Helsinki, Finland

Barbara Crandall-Stotler Department of Plant Biology, Southern Illinois University, Carbondale, IL, USA

John V. Freudenstein Department of Evolution, Ecology, and Organismal Biology, The Ohio State University Herbarium, Columbus, OH, USA

Gerhard Gebauer Laboratory of Isotope Biogeochemistry, Bayreuth Center of Ecology and Environmental Research (BayCEER), University of Bayreuth, Bayreuth, Germany

József Geml Naturalis Biodiversity Center, Leiden University, Leiden, The Netherlands

Heiko Hentrich Ecosystem Analysis, Institute for Environmental Sciences, University of Landau, Landau, Germany

Sarah E. Hopkins Institute of Arctic Biology, University of Alaska Fairbanks, Fairbanks, AK, USA

Nicole A. Hynson Department of Botany, University of Hawaii at Manoa, Honolulu, HI, USA

Stephan Imhof Spezielle Botanik und Mykologie, Fachbereich Biologie, Philipps-Universität, Marburg, Germany

Aaron H. Kennedy National Identification Services, USDA-APHIS-PPQ-PHP, Beltsville, MD, USA

Jonathan Kissling Institute of Biology, Evolutionary Botany, University of Neuchâtel, Neuchâtel, Switzerland

Matthew R. Klooster Centre College, Danville, KY, USA

Paul J.M. Maas Naturalis Biodiversity Center, Wageningen University, Wageningen, The Netherlands

Hiltje Maas-van de Kamer Naturalis Biodiversity Center, Wageningen University, Wageningen, The Netherlands

Thomas P. Madsen Department of Integrative Biology, University of California, Berkeley, Berkeley, CA, USA

Hugues B. Massicotte Ecosystem Science and Management Program, University of Northern British Columbia, British Columbia, Canada

Lewis H. Melville Department of Molecular and Cellular Biology, University of Guelph, Guelph, ON, Canada

Constantijn B. Mennes Naturalis Biodiversity Center, Leiden University, Leiden, The Netherlands

Vincent S.F.T. Merckx Naturalis Biodiversity Center, Leiden University, Leiden, The Netherlands

Yuki Ogura-Tsujita Department of Chemical and Biological Sciences, Japan Women's University, Tokyo, Japan

Kabir G. Peay Department of Biology, Stanford University, Stanford, CA, USA

R. Larry Peterson Department of Molecular and Cellular Biology, University of Guelph, Guelph, ON, Canada

Melanie Roy Centre d'Ecologie Fonctionnelle et Evolutive (CEFE-CNRS), Montpellier, France

Laboratoire Evolution et Diversite Biologique (UMR5174 EDB), Universite de Toulouse, Toulouse, France

Paula J. Rudall Royal Botanic Gardens, Kew, Richmond, Surrey, UK

Marc-André Selosse Centre d'Ecologie Fonctionnelle et Evolutive (CNRS, UMR 5175), Montpellier, France

Erik F. Smets Naturalis Biodiversity Center, Leiden University, Leiden, The Netherlands

Laboratory of Plant Systematics, KU Leuven, Leuven, Belgium

Chelsea D. Specht Department of Plant and Microbial Biology, University of California Berkeley, Berkeley, CA, USA

Raymond E. Stotler Department of Plant Biology, Southern Illinois University, Carbondale, IL, USA

D. Lee Taylor Institute of Arctic Biology, University of Alaska, Fairbanks, AK, USA

Richard J. Waterman Department of Animal and Plant Sciences, University of Sheffield, Western Bank, Sheffield, UK

Norman Wickett Department of Biology, The Pennsylvania State University, University Park, PA, USA

Mycohetrotrophy: An Introduction

1

Vincent S.F.T. Merckx

The snow plant (*Sarcodes sanguinea*) is more admired by tourists than any other in California. It is red, fleshy and watery and looks like a gigantic asparagus shoot. Soon after the snow is off the round it rises through the dead needles and humus in the pine and fir woods like a bright glowing pillar of fire. In a week or so it grows to a height of eight or twelve inches with a diameter of an inch and a half or two inches; then its long fringed bracts curl aside, allowing the twenty- or twenty-five-lobed, bell-shaped flowers to open and look straight out from the axis. It is said to grow up through the snow; on the contrary, it always waits until the ground is warm, though with other early flowers it is occasionally buried or half-buried for a day or two by spring storms. The entire plant—flowers, bracts, stem, scales, and roots—is fiery red. Its color could appeal to one's blood. Nevertheless, it is a singularly cold and unsympathetic plant. Everybody admires it as a wonderful curiosity, but nobody loves it as lilies, violets, roses, daisies are loved. Without fragrance, it stands beneath the pines and firs lonely and silent, as if unacquainted with any other plant in the world; never moving in the wildest storms; rigid as if lifeless, though covered with beautiful rosy flowers.

John Muir (1912)

V.S.F.T. Merckx (✉)
Naturalis Biodiversity Center, Leiden University,
P.O. Box 9514, 2300 RA Leiden, The Netherlands
e-mail: vincent.merckx@naturalis.nl

1.1 Motivation

"Mycoheterotrophy" describes a plant's ability to obtain carbon from fungi. Many plants are capable of mycoheterotrophy; some closely related others are not. Plants with the ability for mycoheterotrophy ("mycoheterotrophic plants"), particularly those that completely depend on fungal carbon during their entire life cycle ("fully mycoheterotrophic plants" see definitions below), have attracted the attention of biologists for centuries. Studies on their unconventional mode of life have led to novel perspectives in ecology and evolution. Since the term mycoheterotrophy (as "myco-heterotrophy") was coined by Jonathan Leake in 1994, scientific research on the topic has increased considerably (Fig. 1.1). Leake's groundbreaking review on mycoheterotrophic plants unfortunately preceded several technological advances that revolutionized the field, and only few review papers have appeared since. Most notable is the *New Phytologist* Tansley review by Martin Bidartondo (2005) and several relevant chapters in *Mycorrhizal Symbiosis* by Sally Smith and David Read (2008). However, these works mainly focus on the ecological and physiological aspects of mycoheterotrophy, and to date, a single volume has not been dedicated to all aspects of mycoheterotrophy. I hope that this book's multidisciplinary approach in discussing mycoheterotrophy will appeal to scientists and students who wish to understand the biology of mycoheterotrophic plants and that this overview will contribute to an enduring interest in these extraordinary plants.

V.S.F.T. Merckx (ed.), *Mycoheterotrophy: The Biology of Plants Living on Fungi*,
DOI 10.1007/978-1-4614-5209-6_1, © Springer Science+Business Media New York 2013

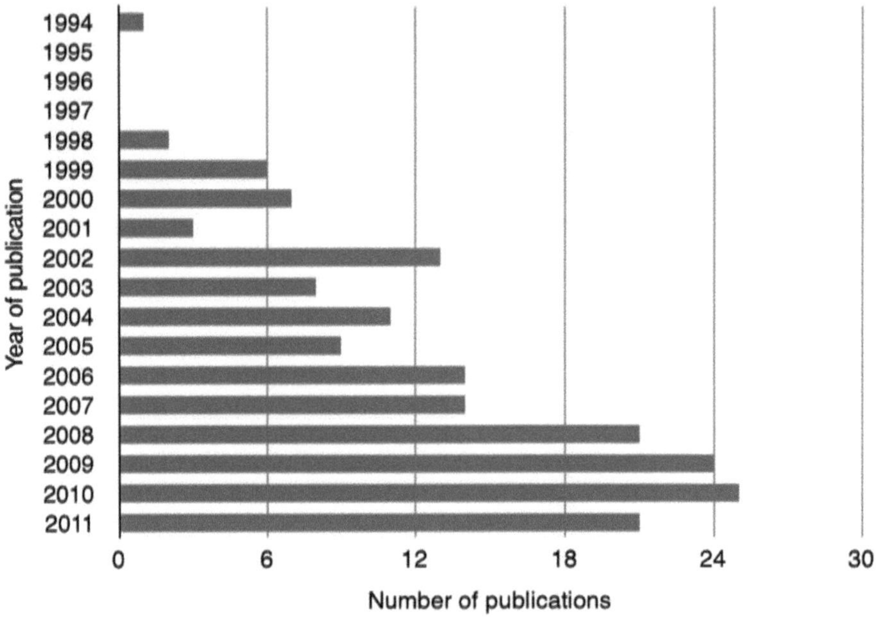

Fig. 1.1 Number of papers published each year since 1994 that are labeled with the topic "myco-hetero*" or "mycohetero*" in Thomson Reuters' Web of Science (as of January 2012)

1.2 Mycorrhizal Symbiosis

1.2.1 Mycorrhizas

Mycoheterotrophy cannot be discussed or appreciated without a basic understanding of the mycorrhizal symbiosis. The term mycorrhiza is derived from the Greek words for "fungus" and "root" and implies the association of specialized soil fungi ("mycorrhizal fungi") with plant roots. In general, this association is mutualistic because both partners benefit: mycorrhizal fungi improve the nutrient uptake of their host plants, and in return they receive photosynthetically fixed carbon that is essential for growth and reproduction of the fungi. The mycorrhizal association is probably the most important symbiosis in nature and plays an essential role in the maintenance of most terrestrial ecosystems such as grasslands and forests. Over 90% of all plant species form mycorrhizas, including most crops. Despite the literal meaning "fungus roots," mycorrhizas also occur in "primitive" plants that do not have true roots such as liverworts and hornworts. Moreover, fossil evidence indicates that the earliest land plants were also associated with hyphal fungi, and it is now generally accepted that the colonization of land by plants relied on a symbiotic association with fungi.

The main diagnostic criteria for the types of mycorrhizas formed in nature are the identity of the fungi engaged in the symbiosis and the morphology at the symbiotic interface of plant and fungus. The two dominant types of mycorrhizas are the arbuscular mycorrhizas (AM) that involve nearly all members of the Glomeromycota and a wide variety of host plants, and the ectomycorrhizas (EM) that involve some members of the Basidiomycota and Ascomycota and several woody plants.

For a comprehensive synthesis on the subject of mycorrhizas, I strongly recommend the book *Mycorrhizal Symbiosis* by Sally Smith and David Read (2008). Here I provide a basic overview of a few aspects of the mycorrhizal symbiosis that are essential in the context of mycoheterotrophy.

1.2.2 Arbuscular Mycorrhizas

Arbuscular mycorrhizas are the most common mycorrhizal type. They are formed by fungi of the phylum Glomeromycota and the majority of land plant species. The arbuscular mycorrhizal symbiosis is extremely ancient, and AM fungi were probably important in the colonization of land by plants, although some evidence suggests that some members of the Mucoromycotina predate the Glomeromycota as the earliest mycorrhizal fungi (Bidartondo et al. 2011). The name "arbuscular" is derived from the characteristic treelike structures, the arbuscules, which occur within the cortical cells of many plant roots and also some mycothalli colonized by AM fungi (Smith and Read 2008). Arbuscular mycorrhizal fungi are obligate symbionts: they cannot survive as free-living organisms. They are also puzzling organisms for taxonomists because they are asexual, multinucleate, and difficult to cultivate, and therefore, species diagnosis and identification remains under debate. However, under the current morphological and molecular concepts, there are no more than 300 species of AM fungi known (Öpik et al. 2010; Krüger et al. 2011). This number strongly contrasts with the number of potential host plant species: perhaps 80–90% of all land plant species are able to form arbuscular mycorrhizas. This illustrates the potential for promiscuity in the arbuscular mycorrhizal symbiosis. Indeed, studies show that an arbuscular mycorrhizal plant typically associates simultaneously with multiple AM fungi, and an AM fungus often associates simultaneously with multiple plants (Giovannetti et al. 2004). This allows for the formation of mycorrhizal networks, linking plants of the same or different species by a shared mycorrhizal fungus. These arbuscular mycorrhizal networks are essential for the existence of mycoheterotrophy because they allow for physiological continuity between an autotrophic plant, its arbuscular mycorrhizal fungus, and a mycoheterotroph (Chap. 5; Fig. 1.2). Mycoheterotrophic interactions through AM fungi occur in lycophytes, ferns, and angiosperms. Over 230 species of fully mycoheterotrophic angiosperms are dependent on AM fungi.

1.2.3 Ectomycorrhizas

Trees in the families Pinaceae, Fagaceae, Dipterocarpaceae, Myrtaceae, and Fabaceae found in both temperate and tropical forests associate with hundreds of ectomycorrhizal fungus species of Basidiomycota and Ascomycota (Bonfante and Genre 2010). Therefore, EM fungi are essential components of the world's forests. The ectomycorrhizal habit has evolved multiple times independently in the evolution of land plants (Bruns and Shefferson 2004). And while the EM symbiosis is considerably younger than the AM symbiosis, fossil, biogeographical, and molecular clock data suggests that EM associations have a long history (Le Page et al. 1997; Moyersoen 2006; Hibbett and Matheny 2009). Over 7,000 species of fungi are known to form ectomycorrhizas, most of them belong to the Basidiomycota, but the actual EM fungi diversity may be considerably higher (Smith and Read 2008; Tedersoo et al. 2010). An EM root is characterized by the presence of three structural components: a mantle of fungal tissue which encloses the root, a labyrinthine inward growth of fungal hyphae between the epidermal and cortical cells (Hartig net), and an outwardly growing system of hyphal elements which extends into the soil (extraradical of external mycelium) (Smith and Read 2008). There may be considerable variation in morphology and development of these structural elements. In general, the EM symbiosis is considered to show a low level of specificity, and it has been demonstrated that an individual tree may have 15 or more different fungal EM partners simultaneously (Saari et al. 2005). Analogous with the AM symbiosis, this allows for the formation of common mycorrhizal networks, linking plants by a shared ectomycorrhizal fungus. Besides ectomycorrhizas, Ascomycota and Basidiomycota are known to form other types of mycorrhizal structures, often categorized as ectendo-, ericoid, arbutoid, and orchid mycorrhizas. However, the type of mycorrhiza formed can be influenced by the identity of both plant and fungus: it is known that the same fungus can form different types of mycorrhiza depending on

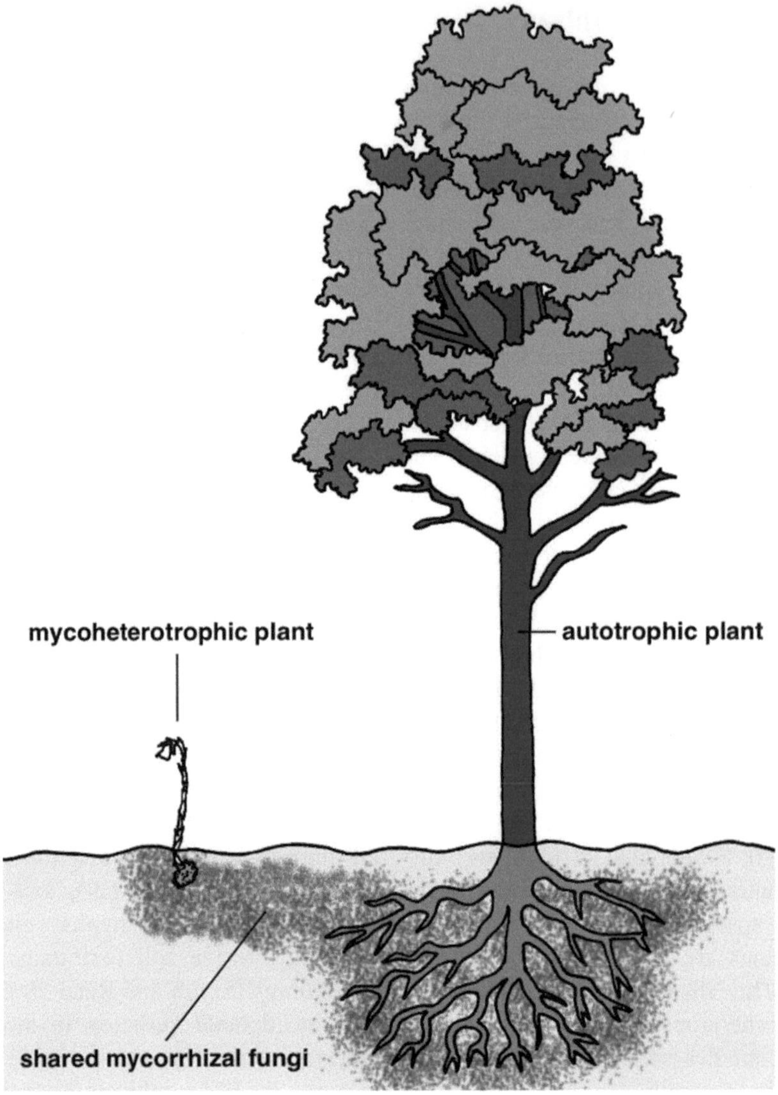

Fig. 1.2 Many fully mycoheterotrophic plants obtain carbon from surrounding autotrophic plants through shared mycorrhizal fungi

the identity of the plant associate. The mycorrhizas involved in the mycoheterotrophic interactions of Monotropoideae, Pyroleae (both Ericaceae), and Orchidaceae have been classified as monotropoid, arbutoid, and orchid mycorrhizas, respectively. Yet, with the exception of mycorrhizal associations between orchids and saprotrophic fungi (SAP), it has been demonstrated that in all of these mycoheterotrophic interactions, the fungi involved simultaneously form ectomycorrhizas with surrounding green plants (to establish the tripartite interaction necessary to sustain the mycoheterotrophic interaction). Therefore, here we group all these interactions as ectomycorrhizas. Mycoheterotrophic interactions through EM fungi occur in liverworts and angiosperms. One liverwort species and at least 48 species of angiosperms (all in Orchidaceae and Ericaceae) are fully dependent on EM fungi for their entire life cycle, although in some cases EM and SAP fungi co-colonize the roots of fully mycoheterotrophic plants (Table 7.1).

1.2.4 Mycorrhizal Associations with Saprotrophic Fungi

The Orchidaceae is one of the largest and most diverse plant families and contains an estimated 22,000 species (Stevens 2012). Almost all orchids produce extremely small seeds ("dust seeds") that contain very few reserves. Consequently they are dependent on carbon and other nutrients provided by mycorrhizal fungi during their early developmental stages ("symbiotic germination") and can be classified as initial mycoheterotrophs. The fungi produce intracellular coils in the embryos of seedlings and in the rhizomes or roots of adult plants (Smith and Read 2008). Fungi isolated from the seedlings of orchids were initially classified under the Basidiomycota genus *Rhizoctonia* (e.g., Bernard 1899; Burgeff 1909). However, molecular phylogenetic analyses demonstrate that rhizoctonia-forming fungi are polyphyletic and phylogenetically spread over the Basidiomycota orders Ceratobasidiales, Tulasnellales, and Sebacinales (Taylor et al. 2002). Rhizoctonia-forming fungi isolated from orchids have been shown to be saprotrophic and are thus able to obtain carbon and other nutrients from decaying matter (Smith and Read 2008). Associations with these SAP fungi are found in all major orchid lineages including the first-diverging Apostasioideae (Kristiansen et al. 2004; Yukawa et al. 2009). An initial dependency on rhizoctonia-forming fungi is thus likely the ancestral state of Orchidaceae, and it has been hypothesized that the unique ability of orchids to recruit free-living SAP fungi into novel mycorrhizas may have led to dramatic expansion of their potential habitat and has triggered their radiation (Ogura-Tsukita et al. 2009). During orchid evolution, many mycoheterotrophic lineages have switched to an association with EM fungi (Chap. 5). However, compared to the total number of species in Orchidaceae, EM orchids represent only a small minority of species. All other orchid species are mycoheterotrophic on SAP fungi at least during their initial development, and therefore, next to AM and EM fungi, SAP fungi are the third source of mycoheterotrophy in plants.

1.2.5 Mutualism–Parasitism Continuum of the Mycorrhizal Symbiosis

Typically the mycorrhizal symbiosis is a mutualistic interaction: both the plant and the mycorrhizal fungus benefit from the association. In a simple model this interaction can be classified as a "win-win" situation. However, evolutionary theory suggests that mutualisms are best viewed as "reciprocal exploitations" that nonetheless provide net benefits to each partner (Herre et al. 1999). This view stresses the disruptive potential of conflicts of interests among the partners and advances the prediction that mutualisms are vulnerable to exploitation (Bronstein 2001; Sachs and Simms 2006). It also emphasizes the dynamic nature of symbiotic interactions: the outcome may vary considerably depending on the context. Thus rather than classifying symbiotic interactions into distinct categories (e.g., mutualism, parasitism, commensalism), they should be viewed as dynamic points along a continuum (Bronstein 1994).

The mycorrhizal symbiosis can be envisioned as such a continuum, in which the mutualistic plant–fungus interaction is the midpoint, and exploitation of a plant by a mycorrhizal fungus, or vice versa, are the endpoints (Egger and Hibbett 2004; Fig. 1.3). Costs and benefits of the mycorrhizal interaction are difficult to determine and may be context dependent (Johnson et al. 1997). However, there is considerable circumstantial evidence for the existence of exploitative strategies by mycorrhizal fungi (e.g., Klironomos 2003; Reynolds et al. 2005; Bever et al. 2009). Mycoheterotrophy is an exploitative strategy that represents the other end of the plant–fungus mycorrhizal continuum, in which the plant exploits its associated mycorrhizal fungi to obtain carbon and other nutrients (Fig. 1.3). The continuum between mycorrhizal mutualism and mycoheterotrophy is dynamic, and shifts along the continuum can occur at a developmental, ecological, or evolutionary timescale (Chaps. 5 and 8). For example, a plant may rely on mycoheterotrophy at the initial stage of its life cycle but develop into an autotrophic mature plant

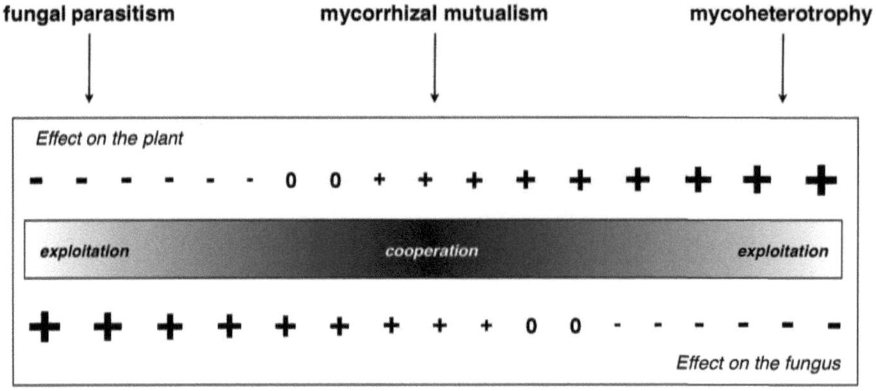

Fig. 1.3 Representation of the symbiotic exploitation–cooperation continuum of the mycorrhizal symbiosis based mutual effect of the interaction on the plant and fungal partners. Three potential outcomes of the interaction (fungal parasitism, mycorrhizal mutualism, and mycoheterotrophy) are indicated above. Figure adapted from Bronstein (1994) and Egger and Hibbett (2004)

(initial mycoheterotrophy) thus shifting from one point in the symbiotic continuum to another during its development. Some plants are able to combine autotrophy and mycoheterotrophy at maturity ("partial mycoheterotrophy," see further). However, where partial mycoheterotrophy can be placed between mycorrhizal mutualism and mycoheterotrophy depends on the context. It is possible that a partially mycoheterotrophic plant exploits its associated mycorrhizal fungi to the same extent as a fully mycoheterotrophic species. But theoretically, a partially mycoheterotrophic plant that acquires a significant amount of its carbon through photosynthesis will demand less carbon from its mycorrhizal fungi than a full mycoheterotrophic plant with a similar net carbon requirement. Stable isotope signatures of partially mycoheterotrophic plant species have revealed intraspecific variation of carbon uptake from fungi depending on light availability (Gebauer 2005; Zimmer et al. 2007; Preiss et al. 2010; Chap. 8). Thus it seems that partially mycoheterotrophic plants do not always exploit their mycorrhizal fungi fully. This suggests that, at least for some plant species, there is a positive correlation between the amount of carbon gained through mycorrhizal fungi and the degree of exploitation of their mycorrhizal fungi (Fig. 1.4).

1.3 A Short History of Research on Mycoheterotrophy

It would require a separate book to provide a detailed historical overview of the pioneering work of early naturalists that has provided the foundations of our understanding of mycoheterotrophy. Therefore, I have limited myself to a rather personal account, highlighting a few of the turning points that have revolutionized the field, with an emphasis on more recent developments. More comprehensive overviews of the early research on mycoheterotrophic plants are provided by Rayner (1927) and more recently by Leake (1994) and Bidartondo (2005).

The story of scientific research on mycoheterotrophy is tightly linked with advances in technology and methodology and begins in the nineteenth century. A major controversy in the 1840s marks the start of a scientific journey that still continues today: naturalists debated over the question of whether *Hypopitys monotropa*[1] (Ericaceae) is parasitic on the roots of beech trees

[1]Based on recent genetic evidence, we place *Monotropa hypopitys* its own genus, *Hypopitys*, with the single species *Hypopitys monotropa* (see Chap. 2).

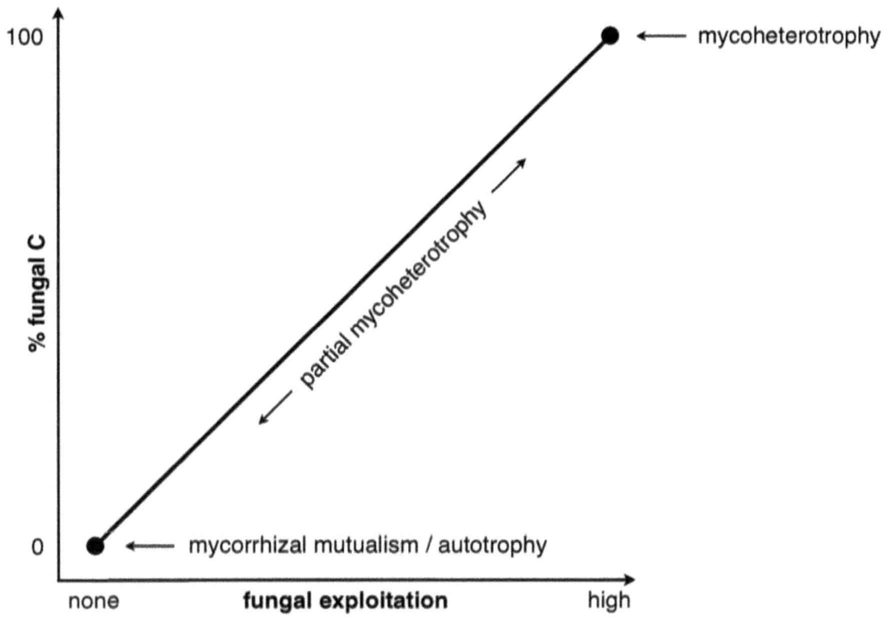

Fig. 1.4 Hypothetical model showing plant dependence on fungal carbon (C) as a function of the amount of fungal exploitation (*solid line*). When a mycorrhizal plant is fully autotrophic, it does not exploit its mycorrhizal fungi for carbon. On the other hand, when a plant completely depends on fungal carbon, fungal exploitation is high. Consequently, increased levels of partial mycoheterotrophy may relate to increased levels of fungal exploitation

or not. In the heated discussion that followed, it became clear—mainly through microscope observations—that the hairy fibers that link the roots of *Hypopitys* to those of surrounding trees are not parasitic plant haustoria but fungi (Rylands 1842). Around the same time, the first description of fungal infections in the roots of "saprophytic" orchids appeared (Reissek 1847; Schacht 1854; Prillieux 1856; Reinke 1873; Drude 1873). Pfeffer (1877) suggested that the observed fungus also penetrated the soil and brought nutrients to the orchid. An important next step was taken by Franz Kamienski, whose detailed investigation resulted in a breakthrough paper in which he postulates that *H. monotropa* lives on a fungus that is connected to tree roots and provides the first unambiguous description of a mycorrhiza (Kamienski 1882). To his merit, he explicitly discusses his observations in the context of the then recent definition of symbiosis, of which both antagonistic and mutualistic examples were known (de Bary 1879). But Kamienski writes: "It is not necessary to prove that those two forms of symbiosis are but extremes and that between them are to be found an infinite number of intermediary forms [...]," and he concludes that the symbiosis between *Hypopitys* and its fungus is an "example of the most striking of the 'mutualistic' union of two vegetative organisms" (Kamienski 1882 *fide* Berch et al. 2005). Daniel MacDougal went on to suggest that many achlorophyllous plants gain "complex substances" by a symbiotic association with root-colonizing fungi, a concept he described as "symbiotic saprophytism" (MacDougal 1899). Francke (1934) observed fungal hyphae release their contents into root cells of *H. monotropa* and successfully grew *Hypopitys* seedlings from seeds by inoculating them with the mycorrhizal fungus extracted from mature plants.

The rich tradition of microscopy work on fungus–plant interactions of mycoheterotrophic plants is still continued today, now aided by powerful electron microscopes (e.g., Imhof 1999; Massicotte et al. 2005; Domínguez et al. 2009). Nevertheless, many aspects of the physiology of the mycoheterotrophic fungus interaction still remain unclear (see Chap. 4).

In the 1950s, research on mycoheterotrophy entered a new era with the experimental fieldwork of Erik Björkman (1960). He observed that physical separation of *H. monotropa* from tree roots resulted in a reduced growth of the former, and he was able culture a fungus from *Hypopitys* roots that formed ectomycorrhizas with pine roots. More importantly, he used radioactive-labeled isotopes to reveal that more carbon and phosphorus are transported from *Picea* to *Hypopitys* than to any other neighboring plant. This was the first "direct" demonstration of a fully mycoheterotrophic plant being energetically dependent on surrounding trees through shared mycorrhizal fungi. Radioactive tracer experiments have later been repeated to study mycoheterotrophic interactions, although for obvious reasons mostly in vitro rather than in natural habitats (McKendrick et al. 2000a; Bidartondo et al. 2003). Recently a series of radioactive-labeling experiments were used to show that the green orchid *Corallorhiza trifida* actually derives most of its carbon through mycoheterotrophy (Cameron et al. 2008, 2009).

The development of DNA sequencing and amplification techniques in the 1970s and 1980s revolutionized the field of biology. Ecologists were quick to adapt these new techniques for the identification of mycorrhizal fungi whose characterization had been hampered by their undifferentiated morphologies and difficulties in culturing them (e.g., Simon et al. 1993). Among the first to identify the fungi in the roots of fully mycoheterotrophic plants with molecular methods were Ken Cullings and colleagues, who sequenced the ectomycorrhizal fungi associated with the roots of monotropes (Cullings et al. 1996). This work was further elaborated by Kretzer et al. (2000), Bidartondo et al. (2000), and Bidartondo and Bruns (2001, 2002). Almost simultaneously, the first reports on the molecular identification of fully mycoheterotrophic orchids were published (Taylor and Bruns 1997; McKendrick et al. 2000b, 2002), followed by molecular studies on mycoheterotrophic plants living on arbuscular mycorrhizal fungi (Yamato 2001; Bidartondo et al. 2002). These and subsequent studies confirmed that most full mycoheterotrophs obtain carbon from surrounding green plants, through shared ectomycorrhizal or arbuscular mycorrhizal fungi. It was already known for a long time that some fully mycoheterotrophic orchids associate with litter- and wood-decaying ("saprotrophic") fungi (e.g., Kusano 1911; Hamada 1939), and these fungi were first identified with molecular methods by Yamato et al. (2005), Ogura-Tsujita and Yukawa (2008), Ogura-Tsujita et al. (2009), and Martos et al. (2009). The molecular identification of the fungi associated with fully mycoheterotrophic plants revealed an important and novel aspect of the mycoheterotrophic interaction: while autotrophic plants typically associate with multiple distantly related fungi and a mycorrhizal fungus often associates simultaneously with distantly related plants, fully mycoheterotrophic plants frequently show high specificity toward narrow lineages of fungi (Chap. 7).

The development of DNA sequencing tools also caused a revolution in the unraveling of the evolutionary relationships of plants (e.g., Chase et al. 1993). Due to high sequence divergences, the evolutionary relationships of mycoheterotrophic plants were often difficult to infer, yet despite this difficult start, phylogenetic studies based on DNA data soon proved to be extremely valuable for resolving lasting problems about the taxonomic position of "difficult" mycoheterotrophic plant lineages (e.g., Molvray et al. 2000). In many cases, this led to dramatic changes in our understanding of mycoheterotrophic plant relationships (Caddick et al. 2000, 2002; Cameron et al. 2003). Subsequently, phylogenetic hypotheses based on DNA data were successfully used to study divergence times and biogeographical scenarios of mycoheterotrophic plant lineages (e.g., Merckx et al. 2008).

Achlorophyllous mycoheterotrophic plants offer excellent opportunities to study genome evolution. But while the first complete plastid genome of a holoparasitic plant (*Epifagus virginiana*) was published in 1990 by Claude dePamphilis and Jeffrey Palmer (dePamphilis and Palmer 1990), it took until 2008 before the first mycoheterotrophic plant genome was sequenced (Wickett et al. 2008). The recent development of high-throughput sequencing methods offers promising new opportunities to study genome evolution and has already resulted in the complete chloroplast genomes of the fully mycoheterotrophic orchids *Rhizanthella gardneri* (Delannoy et al. 2011) and *Neottia nidus-avis* (Logacheva et al. 2011).

In the early 2000s, another approach helped the study of mycoheterotrophy. It was already established that nitrogen (N) and carbon (C) in fungi are isotopically distinct from N and C of accompanying vegetation (Gebauer and Dietrich 1993; Gleixner et al. 1993); thus, the stable isotope's natural abundances of N and C presented tools to study nutrient fluxes between fungi and mycoheterotrophic plants in their natural habitats. The first studies confirmed that achlorophyllous mycoheterotrophic plants are isotopically different from the surrounding green vegetation (Gebauer and Meyer 2003; Trudell et al. 2003). Gebauer and Meyer (2003) also discovered that the carbon stable isotope values of several green orchid species in their study fell between those of autotrophic and achlorophyllous mycoheterotrophic plants and consequently produced the first evidence of partial mycoheterotrophy. More evidence for partial mycoheterotrophy through ectomycorrhizal fungi in temperate orchids was gathered by Bidartondo et al. (2004), Julou et al. (2005), and Abadie et al. (2006). Using stable isotope measurements, partial mycoheterotrophy was soon after discovered in Ericaceae as well (Tedersoo et al. 2007; Zimmer et al. 2007), and stable isotope analyses were subsequently adapted to study full mycoheterotrophic plants living on saprotrophic fungi (Ogura-Tsujita et al. 2009) and arbuscular mycorrhizal fungi (Merckx et al. 2010; Courty et al. 2011) (Chap. 8).

Recently, additional poorly known aspects of mycoheterotrophic plants became the focus of investigations, including population genetics (Taylor et al. 2004; Klooster and Culley 2010; Beatty and Provan 2011; Dowie et al. 2011) and reproductive biology (Klooster and Culley 2009; Hentrich et al. 2010). However, despite the long history of research and the increasing interest in mycoheterotrophic plants, many facets of this extraordinary plant–fungus interaction remain unresolved. More research is awaited on identifying the drivers behind fungal specificity, chemical signaling in the interaction, specific pathways of metabolite transport, evolution of plastid genomes, and morphological and physiological convergences between different types of mycoheterotrophy and their genetic background. Thanks to technological advances, there is no doubt that many of these questions will be successfully answered in the coming years.

1.4 Concepts and Terminology

Nearly all plants are autotrophs: they convert carbon dioxide into organic compounds, especially sugars, using the energy from sunlight in a process known as photosynthesis. Achlorophyllous mycoheterotrophic plants have lost the ability to perform photosynthesis and are able to obtain carbon through a symbiotic association with fungi. The majority of fully mycoheterotrophic plant species associate with fungi that are mycorrhizal with surrounding autotrophic plants (either arbuscular mycorrhizal or ectomycorrhizal fungi). Typically, mycorrhizal fungi receive carbon from autotrophic plants in exchange for water and soil minerals. The fungal carbon on which these mycoheterotrophs rely thus ultimately comes from autotrophic plants (Fig. 1.2). At least a few species of mycoheterotrophic orchids are known to associate with saprotrophic fungi, which obtain carbon from dead or decaying organic matter. Mycoheterotrophy occurs in most major groups of land plants, including liverworts, lycophytes, ferns, angiosperms, and perhaps gymnosperms as well (Fig. 1.5). There are about 23,000 species of land plants that rely on a mycoheterotrophic interaction at some stage in their life cycle. Most are orchids that rely on fungal carbon during the early stages of their development. At least 514 species of angiosperms and a single liverwort species entirely depend on fungal carbon during their complete life cycle ("full mycoheterotrophs"; see definitions below). Fully mycoheterotrophic plants have at least 46 independent origins in land plant evolution (see Chap. 5).

Because it was assumed that they lived directly on soil organic matter, mycoheterotrophic plants

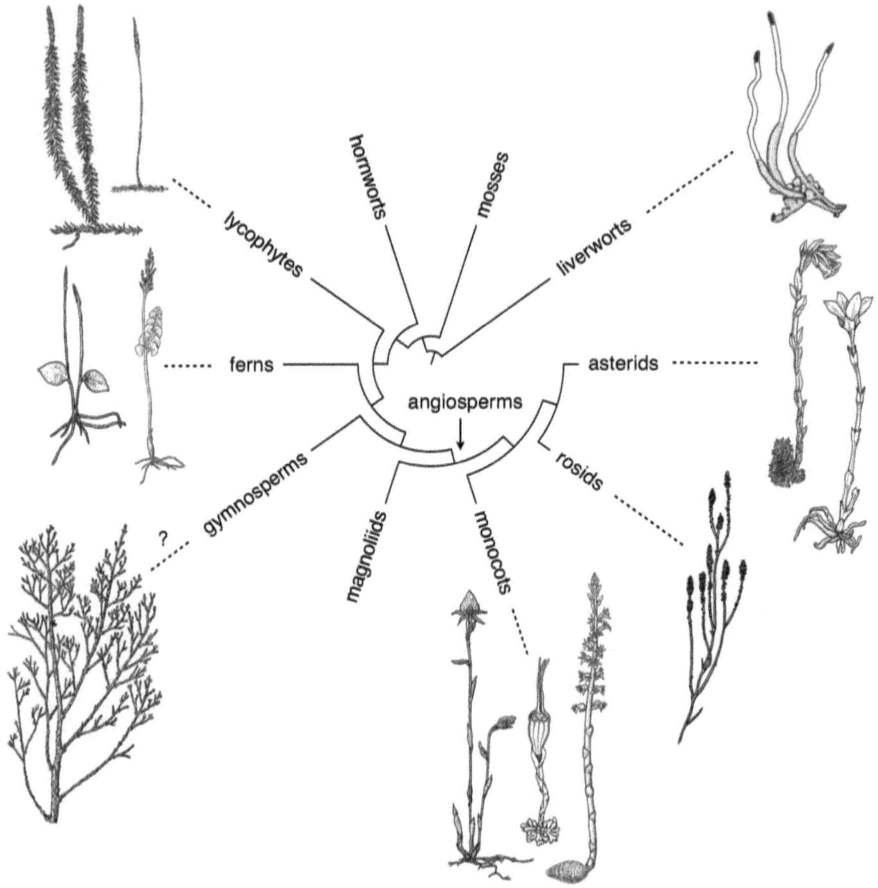

Fig. 1.5 Simplified land plant phylogeny. For each land plant group that contains mycoheterotrophic species, a few examples are shown (not drawn to scale). For lycophytes and ferns, autotrophic sporophytes are shown instead of the mycoheterotrophic gametophytes. For clarity, a few angiosperm lineages were omitted; see Fig. 5.4 for a complete angiosperm tree

were long regarded as "saprophytes" (a term which still appears in new literature from time to time). However, since the pioneering work of nineteenth-century naturalists, evidence from experimental, physiological, and molecular studies has continuously demonstrated that mycoheterotrophs depend on an association with fungi. In contrast, there is no evidence to date that direct carbon transfer from dead organic material to plants exists. "The myth of saprophytism" (Leake 2005) is thereby shattered, and we can advise strongly against the use of the term "saprophyte," as this misrepresents the mode of life of these remarkable plants.

The persistence of the incorrect term "saprophyte" highlighting the need for a consistent terminology based on clear definitions is necessary for meaningful discussion of mycoheterotrophic plants. Unfortunately, a unifying terminology has never been established, and in the current literature, different terms for the same phenomena compete with each other. I made an attempt to create a simple and consistent terminology that we will use throughout this book and that hopefully will be adapted by other authors as well. The first phenomenon we need to define is mycoheterotrophy itself:

"Mycoheterotrophy" is the ability of a plant to obtain carbon from fungi.

An important aspect of this definition is that it is solely based on the plant's ability to obtain

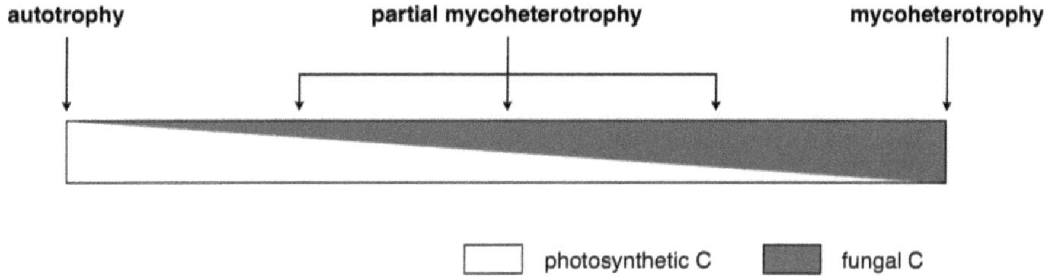

Fig. 1.6 Schematic representation of the trophic strategies autotrophy, mycoheterotrophy, and partial mycoheterotrophy based on the carbon (C) source a plant uses

carbon from root-associated fungi. It is likely that the metabolite fluxes of a mycoheterotrophic plant–fungus interaction differ from a normal mutualistic (mycorrhizal) plant–fungus interaction for other nutrients than carbon as well. For example, stable isotope analyses have shown that many mycoheterotrophs are distinctly enriched in ^{15}N compared to autotrophic plants growing at the same localities. However, ^{15}N enrichment is not a universal feature for mycoheterotrophic plants, and the extent of ^{15}N enrichment is not always linearly related to the extent of heterotrophic carbon gain (Leake and Cameron 2010).

Mycoheterotrophy is sometimes termed "epiparasitism" as well (e.g., Björkman 1960). This term stresses the fact that carbon can be received from green plants through a common mycorrhizal association but excludes dependence on saprotrophic fungi. Also, it remains unknown whether mycoheterotrophy has measurable costs to the green plants that supply carbon to the mycorrhizal fungi targeted by mycoheterotrophic plants. Therefore, "mycoheterotrophy" should be preferred over "epiparasitism."

Mycoheterotrophy is a trophic strategy that contrasts with autotrophy.[2] However, some plant species have been shown to be able to simultaneously combine autotrophy and mycoheterotrophy:

"Partial mycoheterotrophy" is the ability of a plant to obtain carbon simultaneously through autotrophy and mycoheterotrophy.

Thus autotrophy and mycoheterotrophy are the extreme ends of a continuum of trophic strategies that can occur in plants (Fig. 1.6). All intermediate strategies between autotrophy and mycoheterotrophy are designated as partial mycoheterotrophy, even though the relative amount of carbon received through either autotrophy or mycoheterotrophy may differ considerably. Partial mycoheterotrophy has also been termed "mixotrophy" (e.g., Selosse and Roy 2009), but we prefer "partial mycoheterotrophy" because "mixotrophy" is already used in a more general and different context (e.g., bacteria (Eiler 2006), protists (Thingstad et al. 1996), and sea anemones (Bachar et al. 2007)).

Autotrophy, mycoheterotrophy, and partial mycoheterotrophy are different strategies used by plants to obtain complex organic compounds (carbohydrates). It is now known that some plants can change their trophic strategy during their development. For example, the initial developmental stage of all orchids is a nonphotosynthetic protocorm that relies on mycoheterotrophy (Alexander and Hadley 1985; Leake 1994; Rasmussen 1995; Rasmussen and Whigham 1998). Nevertheless, most orchids lose their dependence on fungi as a source of carbon and develop into mature plants that solely rely on autotrophy. In addition, there is evidence that some plant species show plasticity in trophic strategy in relation to the environmental conditions in which they are growing. For example, research by Katja Preiss and colleagues has demonstrated that the orchid species *Cephalanthera damasonium* and *C. rubra* strongly supplement their carbon gain through photosynthesis by organic carbon from fungal partners under low-light conditions but become almost completely autotrophic when they are

[2] In a broad sense, autotrophy includes both phototrophy, in which light is used as an energy source (photosynthesis), and lithotrophy (or chemoautotrophy), in which inorganic compounds are oxidized (chemosynthesis). In the context of plants, autotrophy is restricted to phototrophy.

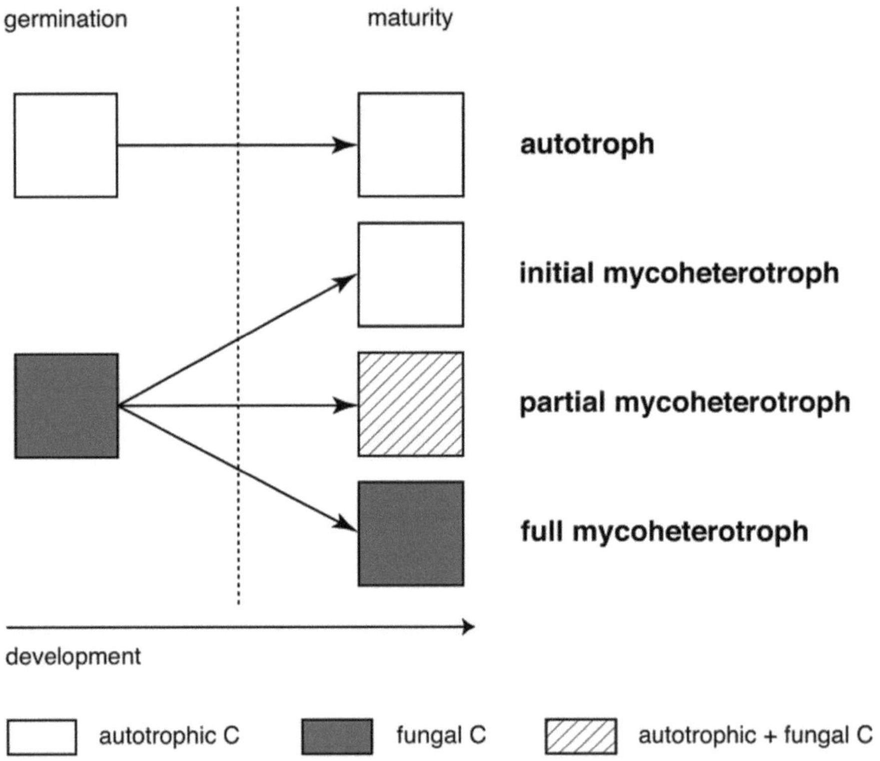

Fig. 1.7 Schematic representation of the different types of mycoheterotrophic plants known, based on life stage and their primary source of carbon (C)

exposed to sufficiently high irradiances (Preiss et al. 2010).

This trophic plasticity hampers the assignment of a plant species to a particular trophic category. Yet, for a comprehensive discussion of mycoheterotrophy, the designation of plant species according to their trophic capabilities is required. Therefore, I propose the following trophic categories, in which the developmental context plays an important role (Fig. 1.7).

A "fully mycoheterotrophic" plant ("full mycoheterotroph") solely depends on fungal carbon during its entire life cycle.

A fully mycoheterotrophic plant thus derives all of its carbon from fungi and does not need a functional photosynthetic apparatus. Plants that lack visible traces of chlorophyll and do not have a direct connection with autotrophic plants—and thus are not holoparasites—are putative fully mycoheterotrophic plants. Over 500 species of land plants fall into this category, the vast majority being angiosperms (Chap. 2).

An "initially mycoheterotrophic" plant ("initial mycoheterotroph") is fully dependent on associated fungi for its carbon supply during the early stages of development.

An initially mycoheterotrophic plant species relies on mycoheterotrophy in the beginning of its life cycle. In a broad sense, all full mycoheterotrophs are initial mycoheterotrophs as well, but we propose to use the term particularly for species that depend on autotrophy or partial mycoheterotrophy at maturity. Hence all orchids, except those that are fully mycoheterotrophic, are initial mycoheterotrophs (>20,000 spp.). Species of Pyroleae (Ericaceae) are probably also initial mycoheterotrophs (Smith and Read 2008; Eriksson and Kainulainen 2011). Other plant species that produce small dustlike seeds with limited nutritional reserves (e.g., Rubiaceae, Buddlejaceae,

Gesneriaceae) are putative initial mycoheterotrophs as well (Eriksson and Kainulainen 2011). In their early development, the sporophytes of lower plants are "temporary parasitic on the gametophyte" (Leake et al. 2008). If those gametophytes depend on mycoheterotrophy, the sporophytes of these species (Chap. 2) can be considered initial mycoheterotrophs as well.

A "partially mycoheterotrophic" plant ("partial mycoheterotroph") combines autotrophy and mycoheterotrophy to obtain carbon during at least one stage of its life cycle.

By definition, partial mycoheterotrophic plants have retained a functional photosynthetic apparatus. Because partial mycoheterotrophs retain chlorophyll, their trophic life strategy can only be detected by examining their physiologies (Chap. 8). Partial mycoheterotrophy has been shown to exist in a few species of Orchidaceae, Gentianaceae, and Ericaceae but may be present in other plants families as well. Green plant species that are closely related to fully mycoheterotrophic species or can survive in extreme low-light conditions (e.g., forest understory habitats) are prime candidates for undiscovered partial mycoheterotrophy. The dependence on fungal carbon can greatly differ between different partially mycoheterotrophic species (Gebauer and Meyer 2003) and between specimens of the same species that grow in different light conditions (Preiss et al. 2010). Even seasonal fluctuations occur within plant populations (Hynson et al. 2011).

Note that the terminology above is not necessarily mutually exclusive: all species of Orchidaceae are initial mycoheterotrophs, and they can develop into either autotrophic, partially mycoheterotrophic, or fully mycoheterotrophic mature plants. In the latter case, however, the term "full mycoheterotroph" is preferred.

1.5 Parasitic Plants

Fully mycoheterotrophic plants are often confused with parasitic plants. However, parasitic plants are a distinct category of achlorophyllous plants. These include holoparasitic plants, which obtain carbon from autotrophic plants through a direct physical attachment. The enigmatic *Rafflesia arnoldii* (Rafflesiaceae), which produces the biggest flower in the world, is one such example (Nais 2001). There are about 390 species of holoparasitic plants. In addition, about 4,100 eudicot plant species are hemiparasites: they have retained the ability to photosynthesize, but they primarily absorb water and nutrients from their host plants (Heide-Jørgensen 2008). The parasitic mode of life has evolved at least 11 times independently in the evolution of eudicots (Barkman et al. 2007). A single gymnosperm (*Parasitaxus usta*) may be a holoparasite as well, although there are indications that a fungus is involved in the interaction, and therefore, the species is sometimes regarded as a mycoheterotroph (see Chap. 2).

Mycoheterotrophy and holoparasitism represent distinct evolutionary pathways toward heterotrophy in plants. Yet there is often a striking morphological convergence between mycoheterotrophs and holoparasites: in general, both have highly reduced leaves, contain little or no chlorophyll, and produce prodigious numbers of seeds that cannot establish in absence of a host (Cameron and Leake 2007). Therefore, it is not surprising that Linnaeus initially considered *H. monotropa* to be a species of *Orobanche* (Leake 1994)!

References

Abadie J, Püttsepp Ü, Gebauer G, Faccio A, Bonfante P, Selosse MA (2006) *Cephalanthera longifolia* (Neottieae, Orchidaceae) is mixotrophic: a comparative study between green and nonphotosynthetic individuals. Can J Bot 84:1462–1477

Alexander C, Hadley G (1985) Carbon movement between host and mycorrhizal endophyte during development of the orchid *Goodyera repens*. New Phytol 101:657–665

Bachar A, Achituv Y, Pastemak Z, Dubinsky Z (2007) Autotrophy versus heterotrophy: the origin of carbon determines its fate in a symbiotic sea anemone. J Exp Mar Biol Ecol 349:295–298

Barkman TJ, McNeal JR, Lim S-H, Coat G, Croom HB, Young ND, dePamphilis CW (2007) Mitochondrial DNA suggests at least 11 origins of parasitism in angiosperms and reveals genomic chimerism in parasitic plants. BMC Evol Biol 7:248

Beatty GE, Provan J (2011) High clonal diversity in threatened peripheral populations of the yellow bird's nest (*Hypopitys monotropa*; syn. *Monotropa hypopitys*). Ann J Bot 107:663–670

Berch SM, Massicotte HB, Tackaberry LE (2005) Re-publication of a translation of 'The vegetative organs of *Monotropa hypopitys* L.' published by F. Kamienski in 1882, with an update on *Monotropa* mycorrhizas. Mycorrhiza 15:323–332

Bernard N (1899) Sur la germination du *Neottia nidus-avis*. C R Hebd Seances Acad Sci 128:1253–1255

Bever JD, Richardson SC, Lawrence BM, Holmes J, Watson M (2009) Preferential allocation to beneficial symbiont with spatial structure maintains mycorrhizal mutualism. Ecol Lett 12:13–21

Bidartondo MI (2005) The evolutionary ecology of myco-heterotrophy. New Phytol 167:335–352

Bidartondo MI, Burghardt B, Gebauer G, Bruns TD, Read D (2004) Changing partners in the dark: isotopic and molecular evidence of ectomycorrhizal liaisons between forest orchids and trees. Proc R Soc B 271:1799–1806

Bidartondo MI, Bruns TD (2001) Extreme specificity in epiparasitic Monotropoideae (Ericaceae): widespread phylogenetic and geographic structure. Mol Ecol 10:2285–2295

Bidartondo MI, Bruns TD (2002) Fine-level mycorrhizal specificity in the Monotropoideae (Ericaceae): specificity for fungal species groups. Mol Ecol 11:557–569

Bidartondo MI, Bruns TD, Weiß M, Sérgio C, Read DJ (2003) Specialized cheating of the ectomycorrhizal symbiosis by an epiparasitic liverwort. Proc R Soc Lond B 270:835–842

Bidartondo MI, Kretzer AM, Pine EM, Bruns TD (2000) High root concentration and uneven ectomycorrhizal diversity near *Sarcodes sanguinea* (Ericaceae): a cheater that stimulates its victims? Am J Bot 87:1783–1788

Bidartondo MI, Read DJ, Trappe JM, Merckx V, Ligrone R, Duckett JG (2011) The dawn of symbiosis between plants and fungi. Biol Lett 7:574–577

Bidartondo MI, Redecker D, Hijiri I, Wiemken A, Bruns TD, Dominguez LS, Sérsic A, Leake JR, Read DJ (2002) Epiparasitic plants specialized on arbuscular mycorrhizal fungi. Nature 419:389–392

Björkman E (1960) *Monotropa hypopitys* L.—an epiparasite on tree roots. Physiol Plant 13:308–327

Bonfante P, Genre A (2010) Mechanisms underlying beneficial plant-fungus interactions in mycorrhizal symbiosis. Nat Commun 1:48

Bronstein JL (1994) Conditional outcomes in mutualistic interactions. Trends Ecol Evol 9:214–217

Bronstein JL (2001) The exploitation of mutualisms. Ecol Lett 4:277–287

Bruns TD, Shefferson RP (2004) Evolutionary studies of ectomycorrhizal fungi: recent advances and future directions. Can J Bot 82:1122–1132

Burgeff H (1909) Die wurzelpiltze der orchideen. Gustav Fisher, Jena, Germany

de Bary HA (1879) Die erscheinung der Symbiose. Trübner, Strassburg, Germany

Caddick LR, Rudall PJ, Wilkin P, Chase MW (2000) Yams and their allies: systematics of Dioscoreales. In: Wilson KJ, Morrison DA (eds) Monocots: systematics and evolution. CSIRO Publishing, Melbourne, pp 475–487

Caddick LR, Rudall PJ, Wilkin P, Hedderson TA, Chase MW (2002) Phylogenetics of Dioscoreales based on combined analyses of morphological and molecular data. Bot J Linn Soc 138:123–144

Cameron DD, Johnson I, Read DJ, Leake JR (2008) Giving and receiving: measuring the carbon cost of mycorrhizas in the green orchid, Gooyera repens. New Phytol 180:176–184

Cameron DD, Leake JR (2007) A different kind of parasitic plant: a brief history of mycoheterotrophy and parasitism. Haustorium 50:4–6

Cameron DD, Preiss K, Gebauer G, Read DJ (2009) The chlorophyll-containing orchid *Corallorhiza trifida* derives little carbon through photosynthesis. New Phytol 183:358–364

Cameron KM, Chase MW, Rudall P (2003) Recircumscription of the monocotyledonous family Petrosaviaceae to include *Japonolirion*. Brittonia 55:214–225

Chase MW, Soltis DE, Olmstead RG, Morgan D, Les DH, Mishler BD, Duvall MR, Price RA, Hills HG, Qiu Y-L, Kron KA, Rettig JH, Conti E, Palmer JD, Manhart JR, Sytsma KJ, Michaels HJ, Kress WJ, Karol KG, Clark WD, Hedrén M, Gaut BS, Jansen RK, Kim K-J, Wimpee CF, Smith JF, Furnier GR, Strauss SH, Xiang Q-Y, Plunkett GM, Soltis PS, Swensen SM, Williams SE, Gadek PA, Quinn CJ, Equiarte LE, Golenberg E, Learn GH, Graham SW, Barrett SCH, Dayanandan S, Albert VA (1993) Phylogenetics of seed plants: an analysis of nucleotide sequences from the plastid gene *rbcL*. Ann Missouri Bot Gard 80:528

Courty P-E, Walder F, Boller T, Ineichen K, Wiemken A, Rousteau A, Selosse M-A (2011) C and N metabolism in mycorrhizal networks and mycoheterotrophic plants of tropical forests: a stable isotope analysis. Plant Physiol 156:952–961

Cullings KW, Szaro TM, Bruns TD (1996) Evolution of extreme specialization within a lineage of ectomycorrhizal epiparasites. Nature 379:63–66

Delannoy E, Fujii S, Colas des Francs C, Brundrett M, Small I (2011) Rampant gene loss in the underground orchid *Rhizanthella gardneri* highlights evolutionary constraints on plastid genomes. Mol Bio Evol 28:2077–2086

dePamphilis CW, Palmer JD (1990) Loss of photosynthetic and chlororespiratory genes from the plastid genome of a nonphotosynthetic plant. Nature 348:337–339

Domínguez LS, Melville L, Sérsic A, Faccio A, Peterson RL (2009) The mycoheterotroph *Arachnitis uniflora* has a unique association with arbuscular mycorrhizal fungi. Botany 87:1198–1208

Dowie NJ, Hemenway JJ, Trowbridge SM, Miller SL (2011) Mycobiont overlap between two mycoheterotrophic genera of Monotropoideae (*Pterospora andromedea* and *Sarcodes sanguinea*) found in the Greater Yellowstone Ecosystem. Symbiosis 54:29–36

Drude O (1873) Die Biologie von *Monotropa Hypopitys* L. und *Neottia-nidus-avis* L. unter vergleichender Hinzuziehung anderer Orchideen. Unversität Göttingen Preisgerkrönte Schrift, Göttingen, Germany

Egger KN, Hibbett DS (2004) The evolutionary implications of exploitation in mycorrhizas. Can J Bot 82:1110–1121

Eiler A (2006) Evidence for the ubiquity of mixotrophic bacteria in the upper ocean: implications and consequences. Appl Environ Microbiol 72:7431–7437

Eriksson O, Kainulainen K (2011) The evolutionary ecology of dust seeds. Perspect Plant Ecol Evol Systemat 13:73–87

Francke H-L (1934) Beiträge zur Kenntnis der Mykorrhiza von *Monotropa hypopitys* L. Analyse und Synthese der Symbiose. Flora 129:1–52

Gebauer G (2005) Partnertausch im dunklen Wald–Stabile IsotopegebenneueEinblickeindasErnährungsverhalten von Orchideen. In Rundgespräche der Kommission für Ökologie, vol 30 (ed. Bayerische Akademie der Wissenschaften). Verlag Dr. Friedrich Pfeil, München, Germany, p 55–67

Gebauer G, Dietrich P (1993) Nitrogen isotope ratios in different compartments of a mixed stand of spruce, larch and beech trees and of understorey vegetation including fungi. Isotopenpraxis 29:35–44

Gebauer G, Meyer M (2003) 15N and 13C natural abundance of autotrophic and myco-heterotrophic orchids provides insight into nitrogen and carbon gain from fungal association. New Phytol 160:209–223

Giovannetti M, Sbrana C, Avio L, Strani P (2004) Patterns of below-ground plant interconnections established by means of arbuscular mycorrhizal networks. New Phytol 164:175–181

Gleixner G, Danier H-J, Werner RA, Schmidt H-L (1993) Correlations between the 13C content of primary and secondary plant products in different cell compartments and that in decomposing basidiomycetes. Plant Physiol 102:1287–1290

Hamada M (1939) Studien über die Mykorrhiza von *Galeola septentrionalis* Reichb. F.—ein neuer Fall der Mykorrhiza-Bildung durch intraradicale Rhizomorpha. Jap J Bot 10:151–211

Heide-Jørgensen HS (2008) Parasitic flowering plants. Koninklijke Brill NV, Leiden, The Netherlands

Hentrich H, Kaiser R, Gottsberger G (2010) The reproductive biology of *Voyria* (Gentianaceae) species in French Guiana. Taxon 59:867–880

Herre EA, Knowlton N, Mueller UG, Rehner SA (1999) The evolution of mutualisms: exploring the paths between conflict and cooperation. Trends Ecol Evol 14:49–53

Hibbett DS, Matheny PB (2009) The relative ages of ectomycorrhizal mushrooms and their plant hosts estimated using Bayesian relaxed clock analyses. BMC Evol Biol 7:13

Hynson NA, Mambelli S, Amend AS, Dawson TE (2011) Measuring carbon gains from fungal networks in understory plants from the tribe Pyroleae (Ericaceae): a field manipulation and stable isotope approach. Oecologia 169:307–317

Imhof S (1999) Anatomy and mycotrophy of the achlorophyllous *Afrothismia winkleri*. New Phytol 144:533–540

Johnson NC, Graham JH, Smith FA (1997) Functioning of mycorrhizal associations along the mutualism-parasitism continuum. New Phytol 135:575–586

Julou T, Burghardt B, Gebauer C, Berveiller D, Damesin C, Selosse MA (2005) Mixotrophy in orchids: insights from a comparative study of green individuals and nonphotosynthetic individuals of *Cephalanthera damasonium*. New Phytol 166:639–653

Kamienski F (1882) Les organs végétatifs du *Monotropa hypopitys* L. Mémoires de la Société Nationale des Sciences Naturelles et Mathématiques de Cherbourg 24:5–40

Klironomos JN (2003) Variation in plant response to native and exotic arbuscular mycorrhizal fungi. Ecology 84:2292–2301

Klooster MR, Culley TM (2009) Comparative analysis of the reproductive ecology of *Monotropa* and *Monotropsis*: two mycoheterotrophic genera in the Monotropoideae (Ericaceae). Am J Bot 96:1337–1347

Klooster MR, Culley TM (2010) Population genetic structure of the mycoheterotroph *Monotropa hypopitys* L. (Ericaceae) and differentiation between red and yellow color forms. Int J Plant Sci 171:167–174

Kretzer AM, Bidartondo MI, Szaro TM, Grubisha L, Bruns TD (2000) Regional specialization of *Sarcodes sanguinea* on a single fungal symbiont from the *Rhizopogon ellenae* species complex. Am J Bot 87:1778–1783

Kristiansen KA, Freudenstein JV, Rasmussen FH, Rasmussen HN (2004) Molecular identification of mycorrhizal fungi in *Neuwiedia veratrifolia* (Orchidaceae). Mol Phylogenet Evol 33:251–258

Krüger M, Krüger C, Walker C, Stockinger H, Schüßler A (2011) Phylogenetic reference data for systematics and phylotaxonomy of arbuscular mycorrhizal fungi from phylum to species level. New Phytol 193:970–984

Kusano S (1911) *Gastrodia elata* and its symbiotic association with *Armillaria mellea*. J Coll Agr Imp Univ Tokyo 4:1–65

Le Page BA, Currah RS, Stockey RA, Rothwell GW (1997) Fossil ectomycorrhizae from the Middle Eocene. Am J Bot 84:410–412

Leake JR (1994) The biology of myco-heterotrophic ('saprophytic') plants. New Phytol 127:171–216

Leake JR (2005) Plants parasitic on fungi: unearthing the fungi in myco-heterotrophs and debunking the 'saprophytic' plant myth. Mycologist 19:113–122

Leake JR, Cameron DD (2010) Physiological ecology of mycoheterotrophy. New Phytol 185:601–605

Leake JR, Cameron DD, Beerling BJ (2008) Fungal fidelity in the myco-heterotroph-to-autotroph life cycle of Lycopodiaceae: a case of parental nurture? New Phytol 177:572–576

Logacheva MD, Schelkunov MI, Penin AA (2011) Sequencing and analysis of plastid genome in myco-heterotrophic orchid *Neottia nidus-avis*. Genome Biol Evol 3:1296–1303

MacDougal DT (1899) Symbiotic saprophytism. Ann Bot 13:1–47

Martos F, Dulormne M, Pailler T, Bonfante P, Faccio A, Fournel J, Dubois M-P, Selosse M-A (2009) Independent recruitment of saprotrophic fungi as mycorrhizal partners by tropical achlorophyllous orchids. New Phytol 184:668–681

Massicotte HB, Melville LH, Peterson RL (2005) Structural features of mycorrhizal associations in two members of the Monotropoideae, *Monotropa uniflora* and *Pterospora andromedea*. Mycorrhiza 15:101–110

McKendrick SL, Leake JR, Read DJ (2000a) Symbiotic germination and development of myco-heterotrophic plants in nature: transfer of carbon from ectomycorrhizal *Salix repens* and *Betula pendula* to the orchid *Corallorhiza trifida* through shared hyphal connections. New Phytol 145:539–548

McKendrick SL, Leake JR, Taylor DL, Read DJ (2000b) Symbiotic germination and development of myco-heterotrophic plants in nature: ontogeny of *Corallorhiza trifida* and characterization of its mycorrhizal fungi. New Phytol 145:523–537

McKendrick SL, Leake JR, Taylor DL, Read DJ (2002) Symbiotic germination and development of the myco-heterotrophic orchid *Neottia nidus-avis* in nature and its requirement for locally distributed *Sebacina* spp. New Phytol 154:233–247

Merckx V, Chatrou LW, Lemaire B, Sainge MN, Huysmans S, Smets E (2008) Diversification of myco-heterotrophic angiosperms: evidence from Burmanniaceae. BMC Evol Biol 8:178

Merckx V, Stöckel M, Fleischmann A, Bruns TD, Gebauer G (2010) 15N and 13C natural abundance of two mycoheterotrophic and a putative partially myco-heterotrophic species associated with arbuscular mycorrhizal fungi. New Phytol 188:590–596

Molvray M, Kores PJ, Chase MW (2000) Polyphyly of mycoheterotrophic orchids and functional influences of floral and molecular characters. In: Wilson KL, Morrison DA (eds) Monocots: systematics and evolution. CSIRO Publishing, Collingwood, Australia, pp 441–448

Moyersoen B (2006) *Pakaraimaea dipterocarpaceae* is ectomycorrhizal, indicating an ancient Gondwanaland origin of the ectomycorrhizal habit in Dipterocarpaceae. New Phytol 172:753–762

Muir J (1912) The Yosemite. The Century Co, New York, USA

Nais J (2001) *Rafflesia* of the world. Natural History Publications, Kota Kinabalu, Malaysia

Ogura-Tsujita Y, Gebauer G, Hashimoto T, Umata H, Yukawa T (2009) Evidence for novel and specialized mycorrhizal parasitism: the orchid *Gastrodia confusa* gains carbon from saprotrophic *Mycena*. Proc R Soc Lond B 22:761–767

Ogura-Tsujita Y, Yukawa T (2008) High mycorrhizal specificity in a widespread mycoheterotrophic plant, *Eulophia zollingeri* (Orchidaceae). Am J Bot 95:93–97

Öpik M, Vanatoa A, Vanatoa E, Moora M, Davison J, Kalwij JM, Reier Ü, Zobel M (2010) The online database Maarj*AM* reveals global and ecosystemic distribution patterns in arbuscular mycorrhizal fungi (Glomeromycota). New Phytol 188:223–241

Pfeffer W (1877) Über fleischfressende Planzen und über die Ernährung durch Aufnahme organischer Stoffe überhaupt. Landwirtschaftliche Jahrbücher 6:969–988

Preiss K, Adam IK, Gebauer G (2010) Irradiance governs exploitation of fungi: fine-tuning of carbon gain by two partially myco-heterotrophic orchids. Proc R Soc Lond B 277:1333–1336

Prillieux E (1856) De la structure anatomique et du mode de végétation du *Neottia nidus-avis*. Ann des Sci Naturelles Series 4:267–282

Rasmussen HN (1995) Terrestrial orchids: from seed to mycotrophic plant. Cambridge University Press, Cambridge, UK

Rasmussen HN, Whigham DF (1998) The underground phase: a special challenge in studies of terrestrial orchid populations. Bot J Linn Soc 126:49–64

Rayner MC (1927) Mycorrhiza: an account of non-pathogenic infection by fungi in vascular plants and bryophytes. New Phytologist Reprint no. 15. Wheldon & Wesley, London, UK

Reinke J (1873) Zur Kenntnis des Rhizomes von *Corallorhiza* und *Epipogon*. Flora 56:161–167

Reissek S (1847) Über endophyten der Pflanzenzelle, eine gesetzmässige den Samenfaden oder beweglichen Spiralfasern analoge Erscheinung. Naturwissenschaftlige Abhandlungen 1:3146

Reynolds HL, Hartley AE, Vogelsang KM, Bever JD, Schultz PA (2005) Arbuscular mycorrhizal fungi do not enhance nitrogen acquisition and growth of old-field perrenials under low nitrogen supply in glasshouse culture. New Phytol 167:869–880

Rylands TG (1842) On the mode of growth of *Monotropa hypopitys*. Phytologist 16:329–330

Saari SK, Campbell CD, Russell J, Alexander IJ, Anderson IC (2005) Pine microsatellite markers allow roots and ectomycorrhizas to be linked to individual trees. New Phytol 165:295–304

Sachs JL, Simms EL (2006) Pathways to mutualism breakdown. Trends Ecol Evol 21:585–592

Schacht H (1854) Pilzfaden im Innern der Zellen und der Starkmehlkörner vor. Flora 39:618–624

Selosse MA, Roy M (2009) Green plants that feed on fungi: facts and questions about mixotrophy. Trends Plant Sci 14:64–70

Simon L, Bousquet J, Lévesque RC, Lalonde M (1993) Origin and diversification of endomycorrhizal fungi and coincidence with vascular land plants. Nature 363:67–69

Smith SE, Read DJ (2008) Mycorrhizal symbiosis, 3rd edn. Academic, London, UK

Stevens PF (2012) Angiosperm Phylogeny Website. Version 9, June 2008 [and more or less continuously updated since]
Taylor DL, Bruns TD (1997) Independent, specialized invasions of the ectomycorrhizal mutualism by two non-photosynthetic orchids. Proc Natl Acad Sci USA 94:4510–4515
Taylor DL, Bruns TD, Hodges SA (2004) Evidence for mycorrhizal races in a cheating orchid. Proc R Soc Lond B 271:35–43
Taylor DL, Bruns TD, Leake JR, Read DJ (2002) Mycorrhizal specificity and function in myco-heterotrophic plants. In: Van der Heijden MGA, Sanders I (eds) Mycorrhizal ecology. Springer, Berlin, Germany, pp 375–413
Tedersoo L, May T, Smith M (2010) Ectomycorrhizal lifestyle in fungi: global diversity, distribution, and evolution of phylogenetic lineages. Mycorrhiza 20:217–263
Tedersoo L, Pellet P, Kõljalg U, Selosse MA (2007) Parallel evolutionary paths to mycoheterotrophy in understorey Ericaceae and Orchidaceae: ecological evidence for mixotrophy in Pyroleae. Oecologia 151:206–217
Thingstad TF, Havskum H, Garde K, Riemann B (1996) On the strategy of "eating your competitor": a mathematical analysis of algal mixotrophy. Ecology 77:2108–2118
Trudell SA, Rygiewicz PT, Edmonds RL (2003) Nitrogen and carbon stable isotope abundances support the myco-heterotrophic nature and host-specificity of certain achlorophyllous plants. New Phytol 160: 391–401
Wickett NJ, Fan Y, Lewis PO, Goffinet B (2008) Distribution and evolution of pseudogenes, gene losses, and a gene rearrangement in the plastid genome of the nonphotosynthetic liverwort, *Aneura mirabilis* (Metzgeriales, Jungermanniopsida). J Mol Evol 67:111–122
Yamato M (2001) Identification of a mycorrhizal fungus in the roots of achlorophyllous *Sciaphila tosaensis* Makino (Triuridaceae). Mycorrhiza 11:83–88
Yamato M, Yagame T, Suzuki A, Iwase K (2005) Isolation and identification of mycorrhizal fungi associating with an achlorophyllous plant, *Epipogium roseum* (Orchidaceae). Mycoscience 46:73–77
Yukawa T, Ogura-Tsujita Y, Shefferson R, Yokoyama J (2009) Mycorrhizal diversity in *Apostasia* (Orchidaceae) indicates the origin and evolution of orchid mycorrhiza. Am J Bot 96:1997–2009
Zimmer K, Hynson NA, Gebauer G, Allen EB, Allen MF, Read DJ (2007) Wide geographical and ecological distribution of nitrogen and carbon gains from fungi in pyroloids and monotropoids (Ericaceae) and in orchids. New Phytol 175:166–175

Taxonomy and Classification 2

Vincent S.F.T. Merckx, John V. Freudenstein, Jonathan Kissling, Maarten J.M. Christenhusz, Raymond E. Stotler, Barbara Crandall-Stotler, Norman Wickett, Paula J. Rudall, Hiltje Maas-van de Kamer, and Paul J.M. Maas

2.1 Introduction

Fully mycoheterotrophic plants share only one particular feature—the obligation to obtain carbon from fungi. The plants that fall within this definition do not necessarily have to be evolutionarily related and therefore mycoheterotrophic plants consist of a wide variety of taxa. Although mycoheterotrophy is relatively rare in nature, multiple independent origins of the mycoheterotrophic mode of life have produced a remarkable array of mycoheterotrophic species in almost every major group of land plants. Furman and Trappe (1971) and particularly Leake (1994) presented excellent overviews of mycoheterotrophic plant species. Here we provide an updated list of all putatively fully mycoheterotrophic plant species, excluding initial and partial mycoheterotrophs, with details on their morphology, distribution, and ecology. This list covers 17 plant families, 101 genera, and ca. 880 species.

Fern and lycophyte species were included when there is evidence for the lack of chlorophyll in the gametophytic phase. For "higher" plants we considered all leafless, achlorophyllous plant species that are not holoparasitic (i.e., physically connected to a host plant) as "putative" full mycoheterotrophs (but see *Parasitaxus*). In some cases we also included species that retain at least some chlorophyll, but in these cases mycoheterotrophy is suspected on the basis of their extremely reduced leaves. However, to determine the trophic status of a species careful investigation is needed. Since these data are lacking for the great majority of putative mycoheterotrophs, the presence of mycorrhizas and the absence of both chlorophyll and a direct physical link to a host plant are probably the best indications for full mycohetertrophy. Nevertheless, it is important to

V.S.F.T. Merckx (✉)
Naturalis Biodiversity Center, Leiden University, P.O. Box 9514, 2300 RA, Leiden, The Netherlands
e-mail: vincent.merckx@naturalis.nl

J.V. Freudenstein
Department of Evolution, Ecology, and Organismal Biology, The Ohio State University Herbarium, 1315 Kinnear Road, Columbus, OH 43212, USA

J. Kissling
Institute of Biology, Evolutionary Botany, University of Neuchâtel, Rue Emile-Argand 11, 2009 Neuchâtel, Switzerland

M.J.M. Christenhusz
Botanical Garden and Herbarium, Finnish Museum of Natural History, Helsinki University, Unioninkatu 44, PO-Box 7, 00014, Finland

R.E. Stotler • B. Crandall-Stotler
Department of Plant Biology, Southern Illinois University, Carbondale, IL 62901-6509, USA

N. Wickett
Department of Biology, The Pennsylvania State University, University Park, PA 16802-5301, USA

P.J. Rudall
Royal Botanic Gardens, Kew, Richmond, Surrey TW9 3AB, UK

H. Maas-van de Kamer • P.J.M. Maas
Naturalis Biodiversity Center, Wageningen University, Generaal Foulkesweg 37, 6703 BL Wageningen, The Netherlands

V.S.F.T. Merckx (ed.), *Mycoheterotrophy: The Biology of Plants Living on Fungi*,
DOI 10.1007/978-1-4614-5209-6_2,

keep in mind that the "full mycoheterotrophy" status of many species listed here remains speculative until a careful physiological analysis has been carried out.

Partial and initial mycoheterotrophs were not included in our overview, although we tried to mention confirmed partially mycoheterotrophic species where appropriate. Partial mycoheterotrophy has been detected in green Orchidaceae, Ericaceae, and Gentianaceae but may occur in several other plant families (Selosse and Roy 2009). In addition, probably all Orchidaceae species are initial mycoheterotrophs (Leake 1994; Chaps. 1 and 5) and this group includes over 20,000 species. Furthermore, many species that produce small dust-like seeds (e.g., Pyroleae in Ericaceae) may depend on a mycorrhizal fungus during early development and are thus initial mycoheterotrophs (Chap. 1). Currently, we know little about the phylogenetic range of partial and initial mycoheterotrophy in most plant lineages, but the chlorophyllous relatives of full mycoheterotrophs are the prime candidates to discover more partial and initial mycoheterotrophs, and we hope that this overview may be a valuable tool in search of new mycoheterotrophs.

The taxonomic affinities of many groups of mycoheterotrophs have puzzled systematists for almost three centuries. Many mycoheterotrophic plants are rare or at least very difficult to find, and in extreme cases particular species are known only from one or two collections. Obtaining study material is therefore often the first obstacle to be tackled when trying to unravel the evolutionary history of these intriguing plants. In addition, mycoheterotrophic plants have evolved convergent adaptations in their morphology and anatomy as a result of their peculiar mode of life, making identification of the close relatives of mycoheterotrophic plants in many cases a taxonomic challenge. The application of DNA data has offered new opportunities to elucide the relationships of mycoheterotrophic plant groups, although scarcity of study material and elevated DNA substitution rates have prevented an accurate inference of phylogenetic relationships of many mycoheterotrophic lineages to date (see Chap. 5). This overview has been compiled using the latest taxonomic and phylogenetic insights of the groups in question. For information on species numbers and distributions the International Plant Names Index (IPNI 2011), the World Checklist of Selected Plant Families (2011), and Tropicos (2011) were of great value, although in some cases we adopted information from alternative sources. In a few cases the choice between contradicting taxonomical classifications was made entirely arbitrarily. As mentioned above, the taxonomic status of many groups of mycoheterotrophs remains doubtful and it is likely to change, as new data will become available. Moreover, new species and even genera are still being identified and there is no doubt that more species are waiting to be discovered.

2.2 Liverworts

Liverworts (Marchantiophyta) are postulated to be the sister lineage to all other land plants and to have had a Late Ordovician origin (Heinrichs et al. 2007). Together with mosses and hornworts, they comprise a paraphyletic grade of the embryophytes in which the haploid gametophyte is the persistent autotrophic generation and the diploid sporophyte is short-lived, unbranched, and permanently matrotrophic. In liverworts the gametophyte can be either a thalloid or a leafy plant and the sporophyte is enclosed by tissues of the gametophyte until sporogenesis is completed.

Liverworts consist of approximately 8,000 species in nearly 500 genera. The gametophyte phase of many liverworts forms an association with endophytic fungi, including mucoromycetes, glomeromycetes, ascomycetes, and basidiomycetes (Pressel et al. 2010). In contrast to ferns, this association has never been observed in the sporophyte phase (Ligrone et al. 1993). Considering the huge number of taxa and the antiquity of the lineage, it is surprising that only a single fern taxon, namely *Aneura mirabilis* (Malmb.) Wickett & Goffinet (=*Cryptothallus* Malmb.), has ever been demonstrated to be fully mycoheterotrophic. This taxon, described by von Malmborg (1933) as *Cryptothallus mirabilis*, was first recorded from France by Denis (1919) who regarded it as an achlorophyllous form of *Aneura pinguis* and compared it with gametophytes of the clubmosses *Lycopodium selago* and *L. phlegmaria* (=*Huperzia phlegmaria*). Recent molecular analysis by Wickett and Goffinet (2008) has not supported its recognition as an autonomous genus, rather showing it to be *Aneura*, as Denis had originally suggested. It is interesting that *Tulasnella*, the basidiomycete fungal symbiont of the photosynthetic *Aneura pinguis*, is the same genus found in the mycoheterotrophic liverwort *Aneura mirabilis*. Although a second species of *Cryptothallus*, *C. hirsutus*, was described from Costa Rica (Crum and Bruce 1996), this problematic taxon was never proven to be mycoheterotrophic and subsequent efforts to collect this liverwort from the type locality proved to be unsuccessful.

2.2.1 Aneuraceae

Aneuraceae H. Klinggr., Höh. Crypt. Preuss.: 11 (1858).
Riccardiaceae Sanborn, Univ. Oregon Publ., Pl. Biol. Ser. 1(1): 33 (1929).

Persistent gametophyte plants thalloid, with the thallus often fleshy, 1–10 cm long, 0.2–12 mm wide, lacking a strongly differentiated midrib; vegetative branching monopodial or absent; ventral slime papillae 2-celled, in 2 rows or dispersed; oil bodies small and numerous or large and 1 to 3 per cell, finely granular; monoicous or dioicous; antheridia sunken in chambers on abbreviated lateral branches (on the main thallus in *Verdoornia*); archegonia in clusters, with paraphyses, on abbreviated lateral branches (on the main thallus in *Verdoornia*); sporophytes enclosed by a fleshy shoot calyptra or coelocaule; capsules ellipsoidial to cylindric, with the wall 2 cells thick, usually opening by 4 valves, with an apical elaterophore; elaters 1(2)-spiraled; gemmae usually endogenous (exogenous, but rare in *Aneura*).

Number of genera and species—The family comprises 4 genera with fewer than 150 accepted species. *Riccardia* is the largest genus with ca. 100 species while *Aneura* includes 10 to 20 species. *Lobatiriccardia*, with eight species, was formerly regarded as a subgenus of *Riccardia*. *Verdoornia* is a monotypic New Zealand endemic.

Distribution and habitat—Aneuraceae are distributed worldwide, from the High Arctic to Fuegia and the islands of the Sub-Antarctic. Populations can be found from sea level to over 4,000 m elevations, growing on moist rotting wood and bark, wet rocks, and over moist soil or boggy ground. The single mycoheterotrophic taxon, *Aneura mirabilis*, is a northern oceanic disjunct from Greenland and Europe.

Classification—Molecular phylogenetic analyses have consistently resolved two backbone lineages of simple thalloid liverworts, recognized as the subclasses Pelliidae and Metzgeriidae of the Jungermanniopsida (Forrest et al. 2006; He-Nygrén et al. 2006; Crandall-Stotler et al. 2009). Aneuraceae form a monophyletic lineage that is sister to the Metzgeriaceae within the Metzgeriidae. A recent phylogenetic hypothesis of relationships within the family resolves each of the four genera as monophyletic and strongly supports a sister group relationship between *Aneura* and *Lobatiriccardia*, with *Riccardia* and *Verdoonia* forming successive sister groups (Preußing et al. 2010).

Evolutionary history—Members of the Aneuraceae are absent from the fossil record, but calculations of divergence times from chloroplast sequence data estimate an origin of the family in the mid Jurassic, about 170 Ma (Heinrichs et al. 2007). Recent molecular analyses evince considerable cryptic speciation, especially in *Aneura* (Wachowiak et al. 2007), and morphological character reconstructions support recognition of the family as a crown group in liverwort phylogeny (Crandall-Stotler et al. 2005).

Mycoheterotrophy—Associations between liverworts and fungi are very common, and Aneuraceae is no exception. Typically, the fungal partner in these associations is either an ascomycete or glomeromycete (Davis and Shaw 2008; Bidartondo and Duckett 2010) or, in the earliest diverging lineage, a mucoromycete (Pressel et al. 2010); however, Aneuraceae is unique among liverworts in having an intracellular association with the basidiomycete genus *Tulasnella* (Kottke and Nebel 2005). Preußing et al. (2010) suggest that the well developed, tulanelloid mycothallus may have opened up novel habitats, which subsequently contributed to this being the most speciose group of simple thalloid liverworts. In *Riccardia*, the mycothallus, if present, is epidermal, whereas the remaining Aneuraceae are characterized by a parenchymal mycothallus. With a fleshy thallus composed almost entirely of parenchyma, *A. mirabilis* is the most intensely colonized species, and derives its fixed atmospheric carbon from its endophyte by way of either *Pinus* or *Betula* upon which the fungal partner is simultaneously ectomycorrhizal (Bidartondo et al. 2003). In all cases, the fungus enters the thallus at the base of the rhizoids and colonizes the interior of the host cells after penetrating the cell walls; there is no intercellular growth of the fungus (Preußing et al. 2010).

2.2.1.1 *Aneura* (Fig. 2.1a)

Aneura Dumort., Comment. Bot.: 115 (1822).
Cryptothallus Malmb., Ann. Bryol. 6: 122 (1933).

Thalli large, 1–5 cm long, 2–10 mm wide, thick and fleshy, sparingly and very irregularly branched, occasionally simple; oil bodies small, numerous, 6 to 40 per cell; dioicous, heteromorphic, with male plants somewhat smaller; antheridia in 2 to 6 rows on lateral branches; archegonia clustered in small lateral notches of the thallus margin.

The number of accepted *Aneura* species is problematic, ranging from as few as 6 to as many as 20. Apart from *A. mirabilis*, which is easily distinguished from other species by its lack of chlorophyll, morphological diagnoses of *Aneura* haven proven to be difficult to reconcile with patterns of DNA sequence variation; it has often been acknowledged that this genus is in need of intense morphological and molecular study in order to understand species level relationships. *Aneura mirabilis*, which is the only fully mycoheterotrophic member of the genus, is nested within the chlorophyllose *Aneura pinguis* complex (Wickett and Goffinet 2008; Preußing et al. 2010). It is predominantly found in oceanic Europe with a distribution covering the United Kingdom, Germany, France, Portugal, Russia, Sweden and Norway (Schuster 1992). Additionally, a disjunct locality has been reported from the western Greenland (Peterson 1972), and ecological models have predicted that its range should include Spain (Sérgio et al. 2005). In its northern localities, *A. mirabilis* is generally found buried up to 20 cm beneath the surface of *Sphagnum* dominated peat bogs (pH ± 3.8), occurring in proximity to birch trees from which the liverwort derives its fixed atmospheric carbon by way of its mycobiont. It is associated closely with pines in its more southern localities (e.g., Portugal) where it is frequently covered by mats of pleurocarpous mosses. It was initially suggested that *A. mirabilis* was saprophytic, harboring a fungal partner different from surrounding tree roots. However, clear transfer of carbon fixed by the associated tree to the liverwort via the fungal hyphae has been demonstrated (Bidartondo et al. 2003) and fungi with identical genotypes have been collected from *A. mirabilis* and a neighboring tree (Bidartondo and Duckett 2010). It appears that the mycobiont of *A. mirabilis* is species-specific given that it will not invade chlorophyllose *Aneura* (Duckett et al. 2004).

Unlike its sister species *Aneura pinguis*, it appears that *A. mirabilis* develops in response to temperature, rather than day length (Benson-Evans 1961). Fertilization occurs in June or July, followed by the development of the sporophyte, within a protective calyptra, with the seta elongating the following spring (Benson-Evans 1952, 1960). The elongation of the seta pushes the capsule nearer to the surface of the overlaying vegetation, allowing the spores to be dispersed in sufficient light. Various experimental conditions have been tested to determine optimal conditions for spore germination and suggest that diffuse light is required for, and frost exposure increases the rate of germination. Fungi do not colonize either the spores themselves or the sporophytes and gametangia (Ligrone et al. 1993), and spores germinated in culture do not mature beyond a 20- to 30-celled stage that follows the initiation of rhizoids (Benson-Evans 1960). This suggests

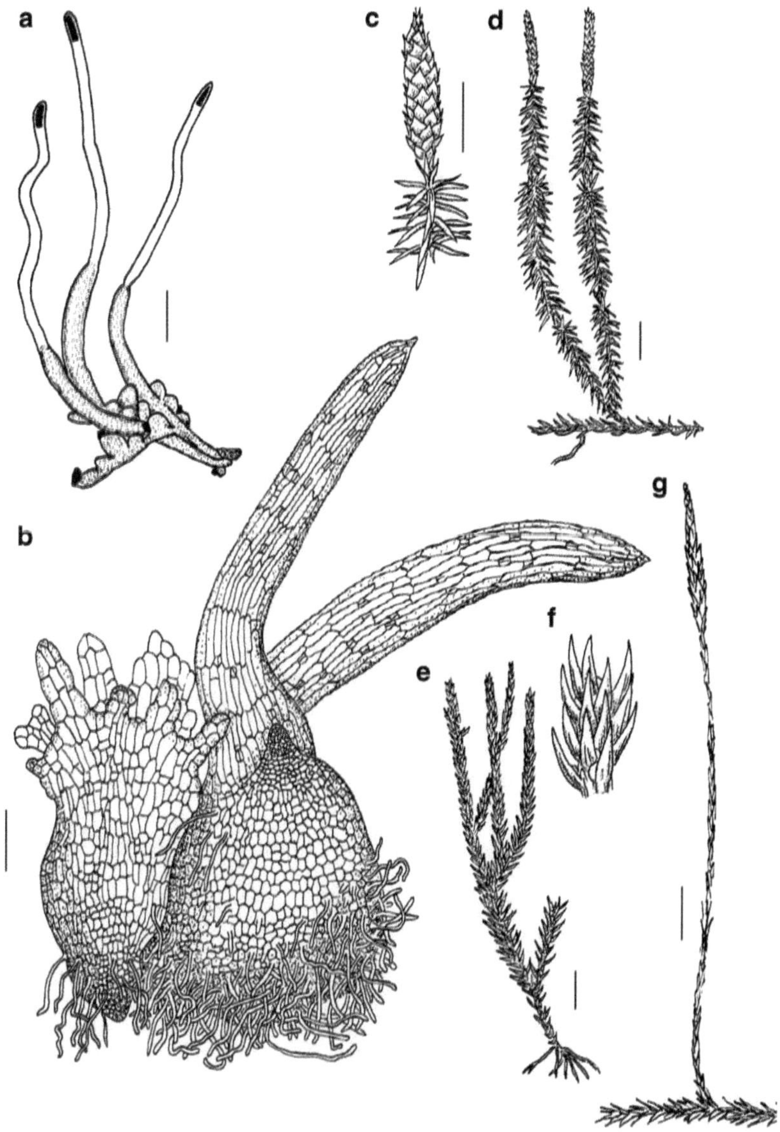

Fig. 2.1 Aneuraceae and Lycopodiaceae. (**a**) Fully mycoheterotrophic gametophyte of *Aneura mirabilis* (Aneuraceae). Drawn from Vanderpoorten and Goffinet (2009). (**b**) Mycoheterotrophic gametophyte (*left*) and emerging sporophyte (*right*) of *Huperzia phlegmaria* (Lycopodiaceae). Redrawn from Haig (2008). (**c**, **d**) Sporophyte of *Lycopodium annotinum*: (**c**) strobilus, (**d**) habit. (**e**, **f**) Sporophyte of *Huperzia selago*: (**e**) habit, (**f**) detail of microphylls. (**g**) Sporophyte of *Lycopodiella appressa*. (**c**–**g**) Redrawn from Wagner and Beitel (1993). Bar = 1 cm, except (**b**) bar = 0.1 cm

Tulasnella must colonize the thallus at this early stage of development. It is important for rhizoids to form pre-colonization as the fungal hyphae penetrate through the rhizoids and proceed to form dense, intracellular coils (Preußing et al. 2010). The liverwort appears to mediate the spread of the fungus, as evidenced by the clear demarcation of the mycothallus (limited to non-reproductive parenchyma), and the ability of a cell to persist (and be subsequently re-colonized) following the death of its intracellular hyphae (Ligrone et al. 1993). While its morphology clearly places it in *Aneura*, *A. mirabilis* is unique in that it is strongly sexually dimorphic and has reticulately ornamented spores that are permanently retained in tetrads (Schuster 1992). Though speculative, it is tempting to invoke the subterranean habitat of *A. mirabilis* as a selective force in the evolution of this divergent spore morphology. Given that the thalli of *A. mirabilis* can be found up to 20 cm beneath the surface of *Sphagnum* peat, it is unlikely that any photosynthesis occurs in this liverwort. Wickett et al. (2008a, b) demonstrated that photosynthesis is likely impossible due to the pseudogenization of, and relaxation of purifying selection on several of the plastid genes that encode structural subunits of the photosynthetic apparatus.

2.3 Ferns and Clubmosses

Of some clubmosses (Lycopodiopsida) and ferns (Polypodiopsida), the sexual stage or gametophyte is achlorophyllous and subterraneous. Many fern and clubmoss gametophytes have mycorrhizal associations as have many fern sporophytes, but in the cases where chlorophyll is absent, the prothallus is entirely dependent on the fungus for survival. These mycoheterotrophic gametophytes are widespread in Lycopodiaceae, Ophioglossaceae, and Psilotaceae, and are also known in *Actinostachys* (Schizaeaceae) and *Stromatopteris* (Gleicheniaceae). In other fern families they are not known to occur.

2.3.1 The Life Cycle of Ferns and Clubmosses

Under normal circumstances there is a regular alternation between a gametophyte (sexual) phase and a sporophyte (asexual) phase. The sporophyte is the dominant generation in ferns (in contrast, the gametophyte is dominant in mosses). It soon becomes independent of the gametophyte and grows to a much greater size and anatomical complexity than the gametophyte. These sporophytes produce haploid spores through meiosis, which, when germinated (usually under warm and moist conditions), form a sexual, free-living, haploid gametophyte. Gametophytes (prothalli) often resemble a liverwort or alga and form the gametes. The male gametes, produced in numbers by antheridia are known as antherozoids or spermatozoids, because they are flagellated and able to swim in water. The female gametes (or egg cells) are non-motile and are borne singly in flask-shaped archegonia. Under humid circumstances a fusion between an egg-cell and an antherozoid may result in the formation of a zygote, containing the combined nuclear material of the two gametes (diploid). The zygote develops through mitotic division into the diploid sporophyte. When mature, the sporophyte produces non-motile, haploid spores that are formed through meiosis, completing the life cycle.

The complexity of the sporophyte in comparison with the gametophyte allows the sporophyte to live under a much wider range of environmental conditions than the gametophyte. The sporophyte is however dependent to grow in places where the gametophyte can survive long enough for fertilization to take place. This is limiting species whose gametophytes are lacking a cuticle and are thus susceptible to dehydration, but not all gametophytes are limited that way: some make associations with fungi, are subterranean, and others are retained within the resistant spore wall (e.g., in heterosporous species).

2.3.2 Gametophytes and Mycoheterotrophy

Gametophytes of ferns are usually chlorophyllous, but in some cases the gametophyte is formed underground and lacks chlorophyll. This is especially the case in Lycopodiaceae (clubmosses) and early branching ferns like Ophioglossaceae and Psilotaceae. These gametophytes are assumed to be mycoheterotrophic, but studies have not been carried out to establish the intricate relationship between the achlorophyllous gametophytes of ferns and clubmosses and the fungus that it associates with.

2.3.3 Clubmosses

The clubmosses (Lycopodiopsida) consists of three families in three orders within the subclass

Lycopodiidae (Christenhusz et al. 2011a) that together form the sister lineage to all other vascular plants and are not closely allied to ferns. Only Lycopodiaceae show mycoheterotrophy at the gametophyte stage. The other two families, Isoëtaceae and Selaginellaceae are both heterosporous and have endosporic gametophytes that do not produce fungal associations. In all families mycorrhizal associations are known to occur at the sporophytic stage.

2.3.4 Lycopodiaceae

Lycopodiaceae P.Beauv. ex Mirb. in Lam. & Mirb., Hist. Nat. Vég. 4: 293 (1802).

Terrestrial or epiphytic plants. Habit erect, trailing or pendent. Stems usually branched, densely clothed nearly throughout with numerous small, simple, 1-nerved leaves, these all similar or dimorphic. Sporophylls similar to other leaves or modified and aggregated to form compact spikes (strobiles). Sporangia axillary, monolocular. Spores all alike, minute, very numerous. Gametophytes (hemi-)mycoheterotrophic, fleshy or tuberous, with or without chlorophyll, monoecious.

Number of genera and species—The Lycopodiaceae comprise 3 genera and ca. 300 species. The largest genus is *Huperzia* with ca. 200 species, followed by *Lycopodium* with ca. 40 species and *Lycopodiella* with ca. 40. Because of the varied growth forms of species within these genera, some authors have adopted a system with more genera, accepting *Phlegmariurus* for epiphytic *Huperzia*, *Palhinhaea* for the *Lycopodiella cernua* complex, and *Diphasiastrum* for *Lycopodium* with flattened branchlets (the cypress clubmosses).

Distribution and habitat—Lycopodiaceae occur worldwide in a variety of habitats. They can be terrestrial or epiphytic, erect or pendent, creeping or climbing plants, and are usually confined to habitats that are at least humid during some part of the year. *Huperzia drummondii* is peculiar in surviving dry periods by forming underground tubers.

Classification—Traditionally Lycopodiaceae was treated as a family of two genera *Lycopodium* and *Phylloglossum* (e.g., Sporne 1962; Bierhorst 1971; Tryon and Tryon 1982). Further division of *Lycopodium* led to as many as 12 genera (Rothmaler 1964; Holub 1983), mainly based on habit morphology. Øllgaard (1987) proposed a consensus classification, which I (MJMC) recommend to be followed, dividing the genus into four genera, but keeping many of the names proposed by Holub at subgeneric levels. Recent molecular studies (Wikström and Kenrick 1997; Wikström et al. 1999) supported Øllgaard's classification in which Lycopodiaceae consist of three major clades: *Lycopodiella* and *Lycopodium* s.s., the pair being sister to an enlarged *Huperzia* (including *Phylloglossum*). To maintain *Phylloglossum*, *Phlegmariurus* will have to be redefined, but in that case it has few morphological characters that separate it from *Huperzia* s.s., and thus inclusion of *Phylloglossum* in *Huperzia* appears to be a better alternative. Further division of *Lycopodiella* and *Lycopodium* is possible (as validated by Holub (1983); Øllgaard (2012)), but has only created confusion in the past, and should therefore be avoided.

Evolutionary history—Lycopods are only a fraction of the present-day vascular plant diversity, but their peak of evolution happened about 300 million years ago (Ma). During the Carboniferous, lycopods were a conspicuous and abundant element of the land flora (Kenrick and Crane 1997). Several groups have now become extinct, and the once dominant clade of rhizomorphic species that included arborescent species of the Carboniferous coal swamps, have been reduced to the few small herbaceous species of *Isoëtes*.

Because of the long fossil record lycopods were recognized as descendents of early divergences in the land plant evolution. Molecular studies have supported this view and place the lycopods as sister to all other vascular plants (Raubeson and Jansen 1992).

Mycoheterotrophy—In some species spores germinate without delay while on the surface of the ground, and then form chlorophyllous gametophytes (*Lycopodiella*). In other species there may be delay of many years in germination, by which time the spores may have been buried and they will become colorless and dependent on a mycorrhizal association. An arbuscular mycorrhizal association appears to occur in all species growing under natural circumstances (Sporne 1962; Winther and Friedman 2008 and references therein).

Photosynthetic (putative partially mycoheterotrophic) gametophytes of *Lycopodiella* and some *Huperzia* are cone-shaped, and have an upper green part, and a colorless lower part with fungal hyphae, whereas in mycoheterotrophic gametophytes the green part is missing and fungal hyphae occur throughout the gametophyte. In both cases the archegonia and antheridia are restricted to the upper part. Epiphytic species of *Huperzia* also have colorless prothalli, but they are slender, branching, and show a pronounced apical growth (Sporne 1962).

2.3.4.1 *Huperzia* (Fig. 2.1b, e, f)

Huperzia Bernh., J. Bot. (Schrader) 1800(2): 126 (1801).
Phylloglossum Kunze, Bot. Zeit. 1 (1843).
Urostachys Herter, Beih. Bot. Centralbl. 39 Abt. 2: 249 (1922).
Phlegmariurus (Herter) Holub, Preslia 36: 17 (1964).

Terrestrial, lithophytic or epiphytic plants. Roots in distal parts of shoots, but sometimes branches rooting near their tips or along prostrate shoots. Sometimes (*H. drummondii*) forming a subterraneous tuber on a leafless geotropic branch. Shoots clustered, dichotomously branching, erect or pendent, clothed with numerous monomorphic or dimorphic leaves, these imbricate or not, sometimes bearing gemmiferous branchlets and gemmae, or all leaves clustered in a basal rosette (*H. drummondii*). Sporangia reniform, borne singly in the axils of undifferentiated or highly differentiated sporophylls. Gametophytes achlorophyllous, mycoheterotrophic (but green in a few species), cylindrical, with pluricellular uniseriate hairs among gametangia.

A genus of over 200 species, but due to numerous species complexes the number is uncertain (Tryon and Tryon 1982). The genus is cosmopolitan extending from the tropics (there mainly epiphytic) to the Arctic and Subantarctic. The peculiar *Huperzia drummondii* (formerly in the monotypic genus *Phylloglossum*) has green mycorrhizal gametophytes and is the only Lycopodiaceae that produces tubers as a survival strategy for dry seasons. Its morphology is highly derived, but phylogenetically it is embedded within the genus *Huperzia* (Wikström and Kenrick 1997). *Huperzia selago* (which is the type of *Huperzia*, and which was found to be sister to *H. drummondii*) can have either colorless or green gametophytes (hemimycoheterotrophic), and can thus be considered a transition from mycoheterotrophic towards independent gametophytes. Most other *Huperzia* (such as the epiphytic *H. phlegmaria*) have mycoheterotrophic gametophytes. Mycoheterotrophic gametophytes of *Huperzia hypogeae* were found to grow with *Glomus* Group A fungi (Winther and Friedman 2008).

2.3.4.2 *Lycopodiella* (Fig. 2.1g)

Lycopodiella Holub, Preslia 36: 20 (1964).
Palhinhaea Franco & Carv., in Carv. Vasc. & Franco, Bol. Soc. Brot. Ser. 2, 41: 24 (1967).
Lateristachys Holub, Folia Geobot. Phytotax. 18(4): 440 (1983).
Pseudolycopodiella Holub, Folia Geobot. Phytotax. 18(4): 441 (1983).

Terrestrial plants with indeterminate main stems (rhizomes) which can be subterraneous, scrambling or creeping, indeterminate. Side branches determinate, simple to much-branched, arising dorsally along main stems. Leaves monomorphic or dimorphic. Strobili erect to nodding or pendulous, terminating simple (rarely forked) branches or much-branched branchlet systems. Sporophylls subpeltate, medially basiscopically winged or with membranes almost enclosing the sporangia. Sporangia axillary or on the base of sporophylls. Spores rugose. Gametophytes green, tuberous, lobed above, living on the surface, hemimycoheterotrophic, lacking pluricellular hairs.

A genus of ca. 40 species widespread in moist-temperate and tropical regions. Especially diverse in the New World.

2.3.4.3 *Lycopodium* (Figs. 2.1c, d and 4.1a)

Lycopodium L., Sp. Pl. 1100 (1753).
Lepidotis P.Beauv. ex Mirb. in Lam. & Mirb., Hist. Nat. Vég. 3: 477; 4: 311 (1802).
Diphasium Rothm., Feddes Repert. Spec. Nov. Regni Veg. 54: 64 (1944).
Diphasiastrum Holub, Preslia 47: 104 (1975).
Lycopodiastrum Holub ex R.D.Dixit, J. Bombay Nat. Hist. Soc. 77(3): 540 (1980 publ. 1981).
Pseudolycopodium Holub, Folia Geobot. Phytotax. 18(4): 441 (1983).
Pseudodiphasium Holub, Folia Geobot. Phytotax. 18(4): 440 (1983).
Austrolycopodium Holub, Folia Geobot. Phytotax. 26(1): 91 (1991).
Dendrolycopodium A.Haynes, Fam. Huperziac. Lycopodiac. New England 84 (2003).
Spinulum A.Haynes, Fam. Huperziac. Lycopodiac. New England 85 (2003).

Terrestrial or lithophytic plants. Main stems trailing along the ground, sometimes climbing, indeterminate, rooting on the underside, usually long-creeping or arching and rooting. Erect determinate shoots scattered along horizontal stems, unbranched or dichotomously or irregularly branched. Leaves spiral to subverticillate, or distinctly flattened and in rows, monomorphic or dimorphic. Strobili erect, sessile or pedunculate or pendulous to nodding. Sporophylls peltate, subpeltate or paleate. Sporangia attached to sporophyll base or axillary, reniform. Gametophytes subterranean, tuberous, mycoheterotrophic, lacking hairs among gametangia.

A genus of ca. 40 species widespread in temperate and tropical (montane) regions.

2.3.5 Ferns

Ferns (Polypodiopsida) or sometimes informally called monilophytes (to include the former "fern allies" Psilotaceae and Equisetaceae) are the second diverging lineage of vascular cryptogams. They consist of 45 families in 11 orders (Christenhusz et al. 2011a). Mycoheterotrophy of the gametophyte is relatively rare and only known with certainty in Ophioglossaceae, Psilotaceae, *Actinostachys* (Schizaeaceae) and *Stromatopteris* (Gleicheniaceae). There may be cases of mycoheterotrophy in Polypodiales (containing the bulk of ferns with ca. 15,000 species), but this is yet to be confirmed. The majority of ferns have chlorophyllous heart- or butterfly-shaped gametophytes, although rarely other shapes (strap-shaped or filamentose) occur. Partly subterraneous kinds are also reported (Sporne 1962), but are extremely rare. Further study on subterraneous gametophytes is needed, but it is complicated because they are difficult to find. Many families are known to have arbuscular mycorrhizal associations (e.g., Marattiaceae, Schizaeaceae, Gleicheniaceae), but while many species seem to be dependent on their fungal association, mycoheterotrophy at the gametophyte phase seems to be occur in only a few species.

2.3.6 Ophioglossaceae

Ophioglossaceae Martinov, Tekhno-Bot. Slovar: 438 (1820).
Helminthostachyaceae Ching, Bull. Fan Mem. Inst. Biol., Bot. 10: 235 (1941).

Terrestrial or epiphytic plants consisting of a fleshy rhizome bearing numerous fleshy roots, and one to several leaves. The growth is not circinnate as in most other ferns but the parts are folded or bent in bud. Leaves erect or pendent, consisting of a petiole bearing at the apex a simple or variously divided blade. Part of the blade may be fertile, and form erect or divergent sporangia-bearing spikes. Sporangia in two rows, naked, each opening by a slit. Gametophytes subterranean, very small, tuber-like, usually without chlorophyll.

Number of genera and species—The Ophioglossaceae comprise four (or five) genera with 60 or more species. *Mankyua* and *Helminthostachys* are monotypic, whereas *Botrychium* includes about 25 species and *Ophioglossum* 25–30.

Distribution and habitat—The Ophioglossaceae occur worldwide from the tropics to the Arctic and Antarctic. They grow epiphytically or in moist woodland or grassland settings.

Classification—The Ophioglossaceae is an early branching lineage of ferns, not closely related to any other group of vascular plants. Together with Psilotaceae it forms the sister group to all other ferns. Stevenson and Loconte (1996) concluded that *Ophioglossum* and *Botrychium* are sister taxa based on their transverse dehiscence of the sporangia. *Helminthostachys* has vertical dehiscence and they therefore considered this genus "ancestral" and segregated it into Helminthostachyaceae. It is however better pertained within Ophioglossaceae, because the genus is certainly allied to *Botrychium* and *Ophioglossum*. The placement of *Mankyua* is uncertain but it has transverse sporangia dehiscense and is thus probably most closely related to *Ophioglossum*.

Evolutionary history—The fossil history of Ophioglossaceae is very limited. Only a single macrofossil is known of the family, which is from the Palaeocene (Rothwell and Stockey 1989), but the family certainly is much older. This lack of fossils is probably due to the soft tissue decaying swiftly and the habitats where these plants grow providing little chance for fossilization. Nevertheless the family must have an ancient origin because of their large number of plesiomorphic characters. Kato (1990) suggested that the "three-dimensional construction of ophioglossoid fertile leaves with epiphyllous sporophores may be a hypothetical archetype for angiosperm carpels with adaxial ovules."

Mycoheterotrophy—The prothallus of all genera is mycorrhizal, the appropriate arbuscular mycorrhizal fungus is needed for the growth of the prothallus (Winther and Friedman 2007; Smith and Read 2008). In most cases the prothallus is subterraneous and lacks chlorophyll, although cases of superficial green prothalli have been reported (Sporne 1962).

Prothalli are tuberous bodies; flattened in *Botrychium*, cilindrical and elongated in *Ophioglossum*, not unlike rhizome parts. Often a large part of the mycorrhizal fungus is located in an enlarged bulbous base in *Ophioglossum*. The antheridia appear first and are deeply sunken, producing very large numbers of antherozoids. The archegonium is stalked in *Botrychium* and sunken in *Ophioglossum*. The prothalli of *Helminthostachys* and *Mankyua* are not known, but are presumably similar to those of *Ophioglossum* or *Botrychium*.

2.3.6.1 *Botrychium* (Figs. 2.2a and 4.1d)

Botrychium Sw., J. Bot. (Schrader) 1800(2): 8 (1801).
Botrypus Michx., Fl. Bor.-Amer. 2: 274 (1803).
Sceptridium Lyon, Bot. Gaz. 40: 457 (1905).
Japanobotrychium Masam., J. Soc. Trop. Agric. 3: 246 (1931).
Osmundopteris (Milde) Small, Ferns Southeast. States 377, 482 (1938).

Perennial terrestrial ferns. Rhizome subterranean, short, erect, usually unbranched. Leaves 1 or 2, the younger enclosing the buds or succeeding leaves in the sheathing petiole base, consisting of a petiole and a blade divided into a sterile and a fertile part. The sterile blade pinnate to deltate-decompound, the fertile portion being equivalent to the basal two divisions of the blade that have fused and bear sporangia on simple or paniculate branches. Sporangia globose, not immersed in the tissue. Spores trilete, thick-walled, exospore verrucose or reticulate. Gametophytes cylindric or oblong and flattened, unbranched, mycoheterotrophic, entirely without chlorophyll.

A genus of ca. 25 species widely distributed throughout the world in boreal, temperate and tropical regions (in the tropics mostly in mountain areas). The mycoheterotrophic gametophytes and autotrophic sporophytes (which are initially mycoheterotrophic) of *Botrychium lanceolatum* and *B. crenulatum* are associated with *Glomus* Group A fungi (Winther and Friedman 2007). The autotrophic sporophyte of *B. virginianum* is able to associate both with *Glomus* Group A and Gigasporaceae fungi (Kovács et al. 2007).

2.3.6.2 *Helminthostachys* (Fig. 2.2b)

Helminthostachys Kaulf., Enum. Filicum 28 (1824).
Botryopteris C.Presl, Rel. Haenk. 1: 76 (1825), non B.Renault (1875 = Botryopteridaceae fossil).
Ophiala Desv., Mém. Soc. Linn. Paris 6: 195 (1827).

Perennial terrestrial ferns with creeping rhizomes bearing thick fleshy roots. Leaves

Fig. 2.2 Ophioglossaceae, Psilotaceae, Gleicheniaceae, and Schizaeaceae (**a**) Sporophyte of *Botrychium lunaria* (Ophioglossaceae). Redrawn from Wagner and Wagner (1993). (**b**) Sporophyte of *Helminthostachys zeylanica*. Redrawn from Shieh and DeVol (1994). (**c**) Sporophyte of *Ophioglossum nudicaule* (Ophioglossaceae). Redrawn from Wagner and Wagner (1993). (**d**) Sporophyte of *Mankyua chejuense* (Ophioglossaceae). Redrawn from Sun et al. (2001). (**e**) Sporophyte of *Tmesipteris tannensis* (Psilotaceae). Redrawn from McLintock (1966). (**f**) Sporophyte of *Psilotum nudum* (Psilotaceae). Redrawn from Castroviejo (1998). (**g**) Sporophyte of *Stromatopteris moniliformis* (Gleicheniaceae). Redrawn from Diels (1902). (**h**) Sporophyte of *Actinostachys pennula* (Schizaeaceae). Redrawn from Wagner (1993). Bar = 1 cm

consisting of a common basal petiole with two rounded stipules at the base, a palmately divided sterile lamina and an erect sporophore projecting above the lamina and consisting of numerous crowded short lateral sporangiate branches. Sporangia large, globose, opening by a vertical slit. Spores granular, yellow. Gametophytes not sufficiently known.

Helminthostachys is a monotypic genus—*H. zeylanica* being the only species—extending from India and Sri Lanka through Malesia and New Guinea, north to southern Japan and south to Australia and New Caledonia. It grows in forest edges usually in alluvial soils by streams or rivers, or in rich organic soil in swamps (Chinnock 1998b).

2.3.6.3 *Mankyua* (Fig. 2.2d)

Mankyua B.-Y.Sun, M.H.Kim et C.H.Kim, Taxon 50: 1020 (2001, publ. 2002).

Perennial terrestrial ferns. Rhizomes tuberous, short, horizontally creeping, unbranched, Roots fleshy, cylindrical, sparsely branched, without root hairs, producing buds. Leaves usually 1, rarely 2, consisting of a common petiole, a thin sterile blade that is terately divided in several ovate to lanceolate lobes, margins dentate, the segments sessile. Sporophores spike-like, arising from top of a common stalk, placed at the base of the sterile blade, the spikes held above these, the spikes simple or branched. Sporangia sunken in fleshy sporophore, horizontally dehiscent. Spores yellowish white. Gametophytes unknown.

The single species *M. chejuense* is the only species in the genus and is only known from Cheju island south of the Korean Peninsula in a lowland swampy area under evergreen broad-leaved forest (Sun et al. 2001). Only 20 plants are reported and thus the species is of conservation concern.

Mankyua shares fertile characters with *Ophioglossum* in the structure of the sporophyll, but resembles in habit a small *Helminthostachys*.

2.3.6.4 *Ophioglossum* (Fig. 2.2c)

Ophioglossum L., Sp. Pl. 1062 (1753).
Ophioderma (Blume) Endlicher, Gen. 66 (1836).
Rhizoglossum C.Presl, Suppl. Tent. Pteridog. 47 (1845), non Kylin (1924, Rhodophyceae, Delesseriaceae).
Cheiroglossa C.Presl, Suppl. Tent. Pteridogr. 56 (1845).
Cassiopteris H.Karst., Linnaea 20: 437 (1847), nom. inval.
Holubiella Škoda, Preslia 68(4): 345 (1996, publ. 1997).

Perennial or annual, terrestrial or epiphytic herbs. Rhizomes short, usually erect, terminating in an erect exposed bud. Leaves erect or pendent, glabrous, somewhat fleshy or leathery, consisting of a petiole and separate sterile and fertile portions of the blade. Sterile blade simple or palmately lobed, sessile or short-stalked, venation reticulate, the primary areoles enclosing free veinlets and sometimes secondary areoles. Fertile portion (sporophore) one or several, simple, stalked spikes. Sporangia sunken, subglobose, more or less coalescent in two marginal rows. Spores yellowish, thick-walled. Gametophytes small, cylindric or ovoid, simple or branched, mycoheterotrophic, without chlorophyll.

About 25–30 species occurring almost throughout the world, except for very cold Arctic or Antarctic regions. It is a taxonomically difficult genus due to the plasticity of the species and lack of morphological characters. *Ophioglossum reticulatum* is known to be the plant with the highest number of chromosomes of $n=720$ (Khandelwal 1990). The epiphytic *O. palmatum* is sometimes maintained in its separate genus *Cheiroglossa* (Christenhusz et al. 2011a).

2.3.7 Psilotaceae

Psilotaceae J.W.Griff. & Henfr., *Microgr. Dict.*: 540 (1855).
Tmesipteridaceae Nakai, Chosakuronbun Mokuroku [Ord. Fam. Trib. Nov.]: 206 (1943).

Epiphytic, lithophytic or sometimes terrestrial plants. Rhizomes without roots, beset with short hair-like rhizoids. Stems green, simple and provided with small two-ranked leaves or several times dichotomously branched and appearing leafless (the leaves being minute, scale-like and far apart). Sporangia bilocular or trilocular, attached on the adaxial base of minute bifid sporophylls, dehiscing vertically. Spores reniform, all similar. Gametophytes tuberous, subterranean or embedded in humus, mycoheterotrophic.

Number of genera and species—The Psilotaceae consists of two genera and ca. 17 species.

Distribution and habitat—Pantropical, extending into warm temperate areas, epiphytic on trees and tree ferns in peat bogs and in crevices of rocks and on walls.

Classification—The placement of Psilotaceae has always been disputed. Because of the protostele and lack of roots it has a superficial similarity to Psilophytopsida/Rhyniopsida—an extinct group of Devonian plants—but these are more likely to be secondary plesiomorphic characters.

The family consists of two genera that are sister to each other. This clade pairs distantly with Ophioglossaceae as sister to the rest of the ferns (incl. Equisetaceae). The Psilotaceae and Ophioglossaceae share several characters of the sporangia, leaf division into a fertile and sterile part, a reduced root system and mycoheterotrophic gametophytes.

Evolutionary history—Apart from an early Tertiary macrofossil of *Tmesipteris* (Carpenter 1988), the Psilotaceae are without a macrofossil record. The family has traditionally been compared with an early Devonian group of land plants, the Psilophytales (which includes the superficially similar *Rhynia*) or with the lycopods.

Bierhorst (1977) noted affinities with Schizaeales and Gleicheniales. This affiliation has since been rejected (Voirin and Jay 1997; Fineran and Ingerfeld 1985; Brownsey and Lovis 1987). However molecular studies showed Psilotaceae to be sister to Ophioglossaceae (Hasebe et al. 1995).

Mycoheterotrophy—Few botanists have had the good fortune to see living specimens of the gametophyte of either *Psilotum* or *Tmesipteris*, but all remark on their similarity and their resemblance to portions of sporophytic rhizome. They are irregularly dichotomizing colorless, cylindrical structures, covered with rhizoids, and they are packed with mycorrhizal fungus hyphae. Archegonia and antheridia are borne together on the same prothallus (monoicous), but because of their small size it is impossible to distinguish gametophytes from bits of sporophyte rhizomes in the field (Sporne 1962). Both the gametophytes and sporophytes of Psilotaceae form arbuscualr mycorrhizal associations (Wang and Qiu 2006 and references therein).

2.3.7.1 *Psilotum* (Figs. 2.2f and 4.1g)

Psilotum Sw., J. Bot. (Schrader) 1800(2): 8 (1801).
Bernhardia Willd. ex Bernh., J. Bot. (Schrader) 1800(2): 132 (1801).

Erect or pendent plants, epiphytic in crevices of cliffs or old walls or (rarely) terrestrial in peat, humus or on gravel. Rhizome short-creeping, dichotomously branched, beset with small, brown, hair-like rhizoids. Stems loosely clustered, the lower unbranched part more or less elongate, dichotomously branched above into numerous narrow divisions. Leaves alternate, 2- or 3-ranked, minute, gradually tapering to a sharp thin point. Sporangia depressed-globose, sessile, trilocular, 3-lobed. Spores hyaline. Gametophytes subterraneous or in humus, mycoheterotrophic.

A genus of two species: *Psilotum nudum* has a pantropical distribution extending into warm-temperate areas, occurring in a wide range of habitats, whereas *P. complanatum* also has a pantropical distribution, but is restricted to montane rainforests. The mycoheterotrophic gametophytes and autotrophic sporophytes (which are initially mycoheterotrophic) of *Psilotum nudum* are associated with *Glomus* Group A fungi (Winther and Friedman 2009).

2.3.7.2 *Tmesipteris* (Fig. 2.2e)

Tmesipteris Bernh., J. Bot. (Schrader) 1800(2): 131 (1801)

Usually epiphytic plants. Rhizome short-creeping, dichotomously branched, with brown rhizoidal hairs. Shoots pendent or erect, unbranched, or rarely a few branched, beset with leaves. Sterile leaves scale-like at the base of the shoot, large and leaf-like above, decurrent, 1-veined, entire, spirally or distichously arranged. Fertile leaves bifid but otherwise similar in size and form to sterile ones or somewhat smaller, produced basally, in distinct zones, or irregularly along a shoot. Sporangia fused into synangia, these large, bilobed, brown. Gametophytes subterraneous, mycoheterotrophic.

A genus of about 15 species extending from the Philippines to Australia, New Zealand and New Caledonia, and east to Samoa, Fiji, French Polynesia and the Marquesas Islands (Chinnock 1998a).

2.3.8 Gleicheniaceae

Gleicheniaceae C.Presl, Reliq. Haenk. 1: 70 (1825).
Mertensiaceae Corda, Fl. d. Vorwelt: 89 (1845).
Stromatopteridaceae Bierh., Phytomorphology 18: 263 (1968).

Terrestrial ferns with long-creeping rhizomes. Leaves pinnate or more complex, indeterminate, usually pseudodichotomously forked (except *Stromatopteris*), the leaves often branching through axillary buds. Veins free. Soria abaxial, not marginal, exindusiate. Sporangia 5–15(–20), each with a complete transverse medial annulus, opening by a longitudinal slit. Gametophyte green, costate, with club-shaped hairs, or, in the case of *Stromatopteris*, mycoheterotrophic, subterranean and cylindrical.

Number of genera and species—The family consists of six genera (*Dicranopteris*, *Diplopterygium*, *Gleichenella*, *Gleichenia*, *Sticherus*, *Stromatopteris*), with ca. 130 species. Only *Stromatopteris* has mycoheterotrophic gametophytes.

Distribution and habitat—Tropical and Southern Hemisphere, with species reaching Japan. It shows a Gondwana distribution, with centers of diversity in Australasia and South America.

Classification—The Gleicheniaceae (including Stromatopteridaceae) are placed together with the Dipteridaceae and Matoniaceae in the Gleicheniales. The Gleicheniales are placed between Schizaeales and Hymenophyllales among the leptosporangiate ferns, but their exact phylogenetic position is not yet clear.

Evolutionary history—The family is obviously of Antarctic origin, where many fossils dating back to the Cretaceous have been found. Older fossils, such as *Antarctipteris* and *Gleichenipteris*, may be ancestral Gleicheniaceae, but are more likely belonging to extinct lineages of Gleicheniales.

Mycoheterotrophy—Bierhorst (1969) discussed mycoheterotrophy in *Stromatopteris*. Because of the similarities in gametophytes between *Stromatopteris* and *Psilotum* he assumed that the two genera were related. It is now known that the similarities are likely due to convergence rather than to true affinity (Kato 1983). The gametophytes of *Stromatopteris* are subterranean, cylindric, non-green, mycoheterotrophic, bearing rhizoids and superficial gametangia and resemble the rhizome of the sporophyte. The antheridia are large, many-celled and produce numerous spermatozoids. The archegonia are long-necked and variously oriented.

2.3.8.1 *Stromatopteris* (Fig. 2.2g)

Stromatopteris Mett., Ann. Sci. Nat. IV 15: 84 (1861).

Rhizomes creeping, horizontal, protostelic, with vertical, slender unequally dichotomous branches bearing the leaves. Young parts scaly. Leaves several, rigidly erect; petiole long, dark. Lamina erect, 1-pinnate, with numerous, imbricate, simple, ovate-rounded, entire or slightly lobed, coriaceous pinnae, adnate to the dark, sulcate rachis, the margin more or less revolute. Venation obscure, anadromous, free. Sori usually 1 per pinna, roundish, consisting of 15–20 large sporangia, intermingled with small irregularly shaped scales. Sporangial stalk uniseriate, annulus oblique to nearly transverse. Spores monolete, ellipsoidal, rugulose-reticulate.

A genus with a single species *Stromatopteris moniliformis*, endemic to New Caledonia, where it is frequent on serpentine, ultrabasic soils in the southern third of the island, mostly in open places and macchia-like vegetation (Brownlie 1969).

2.3.9 Schizaeaceae

Schizaeaceae Kaulf., *Wesen Farrenkr.*: 119 (1827).

Terrestrial ferns with short-creeping or erect rhizomes. Leaves erect, simple or fan-shaped, lamina and petiole hardly distinguised in some species. Blades flabellate and entire or dichotomously incised to strongly reduced, straplike and similar to the green petiole. Veins dichotomous, free. Sporangia sessile, on marginal, compact, pectinate-pinnate or pseudodigitate, branched or unbranched projections at blade tips, not in discrete sori, exindusiate; spores bilateral, monolete. Gametophytes filamentose, partly green with special rhizoid-bearing cells, antheridia on short branches and archegonia on the filaments or on anchegoniophores (in *Schizaea*), or subterranean, fleshy, tuber-like, achlorophyllous and mycoheterotrophic (in *Actinostachys*).

Number of genera and species—The small family Schizaeaceae is now confined to the two genera *Actinostachys* and *Schizaea*, with a total of ca. 30 species.

Distribution and habitat—Pantropical America and Asia, and southern warm-temperate (in Africa only in the South), one outlying species (*Schizaea pusilla* Pursh) occurs in the temperate zone from New Jersey to Newfoundland. The species are always terrestrial, often on substrates poor in minerals, or on decaying wood.

Classification—Formerly the genera *Lygodium* (Lygodiaceae) and *Anemia* (Anemiaceae) were included in Schizaeaceae, but they diverge significantly and are thus placed into their own families within Schizaeales.

Evolutionary history—*Schizaeopsis*, a Cretaceous fossil is the oldest one assigned to this lineage (Wikström et al. 2002).

2.3.9.1 *Actinostachys* (Fig. 2.2h)

Actinostachys Wallich, Num. List 1. 1829.

Actinostachys differs from *Schizaea* in having a pseudodigitate fertile segment, sporangia in four rows (instead of two) and tuberous mycoheterotrophic gametophytes. Because of the similarities (the digitate fertile segments are essentially pinnate as in *Schizaea*), the genus *Actinostachys* is sometimes considered to be part of *Schizaea*. The differences are clear-cut and thus it is maintained at the level of genus.

2.4 Gymnosperms

The existence of mycoheterotrophy in Gymnosperms is under debate and the discussion revolves around the enigmatic plant *Parasitaxus usta*, which is a member of the Podocarpaceae and has a purple reddish appearance. The plant contains chlorophyll but is incapable of measurable photosynthetic electron transport (Feild and Brodribb 2005). *Parasitaxus usta* is always found sprouting from roots and (rarely) trunks of another podocarp, *Falcatifolium taxoides*, and is therefore often regarded as a holoparasitic plant (de Laubenfels 1959; Köpke et al. 1981). However, a typical haustorium is not formed and the connection to the "host" has been described as an "obligate root graft" (Köpke et al. 1981). Moreover, both *P. usta* and *F. taxoides* are associated with arbuscular mycorrhizal fungi, which are closely associated with the *Parasitaxus–Falcatifolium* union (Woltz et al. 1994; Feild and Brodribb 2005). Stable carbon isotopic measurements indicate that carbon transport from *F. taxoides* to *P. usta* most likely involves this fungal partner (Feild and Brodribb 2005). The situation seems unique in land plants (but see de Vega et al. 2010) and differs from "normal" mycoheterotrophic plants that are not directly linked to another plant (but see *Exochaenium* in Gentianaceae). Despite its doubtful classification as a mycoheterotrophic plant, we included this species in our overview.

In addition to *P. usta*, achlorophyllous specimens of *Sequoia sempervirens* (Taxodiaceae) are known. These plants obtain nutrients by grafting their root system with that of surrounding autotrophic specimens (Davis and Holderman 1980). Because there seem to be no fungi involved in the interaction these "albino" redwood trees are probably not mycoheterotrophic and can best be categorized as parasitic plants.

2.4.1 Podocarpaceae

Podocarpaceae Endl., Synopsis Coniferarum 203 (1847).

Phyllocladaceae Bessey, Nebraska Univ. Stud. 7: 325 (1907).
Phyllocladaceae E.L.Core ex H.Keng, Taiwania 18: 142 (1973), nom, illeg.
Pherosphaeraceae Nakai, Tyosen-Sanrin158: 15 (1938).
Nageiaceae D.Z.Fu, Acta Phytotax. Sin.: 522 (1992).
Acmopylaceae Melikian & A.V.Bobrov, Proc. Intern. Conf. Plant Anat. Morph. (St. Petersburg) 1997: 93 (1997).
Saxegothaeaceae Gaussen ex Doweld & Reveal, Phytologia 84: 365 (1999).
Microcachrydaceae Doweld & Reveal, Phytologia 84: 365 (1999).
Bracteocarpaceae Melikian & A.V.Bobrov, Bot. Zhurn. (Moscow & Leningrad) 85: 60 (2000).
Dacrycarpaceae Melikian & A.V.Bobrov, Bot. Zhurn. (Moscow & Leningrad) 85: 59 (2000).
Falcatifoliaceae Melikian & A.V.Bobrov, Bot. Zhurn. (Moscow & Leningrad) 85: 61 (2000).
Halocarpaceae Melikian & A.V.Bobrov, Bot. Zhurn. (Moscow & Leningrad) 85: 60 (2000).
Lepidothamnaceae Melikian & A.V.Bobrov, Bot. Zhurn. (Moscow & Leningrad) 85: 63 (2000).
Microstrobaceae Doweld & Reveal, Novon 11: 396 (2001).
Parasitaxaceae Melikian & A.V.Bobrov, Bot. Zhurn. (Moscow & Leningrad) 85: 61 (2000).
Prumnopityaceae Melikian & A.V.Bobrov, Bot. Zhurn. (Moscow & Leningrad) 85: 58 (2000).

Evergreen shrubs or trees, usually with straight trunk and more or less horizontal branches. Branching typically with extra, weaker branches along the trunk between the main tiers of 3 and 5 major branches. Leaves usually spirally arranged, or in pairs radiating around the twig, or arranged distichously on more or less flat rows on either side of predominantly horizontal branchlets, needle-like, or broader leaves, usually with a single vein. Plants monoecious or dioecious. Pollen cones usually catkin-like; stamens numerous, close together, imbricate, each with two sporangia; pollen grains usually winged. Female cones maturing in 1 year, much reduced to a few fleshy bracts or scales, pendant, usually borne on a thin peduncle, containing a single inverted ovule. Seeds completely covered by a fleshy structure referred to as an epimatium, wingless. Epimatium and integument sometimes connate and forming a leathery testa. Cotyledons 2, with 2 parallel vascular bundles.

Number of genera and species—Nine genera and ca. 180 species (Christenhusz et al. 2011b). The sole species in *Parasitaxus* is entirely reddish purple and has a heterotrophic mode of life. All other Podocarpaceae species are evergreen trees or shrubs (Eckenwalder 2009).

Dictribution and habitat—Podocarpaceae are widespread in the southern hemisphere and reach their northern distribution limits in Mexico, the West Indies (25°N), East Africa (35°N) and Japan (35°N) (Eckenwalder 2009). Most members of the family are trees native to wet tropical or subtropical forests. A few are small trees or shrubs native to forest understory environments (Earle 1997).

Classification—There has been some debate about the position of the genus *Phyllocladus* either within the Podocarpaceae or as an independent family Phyllocladaceae (Tomlinson et al. 1997; Bobrov et al. 1999; Sinclair et al. 2002). Recent molecular analyses link *Phyllocladus* to Podocarpaceae and suggest that *Phyllocladus* forms a distinct lineage that diverged early in the evolutionary history of Podocarpaceae (Wagstaff 2004). Christenhusz et al. (2011b) classified Podocarpaceae in Araucariales and provided a full synonymy.

Evolutionary history—The current distribution of Podocarpaceae and their fossil record suggests that the family had an extensive distribution over southern Gondwana but also occurred in Laurasia. South America and Antarctica are possibly the cradle of much of the modern Podocarpaceae diversity, and while Malesia and Australasia have the greatest diversity of living podocarps these distributions are regarded as the result of secondary radiations (Mill 2003).

Ecology—All investigated Podocarpaceae are associated with arbuscular mycorrhizal fungi (Wang and Qiu 2006). Like all conifers, Podocarpaceae are wind-pollinated. Podocarpaceae seeds are associated with fleshy parts, and are presumably dispersed by birds. Seed dispersal by streams and rivers has been suggested as well (Page 1990).

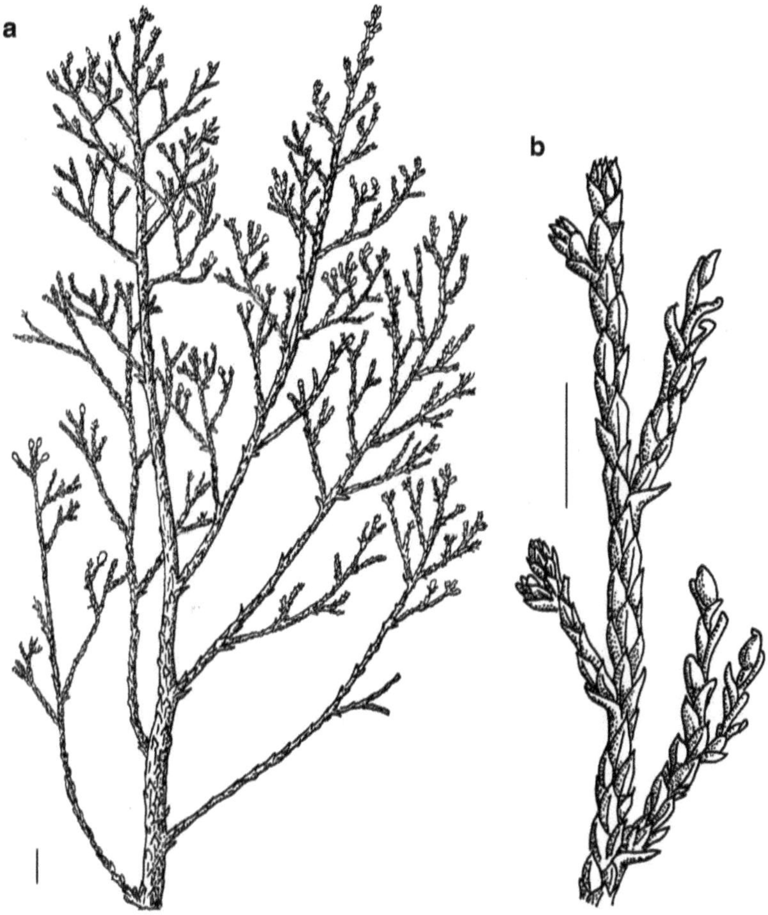

Fig. 2.3 *Parasitaxus usta* (Podocarpaceae): (**a**) habit, (**b**) branch with pollen cones. Redrawn from de Laubenfels (1972). Bar = 1 cm

2.4.1.1 *Parasitaxus* (Fig. 2.3)

Parasitaxus de Laub., Fl. N. Caledonia 4: 44 (1972).

Coral-like, red or purple shrub, 1–1.8(-3) m tall. Roots absent. Stems erect, repeatedly branched, emerging from the roots or lower stems of the host plant. Leaves dense, spirally inserted, scale-like. Plants monoecious. Pollen cones single, at the end of branchlets or in the axils of foliage leaves. Seed cones on a short, leafy stalk to 5 mm long. Combined seed coat and epimatium fleshy, becoming wrinkled when dry, pale bluish white with a thick coating of wax over a reddish purple skin, nearly spherical, 2.5–4 mm in diameter, with a tiny beak at the tip that becomes more pronounced while drying.

Parasitaxus contains a single species, *P. usta*. Its distribution is slightly more restricted than its host plant *Falcatifolium taxoides* (Podocarpaceae), but both species are found in wet montane cloud forests on serpentine soils across New Caledonia and on Île des Pins (Feild and Brodribb 2005; Eckenwalder 2009). Morphological observations suggest that arbuscular mycorrhizal fungi are involved in the interaction between *F. taxoides* and *P. usta* (Woltz et al. 1994), but molecular identification studies are needed to confirm this. Pollination and seed dispersal mechanisms remain to be studied.

2.5 Angiosperms

Mycoheterotrophy occurs in ten families of angiosperms and in terms of species numbers most angiosperms capable of mycoheterotrophy are monocots. We identified ca. 515 angiosperm species that are putatively fully mycoheterotrophic, 468 in monocots (in 7 families) and 47 in eudicots (in 3 families). Full mycoheterotrophy has evolved at least 45 times independently in angiosperms (Chap. 5). The existence of partial mycoheterotrophy has been demonstrated in several species of Orchidaceae and Pyroleae (Ericaceae). Probably all orchid species (>20,000 spp.) are initial mycoheterotrophs and initial mycoheterotrophy may also occur in *Pyrola* (Ericaceae) (Smith and Read 2008) and all other plant species that produce dust-like seeds as well (Eriksson and Kainulainen 2011; Chap. 5).

2.5.1 Petrosaviaceae

Petrosaviaceae Hutch., Fam. Fl. Pl. 2: 36 (1934).
Petrosavieae Engl. In Engl. & Prantl, Nat. Pflanzenfam. Nachtr.: 71,72 (1897).
Miyoshiaceae Nakai, J. Jap. Bot 17: 190 (1941).
Japonoliriaceae Takht., Bot. Zhurn. (Moscow & Leningrad) 79: 97 (1994).
Japonolireae (Takht.) M. N. Tamura in Kubitski, Fam. & Gen. Vasc. Pl. 3: 390 (1998).

Achlorophyllous and mycoheterotrophic (*Petrosavia*) or autotrophic (*Japonolirion*) herbs. Rhizome slender, creeping to erect, with scale-like leaves. Roots filiform. Stems erect, simple. Leaves cauline, reduced to scales and distichous (*Petrosavia*) or basal, linear, and spiral (*Japonolirion*). Inflorescence a terminal bracteate raceme, sometimes corymbiform; each flower subtended by a well-developed bract and possessing a single bracteole in the same radius as one of inner tepals. Flowers bisexual, actinomorphic. Tepals 6, in 2 whorls, those of the outer whorl smaller than those of the inner whorl; tepals erect to patent, persistent, basally connate (*Petrosavia*) or free (*Japonolirion*). Stamens 6; filaments linear, adnate to base of tepals (*Petrosavia*) or free (*Japonolirion*); anthers ovoid, dorsifixed (*Petrosavia*) or basifixed (*Japonolirion*), introrse. Ovary half-inferior (*Petrosavia*) or superior (*Japonolirion*), tricarpellate; carpels in *Petrosavia* postgenitally fused only near their bases (also fused peripherally due to formation of a semi-inferior ovary); carpels in *Japonolirion* postgenitally connate up to the stylar region; ovules numerous (*Petrosavia*) or 4–5 per carpel (*Japonolirion*). Septal nectaries present in both genera; in *Japonolirion* mostly located below the level of the ovary locules (infralocular). Styles 1 per carpel, short, stigmata subcapitate (*Petrosavia*) or linear and recurved (*Japonolirion*). Fruit dry, composed of 3 horizontally patent capsules dehiscing longitudinally and septicidally to the upper side (*Petrosavia*) or 1 ellipsoid, 3-carpellate capsule, dehiscing septicidally (*Japonolirion*). Seeds ellipsoid to broadly ellipsoid, more or less winged (*Petrosavia*) or wingless (*Japonolirion*) (Tamura 1998; Rudall 2002; Cameron et al. 2003; Remizowa et al. 2006a, b).

Number of genera and species—Petrosaviaceae comprise two genera (*Japonolirion* and *Petrosavia*) and four species. All three species of *Petrosavia* are fully mycoheterotrophic.

Distribution and habitat—Petrosaviaceae occur in Southeast Asia and Japan (Cameron et al. 2003). The only species of *Japonolirion* is restricted to serpentine swamps in central and northern Japan (Tomimatsu et al. 2004; Tamura 1998), while *Petrosavia* occurs in forests in Southeast Asia and southern Japan (Chen and Tamura 2000; Ohashi 2000).

Classification—Both genera of Petrosaviaceae have been placed in various groups. For *Petrosavia*, relationships with Melanthiaceae (Beccari 1871), Liliaceae (Engler 1888), Alismatales (Hutchinson 1959), Tofieldiaceae and Nartheciaceae (Tamura 1998), Triuridaceae (Cronquist 1981, 1988), and other families have been suggested (see Cameron et al. 2003 for an overview). *Japonolirion* has been placed in Liliaceae (Nakai 1930; Ohwi 1965), Melanthiaceae (Brummitt 1992; Mabberley 1997), Nartheciaceae (Tamura 1998), and Japonoliriaceae (Takhtajan 1996, 1997). Molecular phylogenetic analyses using plastid and/or nuclear DNA sequences suggest that

Petrosavia is the sister-group of *Japonolirion* (Caddick et al. 2000b; Chase et al. 2000; Fuse and Tamura 2000; Soltis et al. 2000) and that these two genera are phylogenetically isolated within the monocots, not closely related to any of the aforementioned families. Based on this molecular evidence, Cameron et al. (2003) proposed recognition of the family Petrosaviaceae, consisting of *Petrosavia* and *Japonolirion*, and its placement in its own order, Petrosaviales, which was adopted by the APG classification (APG 2009). Although a close relationship between *Petrosavia* and *Japonolirion* was unexpected due to their different mode of life, there are many similarities in their morphology (Cameron et al. 2003; Remizowa et al. 2006a, b; Tobe 2008; Tobe and Takahashi 2009).

Evolutionary history—Petrosaviaceae are sister to the liliid/commelinid clade (Chase 2004; Tamura et al. 2004; Chase et al. 2006; Givnish et al. 2006; Graham et al. 2006). Molecular clock analyses suggest that the Petrosaviaceae diverged during the Early Cretaceous, and the split between *Japonolirion* and *Petrosavia* also occurred in the Early Cretaceous (Janssen and Bremer 2004; Magallón and Castillo 2009).

Ecology—*Petrosavia* is associated with arbuscular mycorrhizal fungi (Yamato et al. 2011b). *Petrosavia* is capable of both cross-pollination and self-pollination (Takahashi et al. 1993). Seed dispersal agents are unknown.

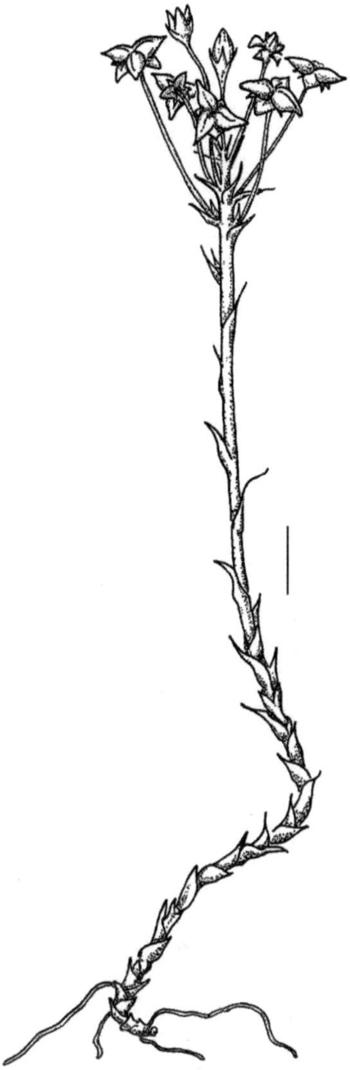

Fig. 2.4 *Petrosavia stellaris*. Redrawn from Tamura (1998). Bar = 1 cm

2.5.1.1 *Petrosavia* (Figs. 2.4 and 2.5a)

Petrosavia Becc., Nuov. Giorn. Bot. Ital. 3: 7 (1871).
Protolirion Ridl., Ann. Bot. 9: 56 (1895).
Miyoshia Makino, Bot. Mag. 17: 144 (1903).

Mycoheterotrophic herbs, cream-colored, 4–30 cm tall. Leaves cauline, reduced to scales and distichous. Inflorescence a sometimes corymbiform raceme, 4- to more than 25-flowered. Flowers funnel-shaped, small, white to yellow, arising from the axil of a small bract, often subtended by a bracteole. Tepals 6, basally connate, inner ones larger than outer ones. Stamens 6, filaments adnate to base of the tepals, anthers dorsifixed. Ovary half-inferior, carpels 3, connate for 1/4–1/2 their length and distally apocarpous, with three separate styles, sometimes only basally connate; ovules numerous. Septal nectaries present. Stigmata subcapitate or slightly 2-cleft. Fruit almost apocarpous, composed of three horizontally patent capsules dehiscing to the upper side. Seeds brown, more or less winged.

Fig. 2.5 Fully mycoheterotrophic species in monocots: (**a**) *Petrosavia stellaris* (Petrosaviaceae) pictured in Borneo. (**b**) *Campylosiphon congestus* (Burmanniaceae) from Cameroon. (**c**) Male plant of *Kupea martinetugei* (Triuridaceae) growing at Diongo Community forest in Cameroon. (**d**) *Gymnosiphon longistylus* (Burmanniaceae) from Cameroon. (**e**) Flower of *Geosiris aphylla* (Iridaceae). Photo by Ehoarn Bidault. (**f**) *Thismia rodwayi* (Thismiaceae) growing on Mount Wellington, Tasmania. (**g**) *Cephalanthera austinieae* (Orchidaceae) photographed in northern California. (**h**) *Afrothismia hydra* (Thismiaceae) from Korup National Park in Cameroon.

Petrosavia comprises three species: *P. sinii*, *P. sakuraii*, and *P. stellaris*, although some authors reduce *P. sinii* to *P. sakuraii* (Ohashi 2000; Cameron et al. 2003; Tobe and Takahashi 2009). *Petrosavia sinii* is endemic to China, *P. sakuraii* is fairly widespread in Southeast Asia (with records from China, Taiwan, Indonesia, Japan, Myanmar, and Vietnam), while *P. stellaris* occurs in Malesia (Chen and Tamura 2000; Ohashi 2000). *Petrosavia* grows in rain forests, mixed forests, bamboo forests, and coniferous forests, from sea level up to 1,800 m (Takahashi et al. 1993; Ohashi 2000; Chen and Tamura 2000).

The pollination biology of *P. sakuraii* was studied by Takahashi et al. (1993). They concluded that the plants are primarily self-pollinating (including insect mediated self-pollination), but cross-pollination by bees and other insects may occur as well. *Petrosavia sakuraii* is associated with a narrow clade of arbuscular mycorrhizal fungi (Yamato et al. 2011b). Dispersal agents remain unknown.

2.5.2 Burmanniaceae

Burmanniaceae Blume, Enum. Pl. Javae 27 (1827).

Achlorophyllous, mycoheterotrophic or autotrophic herbs. Rhizome cylindrical, rarely tuberous or absent, densely covered with scale-like leaves and filiform roots. Stems erect, usually unbranched, leaves alternate, sessile, simple, entire, in achlorophyllous species small and scale-like, in autotrophic species larger and often rosulate. Inflorescence a terminal, bracteate, usually bifurcate, 1-many-flowered cyme, or reduced to a single flower. Flowers syntepalous, actinomorphic, variously colored, campanulate, funnel-shaped, salverform, or tubular. Flower tube wingless, 3- or 6-ribbed, or broadly 3-winged, sessile or pedicellate. Tepals 6, free, entire or sometimes 3-lobed. Stamens 3, erect, inserted in the flower tube just below and opposite the inner tepals, without interstaminal lobes; thecae superposed, transversely dehiscent, connective often bearing apical and basal appendages. Ovary inferior, 1- to 3-locular, with parietal or axile placentation, nectarial glands often present in the septa or on top of the ovary; style equaling the flower tube; stigma variously shaped, sometimes provided with tortuous, filiform appendages; ovules numerous, anatropous. Fruit a capsule, longitudinally or transversely dehiscent, crowned by various flower remnants. Seeds numerous, dust-like, fusiform to subglobose.

Number of genera and species—Burmanniaceae consist of eight genera and ca. 96 species. With respectively 56 and 32 species *Burmannia* and *Gymnosiphon* are the largest genera. The other genera comprise one or two species only. All species, except for 37 species of *Burmannia*, are fully mycoheterotrophic.

Distribution and habitat—Burmanniaceae have a pantropical distribution. The distributions of *Apteria* and a few *Burmannia* species extend into the subtropics. Fully mycoheterotrophic Burmanniaceae species mainly occur in evergreen forest, but *Apteria* sometimes grows in wet savannas. Most species grow at low elevations, but some species occur at 2,000 m and above. Chlorophyllous *Burmannia* species prefer wet savannas or swamps, or grow occasionally in gallery or savanna forests (Maas et al. 1986).

Classification—Burmanniaceae were traditionally associated with the orchids (Lindley 1846; Karsten 1858; Engler 1888; Cronquist 1970; Rübsamen 1986) or other mycoheterotrophic families such as Corsiaceae and Geosiridaceae (now in Iridaceae) (Cronquist 1970; Dahlgren et al. 1985). However, use of molecular data, has shed new light on the position of Burmanniaceae among the monocots. In a phylogenetic analysis of 172 monocot *rbcL* sequences, a *Burmannia* species was sister to *Dioscorea* and *Tacca* (Chase et al. 1995). All subsequent molecular analyses with additional data and sampling recovered a monophyletic family of Burmanniaceae sister to Dioscoreaceae, and therefore part of Dioscoreales (Caddick et al. 2000b, 2002a, b; Soltis et al. 2000; Davis et al. 2004). Only the 26S rDNA analyses by Neyland (2002) and Neyland and Hennigan

(2003) revealed a different hypothesis, with Burmanniaceae (and Corsiaceae) sister to almost all other monocot groups. Analyses using sequence data from the nuclear and mitochondrial genome suggest that Thismiaceae are not part of Burmanniaceae (Merckx et al. 2006; 2009).

Evolutionary history—Burmanniaceae are the second diverging lineage in Dioscoreales (Merckx et al. 2008, 2009). There are no known Burmanniaceae fossils, but molecular clock analyses indicate that the family originated in the Cretaceous (ca. 100–120 Ma). The extant lineages share a common ancestor origniated in West Gondwana during the Late Cretaceous. The diversification rate in Burmanniaceae increased during the warm Eocene, when *Burmannia* and *Gymnosiphon* were able to migrate from the New to the Old World supposedly via boreotropical migration routes (Merckx et al. 2008).

Ecology—The mycorrhizas of only few species of Burmanniaceae have been studied. Morphological observations suggest that they are growing with arbuscular mycorrhizal fungi (Van der Pijl 1934; Terashita and Kawakami 1991; Imhof 1999c). Molecular sequencing detected *Glomus* Group A fungi, and sometimes also Acaulosporaceae fungi, in the roots of *Apteria*, *Burmannia*, *Campylosiphon*, *Gymnosiphon*, and *Hexapterella* species (Leake 2005; Franke et al. 2006; Merckx and Bidartondo 2008; Courty et al. 2011; Merckx et al. 2012). Except for a few *Burmannia* species, pollination has not been studied. The colorful, variously shaped flowers with septal nectaries, indicate insect pollination (Henderson and Stevenson 2004), but in some species of *Burmannia* there is strong evidence for selfpollination (Ernst and Bernard 1912; Wood 1983; Zhang and Saunders 1999, 2000). The tiny seeds are probably dispersed by wind or water (Maas-van de Kamer 1998).

2.5.2.1 *Burmannia* (Figs. 2.6e, f and 4.7g)

Burmannia L., Sp. Pl.: 287 (1753)
Vogelia J.F.Gmel., Syst. Nat. 2: 107 (1791).
Tripterella Michx., Fl. Bor.-Amer. 1: 19 (1803).
Maburnia Thouars, Gen. Nov. Madagasc.: 4 (1806).
Gonianthes Blume, Catalogus: 19 (1823).
Gonyanthes Nees, Ann. Sci. Nat. (Paris) 3: 369 (1824), orth. var.
Tetraptera Miers in J.Lindley, Veg. Kingd., ed. 2.: 172 (1847).
Tripteranthus Wall. ex Miers in J.Lindley, Veg. Kingd., ed. 2.: 172 (1847).
Cryptonema Turcz., Bull. Soc. Imp. Naturalistes Moscou 21(1): 590 (1848).
Nephrocoelium Turcz., Bull. Soc. Imp. Naturalistes Moscou 26(1): 287 (1853).

Mycoheterotrophic, or autotrophic herbs, 5–50 cm tall. Rhizome mostly absent, roots filiform. Leaves green and often rosulate, or without chlorophyll and scale-like. Inflorescence a 1-many- flowered, bifurcate cyme. Flowers erect, often 2-colored, white, yellow, and/or blue, tubular to salverform. Flower tube 3-winged, or 3- or 6-ribbed. Tepals 6, entire, inner tepals smaller than outer ones. Stamens 3, sessile. Ovary 3-locular, with axile placentation, septal nectaries sometimes present; style 3-branched at the apex. Fruit erect, dehiscing longitudinally, transversely, or irregularly, crowned by the persistent perianth. Seeds brown, ellipsoid or rarely narrowly fusiform.

The genus *Burmannia* comprises 19 achlorophyllous mycoheterotrophic species and 37 chlorophyllous species. Some chlorophyllous species are rather robust and have numerous well-developed leaves (e.g., *B. longifolia*, *B. kalbreyeri*, *B. foliosa*). These species are supposedly fully autotrophic. Other chlorophyllous species have reduced vegetative parts and are presumably partial mycoheterotrophs (Leake 1994; but see Merckx et al. 2010b). Full mycoheterotrophy has evolved many times independently in the genus (Merckx et al. 2008). *Burmannia* has a widespread distribution in the tropical and subtropical parts of both the Old and the New World. Nineteen chlorophyllous and one mycoheterotrophic species are found in Central and South America. In Africa (including Madagascar) four chlorophyllous species and one mycoheterotrophic species occur. In Asia, 15 species are chlorophyllous and 17 species are fully mycoheterotrophic. Mycoheterotrophic species grow in rain forests, while chlorophyllous species prefer wet grasslands and swamps (Maas-van de Kamer 1998). All species are terrestrial, except for the chlorophyllous *B. kalbreyeri* from Central and South America, which is an epiphyte growing on various trees

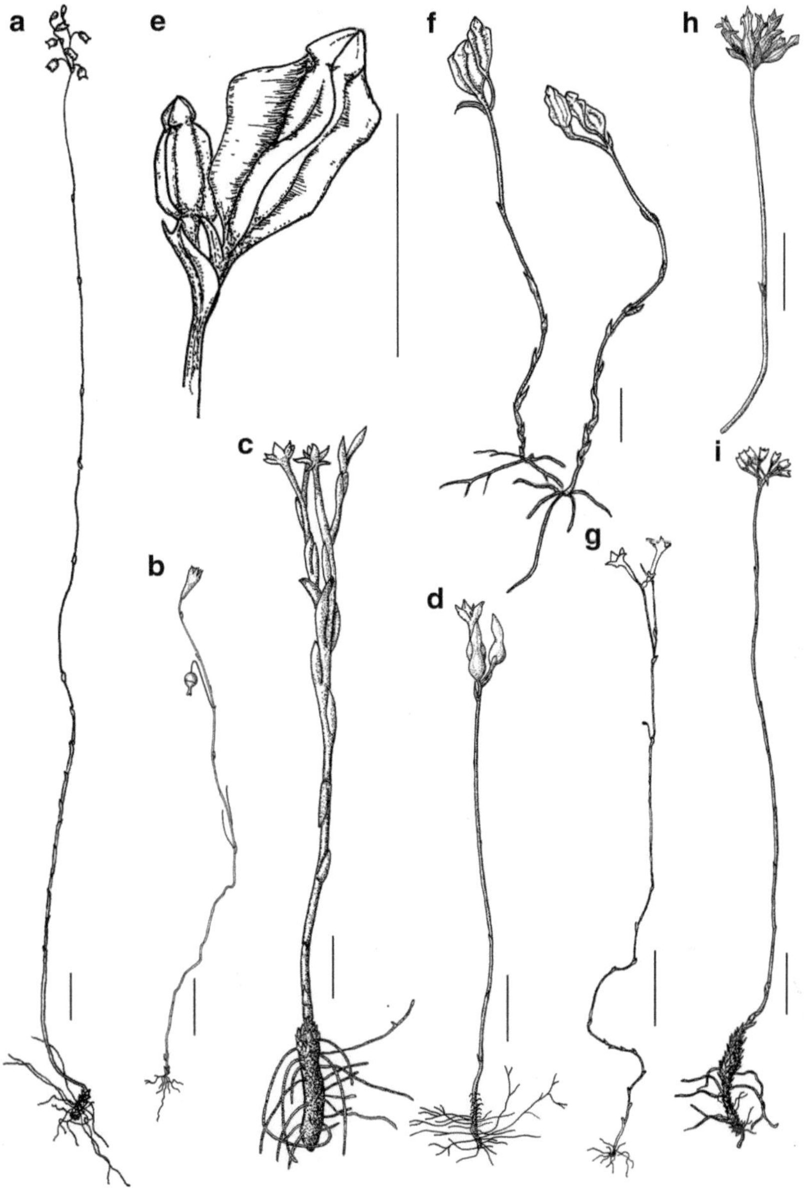

Fig. 2.6 Burmanniaceae. (**a**) *Dictyostega orobanchoides*. (**b**) *Apteria aphylla*. (**c**) *Campylosiphon purpurascens*. (**d**) *Hexapterella gentianoides*. (**e**, **f**) *Burmannia cryptopetala*: (**e**) flower, (**f**) habit. (**g**) *Gymnosiphon divaricatus*. (**h**) *Marthella trinitatis*. (**i**) *Miersiella umbellata*. Bar = 1 cm. Redrawn from Maas et al. (1986), except (**e**, **f**) redrawn from Hsu et al. (2005)

(Maas et al. 1986). *Burmannia* species generally occur at low elevations, but a few species prefer higher elevations of 1,000 m and higher (Jonker 1938; Maas et al. 1986). The Peruvian species *Burmannia stuebellii* has been found up to 4,100 m (León 2006).

Burmannia species are growing with arbuscular mycorrhizal fungi (Van der Pijl 1934; Terashita and Kawakami 1991; Imhof 1999c) belonging to the *Glomus* Group A clade (Franke et al. 2006; Merckx and Bidartondo 2008; Merckx et al. 2010b). Both self-pollination and cross-pollination have been suggested for species of *Burmannia*. Many species of *Burmannia* have strongly colored flowers with prominent wings, presumably to attract pollinators (Maas et al. 1986). Septal nectaries are present (Maas et al. 1986; Caddick et al. 2000a) and the stamens often have glandular apical appendages (Maas et al. 1986). Kato (1996) and Momose et al. (1998) reported that the flowers of *B. lutescens* were visited by mosquitos, suggesting cross-pollination. However, cleistogamy has been observed in the chlorophyllous species *B. capitata* with anther dehiscence in pre-anthesis flowers, and the presence of germinating pollen on the stigmas (Wood 1983). Premature opening of the anthers in pre-anthesis flowers, followed by pollen germination in situ within the anther, and subsequent penetration of the stigma by the pollen tube has furthermore been recorded for the mycoheterotrophic species *B. candida* (Ernst and Bernard 1912), *B. championii* (Ernst and Bernard 1912), and *B. stuebelii* (Spitmann 1975 *fide* Zhang and Saunders 2000). Zhang and Saunders (1999) note that the throat of *B. larseniana*, as with many species in the genus, is blocked by the three stigmatic branches, preventing cross-pollination. The only detailed study on the pollination biology of a *Burmannia* species, concluded that the mycoheterotrophic *B. wallichii* is primarily selfing (Zhang and Saunders 2000). *Burmannia* species have tiny dust-like seeds, which are presumably dispersed by wind or water (Maas et al. 1986).

2.5.2.2 *Campylosiphon* (Figs. 2.5b and 2.6c)

Campylosiphon Benth. in Hooker, Ic. Pl. Ser. 3 14(4): 65 (1882).
Dipterosiphon Huber, Bol. Mus. Paraense Hist. Nat. Ethnogr. 2: 502 (1898).

Mycoheterotrophic herbs, up to 35 cm tall. Rhizome tuberous, cylindrical. Leaves scale-like. Inflorescence a bifurcate 1-many-flowered cyme, flowers sessile or pedicellate. Flowers erect, salverform, blue to white. Flower tube narrowly 3-ribbed. Tepals 6, entire, subequal in length. Stamens 3, sessile. Ovary 1-locular, with 3 parietal placentas in its upper part, 3-locular with axile placentation in its lower part, septal nectaries present; style 3-branched at the apex. Fruit erect, dehiscing irregularly by withering of the fruit wall between the ribs, crowned by the marcescent perianth. Seeds brown, flattened, triangular in outline.

Campylosiphon has a disjunct distribution and contains *C. purpurascens*, which is widely distributed in tropical South America (including Colombia, Venezuela, Guyana, Suriname, French Guiana, Brazil, and Peru) (Maas et al. 1986), and *C. congestus* from West Africa (Guinea, Liberia, Ghana, Nigeria, Cameroon, Gabon, Central African Republic, Angola, and DR Congo) (Bamps and Malaisse 1987; Cheek 2006). Growing in rain forests, often along margins of streams and creeks.

Campylosiphon is dependent on arbuscular mycorrhizal fungi from the *Glomus* Group A clade and the Acaulosporaceae (Franke et al. 2006; Merckx et al. 2012). Pollination syndrome and dispersal agents are unknown.

2.5.2.3 *Hexapterella* (Figs. 2.6d and 4.6f, h)

Hexapterella Urb., Symb. Antill. 3(3): 451 (1903).

Mycoheterotrophic herbs, up to 20 cm tall. Rhizome cylindrical, slightly tuberous. Stems purplish. Leaves scale-like. Inflorescence a bifurcate 1–8-flowered cyme. Flowers erect, salverform, white to purple. Flower tube slightly 6-winged or 6-ribbed. Tepals 6, entire, inner ones much smaller than outer ones, sometimes 3-dentate, soon falling off. Stamens 3, filaments present. Upper part of ovary 1-locular, lower part 3-locular, with 3 parietal placentas, 3 septal nectaries present; style 3-branched at the apex. Fruit erect, dehiscent by transverse slits and/or withering of the fruit wall, crowned by the persistent part of the flower tube. Seeds brown, subglobose to ovoid.

Hexapterella contains two species, *H. steyermarkii* from Venezuela and *H. gentianoides* occurring in lowland forests in Trinidad and northern South America (Colombia, Venezuela, Guyana, Suriname, French Guiana, Brazil) (Maas et al. 1986; Maas and Maas 1989). A specimen of *Hexapterella gentianoides* from French Guiana was found to grow with *Glomus* Group A fungi

(Merckx et al. 2012). Self-pollination seems to occur in *Hexapterella* (Rübsamen 1980). Dispersal agents are unknown.

2.5.2.4 *Dictyostega* (Figs. 2.6a and 4.5a)

Dictyostega Miers, Proc. Linn. Soc. Lond. 1: 61 (1840).

Mycoheterotrophic herbs, 1–50 cm tall. Rhizome cylindric, slightly tuberous. Leaves scale-like. Inflorescence a bifurcate 3-many-flowered cyme. Flowers nodding, tubular, whitish. Flower tube wingless. Tepals 6, entire, inner ones smaller than the outer ones. Stamens 3, sessile. Upper part of ovary 1-locular, with 3 parietal placentas, lower part 3-locular, 3 septal nectaries present, style 3-branched at the apex. Fruit nodding, longitudinally dehiscent, crowned by the persistent perianth. Seeds white, narrowly fusiform.

Dictyostega contains a single, morphological variable species: *D. orobanchoides*. Maas et al. (1986) recognize three subspecies. Widely distributed in the Neotropics, from Mexico in the north to southeastern Brazil in the south, but absent from the West Indies. Growing in rain forests up to 2,600 m (Maas et al. 1986). *D. orobanchoides* is growing with various *Glomus* Group A fungi (Imhof 2001; Merckx et al. 2010b). *Dictyostega* is probably self-pollinating (Miers 1841; Warming 1901). Dispersal agents are unknown.

2.5.2.5 *Miersiella* (Fig. 2.6i)

Miersiella Urb., Symb. Antill. 3(3): 439 (1903).

Mycoheterotrophic herbs, 5–20 cm tall. Rhizome cylindrical, slightly tuberous. Leaves scale-like, almost peltate. Inflorescence a contracted, umbelliform, 4–10(-22)-flowered cyme. Flowers erect, tubular, deep lilac to white. Tepals 6, entire, inner tepals smaller than outer ones. Flower tube wingless. Stamens 3, sessile. Ovary 1-locular, with 3 parietal placentas, and three 2-lobed glands on the top of the ovary; style 3-branched at the apex. Fruit erect, dehiscence longitudinally and loculicidally, crowned by the persistent perianth. Seeds brown, narrowly ellipsoid to ovoid.

Miersiella comprises a single species, *M. umbellata*, growing in dense evergreen rain forests in southeastern and eastern Brazil and the Amazonian parts of Colombia, Venezuela, Guyana, and Peru (Maas et al. 1986; Tropicos 2011). Relationship to the other Burmanniaceae genera remains to be inferred. Pollination syndrome, dispersal agents, and mycorrhizal fungi are unknown.

2.5.2.6 *Gymnosiphon* (Figs. 2.5d and 2.6g)

Gymnosiphon Blume, Enum. Pl. Javae 1: 29 (1827).
Cymbocarpa Miers, Proc. Linn. Soc. London 1: 61 (1840).
Ptychomeria Benth. in Hooker's J. Bot. Kew Gard. Misc. 7: 14 (1855).
Benitzia H. Karst., Linnaea 28: 420 (1857).
Desmogymnosiphon Guinea, Ensayo Geobot. Guin. Continent. Espan.: 264 (1946).

Mycoheterotrophic herbs, up to 30 cm tall. Rhizome cylindrical, slightly tuberous. Leaves scale-like. Inflorescence a bifurcate 1–50-flowered cyme. Flowers erect, salverform, white, occasionally partly yellow or blue, flower tube wingless, upper part very soon falling off. Tepals 6, outer tepals mostly 3-lobed, inner tepals very small often somewhat swollen, inserted in the flower tube below the insertion of the outer tepals. Stamens 3, sessile. Ovary 1-locular, with 3 parietal placentas, septal nectaries present; style 3-branched at the apex, each branch with or without 2 apical, tortuous, filiform appendages. Fruit erect, longitudinally, loculicidally, or irregularly dehiscent, crowned by the persistent part of the flower tube. Seeds greyish-black, ellipsoid or fusiform.

Gymnosiphon has a pantropical distribution, with 16 species in the Neotropics, 8 in Africa (including Madagascar and the Comores), and 8 in Asia. The tiny flowers with soon falling upper parts trouble identification and taxonomy in this genus (Jonker 1938). Species of *Gymnosiphon* occur in lowland rain forests, but some species grow in montane forests up to 2,300 m (Maas et al. 1986). Molecular sequence data suggests that *Gymnosiphon* is the sister clade of *Hexapterella* (Merckx et al. 2008).

Glomus Group A and Acaulosporaceae fungi were detected in the roots of *Gymnosiphon* specimens of various species (Leake 2005; Courty et al. 2011; Merckx et al. 2012). Self-pollination occurs (Maas et al. 1986). Dispersal agents are unknown.

2.5.2.7 *Apteria* (Figs. 2.6b and 4.6b)

Apteria Nutt., J. Acad. Nat. Sci. Philadelphia 7: 64 (1834).
Nemitis Raf., Fl. Tellur. 4: 33 (1838).
Stemoptera Miers, Proc. Linn. Soc. London 1: 62 (1840).

Mycoheterotrophic herbs 5–70 cm tall. Rhizome cylindric, slightly tuberous. Leaves scale-like. Inflorescence a 1–5-flowered cyme. Flowers erect to nodding, funnel-shaped to campanulate. Flower tube wingless. Tepals 6, entire, subequal in length. Stamens 3, filament basally decurrent into a crescent-shaped pouch. Ovary 1-locular, with 3 parietal placentas, 3 septal nectaries present; style 3-branched at the apex. Fruit nodding, longitudinally dehiscent, crowned by the persistent perianth. Seeds brown, ellipsoid to subglobose.

Apteria contains a single species: *A. aphylla*. This species has a very wide distribution, and can be found from southern USA and the West Indies in the north to Peru, Bolivia, Paraguay, and South Brazil in the south (Maas et al. 1986). *Apteria aphylla* grows in rain forests, among decaying leaves or on rotten wood, between mosses and shrubs, or sometimes in savannas.

Apteria is able to associate with *Glomus* Group A and Diversisporales mycorrhizal fungi (Courty et al. 2011; Merckx et al. 2012). There is evidence for self-pollination in *Apteria*, perhaps mediated by flower mites (*Frankeliniella* spp.) (Warming 1901; Ernst and Bernard 1912; Uphof 1929; Maas et al. 1986). Dispersal agents are not known.

2.5.2.8 *Marthella* (Fig. 2.6h)

Marthella Urb., Symb. Antill. 3(3): 447 (1903).

Mycoheterotrophic herbs, up to 10 cm tall. Rhizome cylindric, slightly tuberous. Leaves scale-like. Inflorescence a 2–9-flowered contracted, bifurcate cyme. Flowers erect, tubular, yellowish. Flower tube wingless. Outer tepals entire, inner ones absent. Stamens 3, alternating with the tepals, filament basally decurrent into a crescent-shaped pouch. Ovary 1-locular, with 3 parietal placentas and 3 short-stipate, 2-lobed glands on top of the ovary; style 3-branched at the apex. Fruit erect, crowned by the persistent perianth, dehiscence unknown. Seeds brown, ellipsoid to broadly ovoid.

Marthella is a monotypic genus, only known from Mount Tucuche, Trinidad, growing in rain forests on rotten wood and decaying leaves. The only species, *M. trinitatis*, was last collected in 1898 and may be extinct (Maas et al. 1986). Observations on the biology of this species are lacking.

2.5.3 Thismiaceae

Thismiaceae J. Agardh, Theor. Syst. Pl. Fam. Phan. 99 (1858).

Achlorophyllous, mycoheterotrophic herbs. Underground part tuberous, or a cluster of coralloid or vermiform roots, or creeping cylindrical roots, or a rhizome bearing clumps of small root tubercles. Stems unbranched, leaves alternate, simple, sessile, reduced to scale-like. Flowers terminal, solitary or sometimes in few-flowered monochasial inflorescence, or rarely a panicle composed of few-flowered cincinni. Flowers actinomorphic or zygomorphic, variously colored. Flower tube urceolate or cylindric or obconical, sometimes more or less 2-chambered, sometimes bent in the middle. Tepals 6, free or the inner ones connate into a miter. Stamens 6 or rarely 3, inserted opposite the tepals, pendent or reflexed (*Oxygyne*), thecae dehiscing longitudinally to the abaxial side or latrorse (*Tiputinia*). Ovary inferior, 1-locular with 3 parietal placentae, septal nectaries absent; ovules numerous; style short and thick, sometimes with 3 stigmatic branches or stigma capitate or funnelform. Fruit fleshy and cup-shaped, or a dry capsule (*Haplothismia*). Seeds numerous, dustlike.

Number of genera and species—Thismiaceae comprise five genera and ca. 63 species. All species have a fully mycoheterotrophic mode of life. The largest genus is *Thismia*, with ca. 45 species. A remarkable common feature of most Thismiaceae is their extreme scarcity (Stone 1980; Maas et al. 1986; Franke 2004). The majority of species are known exclusively from the type collection, which in some cases was made more than a century ago (Stone 1980; Maas et al. 1986).

Distribution and habitat—Thismiaceae are widely distributed in the tropical regions of the world, but some species are known from subtropical and even temperate areas (Jonker 1938). The genus *Oxygyne* and several species of *Thismia* have disjunct distribution patterns (Stone 1980; Maas et al. 1986; Abe and Akasawa 1989). Most species occur in the leaf litter of dense tropical rainforest and can only be spotted during the flowering and fruiting period when aboveground organs appear (Maas et al. 1986; Franke 2007).

Classification—Not surprisingly, due to the strong reduction of vegetative organs and the rarity of most species involved, Thismiaceae taxonomy has been the subject of much debate. Most classifications included Thismiaceae, as a subtribe "Thismieae," in a broadly defined Burmanniaceae (Miers 1847; Schlechter 1921; Jonker 1938; Maas et al. 1986; Maas-van de Kamer 1998; Caddick et al. 2002b; APG 2009) while other authors favored the recognition of a separate family of Thismiaceae closely related to the mycoheterotrophic Burmanniaceae (Hutchinson 1934, 1959; Dahlgren et al. 1985; Takhtajan 1997; APG 1998). Thismiaceae or Burmanniaceae (including Thismieae) on their part were linked to various other families, including other mycoheterotrophic groups such as Triuridaceae, Geosiridaceae, Corsiaceae, and Orchidaceae (see Maas et al. 1986 for an overview). However, these relationships are now completely discredited based on convergence of character states involved, due to the mycoheterotrophic mode of life (Soltis et al. 2005). DNA-based phylogenetic analyses place Thismiaceae in Dioscoreales (Caddick et al. 2000b; Caddick et al. 2002a; Davis et al. 2004), but outside Burmanniaceae (Merckx et al. 2006, 2009). Nuclear and mitochondrial DNA data suggest that Thismiaceae are paraphyletic, due to the inclusion of *Tacca* (Merckx et al. 2009).

Evolutionary history—The paraphyletic status of Thismiaceae suggests that a mycoheterotrophic mode of life has evolved independently in *Afrothismia* and in the common ancestor of the remaining Thismiaceae (Merckx et al. 2009). Thismiaceae are absent from the fossil record but according to molecular clock estimates both lineages originated during the Cretaceous (Merckx et al. 2010a).

Ecology—Thismiaceae are poorly known ecologically. The mycorrhizal fungi of several species of *Afrothismia* and a single species of *Thismia* have been identified as arbuscular mycorrhizal fungi belonging to the *Glomus* Group A clade (Franke et al. 2006; Merckx and Bidartondo 2008; Merckx et al. 2012). Pollination has not been studied in detail, but the colorful, variously shaped flowers of most Thismiaceae, and the presence of glandular tissue in the flowers of some *Thismia* species indicate insect pollination (Vogel 1962; Stone 1980; Maas et al. 1986). The particular floral morphology and odor of *Tiputinia* points to sapromyophily (Woodward et al. 2007). The close proximity of the anthers and the stigma in some *Thismia* species suggests self-pollination. The seeds may be dispersed by wind or water (Stone 1980; Maas et al. 1986).

2.5.3.1 *Afrothismia* (Figs. 2.5h, 2.7f, 4.2, 4.3, and 4.4)

Afrothismia Schltr., Bot. Jahrb. Syst. 38: 138 (1906).

Herbs up to 10 cm tall. Rhizome cylindric with clusters of root tubercles, each ending in a terminal rootlet. Stems usually unbranched. Inflorescence a few-flowered cincinnus. Flowers zygomorphic, often with red and yellow pigmentation. Flower tube urceolate to cylindric, 2-chambered and bent in the middle, with an internal flange at the middle and an annulus in the throat. Tepals 6, free, equal or unequal in size. Stamens 6, inserted in the basal part of the flower tube, reflexed; anthers with connective connivent with stigma. Ovary with 3 parietal placentas basally fused into a sterile central column; style with a funnel-shaped stigma. After flowering the perianth and ovary wall soon falling off, only the placentas with the seeds remaining on top of the lengthened sterile central column. Seeds ellipsoid.

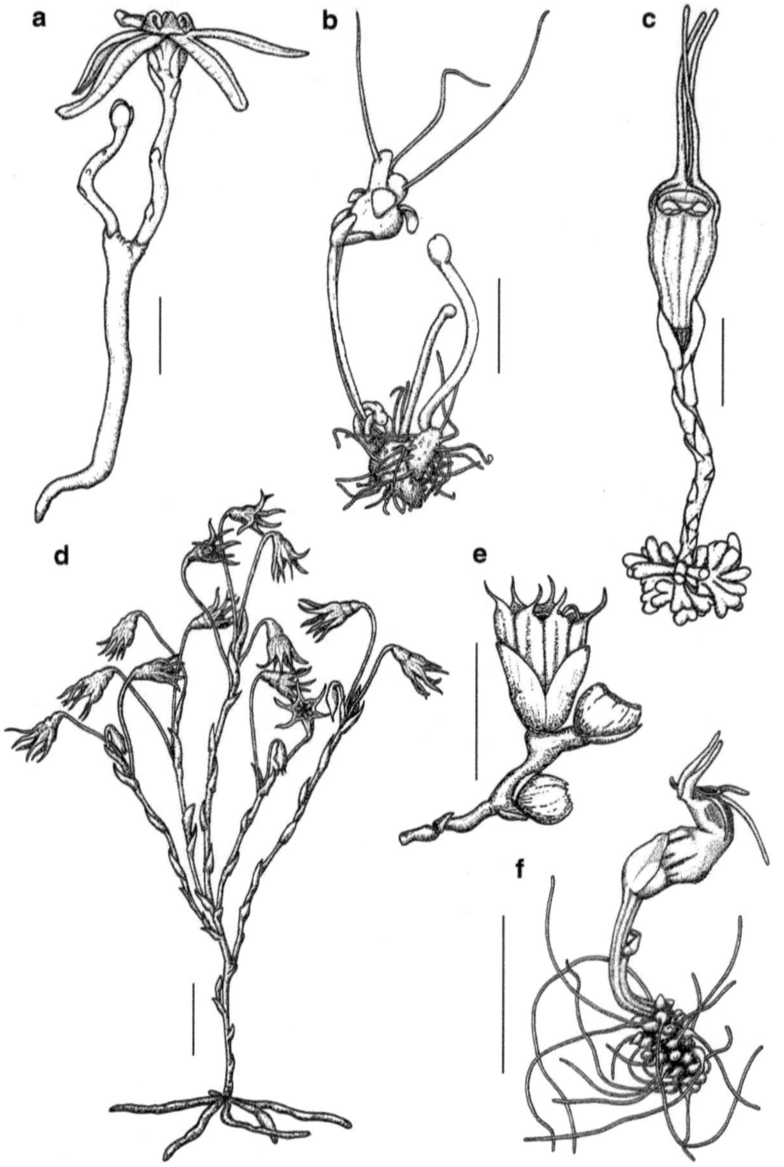

Fig. 2.7 Thismiaceae. (**a**) *Tiputinia foetida*. Redrawn from Woodward et al. (2007). (**b**) *Thismia saülensis*. Redrawn from Maas and Maas (1987). (**c**) *Thismia clavigera*. Redrawn from Chantanaorrapint and Chantanaorrapint (2009). (**d**) *Haplothismia exannulata*. Redrawn from Airy Shaw (1952). (**e**) *Oxygyne hyodoi*. Redrawn from Abe and Akasawa (1989). (**f**) *Afrothismia hydra*. Redrawn from Sainge and Franke (2005). Bar = 1 cm

Currently 12 species of *Afrothismia* are known from tropical Africa, with records from Cameroon, Gabon, Nigeria, Uganda, Malawi, Kenya, and Tanzania (Schlechter 1906; Cowley 1988; Cheek 2003a, 2007, 2009; Maas-van de Kamer 2003; Franke 2004; Franke et al. 2004; Cheek and Jannerup 2005; Sainge and Franke 2005; Sainge et al. 2005; Dauby et al. 2007). Although a few more species from West Africa are awaiting description (Sainge Moses, pers. comm.). The Guineo-Congolian rainforest in southwestern Cameroon is the main center of diversity of the genus (Franke 2007). All species grow exclusively in evergreen rainforest. Species of *Afrothismia* are often found growing with other mycoheterotrophs (Schlechter 1906; Cheek 2003b; Cheek et al. 2003; Sainge et al. 2005). Some species were collected once or are only known from a single locality. The only collection of *A. pachyantha* was made on Mount Cameroon in 1905 by Rudolf Schlechter (Schlechter 1906) and this species is possibly extinct since the type locality has been destroyed by human activity (Franke et al. 2004).

Afrothismia species form very complex mycorrhizas (Imhof 1999a, 2006) with highly specific *Glomus* Group A fungi (Franke et al. 2006; Merckx and Bidartondo 2008). The floral structure of *Afrothismia* species suggests cross-pollination by insects. But there exist only few observations of flower-visiting insects in *Afrothismia*. Engler (1905) mentioned small dipterans he found in the lower part of the perianth tube of *A. winkleri*. Cheek and Williams (1999) reported two dipterans of the same species that left the perianth tube of *A. pachyantha* after a stay of several seconds. Franke (2004) observed a drosophilid fly, which carefully inspected the tepals of an *Afrothismia* flower for several minutes. All these observations strongly suggest myophily. Dispersal agents are unknown.

2.5.3.2 *Thismia* (Figs. 2.5f, 2.7b, c, and 4.10j)

Thismia Griff., Proc. Linn. Soc. London 1: 221 (1845).
Ophiomeris Miers, Proc. Linn. Soc. London 1: 328 (1847).
Sarcosiphon Blume, Mus. Bot. 1: 65 (1850).
Tribrachys Champ. ex Thwaites, Enum. Pl. Zeyl.: 325 (1864)
Myostoma Miers, Trans. Linn. Soc. London 25: 474 (1866).
Bagnisia Becc., Malesia 1: 249 (1878).
Geomitra Becc., Malesia 1: 250 (1878).
Triscyphus Taub., Verh. Bot. Vereins Prov. Brandenburg 36: 66 (1895).
Glaziocharis Taub. ex Warm., Overs. Kongel. Danske Vidensk. Selsk. Forh. Medlemmers Arbeiter 1901: 175 (1902).
Scaphiophora Schltr., Notizbl. Bot. Gart. Berlin-Dahlem 8: 39 (1921).
Mamorea de la Sota, Darwiniana 12: 43 (1960).

Herbs up to 10 cm tall. Underground part tuberous, or creeping cylindrical roots, or a cluster of short hyaline roots. Stems unbranched with few scale-like leaves. Flowers solitary, or rarely in a few-flowered cincinnus, actinomorphic or zygomorphic, variously colored. Flower tube cylindric to urceolate, soon falling off, throat circular, surrounded by an annulus. Tepals 6, often unequal in size, in 2 distinct whorls, the inner whorl sometimes connate forming a miter. Stamens 6, inserted in the throat of the flower tube, pendent, occasionally alternating with interstaminal lobes; connective often with appendages or hairs, connate into a tube with thecae separated, or connective free and thecae united. Ovary with 3 parietal placentas or with 3 free placental columns. Style 3-branched or capitate. Fruit fleshy, cup-shaped. Seeds ellipsoid to ovoid.

Thismia comprises ca. 45 species. The majority of species is known from tropical America and Asia, but some species from Asia extend into the subtropics (southern Japan, Australia, and New Zealand; Thiele and Jordan 2002; Yang et al. 2002) and temperate zones (*T. rodwayi* in Tasmania, Roberts et al. 2003; Wapstra et al. 2005). *Thismia* species are generally known to grow in evergreen forests. A notable exception is *T. americana*. This species was discovered in August 1912 near Chicago (Illinois, USA) in a prairie (Pfeiffer 1914). The population was observed for several subsequent summers, and was probably last seen in 1916. Because the type locality has been destroyed and several intensive searches in the area have failed to rediscover the plant, it is now considered extinct (Lewis 2002). Based on morphological similarities it has been suggested that the closest relative of *T. americana* is *T. rodwayi* from Australia and New Zealand (Jonker 1938; Maas et al. 1986; but see Thiele and Jordan 2002), and thus makes it part of one of the most anomalous disjunction distributions known in flowering plants (Thorne 1972). Phylogenetic analyses based on nuclear and mitochondrial DNA data with a limited *Thismia* sampling suggests that the genus is paraphyletic, but a better sampling is required to investigate the taxonomic status of the genus (Merckx et al. 2006, 2009).

Morphological observations suggest that species of *Thismia* are growing with arbuscular mycorrhizal fungi (Groom 1895; Janse 1897; Pfeiffer 1914; McLennan 1958; Campbell 1968). Indeed, *Glomus* group A fungi were detected in roots of *Thismia rodwayi* (Merckx et al. 2012). Pollination biology was never studied in detail, but both cross-pollination and self-pollination have been hypothesized to occur in *Thismia*. Many species have strongly colored flowers, and tentacle-like tepal tips, which may show the way to enter the flower (Maas et al. 1986). The tips of the tepals, and the base of the perianth of several *Thismia* species are provided with glandular swellings, presumably functioning as osmophores (Vogel 1962). Glandular structures are also present on the anthers of some species (Thiele and Jordan 2002). The stamens of *Thismia* tend to form a funnel or even a connate tube, possibly to guide pollinators down to the stigma, and the flowers often have trap-like structures (Maas et al. 1986; Thiele and Jordan 2002). Moreover, pendent stamens opening towards the wall of the perianth make self-pollination difficult (Maas et al. 1986). *Thismia* flowers generally produce little odor, but Wapstra et al. (2005) reported an odor of rotten fish after boxing the flowers of *T. rodwayi* for a few hours. Miers (1866) reported that the flowers of *T. hyalina* never open and are therefore self-pollinating. Seed dispersal agents are not known, but it is often hypothesized that the dust-like seeds are dispersed by wind or rain-splash (Stone 1980; Maas et al. 1986). It is also possible that the fruits are eaten by birds or mammals (Maas-van de Kamer 1998; Wapstra et al. 2005).

2.5.3.3 *Tiputinia* (Fig. 2.7a)

Tiputinia P.E. Berry & C. Woodw., Taxon 56: 157 (2007).

Herbs, ca. 2 cm tall. Rhizome vertical, cylindric, sympodially branched. Stems unbranched. Flowers solitary, actinomorphic. Flower tube short, obconical. Tepals 6, free, equal in size, olive yellow. Stamens 6, inserted in the throat of the flower tube, alternating with tiny subglobose interstaminal lobes; filaments orange, thick, ascending and then recurved, forming a cage or dome over the throat, orange, with fimbriate appendages; thecae latrorsely dehiscent. Ovary with 3 parietal placentas; style with pyramidal stigma. Fruit and seeds unknown.

Monospecific genus only known from Amazonian Ecuador. Description based on a single specimen of *T. foetida* collected in April 2005 in evergreen forest at the Tiputini Biodiversity Station. Also recorded in Yasuní National Park (A. J. Perez Castañeda pers. comm.). The flowers of *T. foetida* produce a foul, rotten fish-like odor, presumably to attract pollinators (Woodward et al. 2007). Dispersal agents and mycorrhizal fungi unknown.

2.5.3.4 *Haplothismia* (Fig. 2.7d)

Haplothismia Airy Shaw, Kew Bull. 2: 277 (1952).

Herbs, 10–25 cm tall. Underground part a cluster of vermiform tuberous roots. Stems unbranched or branched. Inflorescence a panicle composed of few-flowered cincinni. Flowers nodding, pale brown. Perianth persistent. Flower tube campanulate to funnel-shaped. Tepals 6, free, equal in size. Stamens 6, inserted in the throat of the floral tube, alternating with minute interstaminal lobes; filaments adnate to the flower tube, apical part free and recurved. Ovary with 3 parietal placentas; style with 3-lobed stigma. Fruit a loculidical capsule. Seeds ellipsoid.

Haplothismia is a monospecific genus from India. *H. exannulata* was discovered by A. Abraham and K. C. Jacob in the Western Ghats in 1951 (Airy Shaw 1952). The species was rediscovered at the type locality in 1999 (Sasidharan and Sujanapal 2000). *H. exannulata* is currently known from only two populations in Parabikulam Wildlife Sanctuary, where it occurs in humus rich soil in evergreen rainforest at about 700 m altitude. Flowering and fruiting in October (Sasidharan and Sujanapal 2000). Pollination syndrome, dispersal agents, and mycorrhizal fungi unkown.

2.5.3.5 *Oxygyne* (Fig. 2.7e)

Oxygyne Schltr., Bot. Jahrb. Syst. 38: 140 (1906).
Saionia Hatus., J. Geobot. 24: 2 (1976).

Herbs up to 4 cm tall. Underground part a cluster of vermiform roots. Stems unbranched. Inflorescence a few-flowered cincinnus or reduced to a single terminal flower. Flowers actinomorphic, brown with orange or blue-green, tube funnel-shaped with a well-developed annulus in the throat, upper part of the perianth soon falling off. Tepals 6, free, equal in size. Stamens

3, recurved, inserted opposite and enclosed in the base of the inner tepals, interstaminal lobes absent. Ovary with 3 parietal placentas; style 3-branched. Fruit cup-shaped. Seeds unknown.

Oxygyne is a very rare cryptic genus with a remarkable disjunct distribution. Species of *Oxygyne* differ from other Thismiaceae species by having 3 instead of 6 stamens. The first specimen of *Oxygyne* (*O. triandra*) was discovered by Rudolf Schlechter on Mount Cameroon in September 1905 (Schlechter 1906). A second African species of *Oxygyne* was collected on Mount Etinde, but remains to be described (Cheek et al. 2006). The herbarium specimen *Tisserant 2623* (BM) collected in the Central African Republic probably belongs to *Oxygyne* as well. Other *Oxygyne* species occur in southern Japan, on the islands Shikoku, Okinawa, and Yakushima (*O. hyodoi*, *O. shinzatoi*, and *O. yamashitae*) (Hatusima 1976; Abe and Akasawa 1989; Yahara and Tsukaya 2008). A revision of this genus is required to confirm the relationship between the African and Japanese specimens.

Oxygyne triandra was collected in tropical forest growing with *Afrothismia winkleri*, *A. pachyantha* (Thismiaceae), and *Burmannia hexaptera* (Burmanniaceae) (Schlechter 1906, 1921). The type locality is almost certainly destroyed by human activity and therefore this species might be extinct (Franke et al. 2004). The Japanese species were all collected in evergreen forest and flowered in September and October (Hatusima 1976; Abe and Akasawa 1989; Yahara and Tsukaya 2008; Yokoyama et al. 2008). *O. hyodoi* was found growing together with the mycoheterotrophic plant *Burmannia liukiuensis* (now *B. nepalensis* (Burmanniaceae)) (Abe and Akasawa 1989). Ants and mites were observed visiting the flowers of *O. yamashitae* but did not transfer pollen (Yahara and Tsukaya 2008). Dispersal agents and mycorrhizal fungi unkown.

2.5.4 Triuridaceae

Triuridaceae Gardn., Proc. Linn. Soc. London 19: 160 (1843).
Lacandoniaceae E. Martínez & Ramos, Ann. Missouri Bot. Gard. 76: 128 (1989).

Mycoheterotrophic, dioecious or monoecious herbs, completely white, yellow, purple, brown, or red. Rhizome mostly well-developed, horizontally creeping to vertical, with many scale-like leaves; roots filiform, tuberous, and radiating from the base of the stem, or very rarely coral-shaped, with or without root hairs. Stems mostly unbranched, erect, or decumbent at the base. Leaves few, alternate, sessile, entire, small, and scale-like. Inflorescence a terminal, bracteate few- to many-flowered raceme or spike, in monoecious plants staminate flowers at the top and pistillate flowers at the base of the inflorescence. Flowers unisexual or rarely bisexual (*Lacandonia*, *Sciaphila*), actinomorphic or rarely bilaterally symmetrical (e.g., *Kupea*), white, yellow, purplish, red, or dark brown to black. Tepals 3–10, valvate, equal or rarely unequal, basally connate, often soon reflexed, inner side often densely papillate, apex sometimes with dense tufts of hairs (bearded) or globose knobs or appendages or caudate (with tails to 50 mm long). Bisexual flowers with 2–6 free stamens and ∞ free carpels; staminate flowers with 2–6(-8) stamens, mostly epitepalous (opposite the outer or inner tepals), free or basally connate, sometimes inserted on a central androphore (*Triuris*); anthers 4-locular or sometimes 2- or 3-locular (in *Triuris* and *Sciaphila* respectively), dithecic or rarely monothecic, sessile or filamented, anther dehiscence longitudinal to transversely extrorse or rarely introrse (*Lacandonia*, *Triuris*), staminodes sometimes present (*Seychellaria*); gynoecium rudiments rarely present (*Triuridopsis*). Pollen inaperturate with characteristic spiny-gemmate surface sculpturing. Pistillate flowers with 10-∞ free carpels inserted on the receptacle. Carpels 1-locular with 1 (or 2 in Kupeaeae) basal, anatropous ovule(s), apical part of carpel often papillate, style 1 per carpel, filiform, persistent, basal to lateral or terminal, stigmatic zone papillate, penicillate, or indistinct. Fruit consisting of indehiscent achenes or follicles dehiscent by a longitudinal slit. Seeds 1 (or 2) per carpel, globose to obovoid, small.

Numbers of genera and species—Triuridaceae comprise approximately 50 species in 11 genera and three tribes. With ca. 30 species, *Sciaphila* is the most species-rich genus, other genera include only a few species each. All species of Triuridaceae are fully mycoheterotrophic.

Distribution and habitat—Triuridaceae occur throughout the tropical parts of the Old and the New World and reach the subtropics in Japan. *Sciaphila* has a pantropical distribution, all other genera are confined to one continent. Triuridaceae generally grow in dense and humid forests, between leaf litter, at the base of large trees or along the bank of streams. Less frequently, they are found in temporarily inundated forests, forests on white sand, bamboo thickets, or on termite nests (*Sciaphila purpurea*, *S. arfakiana*). They often grow in close association with other mycoheterotrophic plants of various families (Maas and Rübsamen 1986; Maas-van de Kamer and Weustenfeld 1998).

Classification—Despite recent advances, the close affinities of Triuridaceae remain under dispute. Prior to molecular analyses, Triuridaceae were often linked with other mycoheterotrophs such as Petrosaviaceae and included in Triuridanae (Takhtajan 1997) or Triuridales (Cronquist 1981; Thorne 1992). Molecular phylogenetic analyses placed Triuridaceae with four other families (Cyclanthaceae, Pandanaceae, Stemonaceae, Velloziaceae) in a recircumscribed Pandanales (Chase et al. 2000, 2006; Davis et al. 2004), but its exact position within the order remains to be determined (e.g., Rudall and Bateman 2006). Based on morphological and embryological differences three distinct tribes within Triuridaceae are recognized: Kupeaeae, Sciaphileae, and Triurideae (Cheek 2003b).

Evolutionary history—Based on a robust morphological analysis, Rudall and Bateman (2006) postulated that the family is closely related to (perhaps embedded in) Stemonaceae (including the anomalous Sumatran genus *Pentastemona*); this hypothesis has yet to be tested in a comprehensive molecular analysis. Gandolfo et al. (1998, 2002) reported a series of fossil flowers from the Late Cretaceous (ca. 90 Ma) that show similarities with extant Triuridaceae. A cladistic analysis of 20 morphological characters placed the fossil flower genera *Mabelia* and *Nuhliantha* within Triuridaceae, with affinities to modern Triurideae. Since no vegetative parts were attached to these fossil flowers, it is impossible to determine whether the plants were mycoheterotrophic. Their affinities with extant Triuridaceae remain under debate; the fossil pollen is monosulcate with foveolate exine (Gandolfo et al. 2002), in contrast with the inaperturate spiny-gemmate pollen that characterizes all extant Triuridaceae (Furness et al. 2002; Furness and Rudall 2006; Rudall et al. 2007).

Ecology—The pollination biology of Triuridaceae remains elusive. Floral morphology strongly suggests insect pollination. The family includes protogynous plants, and unisexual flowers, flowers emitting odor and flowers with papillate tepals provided with glandular areas, hairs, or appendages (Maas-van de Kamer 1995; Maas-van de Kamer and Weustenfeld 1998; Rudall 2003, 2008). The morphology and epidermal anatomy of these filamentous structures indicate that they function as osmophores, at least in some species (Rudall 2003). Filamentous osmophores are highly characteristic of sapromyophilous mycoheterotrophs, which need to attract pollinators to otherwise inconspicuous flowers (Vogel 1990). Momose et al. (1998) reported that the flowers of *Sciaphila secundiflora* are pollinated by Calliphoridae flies. However, Márquez-Gúzman et al. (1993) described preanthetic cleistogamy in the bisexual flowers of *Lacandonia*, which are proterandrous. Seed dispersal mechanisms may include zoochory, anemochory, and hydrochory (Maas-van de Kamer 1995; Maas-van de Kamer and Weustenfeld 1998). Some seeds are reported to have a reticulate outer layer (*Lacandonia*, *Sciaphila*, *Soridium*). Root anatomical investigations show that the roots of Triuridaceae are colonized by arbuscular mycorrhizal fungi (Janse 1897; Imhof 1998, 2003, 2004; Yamato 2001). With molecular methods AM fungi were detected in the roots of *Kupea martinetugei*, *Sciaphila ledermannii*, *S. japonica*, and *S. tosaensis* (Yamato 2001; Franke et al. 2006; Merckx and Bidartondo 2008; Yamato et al. 2011a; Merckx et al. 2012).

Tribe KUPEAEAE Cheek
Plants dioecious; flowers unisexual, tepals 4, stamens 4, style terminal, fruit indehiscent, 2-seeded, bilaterally symmetrical. Two genera in tropical Africa.

2.5.4.1 *Kupea* (Figs. 2.5c, 2.8a, b, and 4.10d, e)

Kupea Cheek & S.A. Williams, Kew Bull. 58: 225 (2003).

Mycoheterotrophic, dioecious herbs, up to 10 cm tall. Rhizome horizontally creeping with 5–7 tuberous, hairy roots, radiating from the stem base. Stems usually unbranched. Leaves scale-like. Inflorescence a 20–70-flowered spike; bracts elliptic or absent. Flowers unisexual, pale yellow. Staminate flowers bilaterally symmetrical with 4 strongly unequal patent tepals, upper 3 (narrowly) elliptic, lower one much larger. Stamens 4; anthers 2-locular. Pistillate flowers radially symmetrical with 4 subequal, patent tepals and 25–60 carpels; ovary 1-locular, ovules 2, style terminal, stigmatic zone indistinct. Fruit a 2-seeded, bilaterally symmetric indehiscent achene. Seeds (1-)2 per carpel.

The genus *Kupea* consists of two species. *Kupea martinetugei* is known from several sites in Southwest Province, Cameroon (Mount Kupe and Mount Cameroon) (Cheek et al. 2003; Franke et al. 2004, 2006) and was also collected at sites in the East Province of Cameroon, near Yokadouma (Sainge 1509, 1621, 1624, YA). *Kupea jonii* is only known from the type locality in Kihansi Gorge in Tanzania, where it occurs with *Kihansia lovettii* (Triuridaceae) and *Afrothismia saingei* (Thismiaceae) in evergreen forest (Cheek 2003b; Maas and Maas-van de Kamer 2010). *Kupea martinetugei* associates with arbuscular mycorrhizal fungi from the *Glomus* Group A clade (Franke et al. 2006; Merckx and Bidartondo 2008). *Kupea* possesses two ovules per carpel, in contrast to the single ovule per locule present in all other Triuridaceae except *Kihansia* (Cheek 2003b; Rudall et al. 2007). *Kupea* is also relatively unusual in that the male flowers are zygomorphic and possess a labellum (Rudall et al. 2007). Pollination biology and seed dispersal mechanisms remain to be studied.

2.5.4.2 *Kihansia* (Fig. 2.8g, h)

Kihansia Cheek, Kew Bull. 58: 943 (2003).

Mycoheterotrophic, dioecious herbs, up to 10 cm tall. Rhizome unknown, with 7–12, tuberous, glabrous roots, radiating from the stem base. Stems usually unbranched. Leaves scale-like. Inflorescence a 2–13-flowered spike; rachis with apical sterile part; bracts dimorphic: fertile ones elliptic-rectangular; sterile ones linear. Flowers unisexual, dark brown to black. Tepals 4. Staminate flowers bilaterally symmetrical, perianth flat apart from the central androecial cavity, upper 3 lobes subequal, triangular, the lower lobe about twice as long as the others; anthers white, 2-locular. Pistillate flowers with 4 subequal, patent tepals and 80–100 carpels; ovary 1-locular, ovules 2, style terminal, stigmatic zone indistinct. Fruit a 2-seeded, bilaterally symmetrical, indehiscent achene.

Kihansia includes a single species, *K. lovettii*, known only from the type locality in the Kihansi Gorge in Tanzania. The type locality consists of evergreen tropical forest at an altitude of 720 m. Two other mycoheterotrophs, *Kupea jonii* (Triuridaceae) and *Afrothismia saingei* (Thismiaceae) were found at the same site (Cheek 2003b; Maas and Maas-van de Kamer 2010). Two collections of a presumably new, yet undescribed, *Kihansia* species have been made in southeastern Cameroon in 2005 and 2006 (Thomas and Chuyong 2006). Information about the biology of *Kihansia* is lacking.

Tribe SCIAPHILEAE Miers
Plants monoecious; flowers unisexual, rarely bisexual (*Sciaphila*); tepals 4, 6, or 8; stamens 2–4, or 6; tepals bearded, papillate or with an apical globose knob; style lateral to basal; fruit dehiscent (except *Soridium*). Five genera. Neotropics and Paleotropics.

2.5.4.3 *Seychellaria* (Figs. 2.8c and 4.10a)

Seychellaria Hemsl., Ann. Bot. (London) 21: 74 (1907).

Mycoheterotrophic, monoecious herbs, up to 20 cm tall. Rhizomes horizontally creeping with filiform, hairy roots. Stems unbranched. Leaves

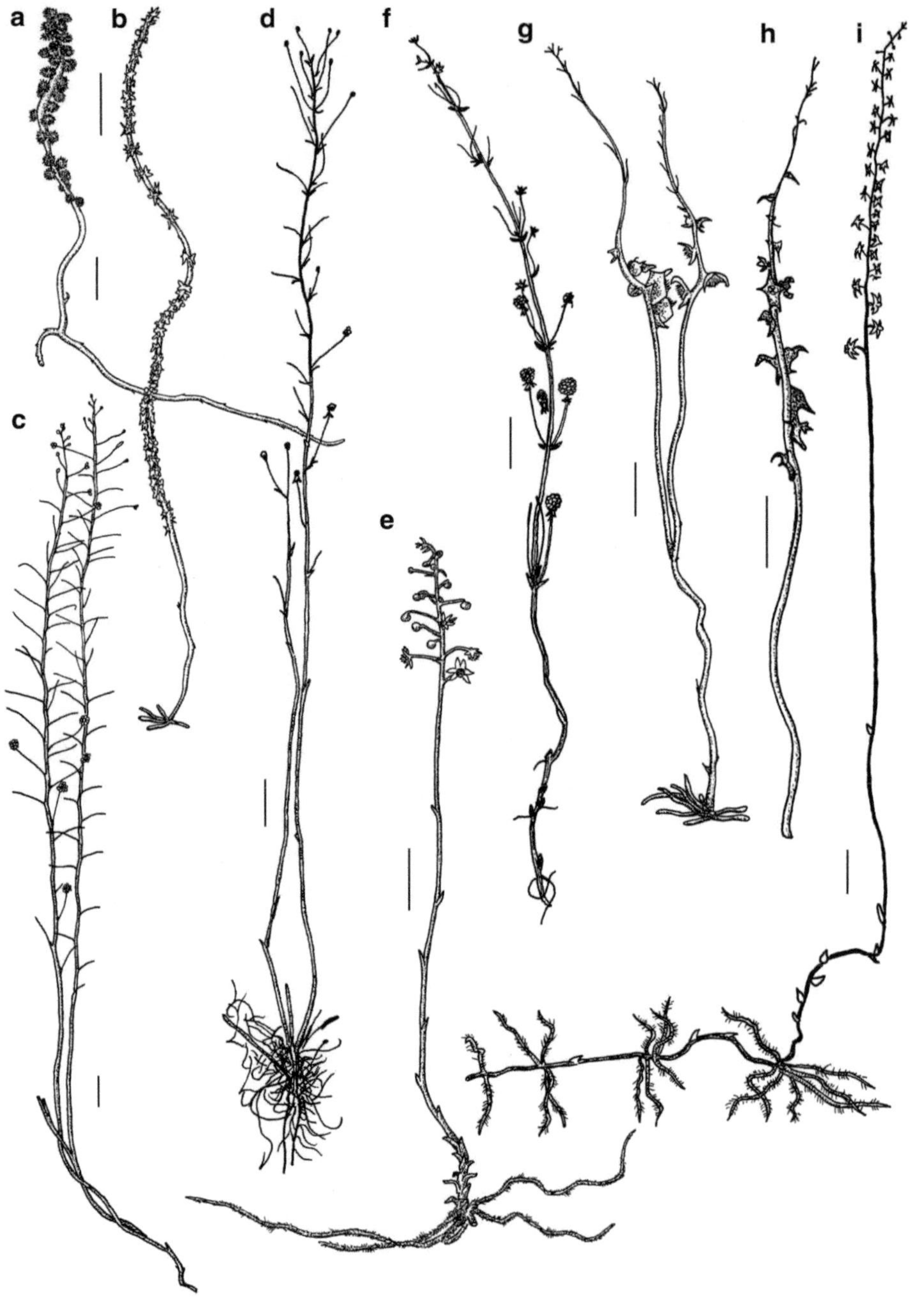

Fig. 2.8 Triuridaceae; Kupeae and Sciaphileae. *Kupea martinetugei*: (**a**) female plant, (**b**) male plant. Redrawn from Cheek et al. (2003). (**c**) *Seychellaria africana*. Redrawn from Vollesen (1982). (**d**) *Andruris australasica*. Redrawn from Giesen (1938). (**e**) *Sciaphila albescens*. Redrawn from Maas and Rübsamen (1986). (**f**) *Hyalisma janthina*. Redrawn from Giesen (1938). *Kihansia lovettii*: (**g**) female plant, (**h**) male inflorescence. Redrawn from Cheek (2003a, b). (**i**) *Soridium spruceanum*. Redrawn from Maas and Rübsamen (1986). Bar = 1 cm

scale-like. Inflorescence an up to 50-flowered raceme, sometimes with 2 flowers per node. Flowers unisexual, whitish, or reddish. Tepals 6, unequal. Staminate flowers with 3 stamens opposite the 3 larger tepals, alternating with 3 staminodes, sometimes connective provided with a long appendage; anthers 4-locular. Pistillate flowers with ∞ carpels; ovary 1-locular, ovule 1, style lateral, stigmatic zone indistinct. Fruit a dehiscent follicle.

Three species of *Seychellaria* are known: *S. madagascariensis* (Madagascar and the Comores) (including *S. perrieri*), *S. thomassettii* (the Seychelles), and *S. africana* (Iringa region in Tanzania). All species occur in rainforest. *Seychellaria* differs from *Sciaphila* mainly by the presence of staminodes in the staminate flowers (Giesen 1938; Perrier de la Bathie 1946; Vollesen 1982), and both genera are probably closely related (Rudall and Bateman 2006). Data on pollination biology, seed dispersal, and mycorrhizae are lacking.

2.5.4.4 *Sciaphila* (Figs. 2.8e, 4.8i, and 4.9a–c)

Sciaphila Blume, Bijdr. Fl. Ned. Ind. 514 (1825).
Aphyleia Champ., Calcutta J. Nat. Hist. 7: 468 (1847).
Lilicella Rich. ex Baill., Bull. Mens. Soc. Linn. Paris 2: 1188 (1895).

Mycoheterotrophic, monoecious herbs, up to 30, very rarely to 150 cm tall (*S. purpurea*). Rhizomes mostly horizontally creeping, with filiform or rarely coral-shaped, hairy to glabrous roots. Stems unbranched or sometimes branched. Leaves scale-like. Inflorescence a 7–55-flowered raceme, sometimes basally branched. Flowers unisexual, sometimes bisexual, whitish, pink, purplish, or red. Tepals (4-)6–8(-10), equal or unequal, inner side papillate, apex sometimes bearded. Staminate flowers with 2–3 or 6 stamens; anthers 3–4-locular. Pistillate flowers with ∞ carpels; carpels 1-locular, ovule 1, style (sub) basal to lateral, stigmatic zone papillate or penicillate or indistinct. Fruit a dehiscent follicle.

Sciaphila comprises ca. 29 species. Seven species occur in the Neotropics, two species are known from tropical West Africa, and 19 species occur in tropical Asia (including Japan, New Caledonia, Fiji, India and Sri Lanka) (van de Meerendonk 1984; Maas and Rübsamen 1986; Cheek 2006). The fungal associates of *S. japonica*, *S. ledermannii*, and *S. tosaensis* have been identified as glomeromycetes belonging to *Glomus* Group A and Acaulosporaceae (Yamato 2001; Franke et al. 2006; Merckx and Bidartondo 2008; Yamato et al. 2011a; Merckx et al. 2012). There is only a single report on the pollination biology of *Sciaphila*, which states that the flowers of *Sciaphila secundiflora* are pollinated by Calliphoridae flies (Momose et al. 1998). Seed dispersal mechanisms in *Sciaphila* remain unstudied.

2.5.4.5 *Hyalisma* (2.8f)

Hyalisma Champ., Calcutta J. Nat. Hist. 7: 466 (1847).

Mycoheterotrophic, monoecious herbs, up to 20 cm tall. Rhizomes unknown, roots filiform radiating from the base of the stem mostly horizontally creeping, with filiform, hairy to glabrous roots. Stems unbranched or sometimes branched. Leaves scale-like. Inflorescence an up to 20-flowered raceme, sometimes basally branched, with several flowers per node. Flowers unisexual, purplish. Tepals 8, equal. Staminate flowers with 4 stamens; anthers 4-locular. Pistillate flowers with ∞ carpels; ovary 1-locular, ovule 1, style (sub)basal to lateral, stigmatic zone indistinct. Fruit a dehiscent follicle.

Hyalisma comprises a single species, *H. janthina*, from south India and Sri Lanka. Mycorrhiza, pollination, and seed dispersal mechanisms remain unstudied (Maas-van de Kamer and Weustenfeld 1998).

2.5.4.6 *Andruris* (Fig. 2.8d)

Andruris Schltr., Bot. Jahrb. Syst. 49: 71 (1912).
Parexuris Nakai & F. Maek., Iconogr. Pl. Asiae Orient. 1: 23 (1936), partly.

Mycoheterotrophic, monoecious herbs, up to 25 cm tall. Rhizomes mostly horizontally creeping, with filiform, hairy to glabrous roots. Stems unbranched or sometimes branched. Leaves scale-like. Inflorescence a 5–50-flowered raceme, sometimes basally branched. Flowers unisexual, whitish, pink, purplish, or red. Tepals (4-)6, unequal, apex sometimes with terminal knobs. Staminate flowers with 3 stamens opposite the larger tepals; anthers 4-locular, connectives with a long subulate appendage. Pistillate flowers with ∞ carpels; ovary 1-locular, ovule 1, style (sub) basal to lateral, stigmatic zone indistinct. Fruit a dehiscent follicle.

Andruris comprises five species from Malesia, Polynesia, Micronesia, eastern India, southern Japan and northeastern Australia. Mycorrhiza, pollination, and seed dispersal mechanisms in *Andruris* remain unstudied (Maas-van de Kamer and Weustenfeld 1998).

2.5.4.7 *Soridium* (Fig. 2.8i)

Soridium Miers, Proc. Linn. Soc. London 2: 74 (1850).

Mycoheterotrophic, monoecious herbs, up to 30 cm tall. Rhizomes horizontally creeping, with filiform, hairy roots. Stems unbranched. Leaves scale-like. Inflorescence a 10–50-flowered raceme. Flowers unisexual, white. Tepals 4, inner side papillate. Staminate flowers with 2(-3) stamens; anthers 2-locular. Pistillate flowers with 25–40 carpels; carpels 1-locular, ovule 1, style lateral, stigmatic zone penicillate. Fruit an indehiscent achene.

Soridium consists of a single species, *S. spruceanum*. The genus is confined to Central America and northern South America (Maas and Rübsamen 1986). Observations about the mycorrhizal fungi, pollination biology, and seed dispersal are lacking.

TRIURIDEAE Miers
Plants dioecious, flowers unisexual, rarely bisexual (*Lacandonia*); tepals 3 or 6; stamens 3 or 6; tepals caudate or appendaged (*Triuridopsis*); style (sub)terminal or lateral (*Peltophyllum*); fruit indehiscent. Four genera in the Neotropics.

2.5.4.8 *Peltophyllum* (Fig. 2.9a, b)

Peltophyllum Gardn., Proc. Linn. Soc. London 1: 176 (1843).
Hexuris Miers, Proc. Linn. Soc. London 2: 72 (1850).

Mycoheterotrophic, dioecious herbs, up to 10 cm tall. Rhizomes vertical with filiform, glabrous roots. Stems unbranched. Leaves scale-like. Inflorescence a 6–16-flowered raceme. Flowers unisexual, yellowish white. Tepals (3-)6(-8), horizontally patent, apex caudate. Staminate flowers with 3 stamens; anthers 4-locular. Pistillate flowers with ∞ carpels; carpels 1-locular, ovule 1, style lateral, stigmatic zone indistinct. Fruit an indehiscent achene.

Peltophyllum consists of two species: *P. luteum* and *P. caudatum*. The former occurs in southeastern Brazil, northern Argentina, and southern Paraguay, but also in Guyana, in evergreen forests. The latter species is known from a single collection from Alto Macahé, Rio de Janeiro, Brazil, where it was found in the shade of large trees near a river, growing in leaf mold (Maas and Rübsamen 1986). Mycorrhizal associates, pollination syndrome, and seed dispersal of *Peltophyllum* remain to be studied.

2.5.4.9 *Lacandonia* (Fig. 2.9g)

Lacandonia E. Martínez & Ramos, Ann. Missouri Bot. Gard. 76: 128 (1989).

Mycoheterotrophic dioecious herbs, up to 10 cm tall. Rhizome horizontally creeping, with filiform, hairy roots. Stems unbranched. Leaves scale-like. Inflorescence a 3–7(-13)-flowered raceme. Flowers bisexual, sometimes unisexual, whitish. Tepals (4-)6, inner side papillate, apex caudate. Stamens (2-)3(-4); anthers 2(-3) locular. Gynoecium composed of ∞ carpels surrounding the stamens; carpels 1-locular, ovule 1, style subterminal, stigmatic zone indistinct. Fruit an indehiscent achene.

Lacandonia includes two species. *Lacandonia schismatica* occurs in scattered populations in the Lacandon rainforest in Mexico at an elevation of about 200 m (Vergara-Silva et al. 2003). Recently, a second species of *Lacandonia*, *L. brasiliana*, was described from material collected in the Atlantic rainforest in Brazil (Melo and Alves 2012). Bisexual flowers of *Lacandonia* have stamens borne inside the carpels, a characteristic unique in angiosperms (Ambrose et al. 2006; Rudall 2008). Based on this feature the genus was originally placed in its own family, Lacandoniaceae (Martínez and Ramos 1989). The flowers of *L. schismatica* are cleistogamous (Márquez-Gúzman et al. 1993). Mycorrhiza, pollination, and seed dispersal remain unstudied.

Fig. 2.9 Triuridaceae; Triurideae. *Peltophyllum luteum*: (**a**) male plant, (**b**) female plant. Redrawn from Maas and Rübsamen (1986). *Triuridopsis intermedia*: (**c**) male plant, (**d**) female plant. Redrawn from Franke et al. (2000). *Triuris hyalina*: (**e**) male plant, (**f**) female plant. Redrawn from Maas and Rübsamen (1986). (**g**) *Lacandonia schismatica*. Redrawn from Martínez and Ramos (1989). Bar = 1 cm

2.5.4.10 *Triuridopsis* (Fig. 2.9c, d)

Triuridopsis H. Maas & Maas, Pl. Syst. Evol. 192: 257 (1994).

Mycoheterotrophic, dioecious herbs, up to 12 cm tall. Rhizomes horizontally creeping, each node provided with 2 filiform, glabrous roots. Stems unbranched. Leaves scale-like. Inflorescence a 1–12-flowered raceme. Flowers unisexual, white. Tepals 3(–4), with a subapical reflexed appendage. Staminate flowers with 3 stamens with bithecal, 4-locular anthers or 6 stamens with monothecal, 2-locular anthers; center of the (staminate) flowers provided with a subulate projection. Pistillate flowers with reflexed tepals and ∞ carpels; carpels 1-locular, ovule 1, style terminal, stigmatic zone indistinct. Fruit an indehiscent achene.

Triuridopsis consists of two species: *T. peruviana* from Iquitos, Loreto, Peru, and *T. intermedia* from La Paz, Bolivia (Maas-van de Kamer and Maas 1994; Franke et al. 2000). The genus is probably closely related to *Triuris* (Maas-van de Kamer and Maas 1994). Mycorrhizal associates, pollination syndrome, and seed dispersal of *Triuridopsis* remain to be studied.

2.5.4.11 *Triuris* (Figs. 2.9e, f, and 4.8b)

Triuris Miers, Proc. Linn. Soc. London 1: 96 (1841).

Mycoheterotrophic, dioecious herbs, up to 20 cm tall. Rhizomes vertical, with filiform, glabrous to hairy roots. Stems unbranched. Leaves scale-like. Inflorescence a 1–4-flowered raceme. Flowers unisexual, white to brown. Tepals 3, soon reflexed, apex long-caudate, the tails in bud rolled inwards like a watch spring. Staminate flowers with 3 stamens with 4-locular anthers alternating with the tepals or with 6 stamens with 2-locular anthers, androphore large, fleshy, conical to deltoid, the stamens inserted at its base. Pistillate flowers with ∞ carpels; ovary 1-locular, ovule 1, style terminal, stigmatic zone indistinct. Fruit an indehiscent achene.

Three species are known. *Triuris hyalina* is widespread from Central America (Guatemala) in the North to southeastern Brazil in the South. *Triuris hexophthalma* and *T. alata* Brade are only known from the type locality in the Pakaraima Mountains, Guyana, and in Itatiaia, Rio de Janeiro, Brazil, respectively (Brade 1943; Maas et al. 1986). Mycorrhizal associates, pollination syndrome, and seed dispersal of *Triuris* remain to be studied.

2.5.5 Corsiaceae

Corsiaceae Becc., Malesia 1: 238 (1878) as "1877."

Mycoheterotrophic herbs, up to 30 cm tall. Underground part a rhizome or a cluster of tuberous roots. Stems simple, erect. Leaves reduced to a few amplexicaul sheaths. Flowers terminal, solitary, bisexual or unisexual, zygomorphic. Tepals 6, in 2 whorls, outer median tepal ("labellum") much larger than the other ones, covering the reproductive parts of the flower, sometimes with a large, glandular, basal callus; other 5 tepals filiform. Stamens 6, in 2 whorls, filaments short, anthers 2-celled, dorsifixed, extrorsely dehiscing by longitudinal slits. Ovary inferior, 1-locular with 3 parietal placentas or 3-locular with 3 axile placentas, ovules numerous; style absent, stigmas 3, sessile and connate, or style(s) present and short, either 1 with a 3-lobed stigma or 3 each with a stigma; septal nectaries absent. Fruit a capsule, dehiscing by 3 valves. Seeds numerous, very small, winged, pendent.

Number of genera and species—Corsiaceae comprise three genera and 27 species. All species are fully mycoheterotrophic. The largest genus is *Corsia* (25 species).

Distribution and habitat—Corsiaceae have a remarkable disjunct distribution, occurring in tropical and subantarctic South America (*Arachnitis*), China (*Corsiopsis*), and tropical Australasia (*Corsia*).

Classification—Corsiaceae were formerly linked to or included in Burmanniaceae (Beccari 1878; Bentham 1883; Engler 1888; Hutchinson 1959; Dahlgren et al. 1985). Based on mitochondrial *atpA* sequence data *Arachnitis* is tentatively included in Liliales (Davis et al. 2004; Fay et al. 2006), a relationship that is supported by nuclear 18S rDNA data analysis (Chase et al. 2006). Neyland and Hennigan (2003), using partial sequences for 26S rDNA alone, suggested that Corsiaceae may be polyphyletic. In their analysis, *Corsia* was placed within Liliales, whereas *Arachnitis* was related to *Thismia* (Thismiaceae). Floral anatomy and pollen morphology suggest that *Corsia* is related to Campynemataceae (Liliales) or *Thismia* (Rudall and Eastman 2002).

Evolutionary history—Due to the uncertain systematic relationships the evolutionary history of Corsiaceae remains unclear.

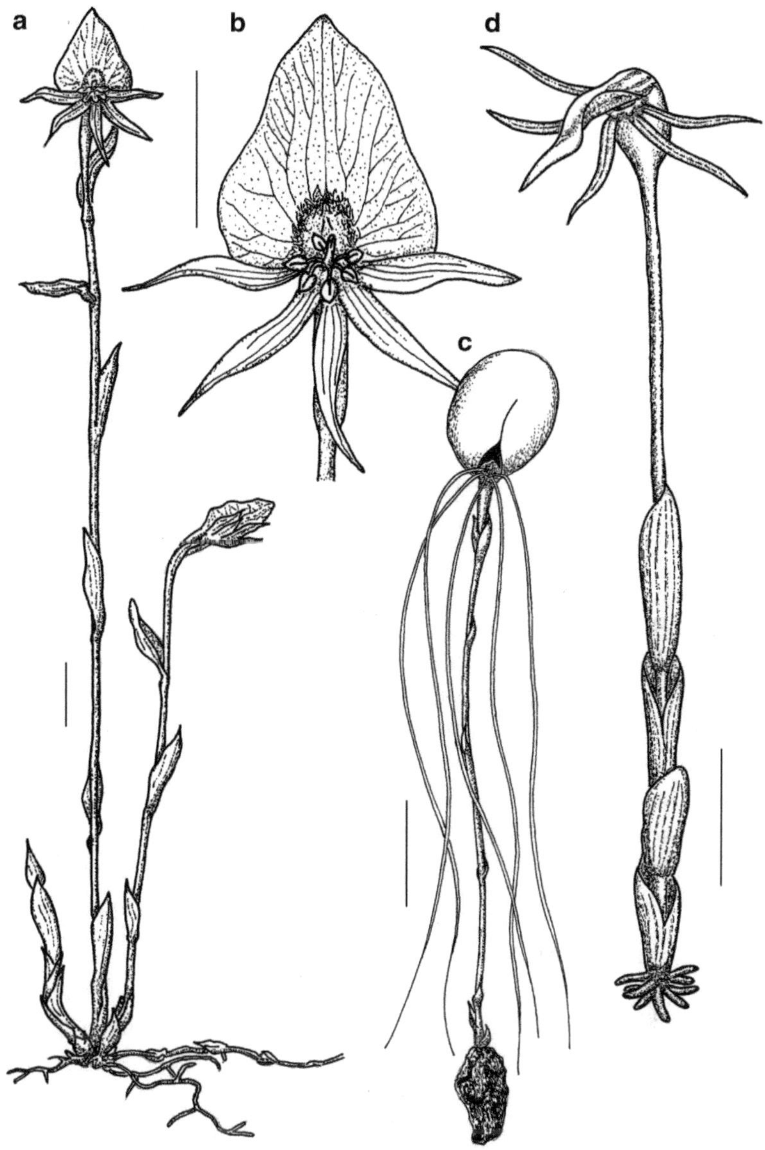

Fig. 2.10 Corsiaceae. *Corsia pyramidata*: (**a**) habit, (**b**) flower. Redrawn from Cribb (1985). (**c**) *Corsiopsis chinensis*. Redrawn from Zhang et al. (1999). (**c**) *Arachnitis uniflora*. Redrawn from Dimitri (1972). Bar = 1 cm

Ecology—Specimens of *Arachinitis* form arbuscular mycorrhizas and are associated with a narrow lineage within the *Glomus* group A clade (Bidartondo et al. 2002). The mycorrhizal fungi of *Corsia* and *Corsiopsis* are not known, but these genera are probably associated with arbuscualar mycorrhizal fungi as well. Pollination syndrome and dispersal agents remain poorly studied. Rudall and Eastman (2002) noted that the flowers of *Corsia* are protandrous. Moreover, during anthesis the flowers of *Arachnitis* grow considerably and show great morphological plasticity (Minoletti 1986; Ibisch et al. 1996). These observations suggest cross-pollination in Corsiaceae.

2.5.5.1 *Arachnitis* (Figs. 2.10 and 4.11a)

Arachnitis Phil., Bot. Zeitung (Berlin) 22: 217 (1864).

Mycoheterotrophic herbs, up to 30 cm tall. Underground part a cluster of tuberous roots. Stems reddish. Flowers bisexual or unisexual.

Tepals whitish yellow to violet or dark red, the 3 inner and 2 outer lateral ones filiform and spreading, the median outer tepal narrowly ovate, basally concave and covering the reproductive parts of the flower, apical part pendent. Stamens free, soon falling off. Ovary 1-locular, styles 3, each with a stigma. Fruit globose, pendent when ripe, dehiscing at the top by 3 horizontally splitting valves. Seeds ovoid.

Arachnitis comprises a single species with a remarkably wide distribution. *A. uniflora* is known from humid subantarctic *Nothofagus* forests in Argentina and Chile, and subhumid and humid tropical Andean forests in Bolivia (Ibisch et al. 1996; Neinhuis and Ibisch 1998). *Arachnitis* is also found on the treeless Falkland Islands, growing "in sand amongst rocks" (Cribb et al. 1995), occurring between sea level and 1,000 m. Some authors recognize a second species, *A. quetrihuensis*, based on differences in flower proportions (Dimitri 1972; Neinhuis and Ibisch 1998), but given the substantial morphological variability of *Arachnitis* over its wide distribution, we consider *A. quetrihuensis* conspecific to *A. uniflora* (see also Ibisch et al. 1996).

Bidartondo et al. (2002) sequenced the fungal symbionts of eight individuals of *A. uniflora* from three populations in subantarctic forests in Argentina and found that the plants form arbuscular mycorrhizae and are specialized to a narrow lineage within the *Glomus* group A clade. The pollination biology of *Arachnitis* has not been studied in the field, but Vogel (1978) suggests that *Arachnitis* may be pollinated by fungus gnats.

2.5.5.2 *Corsia* (Fig. 2.10a, b)

Corsia Becc., Malesia 1: 238 (1878) as "1877."

Mycoheterotrophic herbs, up to 30 cm tall. Rhizome cylindric, more or less horizontal; roots filiform. Stems often purplish or reddish brown. Flowers bisexual, nodding at anthesis. Tepals reddish, the 3 inner and 2 outer lateral ones filiform, spreading to pendent, the median outer tepal ovate and often cordate, with a large, glandular, basal callus, covering the reproductive parts of the flower as an umbrella-like structure. Stamens connate at the base with each other and the base of the style. Ovary 1- or 3-locular, style 1, with 3 thick, short stigmas; septal nectaries absent. Fruit fusiform, longitudinally dehiscing by 3 valves, the valves curving downwards exposing the 3 erect placentas carrying the seeds. Seeds fusiform, pendent.

Corsia contains approximately 25 species. Most species are endemic to New Guinea, but at least two are found on the Solomon Islands and one species occurs in northern Australia (Van Royen 1972; Jones and Gray 2008). *Corsia* grows in the upper parts of lowland forests and in montane forests between 900 and 2,300 m (Rübsamen 1986). The species typically grow in damp places and are often found growing together with *Sciaphila* spp. (Triuridaceae) and *Burmannia* spp. (Burmanniaceae) (Van Royen 1972). Flowers of *Corsia* are protandrous according to Smith (1909) and produce nectar (Beccari 1878), suggesting pollination is mediated by insects. Seed dispersal mechanisms and identity of mycorrhizal fungi remain unknown.

2.5.5.3 *Corsiopsis* (Fig. 2.10c)

Corsiopsis D.X. Zhang, R.M.K. Saunders and C.M. Hu, Syst. Bot. 24: 313 (1999).

Mycoheterotrophic herbs, up to 6 cm tall. Rhizome vertical, ellipsoid-obovoid. Stems white. Flowers unisexual. Tepals white, the 3 inner and 2 outer lateral ones filiform, pendent, the median outer tepal broadly ovate, erect, inflated into a bowl-shaped structure, covering the reproductive parts of the flower. Stamens free, each with an obtuse, apical extension of the connective. Ovary of female flowers narrowly ellipsoid, 1-locular, stigmas 3, sessile, connate. Fruit and seeds unknown.

Corsiopsis comprises a single species, *C. chinensis*, which is only known from a single collection from Guangdong Province, China made in 1974 (Zhang et al. 1999). Data on pollination syndrome, dispersal agents, and the identity of mycorrhizal fungi are lacking.

2.5.6 Orchidaceae

Orchidaceae Juss., Gen. Pl. 64–65 (1789).

Epiphytic, terrestrial, lithophytic, or rarely aquatic or subterranean herbs, usually green and photosynthetic, some without chlorophyll and putatively fully mycoheterotrophic. Roots subterranean or aerial, thickened, when epiphytic provided with a multilayered epidermal velamen, sometimes with tubers. Stems elongate to shortened, rarely vining (as in *Vanilla*), thickened in many species, in which case either forming a one- to several-nodal pseudobulb or subterranean corm, predominantly exhibiting sympodial growth although monopodial in some groups. Leaves membranous to thickened or terete, plicate to conduplicate, sometimes reduced to sheaths, absent in fully mycoheterotrophic species. Inflorescences elongate to condensed racemes, spikes or panicles, terminal or lateral, numerous-flowered to solitary. Flowers usually zygomorphic, frequently resupinate via a twisting of the pedicellate ovary. Sepals free, sometimes variously connate, often colored like the petals. Uppermost petal (lowermost in resupinate species) usually modified and enlarged relative to the lateral petals. Functional anthers 1–3, most frequently one, filaments and styles united to form a gynostemium (column). Column short to elongate, occasionally prolonged at base and united with sepals to form a foot, sometimes also forming a spur or mentum. Pollen usually aggregated into masses (pollinia), sometimes forming hard, bony masses, sometimes with stalks that affix the pollinia to insects, pollinia 2–8. Stigma borne on the adaxial side of the column, often below a sticky mass (viscidium) that aids in attachment of pollinia to insects. Ovary inferior, usually unilocular with parietal placentation, but trilocular and axile in some. Ovules small, up to a million or more per flower in some species. Fruits capsular except in some *Vanilla* where an indehiscent structure may be produced. Seeds minute, without differentiated embryo or endosperm.

Number of genera and species—Orchidaceae are usually considered to be the largest family of flowering plants with ca. 22,000 species in about 880 genera (Stevens 2001). Circa 235 species in 43 genera are leafless and are putative full or nearly full mycoheterotrophs. The largest genera of full mycoheterotrophs are the Old World *Aphyllorchis* (33 species) and *Gastrodia* (22 species). Partial mycoheterotrophy has been detected in many green-leaved species (e.g., *Cephalanthera* spp., *Cheirostylis montana*, *Cymbidium* spp., *Epipactis* spp., *Ophrys insectifera*, *Platanthera bifolia*) and may be relatively common in terrestrial orchids (e.g., Gebauer and Meyer 2003; Bidartondo et al. 2004; Julou et al. 2005; Tedersoo et al. 2007; Cameron et al. 2009; Roy et al. 2009a; Liebel et al. 2010; Motomura et al. 2010; Preiss et al. 2010; Girlanda et al. 2011). Occasionally achlorophyllous "albino" individuals are found in some otherwise partially mycoheterotrophic species, notably in *Epipactis* and *Cephalanthera* (Salmia 1989; Selosse et al. 2004; Abadie et al. 2006). Some terrestrial orchid species have separate vegetative and leafless flowering stages, and have been misinterpreted as mycoheterotrophs (Chen and Luo 2002).

Distribution and habitat—Orchidaceae have a worldwide distribution, occurring in almost every habitat on the planet and absent only from the polar regions and the driest of deserts (Chase 2005). The great majority are to be found in the tropics, mostly in Southeast Asia and in the Neotropics, and their diversity peaks in montane tropical regions where abundant rainfall allows for the maximum growth of epiphytes. Most fully mycoheterotrophic orchids are found in the tropics, but their diversity has a highly uneven distribution. The vast majority of tropical mycoheterotrophic orchids occur in Southeast Asia and adjacent Australasia. In contrast, the floras of tropical Africa and particularly the Neotropics are surprisingly poor in fully mycoheterotrophic Orchidaceae. A majority of orchids are perennial epiphytes, which grow anchored to trees or shrubs in the tropics and subtropics. Other species are terrestrial or lithophytes, growing on rocks or very rocky soil. Nearly all temperate orchids are terrestrial. All mycoheterotrophic orchids are terrestrial, although some species, such as

Erythrorchis cassythoides, are climbers (Dearnaley 2006).

Classification—Recent phylogenetic analyses suggest that the Orchidaceae are sister to the remainder of the Asparagales (Givnish et al. 2006; Graham et al. 2006; Pires et al. 2006). Traditionally, classification of Orchidaceae has been based on the construction of the fused gynoecium and androecium ("column" or "gynostemium"), which is, in its details, unique to the family. The number of anthers has been the primary trait emphasized, which has resulted in the family being split into three main groups, often recognized as subfamilies. Five subfamilies are currently recognized. *Apostasia* and *Neuwiedia*, two Southeast Asian genera that comprise Apostasioideae, have been sometimes viewed as orchid relatives and placed in a separate family, Apostasiaceae, believed to be more closely related to other families such as Hypoxidaceae (Hutchinson 1959) than to Orchidaceae. Vanilloideae is the most recently recognized subfamily, having been resolved by molecular data and thereby clarifying a long-standing uncertainty in the relationships of its species based on their unusual combination of primitive and advanced morphological features. The remainder of the family comprises Cypripedioideae with their distinctive slipper-shaped labellum, Orchidoideae, which contains most of the temperate species, and Epidendroideae, which contains the great majority of the family and is primarily tropical and epiphytic.

Evolutionary history—The family's placement as sister to the remainder of the Asparagales suggests a relatively ancient origin for Orchidaceae; Chase (2001) suggested that the family might date from approximately 110 Ma by relating its phylogenetic placement to other groups. Orchid fossils are rare and usually consist of pollinaria. A 15–20 Ma fossil pollinarium was recently used to date the family at 76–84 Ma (Ramirez et al. 2007). The family was undoubtedly primitively terrestrial, with multiple derivations of epiphytism, primarily in the large subfamily Epidendroideae. Containing ca. 80% of the family's diversity, Epidendroideae comprise most of the epiphytes and exhibits the most advanced pollinarium morphologies related to specialized pollination strategies. The majority of mycoheterotrophs are members of Epidendroideae. The diversity of leafless species in the family represents an estimated minimum of 30 independent shifts to heterotrophy (Freudenstein and Barrett 2010).

Ecology—Orchids are well-known for their pollination specializations, ranging from the perfume-collecting euglossine bee syndrome found in many neotropical species to the pseudocopulatory syndrome of genera such as *Ophrys* and *Chiloglottis*. Most orchids outside of Apostasioideae and Cypripedioideae disperse their pollen in masses (pollinia). Small seeds without endosperm, reliance on fungi for germination, and pollen aggregated into pollinia together form a highly specialized strategy for orchids, in which in order to produce the large numbers of highly mobile seeds required to ensure that some may find a suitable fungus, large numbers of pollen grains are also necessary. Although it might be possible to achieve this with granular pollen, pollen masses provide an "all-or-nothing" strategy in which large numbers of ovules are fertilized or none at all.

As far as is known, all orchids depend on a mycoheterotrophic interaction with a symbiotic fungus for germination (initial mycoheterotrophy) (Leake 1994). In most orchids, particularly epiphytic species, this dependence appears to be required only during early seedling development prior to photosynthesis. Leafless epiphytes, such as *Dendrophylax*, photosynthesize with their roots and thus are not mycoheterotrophic. At least some terrestrial orchids that photosynthesize also obtain carbohydrates from fungi, and so are partially mycoheterotrophic (Rasmussen 1995). In others, the initial completely fungally-dependent phase has been prolonged throughout the plant's life (the fully mycoheterotrophic species). Both roots and rhizomes may be used to interact with fungi. Coralloid rhizomes, which characterize some full mycoheterotrophs, may be viewed as a paedomorphic extension of the protocorm stage that facilitates fungal interaction in the mature plant (Rasmussen 1995). Two distinct types of orchid mycorrhiza are recognized (Burgeff 1932). In the most common *tolypophagous* type, hyphae infect the rhizome or root, form coils (pelotons) in cortical cells, and are digested. In the infrequent *ptyophagous* type, hyphae that have entered a root experience lysis at the tips and cell contents are released. The latter type is little known and may be confined essentially to the tropics.

Subfamily VANILLOIDEAE Szlachetko

2.5.6.1 *Cyrtosia*

Cyrtosia Blume, Bijdr.: 396 (1825).
Conchoglossum Breda, Gen. Sp. Orchid. Asclep. 4: t. 17 (1830).

Cyrtosia contains seven species, all of which are achlorophyllous and thus putative mycoheterotrophs (Cameron 2003). They are widespread in tropical and subtropical Southeast Asia. *Cyrtosia septentrionalis* has been reported to grow with wood-decaying *Armillaria* fungi (Hamada 1939; Cha and Igarashi 1996; Rasmussen 2002).

2.5.6.2 *Erythrorchis*

Erythrorchis Blume, Rumphia 1: 200 (1837).
Haematorchis Blume, Rumphia 4: t. 200 B (1849).
Ledgeria F. Muell., Fragm. 1: 238 (1859).

Erythrorchis comprises three species, which are full mycoheterotrophs. *Erythrorchis altissima* ranges from Indonesia to the Philippines, *Erythrorchis ochobiensis* from Japan through Taiwan to Vietnam, Cambodia, Laos, and Thailand. *Erythrorchis cassythoides* occurs in eastern Australia. *Erythrorchis ochobiensis* is reported to form mycorrhizas with a wide range of wood-rotting and ectomycorrhizal fungi (Umata 1995, 1997a, b, 1998a, b), and the roots of *E. cassythoides* are also colonized by both ectomycorrhizal and saprotrophic fungi (Dearnaley 2006).

2.5.6.3 *Galeola*

Galeola Lour., Fl. Cochinch. 2: 520 (1790).
Pogochilus Falc., J. Bot. (Hooker) 4: 73 (1842).

Galeola includes six fully mycoheterotrophic species. The genus has a remarkably widespread distribution with *G. humblotii* occurring in Madagascar and the Comores and all other species growing in tropical and subtropical Southeast Asia. *Galeola septentrionalis* is associated with species of *Armillaria* fungi (Terashita 1996).

2.5.6.4 *Lecanorchis*

Lecanorchis Blume, Mus. Bot. Lugd. Bat. 2: 188 (1856).

Lecanorchis includes approximately 20 species. All species are fully mycoheterotrophic. The genus is quite diverse in Japan, but extends widely in tropical and subtropical Southeast Asia (Hashimoto 1990).

2.5.6.5 *Pseudovanilla*

Pseudovanilla Garay, Bot. Mus. Leafl. 30: 234 (1986).

Pseudovanilla includes about eight species from Malesia and the Pacific islands. All species have reduced leaves and stems that are orange to yellow when young but are green when mature. Their (partial) mycoheterotrophic status remains to be investigated in detail.

Subfamily ORCHIDOIDEAE Lindley

2.5.6.6 *Arthrochilus*

Arthrochilus F. Muell., Fragm. 1: 42 (1858).
Drakaea Lindl. Sect. *Akaedra* Schltr., Bot. Jahrb. Syst. 45: 383 (1911).

Limited to eastern Australia (including Tasmania) and southern Papua New Guinea, this genus of ten species contains a single leafless species, *A. huntianum*, which has the typical tubers of the genus reduced to protocorm-like structures.

2.5.6.7 *Brachycorythis*

Brachycorythis Lindl., Gen. Sp. Orchid. Pl.: 363 (1838).
Schwartzkopffia Kraenzl., Bot. Jahrb. Syst. 28: 177 (1900).
Phyllomphax Schltr., Repert. Spec. Nov. Regni Veg. Beih. 4: 118 (1919).
Diplacorchis Schltr., Beih. Bot. Centralbl. 38(2): 127 (1921).

Gyaladenia Schltr., Beih. Bot. Centralbl. 38(2): 124 (1921).
Afrorchis Szlach., Richardiana 6: 82 (2006).

Brachycorythis includes ca. 35 species and is distributed from tropical and southern Africa to Madagascar and Southeast Asia. Most species are autotrophic with green leaves, except for *B. pumilio* from tropical West Africa and *B. lastii* from tropical East Africa, which are both achlorophyllous (Summerhayes 1955).

2.5.6.8 *Burnettia*

Burnettia Lindl., Gen. Sp. Orchid. Pl.: 517 (1840).

A monospecific genus comprising the mycoheterotrophic *B. cuneata*, from southeast Australia and Tasmania.

2.5.6.9 *Chamaegastrodia*

Chamaegastrodia Makino & Maek., Bot. Mag. (Tokyo) 49: 596 (1935).

Chamaegastrodia comprises three species (Govaerts et al. 2011), which are distributed from Assam to Japan. All species lack chlorophyll and are therefore considered to be fully mycoheterotrophic. *C. shikokiana* has been demonstrated to grow with ectomycorrhizal Ceratobasidiaceae fungi (Yagame et al. 2008).

2.5.6.10 *Corybas*

Corybas Salisb., Parad. Lond.: t. 83 (1807).

Corybas has a widespread distribution that ranges over temperate Asia, Southeast Asia, Australasia, temperate Australia and New Zealand, and the Pacific Islands. The genus comprises ca. 50 species. *C. cryptanthus* from the North Island of New Zealand is leafless and lacks chlorophyll and is therefore a putative full mycoheterotroph (Moore and Edgar 1970). Specimens of *C. cheesemanii* are sometimes achlorophyllous as well.

2.5.6.11 *Cryptostylis*

Cryptostylis R.Brown, Prodr. Fl. Nov. Holl.: 317 (1810).
Chlorosa Blume, Bijdr. 8: 420 (1825).
Zosterostylis Blume, Bijdr. 8: 418 (1825).

A genus of 25 species from tropical and subtropical Asia and the Southwest Pacific. One species from southeast Australia, *C. hunteriana*, is leafless but the stems are green.

2.5.6.12 *Cystorchis* (Fig. 2.11e)

Cystorchis Blume, Fl. Javae ser. 2. 1: 73. t. 24 (1858).

Cystorchis from Southeast Asia and the Pacific islands comprises ca. 20 species, three of which are achlorophyllous: *C. aphylla* from Southeast Asia, *C. saprophytica* from Borneo, and *C. peliocaulos* from New Guinea.

2.5.6.13 *Danhatchia*

Danhatchia Garay & Christenson, Orchadian 11: 469 (1995).

Previously placed in *Yoania* (Epidendroideae), *Danhatchia* was recognized as distinct based on floral structure. The single fully mycoheterotrophic species, *D. australis*, occurs in New Zealand but has recently been discovered in New South Wales, Australia as well.

2.5.6.14 *Degranvillea*

Degranvillea Determann, Amer. Orchid Soc. Bull. 54: 174 (1985).

The rare *Degranvillea dermaptera* is the sole species in this genus and is only known from French Guiana. It bears a coralloid rhizome.

2.5.6.15 *Odontochilus*

Odontochilus Blume, Fl. Javae, n.s., 1: 69 (1858).
Evrardia Gagnep., Bull. Mus. Natl. Hist. Nat., II, 4: 596 (1932).
Evrardianthe Rauschert, Feddes Repert. 94: 433 (1983).
Evrardiana Aver., Bot. Zhurn. (Moscow & Leningrad) 73: 432 (1988).

Four of 40 *Odontochilus* species are putative full mycoheterotrophs—*O. saprophyticus* from Hainan and S Vietnam, *O. poilanei* from China, Myanmar, Thailand, Vietnam, and Japan, *O. asraoa* from Assam and Nepal, and *O. guangdongensis* from southern China.

Fig. 2.11 A few examples of fully mycoheterotrophic Orchidaceae. (**a**) *Gastrodia grandilabris*. (**b**) *Cephalanthera exigua*. (**c**) *Tropidia saprophytica*. (**d**) *Epipogium roseum*. (**e**) *Cystorchis aphylla*. (**f**) *Platanthera saprophytica*. Redrawn from Wood et al. (2011), except for (**b**) redrawn from Pedersen et al. (2009). Bar = 1 cm

2.5.6.16 *Platanthera* (Fig. 2.11f)

Platanthera Rich., De Orchid. Eur.: 26 (1817).
Lysias Salisb., Trans. Hort. Soc. London 1: 288 (1812).
Sieberia Spreng., Anleit. Kenntn. Gew., ed. 2, 2(1): 282 (1817).
Mecosa Blume, Bijdr.: 403 (1825).
Diplanthera Raf., Herb. Raf.: 73 (1833).
Tulotis Raf., Herb. Raf.: 70 (1833).
Perularia Lindl., Edwards's Bot. Reg. 20: t. 1701 (1835).
Blephariglottis Raf., Fl. Tellur. 2: 38 (1837).
Conopsidium Wallr., Linnaea 14: 147 (1840).
Diphylax Hook.f., Hooker's Icon. Pl. 19: t. 1865 (1889).
Limnorchis Rydb., Mem. New York Bot. Gard. 1: 104 (1900).
Lysiella Rydb., Mem. New York Bot. Gard. 1: 104 (1900).
Gymnadeniopsis Rydb. in N.L.Britton, Man. Fl. N. States: 293 (1901).
Piperia Rydb., Bull. Torrey Bot. Club 28: 269 (1901).

Dithrix (Hook.f.) Schltr., Notizbl. Bot. Gart. Berlin-Dahlem 9: 583 (1926).
Pseudodiphryllum Nevski in V.L.Komarov (ed.), Fl. URSS 4: 752 (1935).
Tsaiorchis Tang & F.T.Wang, Bull. Fan Mem. Inst. Biol. 7: 131 (1936).
Fimbriella Farw. ex Butzin, Willdenowia 11: 323 (1981).
× *Platanthopsis* P.M.Br., N. Amer. Native Orchid J. 8: 37 (2002).
× *Blepharopsis* Efimov, Novosti Sist. Vyssh. Rast. 40:48 (2008 publ. 2009).

Broadly distributed from North America through Eurasia, including Southeast Asia, this genus of ca. 200 species includes a single fully mycoheterotrophic species, *P. saprophytica* from Borneo. This species is entirely whitish in color, except for a purple margin to the lip (Wood et al. 2011).

2.5.6.17 *Platythelys*

Platythelys Garay, Bradea 2: 196 (1977).

A New World genus of about ten species distributed from southeastern USA to Brazil and Argentina. One species, *P. pedicellata*, has leaves reduced to bracts and may be mycoheterotrophic.

2.5.6.18 *Rhizanthella*

Rhizanthella R.S.Rogers, J. Roy. Soc. Western Australia 15: 1 (1928).
Cryptanthemis Rupp, Proc. Linn. Soc. New South Wales 57: 58 (1932).

Rhizanthella is a genus of extremely rare, fully subterranean mycoheterotrophic orchids that are endemic to Australia. The genus comprises three species: *R. gardneri* from western Australia, and *R. omissa* and *R. slateri* from southeastern Australia. Fungal associates of *R. gardneri* and *R. slateri* have been identified as *Rhizoctonia*-type fungi that most likely belong to the Ceratobasidiales (Basidiomycota) (Bougoure et al. 2009). *R. gardneri* is only found growing adjacent to individual shrubs of species in the *Melaleuca uncinata s.l.* complex (Myrtaceae) in its native habitats (Bougoure et al. 2008). Fungi isolated from *R. gardneri* demonstrated the ability to form ectomycorrhizas with the roots of *Melaleuca uncinata s.l.* individuals (Warcup 1985, 1991; Bougoure et al. 2009), suggesting that *R. gardneri* obtains carbon from *Melaleuca* plants (Bougoure et al. 2009). Carbon transfer between *R. gardneri* and *Melaleuca uncinata* through common ectomycorrhizal fungi has been confirmed by microcosm experiments (Bougoure et al. 2010). Dixon (2003) suggested that seed dispersal of *R. slateri* is carried out by animals.

Subfamily EPIDENDROIDEAE Lindley

2.5.6.19 *Aphyllorchis*

Aphyllorchis Blume, Tab. Pl. Jav. Orchid.: t. 16, f. 77 (1825).
Sinorchis S.C.Chen, Acta Phytotax. Sin. 16: 82 (1978).

Aphyllorchis includes 33 fully mycoheterotrophic species. The genus is widespread in tropical and subtropical Asia. *A. montana* and *A. caudata* are able to associate with a wide range of ectomycorrhizal fungi (Roy et al. 2009a).

2.5.6.20 *Auxopus*

Auxopus Schltr., Westafr. Kautschuk.-Exped. 275 (1900).

The fully mycoheterotrophic *Auxopus* comprises three species—*Auxopus kamerunensis* and *A. macranthus* from tropical West and Central Africa and *A. madagascariensis* from tropical Madagascar.

2.5.6.21 *Cephalanthera* (Figs. 2.5g and 2.11b)

Cephalanthera Rich., Mém. Mus. Hist. Nat. 4: 51 (1818).
Callithronum Ehrh., Beitr. Naturk. 4: 148 (1789).
Lonchophyllum Ehrh., Beitr. Naturk. 4: 148 (1789).
Dorycheile Rchb., Deut. Bot. Herb.-Buch: 56 (1841).
Xiphophyllum Ehrh., Beitr. Naturk. 4: 148 (1789).
Eburophyton A.Heller, Muhlenbergia 1: 48 (1904).
Tangtsinia S.C.Chen, Acta Phytotax. Sin. 10: 194 (1965).

Cephalanthera includes ca. 18 species. Six species lack chlorophyll and are putative full mycoheterotrophs. Stable isotope data suggest that chlorophyllous species are partial mycoheterotrophs (Julou et al. 2005; Abadie et al. 2006; Preiss et al. 2010; Stöckel et al. 2011).

Albino forms of some green species are known: e.g., *C. damasonium*, *C. longifolia* (Julou et al. 2005; Abadie et al. 2006). *Cephalanthera* is widespread in temperate Eurasia, northern Africa, North America, and Southeast Asia. One achlorophyllous species, *C. austinae*, occurs in western North America. The remaining achlorophyllous species grow in Southeast Asia (*C. calcarata*, *C. ericiflora*, *C. exigua*, *C. gracilis*, *C. pusilla*) (Pedersen et al. 2009). *C. austinae* and *C. exigua* have been found to associate with ectomycorrhizal Thelephoraceae fungi (Basidiomycota) (Taylor and Bruns 1997; Roy et al. 2009a).

2.5.6.22 *Corallorhiza* (Figs. 4.12a and 6.2)

Corallorhiza Gagnebin, Acta Helv. Phys.-Math. 2: 61 (1755).
Rhizocorallon Gagnebin, Acta Helv. Phys.-Math. 2: 61 (1755).
Corallorhiza Châtel., Spec. Inaug. Corallorhiza: 5 (1760).
Cladorhiza Raf., Amer. Monthly Mag. & Crit. Rev. 1: 429 (1817).

Corallorhiza, commonly known as "coralroot orchids," includes 12 species (Freudenstein 1997; Barrett and Freudenstein 2011). All species are achlorophyllous except for *C. trifida*, which has green stems and capsules. However, recent research has shown that *C. trifida* derives most of its carbon from ectomycorrhizal Telophoraceae fungi (McKendrick et al. 2000; Cameron et al. 2009). *Corallorhiza trifida* is extremely widespread in the temperate and subarctic Northern hemisphere. The distribution of the remaining species of *Corallorhiza* is limited to North and Central America. *Corallorhiza striata* and *C. bentleyi* are apparently associated with nonoverlapping clades of ectomycorrhizal *Tomentella* fungi (Thelephoraceae; Basidiomycota) (Barrett et al. 2010), while *C. maculata* and the closely related species *C. mertensiana* have been found to associate with nonoverlapping Russulaceae fungi (Basidiomycota) (Taylor and Bruns 1997, 1999; Taylor et al. 2004). *Corallorhiza odontorhiza* and *C. wisteriana* are sister species that occur in Mexico and northward; *C. wisteriana* is distributed across the USA, while *C. odontorhiza* occurs only in the east of the USA. *Corallorhiza odontorhiza* utilizes Thelephoraceae; populations of *C. wisteriana* in the western portion of the distribution utilize Thelephoraceae, while eastern populations utilize Russulaceae. Some polymorphic populations exist in which plants use either fungus (Freudenstein and Barrett, unpubl.).

2.5.6.23 *Cremastra*

Cremastra Lindl., Gen. Sp. Orch. Pl.: 172 (1883).

A genus of three species extending from Nepal and Sikkim through China, Korea, Japan, and Sakhalin. It contains one full mycoheterotroph, *C. aphylla* from Japan (Yukawa 1999).

2.5.6.24 *Cymbidium*

Cymbidium Sw., Nova Acta Regiae Soc. Sci. Upsal. 6: 70 (1799).
Jensoa Raf., Fl. Tellur. 4: 38 (1838) "1836."
Cyperorchis Blume, Rumphia 4: 47 (1849).
Iridorchis Blume, Coll. Orchid.: 90 (1859).
Arethusantha Finet, Bull. Soc. Bot. France 44: 179 (1897).
Pachyrhizanthe (Schltr.) Nakai, Bot. Mag. (Tokyo) 45: 109 (1931).
× *Cyperocymbidium* A.D.Hawkes, Orchid Rev. 72: 420 (1964).
Liuguishania Z.J.Liu & J.N.Zhang, J. S. China Agric. Univ. 19(1): 73 (1998).
Wutongshania Z.J.Liu & J.N.Zhang, J. S. China Agric. Univ. 19(1): 74 (1998).
Cymbidiopsis H.J.Chowdhery, Indian J. Forest. 32: 154 (2009).

Cymbidium includes ca. 52 species (Du Puy and Cribb 2007). One species, *C. macrorhizon*, lacks foliage leaves but has green stems and capsules. Stable isotope data indicate that this species is fully mycoheterotrophic. Related species with green leaves are partial mycoheterotrophs (Motomura et al. 2010). The chlorophyllous species *C. lancifolium* and *C. goeringii* both associate simultaneously with saprotrophic Tulasnellaceae and ectomycorrhizal fungi, whereas *C. macrorhizon* establishes symbiosis exclusively with ectomycorrhizal fungi (Motomura et al. 2010).

2.5.6.25 *Didymoplexiella*

Didymoplexiella Garay, Arch. Jard. Bot. Rio de Janeiro 13: 33 (1954).
Leucolena Ridl., J. Linn. Soc., Bot. 28: 340 (1891).

Didymoplexiella comprises seven fully mycoheterotrophic species that are restricted to Southeast Asia.

2.5.6.26 *Didymoplexis*

Didymoplexis Griff., Calcutta J. Nat. Hist. 4: 383 (1843).
Leucorchis Blume, Mus. Bot. 1: 31 (1849).
Apetalon Wight, Icon. Pl. Ind. Orient. 5: 22 (1851).
Epiphanes Rchb.f. in B.Seemann, Fl. Vit.: 295 (1868).

Didymoplexis includes about 12 species, all of which are full mycoheterotrophs. The genus has a remarkably wide distribution and occurs in tropical Africa (*D. africana*) and from Afghanistan to India, Southeast Asia, northern Australia, New Guinea, and Vanuatu. *Didymoplexis* is probably absent from Madagascar (Cribb et al. 2010). Morphological observations suggest that *D. minor* is associated with saprotrophic *Marasmius* fungi (Burgeff 1932).

2.5.6.27 *Dipodium*

Dipodium R.Br., Prodr. Fl. Nov. Holl.: 330 (1810).
Leopardanthus Blume, Rumphia 4: 47 (1849).
Wailesia Lindl., J. Hort. Soc. London 4: 261 (1849).
Hydranthus Kuhl & Hasselt ex Rchb.f., Xenia Orchid. 2: 20 (1862).
Trichochilus Ames, J. Arnold Arbor. 13: 142 (1932).

Dipodium ("hyacinth orchids") contains ca. 21 species from Southeast Asia, Australia, and the Pacific Islands. While some species are green and leaf-bearing, at last nine species (e.g., *D. variegatum*, *D. roseum*, *D. hamiltonianum*) have green stems but lack foliage leaves and are likely partial mycoheterotrophs. Several studies report that *Dipodium* species associate with ectomycorrhizal Russulaceae fungi (Bougoure and Dearnaley 2005; Dearnaley and Le Brocque 2006).

2.5.6.28 *Epipogium* (Fig. 2.11d)

Epipogium S.G. Gmel. ex Ehrh., Beitr. Naturk. 4: 149 (1789).
Galera Blume, Bijdr.: 415 (1825).
Ceratopsis Lindl., Gen. Sp. Orchid. Pl.: 383 (1840).
Podanthera Wight, Icon. Pl. Ind. Orient. 5: 22 (1851).
Epipogon S. G. Gmel., Fl. Sibirica 1: 11 (1747).
Epipogion St.-Lag., Ann. Soc. Bot. Lyon 7: 144 (1880).

Epipogium comprises 2–3 fully mycoheterotrophic species: *E. aphyllum* ("Ghost Orchid") from temperate Eurasia, the questionably distinct *E. japonicum* from Japan, Taiwan, and S China, and *E. roseum* from tropical Africa, Southeast Asia, New Guinea, Australia, and the Pacific Islands. *Epipogium aphyllum* associates with ectomycorrhizal fungi (Roy et al. 2009b), while populations of *E. roseum* from Japan were found to grow with saprotrophic Coprinaceae fungi (Yamato et al. 2005; Yagame et al. 2007). *E. aphyllum* was shown to obtain carbon and nitrogen through its ectomycorrhizal association (Liebel and Gebauer 2011).

2.5.6.29 *Eulophia*

Eulophia R.Br. ex Lindl., ("*Eulophus*") Bot. Reg. 7: t. 573 (1821).
Wolfia Dennst., Schlüssel Hortus Malab.: 38 (1818).
Lissochilus R.Br., Bot. Reg. 7: t. 573 (1821).
Cyrtopera Lindl., Gen. Sp. Orchid. Pl.: 189 (1833).
Thysanochilus Falc., Proc. Linn. Soc. London 1: 14 (1839).
Hypodematium A.Rich., Tent. Fl. Abyss. 2: 286 (1850).
Orthochilus A.Rich., Tent. Fl. Abyss. 2: 284 (1850).
Pteroglossaspis Rchb.f., Otia Bot. Hamburg.: 67 (1878).
Platypus Small & Nash in J.K.Small, Fl. S.E. U.S.: 329 (1903).
Triorchos Small & Nash in J.K.Small, Fl. S.E. U.S.: 329 (1903).
Smallia Nieuwl., Amer. Midl. Naturalist 3: 158 (1913).
Donacopsis Gagnep., Bull. Mus. Natl. Hist. Nat., II, 4: 593 (1932).
Semiphajus Gagnep., Bull. Mus. Natl. Hist. Nat., II, 4: 598 (1932).

The pantropical *Eulophia* contains ca. 230 species (only two of which are neotropical). Most species are terrestrial, but a few are epiphytes or lithophytes. At least 17 species are leafless or nearly so and probably mycoheterotrophic, including *E. epiphanoides* (southwest Tanzania), *E. galeioides* (tropical Africa), *E. gastrodioides* (Mozambique and Zambia), *E. macrantha* (Malawi and Zimbabwe), *E. richardsiae* (northern

Zambia), and *E. zollingeri* (widespread in tropical and subtropical Asia and Australia). Despite its wide distribution, *E. zollingeri* was found to associate with a narrow lineage of wood-rooting fungi within Coprinaceae (Ogura-Tsujita and Yukawa 2008).

2.5.6.30 *Gastrodia* (Fig. 2.11a)

Gastrodia R.Br., Prodr. Fl. Nov. Holl.: 330 (1810).
Epiphanes Blume, Bijdr.: 421 (1825).
Gamoplexis Falc. ex Lindl., Gard. Chron. 1847: 103 (1847).
Neoclemensia Carr, Gard. Bull. Straits Settlem. 8: 180 (1935).
Demorchis D.L.Jones & M.A.Clem., Orchadian 14(8: Sci. Suppl.): xiii (2004).

Gastrodia comprises ca. 22 achlorophyllous mycoheterotrophic species (Cribb et al. 2010). The center of diversity of the genus is situated in Southeast Asia, and the genus extends to Japan, Siberia, tropical Australia, New Zealand, New Caledonia, the Pacific Islands, Madagascar, the Mascarene Islands, and tropical Africa (Cribb et al. 2010). Dearnaley and Bougoure (2010) identified a number of fungi in the roots of *G. sesamoides*; the most common were saprotrophic members of Marasmiaceae. In addition, stable isotope analysis suggests that *G. sesamoides* obtains most of its carbon from these wood-rotting fungi (Dearnaley and Bougoure 2010). *Gastrodia elata* associates with saprotrophic and parasitic *Armillaria* and *Mycena* fungi (Kusano 1911; Lan et al. 1994; Xu and Fan 2001). *Gastrodia confusa* also associates with saprotrophic *Mycena* fungi (Ogura-Tsujita et al. 2009). *Gastrodia similis* was found to grow mainly with wood-decaying *Resinicium* species (Martos et al. 2009).

2.5.6.31 *Hexalectris* (Fig. 6.1)

Hexalectris Raf., Neogenyton: 4 (1825).

Hexalectris is a New World genus of about eight fully mycoheterotrophic species, occurring throughout most of the southern U.S.A. and Mexico, with concentrations of diversity in the mountainous regions of southwest U.S.A. and northeastern and western Mexico. They often inhabit inhospitable habitats, such as desert canyons, cedar thickets, and tropical dry forests (Kennedy and Watson 2010). *Hexalectris* species are specialized toward different clades of fungi, mainly in Sebacinaceae. These fungi are presumably ectomycorrhizal with surrounding trees (Taylor et al. 2003; Kennedy et al. 2011).

2.5.6.32 *Kalimantanorchis*

Kalimantanorchis Tsukaya, M. Nakajima & H. Okada, Syst. Bot. 36: 52 (2011).

Kalimantanorchis comprises a single species, the achlorophyllous *K. nagamasui* from Borneo (Tsukaya et al. 2011), which is questionably distinct from *Tropidia*.

2.5.6.33 *Limodorum*

Limodorum Boehm., Defin. Gen. Pl.: 358 (1760).
Centrosis Sw., Adnot. Bot.: 52 (1829), nom. illeg.
Jonorchis Beck, Fl. Nieder-Österreich 1: 215 (1890).
Lequeetia Bubani, Fl. Pyren. 4: 57 (1901).

Limodorum comprises three mycoheterotrophic species: *L. arbortivum* from Europe, North Africa, and the Caucasus, *L. rubriflorum* from Turkey, and *L. trabutianum* from the Mediterranean. The plants have small scale-like leaves and a violet stem. Despite the presence of chlorophyll in *L. arbortivum* (Blumenfeld 1935), CO_2 fixation is insufficient to compensate for respiration in adult plants suggesting that *L. arbortivum* is at least partially mycoheterotrophic (Girlanda et al. 2006). C and N stable isotope signatures also support the mycoheterotrophic status of *L. arbortivum* and *L. trabutianum* (Liebel et al. 2010). Both species associate predominantly with fungal symbionts of the genus *Russula*. The associated fungi were found to be ectomycorrhizal with surrounding trees (Girlanda et al. 2006). Paduano et al. (2011) detected differences in cellular response to Russulaceae and *Ceratobasidium* fungi in *Limodorum abortivum*.

2.5.6.34 *Malaxis*

Malaxis Sol. ex Sw., Nov. Gen Sp. Prodr. 8 (1788).
Limnas Ehrh., Beitr. Naturk. 4: 146 (1789).
Achroanthes Raf., Med. Repos. N. York 5: 352 (1808).
Microstylis (Nutt.) Eaton Man. Bot. ed. 3, 115 (1822).

Cheiropterocephalus Barb. Rodr. Gen Sp. Orch. 1: 28 (1877).
Tamayorkis Szlach., Fragm. Phyt. Geobot. Suppl. 3: 121 (1995).

A widespread genus of ca. 300 species, occurring in tropical and temperate regions on all continents except Antarctica. Two species are putative full mycoheterotrophs: *M. aphylla* and *M. saprophyta*.

2.5.6.35 *Neottia*

Neottia Guett., Hist. Acad. Roy. Sci. Mém. Math. Phys. (Paris, 4to) 1750: 374 (1754).
Ophris Mill., Gard. Dict. Abr. ed. 4 (1754).
Nidus Rivinus, Icon. Pl. Fl. Irreg. Hexapet. T. 7. (1764).
Nidus-avis Ortega, Tab. Bot.: 24 (1773).
Cardiophyllum Ehrh., Beitr. Naturk. 4: 148 (1789).
Epipactis Persoon, Syn. Pl. 2: 513 (1807).
Diphryllum Raf., Med. Repos. Ser. 2, 5: 357 (1808).
Listera R.Br. in W.T.Aiton, Hortus Kew. 5: 201 (1813).
Neottidium Schltdl., Fl. Berol. 1: 454 (1823).
Distomaea Spenn., Fl. Friburg. 1: 245 (1825).
Pollinirhiza Dulac, Fl. Hautes-Pyrénées: 120 (1867).
Bifolium Nieuwl., Amer. Midl. Naturalist 3: 128 (1913).
Holopogon Kom. & Nevski in V.L.Komarov (ed.), Fl. URSS 4: 750 (1935).
Diplandrorchis S.C.Chen, Acta Phytotax. Sin. 17(1): 2 (1979).
Archineottia S.C.Chen, Acta Phytotax. Sin. 17(2): 12 (1979).

Neottia is a genus of ca. 60 species with a distribution ranging through the temperate and subarctic Northern Hemisphere. It now contains the leafy species previously placed in *Listera*. Fourteen species are achlorophyllous and presumably fully mycoheterotrophic. These include *N. acuminata*, *N. brevilabris*, *N. camtschatea*, *N. gaudissartii*, *N. listeroides*, *N. megalochila*, *N. microglottis*, *N. nidus-avis*, *N. pantlingii*, *N. papilligera*, *N. smithiana*, *N. taibaishanensis*, *N. tenii*, and *N. ussuriensis* (Govaerts et al. 2011). Natural abundance ^{15}N and ^{13}C data confirms the mycoheterotrophic nature of *N. nidus-avis* (Gebauer and Meyer 2003). *N. nidus-avis* has been found to associate with *Sebacina* fungi that are ectomycorrhizal with surrounding trees, including *Fagus sylvatica* and *Corylus* sp. (McKendrick et al. 2002; Selosse et al. 2002).

2.5.6.36 *Pogoniopsis*

Pogoniopsis Rchb.f., Otia Bot. Hamburg.: 82 (1881).

A rare genus with two mycoheterotrophic species from eastern Brazil (*P. nidus-avis* and *P. schenkii*). Previously placed in Vanilloideae, it appears to fall among the epidendroids (Cameron 2003; pers. comm.).

2.5.6.37 *Risleya*

Risleya King & Pantl., Ann. Roy. Bot. Gard. Calcutta 8: 246 (1898).

Risleya contains a single species, *R. atropurpurea*. This full mycoheterotroph has been recorded from the eastern Himalayas and China (southeastern Sichuan and northwestern Yunnan) (Govaerts et al. 2011).

2.5.6.38 *Silvorchis*

Silvorchis J.J.Sm., Bull. Dép. Agric. Indes Néerl. 13: 2 (1907).

The fully mycoheterotrophic *Silvorchis colorata* is the sole species of the genus. The species is known only from the type specimen, which was collected in Java.

2.5.6.39 *Stereosandra*

Stereosandra Blume, Mus. Bot. Lugd. Bat. 2: 176 (1856).

Stereosandra javanica, a full mycoheterotroph and the sole species of the genus, has a wide distribution in Southeast Asia, having been recorded from the eastern Himalayas, southern China, the Ryukyu Islands, Taiwan, Vietnam, Borneo, Java, Malaysia, Sumatra, the Philippines, New Guinea, the Solomon Islands, and Samoa.

2.5.6.40 *Tropidia* (Fig. 2.11c)

Tropidia Lindl., Edwards's Bot. Reg. 19: t. 1618 (1833).

Decaisnea Lindl. ex Wall., Numer. List: 7388 (1832), nom. inval.
Cnemidia Lindl., Edwards's Bot. Reg. 19: t. 1618 (1833).
Chloidia Lindl., Gen. Sp. Orchid. Pl.: 484 (1840).
Ptychochilus Schauer, Nov. Actorum Acad. Caes. Leop.-Carol. Nat. Cur. 19(Suppl. 1): 431 (1843).
Govindooia Wight, Icon. Pl. Ind. Orient. 6: 34 (1853).
Schoenomorphus Thorel ex Gagnep., Bull. Soc. Bot. France 80: 351 (1933).
Muluorchis J.J.Wood, Kew Bull. 39: 73 (1984).

A genus with ca. 20 species from tropical and subtropical Asia and the Pacific, with one neotropical species. *T. saprophytica* and *T. connata* from Borneo are full mycoheterotrophs.

2.5.6.41 *Uleiorchis*

Uleiorchis Hoehne, Arq. Bot. Estado São Paulo, n.s., f.m., 1: 129 (1944).

Uleiorchis comprises two fully mycoheterotrophic species from tropical South America. *Uleiorchis liesneri* is known from Venezuela and *U. ulei* has been recorded in Central and tropical South America (Costa Rica, Honduras, Panama, French Guiana, Guyana, Venezuela, Colombia, Ecuador, Peru, and Brazil (Born et al. 1999)).

2.5.6.42 *Wullschlaegelia*

Wullschlaegelia Rchb.f., Bot. Zeitung (Berlin) 21: 131 (1863).

Wullschlaegelia comprises two fully mycoheterotrophic species, *W. aphylla* and *W. calcarata* (Born et al. 1999). Both species are widespread in tropical South and Central America and are present in the West Indies. *Wullschlaegelia aphylla* associates with both litter-decaying *Gymnopus* and *Mycena* species (Martos et al. 2009).

2.5.6.43 *Yoania*

Yoania Maxim., Bull. Acad. Imp. Sci. Saint-Pétersbourg, III, 18: 68 (1872).

Yoania comprises four species, all fully mycoheterotrophic: *Y. amagiensis* (Japan), *Y. flava* (Japan), *Y. japonica* (Assam, China, Taiwan, Japan), and *Y. prainii* (eastern India to northern Vietnam).

2.5.7 Iridaceae

Iridaceae Juss., Gen. Pl.: 57 (1789).
Geosiridaceae Jonker, Recueil Trav. Bot. Néerl. 36: 477 (1939).

Herbs, perennial, rarely annual, evergreen, or seasonal. Underground parts a rhizome, bulb, or corm. Stems simple or branched, terete or variously compressed, angled or winged. Leaves basal and cauline, distichous; proximal 2–3 sometimes membranous, not reaching much above ground; others with open or closed sheaths, usually unifacial. Inflorescences umbellate, monochasial cymes (rhipidia), spikes, or solitary flowers; rhipidia enclosed in 2, opposed, usually large, leafy to dry bracts (spathes). Flowers usually pedicellate actinomorphic or zygomorphic, petaloid, with 2 equal or unequal whorls of 3 tepals each. Tepals usually large, showy, free or connate in tube. Stamens (2-)3, inserted at base of outer tepals or in tube, symmetrically arranged or unilateral; filaments free or partly connate; anthers with 2 pollen sacs, extrorse, usually dehiscing longitudinally. Ovary inferior, (1-)3-locular; placentation axile; ovules 2–few, anatropous; style 1, filiform at least proximally, usually 3-branched or 3-lobed. Fruits a capsule, loculicidal, rarely indehiscent, firm to cartilaginous, occasionally woody. Seeds globose to angular or discoid, sometimes broadly winged; seed coat usually dry.

Number of genera and species—Iridaceae contain over 2,000 species in 66 genera (Goldblatt et al. 2008). The family includes two fully mycoheterotrophic species, *Geosiris aphylla* and *G. albiflora* (Goldblatt and Manning 2008, 2010).

Distribution and habitat—Iridaceae have a cosmopolitan distribution, with a main center of diversity in southern Africa. They are particularly species-rich in the Cape region (Davies et al. 2005). Iridaceae are mainly found in open, seasonable habitats, but also occur in forests, savannas, and semi-arid habitats. They grow on a variety of different soil types.

Classification—Iridaceae are part of the monocot order Asparagales (Chase et al. 2000, 2006; Fay et al. 2000; APG 2009). The family consists of seven subfamilies of which the Iridoideae and Crocoideae contain the majority of the species (Goldblatt et al. 2008). Baillon who described the species *Geosiris aphylla* in 1894, already placed it in Iridaceae (Baillon 1894). However, Engler (1897) transferred *Geosiris* to Burmanniaceae without seeing a specimen. Jonker (1939) concluded that *Geosiris* was not a member of Burmanniaceae and erected a new family, Geosiridaceae, to accommodate the species. Jonker considered Geosiridaceae and Iridaceae closely related. Molecular data confirmed the placement of *Geosiris* in Iridaceae, although it is placed in its own subfamily, Geosiridoideae (Goldblatt et al. 2008).

Evolutionary history—Reliable fossils of Iridaceae date back to the Miocene (23–5 Ma), but molecular clock analyses estimated the origin of Iridaceae at 82 Ma (Wikström et al. 2001) and Ma (Janssen and Bremer 2004), respectively. The start of the divergence of the extant crown group is estimated at 61 (Goldblatt et al. 2008) or 96 Ma (Janssen and Bremer 2004). According to most recent phylogenetic hypotheses *Geosiris* is an early diverging lineage within the family (Goldblatt et al. 2008).

Ecology—The variously shaped and colored flowers of Iridaceae are pollinated by various insects (bees, beetles, flies, wasps, moths, and butterflies) and birds (hummingbirds and sunbirds). Many Iridaceae are highly specialized in their pollinator relationships. Various seed dispersal mechanisms are observed in Iridaceae. These include dispersal by wind, water, ants, and birds (Goldblatt and Manning 2008). The mycorrhizas of only few species of Iridaceae have been examined. From these observations it seems that Iridaceae are mainly associated with AM fungi (Wang and Qiu 2006).

2.5.7.1 *Geosiris* (Figs. 2.5e, 2.12, and 4.10h)

Geosiris Baillon, Bull. Mens. Soc. Linn. Paris 2:1149 (1894).

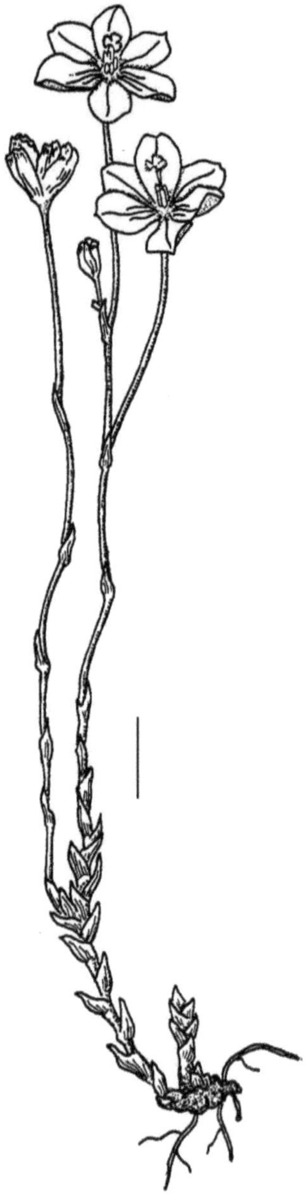

Fig. 2.12 *Geosiris aphylla*. Redrawn from Goldblatt and Manning (2008). Bar = 1 cm

Mycoheterotrophic herbs, up to 12 cm tall. Rhizome short and thick. Roots filiform. Leaves reduced to scales. Inflorescence a binate rhipidium, distorted by crowding of numerous flowers, binate rhipidia few to several. Flowers actinomorphic, without nectaries, lasting a single day. Tepals purple to white, connate at base, spreading. Stamens with free filaments; anthers loculicidal, extrorse. Ovary inferior; style slender, dividing into 3-fringed lobes or apically 3-fid. Fruit a capsule, more or less woody. Seeds minute, dust-like (Goldblatt et al. 2008; Goldblatt and Manning 2008).

Geosiris includes *G. aphylla*, known from evergreen forests in Madagascar. A second species, *G. albiflora*, is endemic to Mayotte (the Comores) (Goldblatt and Manning 2010). A third species has been discovered in Madagascar but remains to be described (Goldblatt and Manning 2010). The identity of the mycorrhizal fungi remains unknown, but like other Iridaceae *Geosiris* is presumably associated with arbuscular mycorrhizal fungi. Pollination syndrome and seed dispersal agents are unknown.

2.5.8 Polygalaceae

Polygalaceae Hoffmans. & Link, Fl. Portug. 1: 62 (1809).

Trees, lianas, shrubs, or perennial as well as annual, rarely mycoheterotrophic herbs. Leaves usually alternate, simple, entire, with pinnate venation. Inflorescence spicate, racemose, or paniculate, sometimes reduced to a single flower, terminal or axillary, bracteate. Flowers bisexual, zygomorphic to actinomorphic. Sepals usually 5, free to more or less connate, lateral sepals often large and petaloid ("wings"). Petals 3 (2 upper ones and 1 lower one) or sometimes 5, imbricate, free, but often all adnate to the staminal tube, the lower petal often boat-shaped and keeled. Stamens (2-)5–8(-10); filaments free or connate into a tube adnate to the petals; anthers basifixed, (2-)4-sporangiate, opening by pores or longitudinal slits. Ovary superior, 2–8-locular, with axile placentation, ovules mostly 1 per carpel, epitropous; style 1, bilobed with 1 stigmatic branch and 1 sterile branch, or stigma capitate. Fruit a loculicidal capsule or a samara, drupe, berry, or nut. Seeds often with stiff hairs.

Number of genera and species—Polygalaceae comprise approximately 21 genera and 1,000 species (Stevens 2001). All species of *Epirixanthes* are fully mycoheterotrophic.

Distribution and habitat—The family of Polygalaceae is cosmopolitan, with its center of diversity in tropical and subtropical areas (Eriksen and Persson 2007).

Classification—Polygalaceae were often considered to be related to Malpighiaceae or Krameriaceae, because of their common bilaterally symmetrical flowers (Cronquist 1981). Analyses of plastid DNA sequences, however, place the family in Fabales (Chase et al. 1993), as the sister group of Surianaceae (Forest et al. 2007; Stevens 2001). Within Polygalaceae four monophyletic tribes are recognized (Xanthophylleae, Polygaleae, Carpolobieae, and Moutabeae) (Eriksen and Persson 2007; Forest et al. 2007). Based on morphology *Epirixanthes* is placed in Polygaleae, but its exact phylogenetic position remains unknown.

Evolutionary history—Molecular clock analyses date the origin of the Polygalaceae to the Late Cretaceous with diversification beginning in the Paleocene (Wikström et al. 2001; Bello et al. 2009).

Ecology—The showy flowers of many Polygalaceae attract various bees and wasps, but self-pollination is also well known. In *Polygala* a pollination mechanism with a movable lower boat-shaped petal is recorded. Species with samaras are dispersed by wind, fleshy-fruited species are vertebrate-dispersed. The loculicidal capsules of *Polygala* release seeds with lobed, aril-like structures that are dispersed over short distances by ants (Judd et al. 1999; Eriksen and Persson 2007). *Polygala* and *Epirixanthes* species associate with arbuscular mycorrhizas (Wang and Qiu 2006; Imhof 2007). Species in other genera are probably arbuscular mycorrhizal as well.

2.5.8.1 *Epirixanthes* (Figs. 2.13a, b, 2.15b, and 4.13c)

Epirixanthes Blume, Catalogus 25 (1823).
Salomonia Loureiro, Fl. Cochinch. 14 (1790).

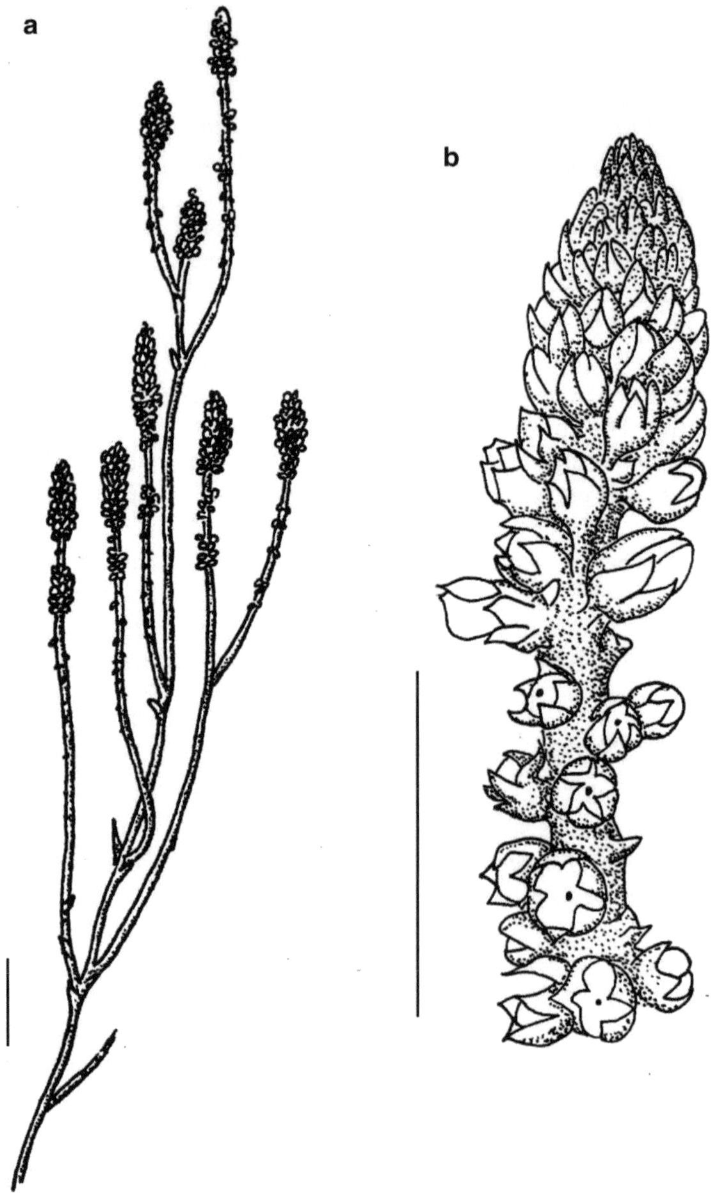

Fig. 2.13 Polygalaceae. *Epirixanthes elongata*: (**a**) habit, (**b**) inflorescence. Redrawn from Hsieh et al. (1995). Bar = 1 cm

Mycoheterotrophic herbs, up to 25 cm tall. Rhizome short. Roots filiform. Stems erect, simple or sparsely branched. Leaves sessile, reduced to scales. Inflorescence spicate, terminal. Flowers very small. Sepals 5, unequal, free or basally connate. Petals 3. Stamens 2–5; filaments united or partly free; anthers introrsely dehiscent by a slit. Ovary 2-locular, compressed; style short and bifurcate toward apex. Fruit indehiscent, with a fleshy pericarp. Seeds ellipsoid, glabrous, nearly without endosperm, with thickened tissue at micropylar end.

Epirixanthes belongs to the tribe Polygaleae. The genus comprises six species (Pendry 2010). It is widely distributed in tropical Asia, including India, Indonesia, Malaysia, Myanmar, Thailand, Vietnam, China, and the Solomon Islands. *Epirixanthes* occurs in the leaf litter of evergreen forest and bamboo groves (Hsieh et al. 1995; Shukun et al. 2008). Root anatomical observations of *Epirixanthes papuana* and *E. elongata* have shown that species of *Epirixanthes* associate with arbuscular mycorrhizas (Imhof 2007). Autogamy has been suggested for *Epirixanthes* (Wirz 1910). Observations on seed dispersal mechanisms are lacking.

2.5.9 Ericaceae

Ericaceae Juss., Gen. Pl.: 159 (1789).

Trees, shrubs, lianas, sometimes epiphytic, occasionally nearly herbaceous, associated with mycorrhizal fungi, some mycoheterotrophic. Leaves simple, alternate, opposite or whorled, margin entire to serrate, occasionally revolute, without stipules. Inflorescence usually a bracteate raceme, flowers sometimes solitary, either terminal or axillary. Flowers usually bisexual, sometimes unisexual in which case the plants are dioecious, actinomorphic to slightly zygomorphic. Sepals 4–5, free to slightly connate. Petals usually 4–5 and connate but sometimes free, often campanulate-urceolate, sometimes funnelform, often pendulous. Stamens 8–10, filaments free or adnate to corolla, sometimes connate, sometimes spurred near junction with anther; anthers becoming inverted, often opening by 2 apical pores, sometimes by slits, in some cases awned or apex narrowed to form a tubule, pollen shed in tetrads, rarely in monads, usually tricolporate, sometimes with viscin threads. Carpels 2–10, usually 5 or 4, ovary superior to inferior, usually with axile or deeply intruded parietal placentation. Style 1, hollow, with fluted cavity, stigma capitate to strongly lobed; ovules 1 to numerous per locule, unitegmic, tenuinucellar. Intrastaminal nectary disc usually present at base of superior ovary or at top of inferior ovary. Fruit a septicidal or loculicidal capsule, berry, or 1-several pitted drupe. Seeds small to minute, embryo developed to undeveloped.

Number of genera and species—Ericaceae comprise approximately 126 genera and 3,995 species (Stevens 2001). Eleven largely monospecific genera comprise fully mycoheterotrophic species. One genus, *Pyrola*, has a single mycoheterotrophic taxon.

Distribution—Cosmopolitan, but uncommon in many lowland tropical and desert regions. The mycoheterotrophic species are primarily temperate and when found in tropical regions are in montane habitats.

Classification—Ericaceae are part of Ericales, which also include Actinidiaceae, Balsaminaceae, Cyrillaceae, Clethraceae, Diapensiaceae, Ebenaceae, Fouquieriaceae, Lecythidaceae, Marcgraviaceae, Mitrastemonaceae, Pentaphylacaceae, Polemoniaceae, Primulaceae, Roridulaceae, Sapotaceae, Sarraceniaceae, Sladeniaceae, Styracaceae, Symplocaceae, Tetrameristaceae, and Theaceae (APG 2009). Ericaceae is the only family among these with mycohetrotrophic species. *Pyrola* has often been placed with three to four other small genera in its own family, Pyrolaceae, and the remaining mycoheterotrophic genera have been segregated as Monotropaceae. Recent phylogenetic analyses (summarized in Kron et al. 2002) have shown these groups to fall within, but near the base of, Ericaceae. Relationships among these basal lineages are still somewhat unclear, but the pyroloids and monotropoids do not appear to form a monophyletic group (Freudenstein and Broe, unpubl.)

Evolutionary history—Ericaceae fall clearly into Ericales, a lower Asterid group that is defined by both molecular and morphological characters (Anderberg 1992; Anderberg et al. 2002). The closest relatives of the family appear to be Clethraceae and Cyrillaceae. Fossils of Ericaceae comprise largely leaf impressions, but seeds and fruits are also known. They are from Tertiary to late Cretaceous, with perhaps the oldest being a charcoalified flower from the Turonian of New Jersey (Nixon and Crepet 1993). The crown node date for Ericales proposed by Bremer et al. (2004) based on molecular dating is ca. 114 Ma, putting Ericaceae somewhat younger than that.

Ecology—Many genera of the family are characterized by urceolate flowers, while others (e.g., *Rhododendron*), have much more open flowers. Anthers that dehisce poricidally are common, bees engaging in vibratory stimulation to release pollen. Bird pollination is also known in some

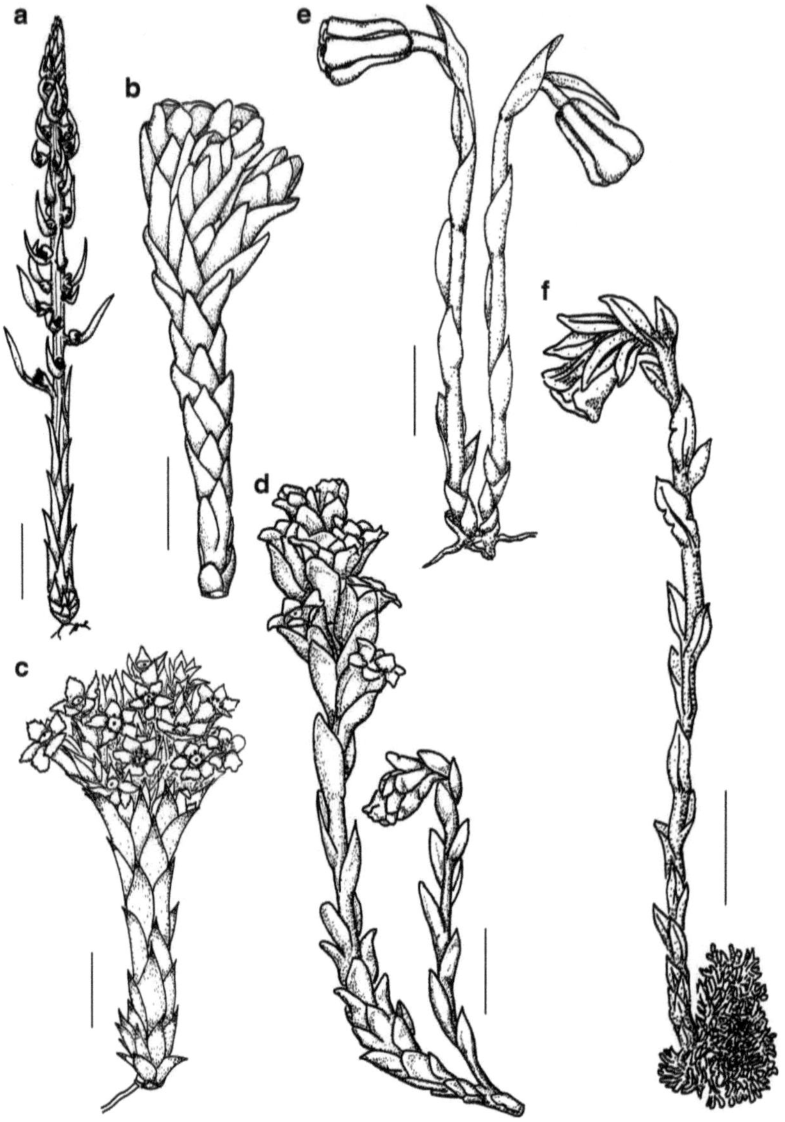

Fig. 2.14 Fully mycoheterotrophic Ericaceae (part 1). (**a**) *Allotropa virgata*. (**b**) *Cheilotheca malayana*. (**c**) *Hemitomes congestum*. (**d**) *Hypopitys monotropa*. (**e**) *Monotropa uniflora*. (**f**) *Monotropastrum humile*. Redrawn from Flora of North America Editorial Committee (2009), except (**b**) redrawn from Hooker (1886), (**d**) redrawn from Fitch (1924), and (**f**) redrawn from Yang et al. (1999). Bar = 3 cm

tropical groups. Often associated with acidic soils, most members of the family are terrestrial, but tropical epiphytes also occur. Members of the family can predominate in acidic bogs and are frequent in arctic regions. Given their frequent occurrence in areas of low nutrient availability, the family is well-known for its mycorrhizal associations. They exhibit three of the fundamental mycorrizal types listed by Smith and Read (2008)—ericoid, arbutoid, and monotropoid. Imhof (2009) examined those types in detail and characterized additional subtypes. Full mycoheterotrophy appears to have arisen twice in the family—once in the monotropoids and a second time in *Pyrola aphylla* in the pyroloids (Freudenstein and Broe, unpubl.). Bidartondo and Bruns (2001, 2002) enumerated the fungal groups with which the monotropoids associate.

2.5.9.1 *Allotropa* (Figs. 2.14a, 2.15d, and 4.16a)

Allotropa Torr. & A. Gray, Pacific Railr. Rep. 6: 81 (1858).

Racemes to 50 cm tall, fleshy, the white axis typically striped longitudinally with red, with

Fig. 2.15 Fully mycoheterotrophic species in eudicots: (**a**) *Voyria clavata* (Gentianaceae). (**b**) A large clump of *Epirixathes* plants (Polygalaceae) from Malaysian Borneo. (**c**) Voyria tenuiflora (Gentianaceae) photographed in French Guiana by Heiko Hentrich. (**d**) *Allotropa virgata* (Ericaceae) in Umqua Forest, Oregon, USA. (**e**) *Sarcodes sanguinea* (Ericaceae) at Lassen Volcanic National Park in California, USA

occasional white bracts that are at first appressed to axis, becoming recurved and eventually purplish with age. Root mass comprising slender brittle roots with occasional shoots, frequently several racemes produced from each mass. Flowers produced on upper half of axis, each subtended by a narrow white bract that is longer than the perianth, and borne on a short pedicel, protogynous, glabrous. Sepals rarely present, but if so reduced to two or more narrowly lanceolate or filiform segments. Petals 5, free, elliptic with rounded to acute apices, margins irregular to erose, white, forming a campanulate corolla. Stamens 10, incurved over ovary and slightly shorter than pistil, filaments white, terete except at base where they are flattened, glabrous; anthers dark red, basifixed, dehiscing by two pore-like slits. Ovary spheroid-slightly ellipsoidal, 5-locular, with axile placentation, glabrous, abruptly narrowed to a short style whose base may be sunken slightly into ovary; stigma discoid, obscurely five-lobed, red; nectary represented by 10 lobes between stamen bases. Fruit a loculicidal capsule. Seeds linear, ca. 1 mm, numerous.

One species, *A. virgata* ("Sugar Stick"), in western North America from the Sierra Nevada in California in the south to southern British Colombia in Canada in the north. Infrequent but sometimes locally abundant, in mixed or evergreen forests, generally flowering from June to August. Flowers may be selfing, although bumblebees have been observed visiting the flowers of *Allotropa* (Wallace 1975). Seeds are presumably dispersed by wind. Clonal reproduction has been reported as well (Lichthardt and Mancuso 1991). Copeland (1938) studied the morphology. Bidartondo and Bruns (2001, 2002) sequenced the ITS region of the fungal symbionts of a number of *Allotropa* plants from California and Oregon and found that all investigated specimens were associated with a narrow clade within the basidiomycete fungal genus *Tricholoma*.

2.5.9.2 *Cheilotheca* (Fig. 2.14b)

Cheilotheca J. D. Hooker in Bentham et J. D. Hooker, Gen. 2: 605, 607 (1876).
Chilotheca T.E. von Post & C.E.O. Kuntze (1903), orth. var.
Wirtgenia H. Andres, Verh. Bot. Vereins Prov. Brandenburg 56: 61 (1914).
Andresia H. Sleumer, Fl. Males., Ser. 1 6: 669 (1967).

Racemes to 10 cm tall, inconspicuous, white to reddish or purplish, stout, fleshy, and densely covered with imbricate bracts. Racemes with 1–few flowers. Root mass shallow, loosely organized, of highly branched, slender, brittle roots, probably perennial like those of all other members of the subfamily. Floral axes emerging from soil erect rather than nodding as in *Monotropa*, with usually a few inflorescences per root mass. Inflorescences with a single flower each are not uncommonly mixed in with racemes. Flowers erect at anthesis on short, stout pedicels, each subtended by a broad-based bract. Bracts broader and generally larger than perianth segments, overlapping with acute thickened apices. Calyx polysepalous; sepals 2–4, sometimes difficult to discern from nearby bracts except for their position on the pedicels, arranged in a lateral pair and a dorsiventral pair; usually the dorsiventral pair will be lacking if there are fwer than 4 sepals; sepals may lack the conspicuously constricted acute apices found on bracts. Corolla polypetalous; petals 3, distinctive, oblong, imbricate over stigma in bud; apices of petals broad, deeply concave and quite thick compared to the lower portions of petals which taper little from a point just below apex. Stamens 6, in 2 series of alternating lengths; filaments flattened, straight, finely pubescent. Anthers in *Cheilotheca khasiana* basifixed linear, dehiscing by longitudinal slits; in *C. malayana* anthers hippocrepiform, dehiscing by a single, somewhat introrse, terminal slit over the connate sacs. Pistil narrow ampulliform with imperceptible external transition area between style and ovary; stigma with 6 low lobes, capitate and slightly umbilicate; style straight, tapering, and stout; ovary slightly pubescent, unilocular; placentation parietal; placentas 6; nectaries represented (not clear in *C. khasiana*) by short lobes projecting downwardly between staminal bases. Seeds numerous, embedded in viscous material within fruit (Reproduced with small changes from Wallace 1975).

Cheilotheca comprises three species, *C. khasiana*, *C. malayana*, and *C. sleumeriana*, which are achlorophyllous and putative full mycoheterotrophs. They are all very poorly known. *Cheilotheca khasiana* is only known from the type locality in western Assam, India. *C. malayana* is restriced to the state of Perak in Malaysia, where it has been found in upper montane dipterocarp and oak-laurel forests (Keng 1974; Wallace 1975). *C. sleumeriana* is based on a single collection from Sumatra (Keng 1974). Keng (1974) merged *Cheilotheca* and *Monotropastrum*, creating a heterogeneous group. Observations on pollination, seed dispersal, and mycorrhiza for these rare species are lacking.

2.5.9.3 *Hemitomes* (Fig. 2.14c)

Hemitomes A. Gray, Pacific Railr. Rep. 6: 80 (1858).
Newberrya Torr., Ann. Lyceum Nat. Hist. New York 8: 55 (1864).

Root mass comprising short brittle roots from which emerge one or more inflorescences. Inflorescence axis covered with imbricate bracts. Racemes congested, to 10 cm tall, variable in form, from a simple structure compressed into a capitate inflorescence at the soil surface to a compound structure with cymulose branches to occasionally a solitary flower. Flowers and inflorescence axis reddish to pink to white or slightly yellowish, each flower subtended by a ciliate-margined bract that is longer than the perianth. Flowers 4–5-merous, protogynous. Pedicels stout, erect. Sepals free, appressed to corolla, glabrous on abaxial surface, with scattered trichomes on adaxial surface, when four of two unequal pairs, with lateral pair keeled, longer than dorsiventral pair, which are flattened. Petals connate, apical lobes narrowly ovate, glabrous abaxially but densely hairy adaxially, slightly saccate at base. Stamens usually 8, in two series with alternating lengths, filaments slender, terete, pubescent; anthers basifixed, without awns or appendages, dehiscing by two elongate slits; adaxial anther sacs smaller and shorter, becoming joined to adaxial sacs at dehiscence, obscured at maturity. Pistil narrowly ampulliform, style pubescent, merging imperceptibly with ovary. Stigma discoid, smooth, unlobed, subtended by a ring of hairs, with a depression in the center, yellow; ovary unilocular, with parietal placentation. Nectaries present as 8 paired lobes projected between staminal bases. Fruit a globose berry with a sticky mass of numerous minute seeds inside. Seeds subovoid, endosperm present.

One species, *Hemitomes congestum* ("Gnome Plant"), in mixed or coniferous forests in western North America. Self-pollination may occur, although bumblebees have been collected visiting the flowers (Wallace 1975). Seed dispersal has not been studied, but the sticky seeds are presumably dispersed by animals. Morphology and anatomy were studied by Copeland (1934, 1941). Populations of *Hemitomes congestum* from Oregon were found to associate with a narrow range of *Hydnellum* fungi (Basidiomycota) (Bidartondo and Bruns 2001).

2.5.9.4 *Hypopitys* (Fig. 2.14d)

Hypopitys J. Hill, Brit. Herbal 221 (1756).
Hypopithys G.A. Scopoli, Fl. Carniol., ed. 2. 1: 285 (1771).
Hypopithis Rafinesque, Med. Repos. ser. 3. 1: 297 (1810).

Racemes slender, to 30 cm tall, fleshy, arising from a mass of brittle roots, 1-several flowered, secund, nodding at anthesis, erect in fruit, cream-colored to tawny brown to reddish, with occasional bracts. Flowers subtended by single pubescent bracts that are elliptic to ovate to narrowly ovate, erose-lacerate. Pedicels slender, finely pubescent. Sepals usually 4–5, oblanceolate, pubescent adaxially or on both surfaces, margin ciliate. Petals rectangular to slightly spatulate, pubescent on adaxial only or on both surfaces, slightly deflexed at apex, narrowed at base by infolding of margins to form a distinct saccate structure, pubescent on one or both surfaces, erose to coarsely toothed. Stamens included, about equaling the length of the pistil, filaments pubescent, somewhat flattened, arranged in two series of alternating lengths. Anthers hippocrepiform, shortened, opening by a terminal slit across the anther sacs. Ovary spheroidal, lobed, pubescent, narrowed abruptly to a cylindrical style that broadens toward apex, bearing a funnelform stigma with undulating inward lobes, with a distinct zone of coarse hairs just below stigma. Nectaries represented by 8–10 paired lobes projecting downward between stamen bases. Fruit a loculicidal capsule. Seeds minute, scobiform.

Hypopitys monotropa has the broadest continuous distribution range of any Monotropoideae, and any mycoheterotrophic plant in general (but it may consist of multiple separate species). In the New World *H. monotropa* occurs from Alaska and British Colombia in the north, throughout northwestern and eastern USA, and into Mexico and Central America in the south. In the Old World, it is found throughout most of Europe and into Central Asia, and east through Afghanistan and along the Himalayas in India and Nepal into China and Japan. *Hypopitys* associates with *Tricholoma* fungi (Bidartondo and Bruns 2001). Bumblebees (*Bombus* spp.) serve as cross-pollination agents in *Hypopitys* (Klooster and Culley 2009).

Although *H. monotropa* and *Monotropa uniflora* have often been placed together in *Monotropa*, molecular phylogenetic studies have demonstrated the paraphyletic nature of *Monotropa* and suggest the presence of cryptic species within *M. hypopithys* (Cullings 2000; Bidartondo and Bruns 2001, 2002; Tsukaya et al. 2008; Freudenstein and Broe, unpubl.).

2.5.9.5 *Monotropa* (Figs. 2.14e and 4.14a)

Monotropa L., Sp. Pl. 1: 387 (1753).
Monotropion Saint-Lager, Ann. Soc. Bot. Lyon vii: 130 (1880).

Inflorescence 1-flowered, slender, often clustered, to 30 cm tall, wholly white to red, arising from a dense mass of highly branched brittle roots, nodding as it emerges from the soil and at anthesis, but erect in fruit, with occasional bracts on the axis. Bracts subtending flowers narrowly elliptic to lanceolate with a lacerate to erose margin. Sepals apparently absent. Petals 5–8, quincuncial, spatulate, apex truncate to rounded, pubescent adaxially, prominently saccate at base. Stamens 8–14, filaments terete above, flattened below, sparsely pubescent. Anthers short, appearing somewhat peltate because of a horizontal orientation at maturity, without awns or appendages, opening by slits that are oriented laterally. Ovary ovoid, appearing as if covered by plates corresponding to the 5–6 locules that will eventually form the capsular valves, glabrous, placentation axile. Style distinct, stout, gradually expanding to a broadened stigma that is lobed-undulating with a prominent central cavity. Nectary present as prominent pairs of fingerlike lobes surrounding bases of alternate stamens. Capsule loculicidal, containing numerous minute scobiform seeds with endosperm.

Monotropa uniflora is widespread in North America, and its distribution extends into Central America and even Colombia. The species is absent from Europe and Central Asia, but occurs in southern China, Japan, northern India, Nepal, and Bhutan. Copeland (1941) studied its morphology and anatomy. *Monotropa uniflora* roots are colonized by a narrow range of Russulaceae fungi (*Russula* or *Lactarius*) (Bidartondo and Bruns 2001, 2002; Bidartondo 2005). Bumblebees (*Bombus* spp.) serve as cross-pollination agents in *Monotropa* (Klooster and Culley 2009).

2.5.9.6 *Monotropastrum* (Fig. 2.14f)

Monotropastrum Andres, in Hand.-Mazz., Symb. Sin. 7: 766 (1936).
Monotropa D. Don, Prodr. Fl. Nepal 151 (1825).
Monotropanthum H. Andres, Feddes Repert. Spec. Nov. Regni Veg. 64: 87 (1961).

Inflorescences arising from masses of congested brittle roots, often several per root mass. Inflorescences scapose or racemose, to 15 cm tall, elongate, fleshy, white, with sterile bracts, emerging from soil with the flowers nodding and nodding at anthesis, but erect in fruit. Flower 3–4 (-5)-merous, white, but reported to be occasionally yellowish or reddish. Sepals free, usually 3, elliptic-ovate, glabrous, appressed to corolla, with entire to lacerate margins, rounded to obtuse to acute at apex. Petals free, 3–4, oblong-obovate, pubescent adaxially, glabrous abaxially, saccate at base, convex, margins entire, apices rounded to truncate. Stamens 6–10, twice as many as petals, in two series of alternating lengths. Filaments pubescent to glabrous. Anthers basifixed, horizontally reniform, with a nearly horizontal dehiscence suture. Ovary globose to ovoid, lacking grooves or plates on sides, finely pubescent, unilocular, placentation parietal. Style arising from the narrowed ovary, short, stout, terminating in a broad funnelform stigma, which may be bluish-black or yellow. Nectaries present as slender projections in pairs around bases of alternating stamens or as a lobed circular structure. Fruit a globose to ovoid berry bearing many minute ovoid seeds.

This genus contains two to three species: *Monotropastrum humile* and its occasional segregate *M. humile var. glaberrima*, and *M. sciaphilum. Monotropastrum humile* is known from montane mixed and deciduous forests in temperate and subtropical Asia (Bhutan, Burma, southern China, India, Japan, Laos, Nepal, Russia, South Korea, Thailand, and Vietnam). *Monotropastrum sciaphilum* is known only from the type collection (as *Eremotropa sciaphila*) from Yunnan, China (Wallace 1987). *Monotropastrum humile* is pollinated by bumblebees (Tanaka 1978) and seeds are dispersed by insects (Ushimaru and Imamura 2002). Populations of *Monotropastrum humile* var. *humile* from Japan and Taiwan were found to associate with basidiomycete fungi within the genus *Russula* (Bidartondo and Bruns 2001; Bidartondo 2005; Yokoyama et al. 2005), while plants referable to *M. humile* var. *glaberrima* were asscoiated with Thelephoraceae (Yokoyama et al. 2005)

2.5.9.7 *Monotropsis* (Figs. 2.16a and 7.3a, b)

Monotropsis Schweinitz ex S. Elliott, Sketch Bot. S. Carolina 1: 478 (1817).
Schweinitzia S. Elliott ex T. Nuttall, Gen. 2: Add. (1817).
Cryptophila W. Wolf, Amer. Midl. Naturalist 8: 115 (1922).

Inflorescences in clusters or solitary, arising from a diffuse mass of roots. Racemes emerging from soil in a somewhat nodding position, secund, with more or less crowded flowers. Axes purplish-pinkish, to 13 cm tall, with elongate, acute, scarious bracts becoming brown at or after anthesis. Flowers subtended by a single membranous bract and often 2–3 smaller bracteoles. Flowers tyically 5-merous, purplish to maroon to white, borne on short pedicels. Sepals free, narrowly lanceolate to ovate, closely appressed to corolla and in *M. odorata* somewhat obscuring it, glabrous, brown-scarious at maturity. Petals connate, lobes free for about 1/3 their length, either purplish-violet with or without white tips, or entirely white, glabrous, slightly saccate at base. Stamens 10, occuring in two series of alternating lengths; filaments slender, flattened, glabrous. Anthers dorsifixed, without awns or appendages, the sacs of equal size, dehiscing with a common slit for each pair of ad- and abaxial sacs. Pistil flask-shaped and glabrous with a subglobose ovary from which a style narrows, capped by a thickened discoid stigma, obscurely lobed and with a central depression. Overy unilocular with parietal placentation. Nectaries comprising 10 downward-pointing pairs of lobes between bases of the short series of stamens. Fruit a globose berry with a sticky mass containing the numerous minute seeds that contain endosperm.

Monotropsis consists of two mycoheterotrophic species, *M. odorata* ("Sweet Pinesap"), which occurs in the Appalachian Mountains of southeastern North America and *M. reynoldsiae*, restricted to northern Florida. *Monotropsis odorata* is found almost exclusively growing in upland, mixed oak–pine forests where it flowers in spring (Jones 2005), while *M. reynoldsiae* occurs in scrub oak hammocks and flowers in December-January. *Monotropsis odorata*, which is relatively rare and easily overlooked, is known for its highly fragrant flowers (Wallace 1975). Copeland (1939) studied the morphology of *M. odorata*. Cross-pollination of *M. odorata* is carried out by bumblebees (*Bombus* spp.) and the seeds are animal dispersed (Klooster and Culley 2009). *Monotropsis odorata* was found to associate with a narrow clade of fungi of the genus *Hydnellum* (Basidiomycota) (Bidartondo and Bruns 2001; Bidartondo 2005). *Monotropsis reynoldsiae* is even more rare and has not always been recognized as distinct from *M. odorata*.

2.5.9.8 *Pityopus* (Figs. 2.16b and 4.17a)

Pityopus Small, N. Amer. Fl. 29: 16 (1914).
Monotropa Eastwood, Bull. Torrey Bot. Club 29: 75, pl. 7 (1902).
Hypopitys A. Heller, Muhlenbergia 9: 68 (1913).

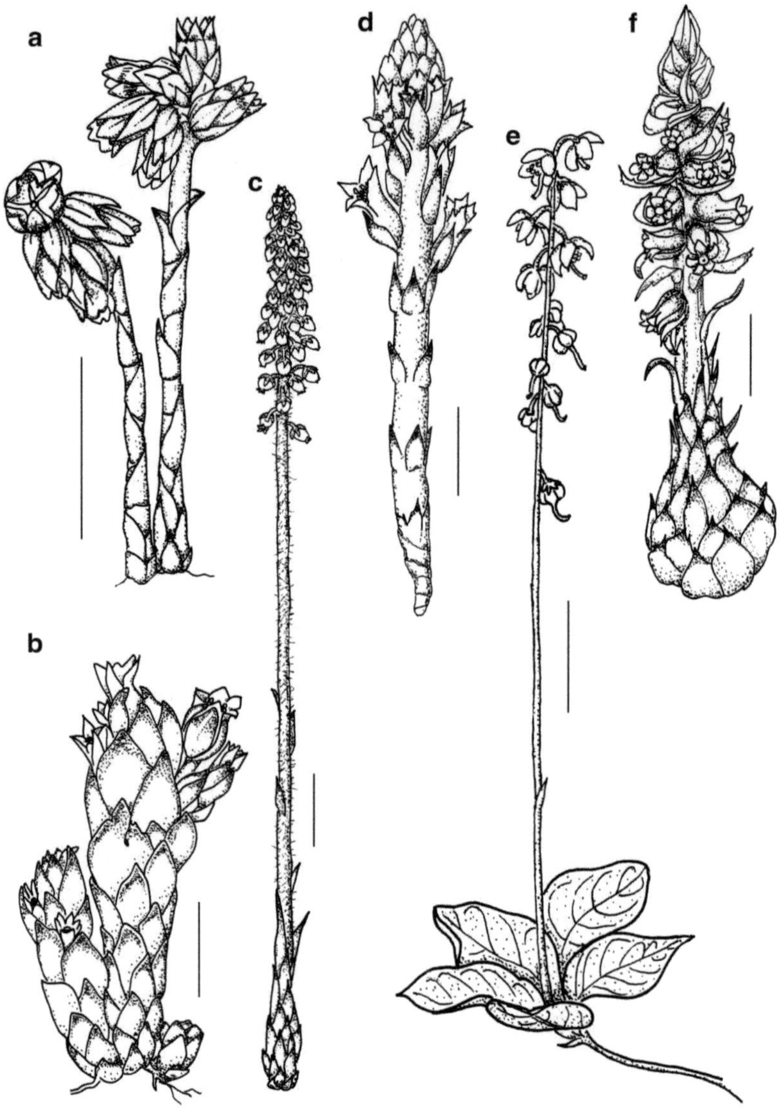

Fig. 2.16 Fully mycoheterotrophic Ericaceae (part 2). (**a**) *Monotropsis odorata*. (**b**) *Pityopus californica*. (**c**) *Pterospora andromedea*. (**d**) *Pleuricospora fimbriolata*. (**e**) *Pyrola picta*. (**f**) *Sarcodes sanguinea*. Redrawn from Flora of North America Editorial Committee (2009). Bar = 3 cm

Root mass comprising a network of slender, brittle, branched roots. Racemes compact, to ca. 10 cm tall, up to several per root mass, sometimes branched below soil level resulting in a congested mass of racemes, usually several-flowered but sometimes reduced to a single flower, often appearing only slightly above soil level, white to yellowish. Flowers subtended by single concave bracts that are longer than perianth segments, erect, 4-merous, terminal flower sometimes 5-merous. Sepals free, narrowly ovate to oblanceovate, appressed to corolla, glabrous; margins ciliate to erose, lateral pair longer than dorsiventral pair. Petals free, 4(-5), rounded and erose at apex, somewhat saccate at base, glabrous abaxially, densely hirsute adaxially. Stamens twice as many as petals, in two series alternating in length, filaments slender, dorsiventrally somewhat flattened, glabrous or somewhat pubescent basally. Anthers hippocrepiform, opening by a single slit curving over the anther apex. Pistil ampulliform. Stigma yellow, slightly lobed, subtended by a ring of dense hairs, style indistinct, gradually tapering from ovary. Ovary unilocular, pubescent, with parietal placentation. Nectaries comprising lobes that project downward between stamen bases. Fruit a globose berry containing a sticky mass with numerous small seeds.

This genus contains a single species, *P. californicus* ("California pinefoot"), growing in moist conferious or mixed forests in California, Oregon, and Washington. It flowers from May to July, but is very difficult to spot and is among the rarest of all North American Monotropoideae. Its morphology was studied by Copeland (1935). Pollination syndrome and dispersal agents have not been studied in detail, athough there is some information that bumblebees may act as cross-pollination agents (Wallace 1975). *Pityopus californicus* associates with several species groups of *Tricholoma* fungi (Basidiomycota) (Bidartondo and Bruns 2002).

2.5.9.9 *Pleuricospora* (Fig. 2.16d)

Pleuricospora A. Gray, Proc. Amer. Acad. Arts 7: 369 (1868).

Racemes one or several from a slender, brittle, diffuse root system, erect, to 15 cm tall, often forming clumps, usually multiflowered but can be reduced to a single flower, cream to yellowish. Bracts imbricate, extending from base of plant to apex, upper subtending individual flowers, entire to erose, becoming brownish with age. Flowers usually 4-merous, protogynous, erect on stout pedicels. Sepals free, glabrous, narrowly ovate, erose. Petals free, narrowly ovate, erose-fimbriate, glabrous, apices acute to rounded, spreading at maturity. Stamens 8, glabrous, filaments flattened with a prominent connective between the slender anther sacs. Anther sacs elongate, dehiscing by 2 long slits, without appendages or awns. Pistil ampulliform, style indistinct, narrowed gradually from the ellipsoid, unilocular ovary and capped by five prominent lobes that form the stigma. Nectaries obscure, represented by 8 low ridges. Fruit a globose whitish berry, fleshy with numerous minute, ovoid seeds.

Pleuricospora consists of a single, fully mycoheterotrophic species, *P. fimbriolata* ("Fringed Pinesap"). *Pleuricospora fimbriolata* grows in mixed or coniferous forests in western North America (California, Oregon, Washington), but there are also doubtful records from Mexico (Wallace 1975). *Pleuricospora* flowers mainly in July and August. Observations on pollination syndrome and seed dispersal mechanism are lacking for this species, but its morphology was studied by Copeland (1937). *Pleuricospora fimbriolata* associates with a narrow lineage of fungi within the basidiomycete genus *Gautieria* (Bidartondo and Bruns 2002).

2.5.9.10 *Pterospora* (Figs. 2.16c and 4.15a)

Pterospora Nuttall, Gen. N. Amer. Pl. 1: 269 (1818).

Racemes to 200 cm, slender, many-flowered, reddish-pink to tawny brown, with a narrow glandular to pubescent bracteate axis arising from a tight mass of heavily branching brittle roots. Flowers urceolate, protogynous, pendent on slender pedicels, each subtended by a narrow, lanceolate, glandular-pubescent bract with ciliate margins that extends as long or longer than the flower. Sepals 5, free, glandular-pubescent, lanceolate, appressed to corolla, reddish-pink to brownish. Petals connate, cream-colored, apices free, reflexed, rounded to blunt, glabrous, margins minutely erose. Stamens 10, in two series of alternating lengths; filaments slender, laterally flattened and thickened, glabrous. Anthers basifixed and essentially horizontal at maturity, with prominent awns extending from the lower portion of proximal anther sacs, dehiscing by lateral slits. Ovary spheroidal, 5-locular, glabrous, from which ascends a stout, straight, glabrous style, ending in a flat, shallowly 5-lobed stigma, placentation axile. Nectaries present as 10 shallow lobes projecting between staminal bases. Fruit an oblate spheroidal loculicidal capsule, brownish at maturity, pendent. Seeds small, each with a thin membranous wing.

Pterospora consists of a single species, *P. andromedea* ("Pinedrops"), which has a widespread distribution in North America. It is most common in western USA and now rare in the east, and its distribution extends to central Mexico. Copeland (1941) studied its morphology. Pollinators are not known. The winged seeds are presumably dispersed by wind. *Pterospora andromedea* relies on a narrow range of mycorrhizal fungi of the basidiomycete genus *Rhizopogon* (Cullings et al. 1996; Bidartondo and Bruns 2001, 2002; Dowie et al. 2011; Hazard et al. 2011). Fungal specialization exceeds the species level: different genotypes of *Pterospora* were found to grow with different fungal lineages, even when growing in sympatry (Bidartondo and Bruns 2002).

2.5.9.11 *Pyrola* (Fig. 2.16e)

Pyrola L., Sp. Pl. 1: 396 (1753).
Braxilia Rafinesque, Aut. Bot. 102 (1840).
Amelia Alefeld, Linnaea 28: 8, 25 (1856).
Thelaia Alefeld, Linnaea 28: 8, 33 (1856).
Erxlebenia Opiz ex Rydberg, N. Amer. Fl. 29: 28 (1914).

Subshrubs to ca. 40 cm in height. Rhizomes slender, creeping, with sparse roots. Racemes arising from a lax basal rosette of leaves. Leaves subtended by single lanceolate bracts, petioles typically longer than blades, base of lamina often decurrent along petiole, ovate-orbicular-reniform-obovate-elliptical, membraneous to coriaceous, glabrous, sometimes reduced or absent. Flowers subtended by single bracts, pedicels distinct, slender. Flowers pendulous, usually somewhat zygomorphic (although essentially actinomorphic in *P. minor*), with style, stamens, and lower petals downcurved. Sepals usually free or united slightly at base, triangular-lanceolate, margins entire to erose, greenish. Petals free or just slightly united at base, usually obovate, apex broadly rounded, concave. Stamens 10, free, filaments slender, broadened toward base, glabrous, anthers inverting late in development, with two prominent to indistinct horns at base, each with a pore or slit. Ovary spherical-oblate, into which is inserted a narrow style, terminated by a capitate-lobed stigma. Style elongate and downcurved in all but *P. minor*, where it is short and straight. Nectaries absent. Fruits loculicial capsules, pendent, valves connected by slender tissue threads when dehisced. Seeds minute, scobiform, with undifferentiated embryo and loose cellular-reticulate testa.

Pyrola comprises ca. 35 species of small, loosely rosette-forming, nearly herbaceous plants that are found circumboreally and south in the New World to Guatemala and in the Old World to Sumatra in montane habitats. Some species, such as *P. chlorantha*, are occasionally almost leafless. Many species are difficult to separate, especially from preserved material. In the western USA, leafless forms can be relatively common. They have sometimes been treated as a form of *P. picta* (Camp 1940); more recent molecular studies by Jolles (2007) suggest that they may warrant the species status, as *P. aphylla*, that they have often been given. Stable isotope analysis showed that *P. aphylla* was highly enriched in ^{13}C, exhibiting a pattern seen in mycoheterotrophs that associate with ectomycorrhizal fungi, while *P. picta* and other green pyroloids were not enriched for ^{13}C compared to autotrophs (Zimmer et al. 2007; Hynson et al. 2009). However, all examined pyroloids, and especially *P. aphylla*, were enriched for ^{15}N, indicating incorporation of nitrogen from fungi. Both *P. aphylla* and *P. picta* were found to associate with a diversity of fungi, mainly ectomycorrhizal (Hynson and Bruns 2009). Morphological and molecular data support a clade comprising *P. chlorantha* and *P. picta* (including *P. aphylla*) (Freudenstein 1999; Liu et al. 2010). Species of *Pyrola* are cross-pollinated by insects, most commonly flies. Seeds are very small ("dust seeds") and presumably dispersed by wind.

2.5.9.12 *Sarcodes* (Figs. 2.15e and 2.16f)

Sarcodes Torrey, Proc. Amer. Assoc. Advancem. Sci. 4: 193 (1851).
Pterosporopsis A. Kellogg, Pacific (San Francisco) 3: 122 (1854).

Racemes arising from a large mass of brittle roots, with a thick axis and large flowers, to 50 cm tall, strikingly red, solitary or clumped, stout, glandular-pubescent. Flowers horizontal to somewhat downfacing, borne on stout pedicels, subtended by lanceolate, ciliate bracts that are longer than the flowers. Flowers red, protogynous, urceolate, with 5 free sepals that are glandular-pubescent, narrowly ovate, and appressed to the corolla. Petals connate, apices free and spreading, glabrous. Stamens 10, included, glabrous, with slender filaments that are flattened near their bases. Anthers dorsifixed, elongate, opening by large terminal slits. Ovary oblate spheroidal, 5-locular, glabrous, with axile placentation, into which is inserted a stout glabrous style. Stigma subcapitate with 5 shallow lobes and a slight central depression. Nectaries present as 10 low lobes between staminal filament bases. Fruit an irregularly dehiscent capsule. Seeds small, within membranous wings.

The bright red species *Sarcodes sanguinea* ("Snow Plant") is the only member of this genus. It grows in mixed or coniferous forests in western North America (California, Nevada, and Oregon) and Mexico (Baja California). Both bumblebees and hummingbirds have been reported visiting the flowers, but self-pollination has been demonstrated as well (Wallace 1975). The seed dispersal mechanism remains unknown, and the species may reproduce by vegetative reproduction as well (Oliver 1890). *Sarcodes sanguinea* plants rely on a relatively narrow range of mycorrhizal fungi of the basidiomycete fungal genus *Rhizopogon* (Kretzer et al. 2000; Bidartondo and Bruns 2001, 2002; Dowie et al. 2011).

2.5.10 Gentianaceae

Gentianaceae Juss., Gen. Pl.: 141 (1789).

Annual, biennial or perennial, glabrous herbs, shrubs, trees, or rarely lianas; autotrophic, a few achlorophyllous and mycoheterotrophic. Stems erect, decumbent, rarely trailing; rhizomes sometimes present. Leaves generally opposite-deccussate, sometimes in a basal rosette, rarely whorled or alternate, simple and entire, sessile to petiolate; stipules generally absent. Sometimes presence of colleters (multicellular glands) in leaf axils. Inflorescence terminal or axillary (dichasial or monochasial cymes, thyrses, verticillasters or having a solitary terminal flower only); flowers actinomorphic, sometimes slighty zygomorphic, usually monomorphic, rarely imperfect; andromonoecious, gynodioecious or dioecious, iso- or heterostylous. Sepals fused, but free in a few genera, usually green, persistent or rarely absent, often keeled or winged. Petals fused, lobes contorted (twisted) to the right while in bud. Stamens isomerous and alternate with petals; anthers basifixed or dorsifixed, free, rarely connate, two thecae dehiscing longitudinally, rarely with terminal pores. Ovary superior, unilocular or bilocular, rarely pseudotetralocular, placentation parietal or axile; ovules few to many/numerous; style present or absent, straight or deflexed to one side; stigma filiform or two-parted (rarely decurrent along carpel sutures), often capitate, funnelform or 2-lobed. Fruit a capsule, occasionally a berry. Seeds usually small, non-arillate.

In spite of the indubitable monophyly of the Gentianaceae in their current definition (Struwe et al. 1994, 2002; Thiv et al. 2002; Yuan et al. 2003), there is no synapomorphic diagnostic feature confined to the entire family (Struwe and Albert 2002) except the presence of a combination of specific secondary metabolites (xanthone and secoiridoids) (Mandal et al. 1992; Rodriguez et al. 1998; Jensen and Schripsema 2002).

Number of genera and species—Gentianaceae comprise 92 commonly accepted genera and over 1,650 species (Struwe and Albert 2002, updated here). Twenty-five species are achlorophyllous and putative full mycoheterotrophs. These species are part of four genera: *Voyria*, *Voyriella*, *Exacum*, and *Exochaenium*. Species of the North American genera *Bartonia* (four spp.) and *Obolaria* (one sp.) are partial mycoheterotrophs (Cameron and Bolin 2010). Partial mycoheterotrophy is suggested to occur in *Curtia tenuifolia* and species of *Neurotheca* as well (Struwe et al. 2002; Molina and Struwe 2009).

Distribution and habitat—Gentianaceae are a cosmopolitan family, absent only from Antarctica. The majority of the species occurs in temperate zones but the mycoheterotrophic species are restricted to rain forests in the Neotropics and Paleotropics. However, some species of *Voyria* also occur in savannas and extend into subtropical Central America.

Classification—Gentianaceae are part of the order Gentianales, which also includes Rubiaceae, Apocynaceae, Gelsemiaceae, and Loganiaceae (APG 2009). The relationships between these families remain largely unclear (Stevens 2001). Gentianaceae comprise six tribes: Saccifolieae, Exaceae, Chironieae, Helieae, Potalieae, and Gentianeae (Struwe and Albert 2002). Several molecular evidences support the following classification: Saccifolieae is sister to the rest of the family, followed by Exaceae, Chironieae, Potalieae, and finally Helieae and Gentianeae (Struwe et al. 2002; Yuan et al. 2003; Kissling et al. 2009).

Evolutionary history—Fossil records for Gentianaceae are scarce (Struwe and Albert 2002; Yuan et al. 2005) and a minimum age of ca. 50 Ma for the family has been estimated with molecular clock analyses (Yuan et al. 2003). Full mycoheterotrophy has evolved at least four times independently in the family: once in Saccifolieae (*Voyriella*), twice in Exaceae (*Exacum* and *Exochaenium*), and a fourth time in the ancestor of *Voyria* (unplaced, but supposedly not closely related to any of the other fully mycoheterotrophic Gentianaceae lineages).

Ecology—Gentianaceae flowers are pollinated by various vectors, including bees, beetles, hummingbirds, moths, and bats (Struwe and Albert 2002). Fruits and seeds are dispersed by animals (including mammals, bats, and birds) or wind or rain (cf. seed dispersal of *Voyria*). Most species of Gentianaceae are probably arbuscular mycorrhizal but non-mycorrhizal Gentianaceae have been reported as well (Wang and Qiu 2006).

2.5.10.1 *Voyria* (Figs. 2.15a, c, 2.17a, 4.18, 4.19, 4.20, and 7.4)

Voyria Aubl., Hist. Pl. Guiane 1:208 (1775).
Humboldtia de Necker, Elem. Bot. 2: 16 (1790).
Lita Schreb., Gen. Pl. 2: 795 (1791).
Leiphaimos Schltdl. & Cham., Linnaea 6: 387 (1831).
Leianthostemon (Griseb.) Miq., Stirp. Surinam. Select. 147 (1851).
Pneumonanthopsis (Griseb.) Miq., Stirp. Surinam. Select. 150 (1851).
Disadena Miq., Stirp. Surinam. Select. 150 (1851).
Biglandularia H. Karst., Linnaea 28: 416 (1857).

Erect, mycoheterotrophic herbs, up to 30 cm tall. Root system star-like with unbranched roots, or small and coral-like or large with repeatedly branched roots. Stems usually simple, less often brancherd, terete, solitary, or a few together. Leaves opposite, somewhat connate at the base, small, scale-like, the lower ones sometimes alternate. Inflorescence a terminal few- to 30- flowered dichasial/bifurcate cyme or the plant having a solitary, terminal flower only. Flowers erect, rarely nodding, (4-)5(-7)-merous, short- or long-pedicellate, actinomorophic. Calyx tubular to campanulate, (4-)5(-7)-lobed, persistent, sometimes provided with discoid scales at the inner base. Corolla salverform to infundibular, variously colored, far exceeding the calyx, marcescent, apical part often papillate inside, tube elongate, lobes (4-)5(-7), contorted, spreading to recurved, rarely erect. Stamens (4-)5(-7), included in the corolla tube, rarely somewhat exceeding the corolla tube, inserted at various levels in the corolla tube, filaments conspicuous or virtually absent, anthers free or often coherent just below the stigma. Ovary 2-carpellate, 1-locular, sometimes borne on a short gynophore, the parietal placentae protruding, the base of the ovary often provided with two opposite glandular marks or ellipsoid glands, sometimes with two distinctly stalked glands, or eglandular; style filiform, gradually widened towards the ovary; stigma infundibular, rotate, or capitate, with undulate margin, or weakly 2-lobed, appendages sometimes present, ovules anatropous. Fruit a capsule, fusiform to globose, septicidally dehiscent, dehiscing entirely or in the middle only, often indehiscent. Seeds numerous, globose to filiform, in some species with two hair-like projections, embryo few-celled, endosperm present (Maas and Ruyters 1986; Franke 2002).

Voyria has a disjunct distribution with 18 species in tropical and subtropical America and one species with a widespread distribution in tropical

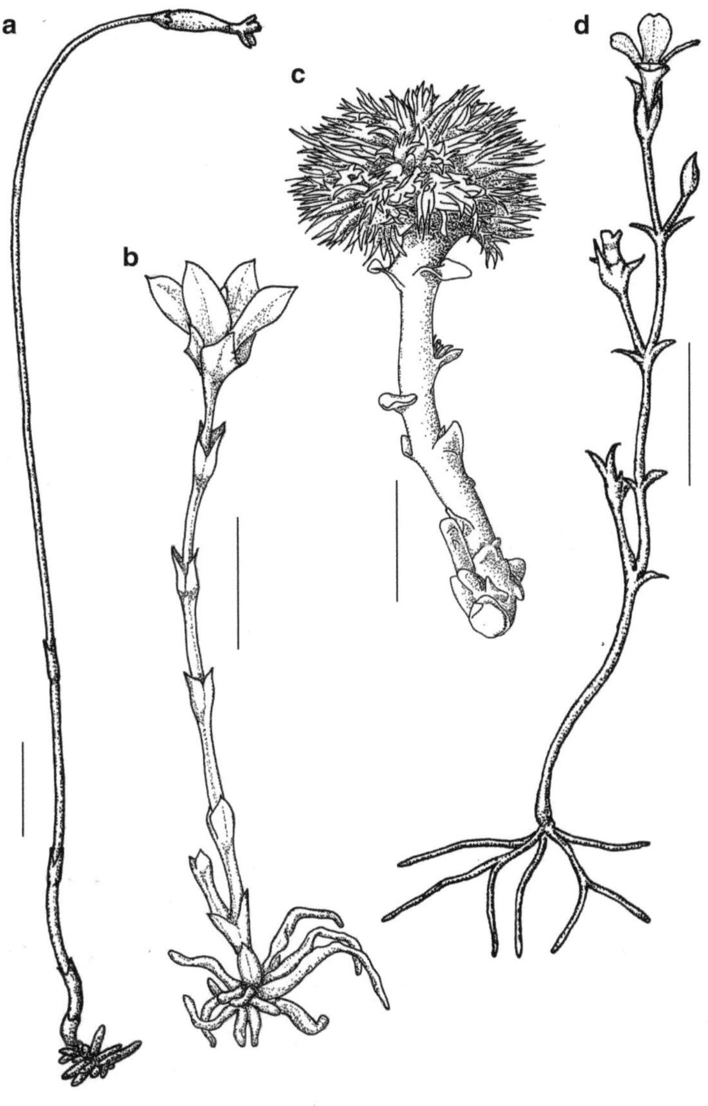

Fig. 2.17 Fully mycoheterotrophic Gentianaceae. (**a**) *Voyria aphylla*. Redrawn from Maas and Ruyters (1986). (**b**) *Exacum tenue*. Redrawn from Struwe and Albert (2002). (**c**) *Voyriella parviflora*. Redrawn from Maas and Ruyters (1986). (**d**) *Exochaenium oliganthum* (drawn after Merckx et al. 103, LV). Bar = 1 cm

West Africa (records from Liberia, Ivory Coast, Ghana, Nigeria, Cameroon, Gabon, and DR Congo) (Raynal-Roques 1967a; Maas and Ruyters 1986; Albert and Struwe 1997). Many neotropical species are widely distributed as well. *Voyria* species occur in various forest types, including lowland rainforest creek forest, swamp forest, montane rainforest hammock forest, and Amazonian caatinga forest. A few species prefer drier vegetation types and grow in white sand savannas and savanna forests. All species are terrestrial, but are sometimes found on dead, decaying logs (Maas and Ruyters 1986). Remarkably, two species have been found growing as epiphytes up to 30 m high on trees (Groenendijk et al. 1997).

Currently DNA data of *Voyria* are lacking, and therefore the phylogenetic relationships of this genus remain to be determined. Certain characteristics (opposite leaves, hypogynous flowers, no latex, no stipules) point to a position close to or in Gentianaceae (Struwe and Albert 2002).

Preliminary analyses based on nuclear and mitochondrial DNA sequences suggest that *Voyria* is an early diverging lineage within Gentianaceae (V. Merckx unpublished results).

Most species of *Voyria* possess brightly colored flowers that emit scent and offer nectar. Consequently, they are generally considered to be cross-pollinated (Maas and Ruyters 1986). Indeed, cross-pollination by butterflies and bees has been observed, although some species may rely on a mixed pollination strategy (individual reproduces both by self-fertilization and out crossing with genetically different individuals) to ensure seed production when pollen transfer by visitors fails (Hentrich et al. 2010). Seed dispersal vectors are poorly studied, but may include water and various animals (Maas and Ruyters 1986; Hentrich et al. 2010). Species of *Voyria* are associated with arbuscular mycorrhizal fungi (Leake 1994; Imhof 1997, 1999b; Imhof and Weber 1997; Franke 2002). Molecular sequencing detected *Glomus* Group A and Diversisporales fungi in the roots of neotropical *Voyria* species (Bidartondo et al. 2002; Merckx et al. 2010b; Courty et al. 2011).

2.5.10.2 *Voyriella* (Fig. 2.17c)

Voyriella Miq., Stirp. Surinam. Select.:146 (1851).
Voyria Aubl. Sect. *Voyriella* Miq., Tijdschr. Wis.-Natuurk. Wetensch. Eerste Kl. Kon. Ned. Inst. Wetensch. 2: 122 (1849).

Erect, mycoheterotrophic herbs, completely white, up to 15 cm tall. Roots filiform. Stems branched or unbranched, fleshy, quadrangular to slightly winged. Leaves opposite, scale-like, small. Inflorescence a terminal or rarely axillary, more or less contracted, 1- to many-flowered bifurcate cyme. Flowers erect, (4-)5(-6)-merous, shortly pedicellate. Sepals almost free, narrowly triangular, persistent, provided with discoid scales at the inner base. Corolla actinomorphic, salverform to tubular, hardly exceeding the calyx, soon falling off, papillate within, lobes small. Stamens (4-)5(-6), included, inserted at various levels in the corolla tube; filaments long or short, anthers free or coherent, introrse, basifixed; connective prolonged beyond the thecae or not, rounded at the base and apex. Ovary bicarpellate, bilocular at the base, unilocular in the upper part, eglandular; placentae parietal. Style filiform, stigma 2-lobed, papillate, ovules numerous, strictly orthotropous. Fruit a capsule, globose to ovoid, indehiscent, provided with a persistent, bifid style. Seeds subglobose, pitted, embryo few-celled, endosperm present.

The genus *Voyriella* consists of one fully mycoheterotrophic species: *Voyriella parviflora* (Maas and Ruyters 1986; Maguire and Boom 1989). Nuclear and chloroplast data suggest that *Voyriella* is a member of the tribe Saccifolieae (Struwe and Albert 2002). *Voyriella parviflora* occurs in lowland forests of northern South America and adjacent Panama. Its main center of distribution is in the Guianas (Maas and Ruyters 1986). *Voyriella parviflora* is autogamous (Oehler 1927). Seed dispersal mechanisms are unknown. *Voyriella parviflora* has been found to associate with *Glomus* Group A fungi (Bidartondo et al. 2002).

2.5.10.3 *Exacum* (Fig. 2.17b)

Exacum L., Sp. Pl. 112 (1753).
Cotylanthera Blume, Bijdr. Fl. Ned. Ind. 707 (1826).

Erect annual herbs to perennial subshrubs, 2 cm to 1 m tall. Stems terete to quandrangular, often with four wings or lines/ribs. Leaves opposite-decussate, rarely verticillate or rosulate, almost leafless in achlorophyllous species. Inflorescence a monochasial or dichasial cyme, sometimes umbel-shaped. Flowers 4- or 5-merous, actinomorphic to often zygomorphic by having the anthers forming a cone above a bent style (enantiostyly). Calyx persistent, each lobe furnished with a keel or a wing that might enlarge in fruit, rarely zygomorphic by having two well-developed wings and three reduced ones. Corolla white to violet, up to 7 cm long, tube short and lobes usually spreading, rarely persistent in fruit. Stamens protruding from the corolla tube, anthers usually connivent around or above the style forming a cone, dehiscent by 1 (in the achlorophyllous species) or 2 apical pores, usually furnished by small papillae on their dorsal sides. Ovary 2-carpellate, 2-locular, placentation axile; style filiform, straight, or curved; stigma small, entire to slightly 2-lobed.

Fruit a capsule, septicidally dehiscent. Seeds numerous, angular, rarely cup-shaped. Testa cells star-shaped or isodiametric. For a detailed description and taxonomy of *Exacum*, see Klackenberg (1985, 2002, 2006)

The genus *Exacum* comprises 68 species (Klackenberg 1985, 2006; Thulin 2001) and shows a typical paleotropical distribution (Klackenberg 1985, 2002; Thulin 2001). *Exacum* has two main centers of diversity, namely Madagascar and the area including Southern India and Sri Lanka. Only a few species occur in Socotra (and the Arabian peninsula), in the Himalayas, Southeast Asia, New Guinea, and in extreme northern Australia. *Exacum* species have a wide spectrum of habitat preferences. Taxa are found from sea level up to the highest mountain tops in Madagascar (ca. 2,800 m elevation), and up to ca. 2,000 m in the Himalayas, South India, and New Guinea. Most species occur in lowland and montane rainforest areas, although they usually grow in full sun (Klackenberg 1985, 1990, 2002). *Exacum* originated in Madagascar and has experienced multiple out-of-Madagascar dispersals (Yuan et al. 2005). The most important is the long-distance dispersal to Sri-Lanka/South-India, which resulted in the extensive radiation of the Socotra-Arabia and other Asian lineages in the northern India Ocean basin regions (Yuan et al. 2005). More recent out-of-Madagascar dispersals include single dispersal of *E. oldenlandioides* to the African mainland, or several dispersals to other islands around Madagascar including the Comores (*E. stenopterum*), or the volcanic island of Mauritius (*E. quinquenervium*) (Klackenberg 1985). The four mycoheterotrophic *Exacum* species occur only in Asia and presumably diversified from an Asian descendant (Klackenberg 2006).

Most *Exacum* species have bright colored enantiostylous flowers suggesting pollination by bees, however no thorough pollination studies have been yet performed on *Exacum*.

Exacum contains four achlorophyllous species (*E. loheri*, *E. nanum*, *E. paucisquamum*, and *E. tenue*) previously placed in the genus *Cotylanthera*. They are distributed from the Himalayas throughout Southeast Asia to New Guinea (Klackenberg 2006). Based on morphological evidences they belong to a small clade of *Exacum* comprising four other tiny chlorophyllous species (Klackenberg 2006) and share the same distribution. The evolution of mycoheterotrophy inside a predominantly chlorophyllous genus has occurred twice within the tribe Exaceae: in *Exacum* in Asia and in *Exochaenium* in Africa. However, yet, few studies have focused on the biology or the evolution of mycoheterotrophy in *Exacum*. Figdor (1897) studied mainly the morphology and anatomy of *E. tenue*, while Oehler (1927) studied its cytology.

2.5.10.4 *Exochaenium* (Fig. 2.17d)

Exochaenium Griseb., DC. Prod. 9: 55 (1845).

Annual, erect or dwarf herbs, rarely achlorophyllous. Stems simple or branched, usually tetragonal, more or less 4-ridged or 4-winged. Leaves well-developed or reduced and scale-like, sessile, opposite, linear-lanceolate to suborbicular. Inflorescence a terminal 1- to many-flowered bifurcate or dichasial cyme. Flowers 5-merous, pedicellate, corolla white, sometimes yellow or salmon, often pendent or inclined, generally with a stylar polymorphism (short- and long-styled flowers). Sepals almost free or forming a short tube, lobes linear-lanceolate to ovate or obovate, dorsally keeled or winged. Corolla tube cylindrical or infundibuliform, the lower portion enlarged in fruit, lobes oblong-obovate, obtuse at the apex or acuminate; filaments filiform, inserted at mid-length of the corolla tube. Stamens included; anthers oblong, basifixed, free or coherent, with a conspicuous apical stipitate gland, with or without two basal minute glands. Ovary ovoid to globose, bilocular, placentation axile, ovules numerous. Style filiform, included in the corolla tube; stigma filiform or clavate, entire or very slightly bilobed, rarely bifid, papillate. Capsule ovoid or obovoid, membranous or coriaceous, septicidally dehiscing by 2 valves. Seeds cubical, black; testa cells star-shaped.

Exochaenium comprises 22 species, all endemic to Africa (Kissling 2012). Most of the species occur on the Katanga plateau (Angola, DR Congo, and Zambia), many extending their distribution to the Sudano-Zambesian domain sensu White (1986). *E. oliganthum*, the single mycoheterotrophic species of the genus, has a remarkable widespread distribution in tropical Africa and has been recorded from Ethiopia, Sudan, Uganda, DR Congo, Equatorial Guinea, Gabon, Central African Republic, Cameroon, Nigeria, Ivory Coast, Guinea-Bissau, Zambia, and Tanzania (Raynal-Roques 1967b; Cheek 2006; Kissling 2012).

The ecology and particular morphology of *E. oliganthum* was described by Raynal-Roques (1967b) and its mycoheterotrophic status has been confirmed. However several individuals determined as *E. oliganthum* have been found to "parasitize" roots of other plant species (mainly Cyperaceae or Poaceae; Nemomissa 2002), but additional research is needed to investigate this claim. Also, within a Zambezi population (*Dessein et al. 499*, NEU) both achlorophyllous and chlorophyllous individuals of this species have been found (Kissling 2012). Fully mycoheterotrophic specimens of *E. oliganthum* associate with arbuscular mycorrhizal fungi from the *Glomus* Group A clade (Franke et al. 2006).

Exochaenium oliganthum is reported to have both underground cleistogamous flowers and aerial chasmogamous flowers (Raynal-Roques 1967b). Heterostyly has consistently been reported for this species (only in the chasmogamous flowers). Its pollination biology has not been studied, but the presence of heterostyly strongly suggests a high outcrossing rate for the chasmogamous flowers, while the morphology of the cleistogamous flowers (i.e., the anthers being compressed on the stigma) suggest selfing. Thus a mixed pollination strategy seems to occur in this mycoheterotrophic species.

Currently no complete phylogeny of the genus exists, however it has been found that *E. oliganthum* is nested deeply inside the genus (Kissling et al. 2009).

Acknowledgements Vincent Merckx, Hiltje Maas-van de Kamer, and Paul Maas assembled the chapter and wrote Section 1 and 4, 5.2–5.5, 5.7–5.8, and 5.10. Raymond Stotler, Barabara Crandall-Stotler, and Norman Wickett contributed Section 2. Maarten Christenhusz wrote Section 3. Paula Rudall contributed to 5.1 and 5.4. John Freudenstein and Vincent Merckx wrote Section 5.6 and 5.9. Jonathan Kissling contributed to Section 5.10.

References

Abadie J, Püttsepp Ü, Gebauer G, Faccio A, Bonfante P, Selosse MA (2006) *Cephalanthera longifolia* (Neottieae, Orchidaceae) is mixotrophic: a comparative study between green and nonphotosynthetic individuals. Can J Bot 84:1462–1477

Abe C, Akasawa Y (1989) A new species of *Oxygyne* (Burmanniaceae) found in Shikoku, Japan. J Jap Bot 64:161–164

Airy Shaw HK (1952) A new genus and species of Burmanniaceae from South India. Kew Bull 2: 277–279

Albert VA, Struwe L (1997) Phylogeny and classification of *Voyria* (saprophytic Gentianaceae). Brittonia 49:466–479

Ambrose BA, Espinosa-Matías S, Vázquez-Santana S, Vergara-Silva F, Martínez E, Márquez-Guzmán J, Alvarez-Buylla ER (2006) Comparative developmental series of the Mexican triurids support a euanthial interpretation for the unusual reproductive axes of *Lacandonia schismatica* (Triuridaceae). Am J Bot 93:15–35

Anderberg AA (1992) The circumscription of the Ericales, and their cladistic relationships to other families of "higher" Dicotyledons. Syst Bot 17:660–675

Anderberg AA, Rydin C, Källersjö M (2002) Phylogenetic relationships in the order Ericales s.l.: analyses of molecular data from five genes from the plastid and mitochondrial genomes. Am J Bot 89:677–687

APG (1998) An ordinal classification for the families of flowering plants. Ann Missouri Bot Gard 85:531–553

APG (2009) An update of the Angiosperm Phylogeny Group classification for the orders and families of flowering plants: APG III. Bot J Linn Soc 161:105–121

Baillon HE (1894) *Geosiris aphylla*. Bull Mens Soc Linn Paris 2:1149

Bamps P, Malaisse F (1987) Extension de l'aire de repartition de *Burmannia congesta* (Wright) Jonker (Burmanniaceae). Bull Jard Bot Nat Belg 57:457–458

Barrett CF, Freudenstein JV (2011) An integrative approach to delimiting species in a rare but widespread mycoheterotrophic orchid. Mol Ecol 20:2771–2786

Barrett C, Freudenstein J, Taylor D, Köljalg U (2010) Rangewide analysis of fungal associations in the fully mycoheterotrophic *Corallorhiza striata* complex (Orchidaceae) reveals extreme specificity on ectomycorrhizal *Tomentella* (Thelephoraceae) across North America. Am J Bot 97:628–643

Beccari O (1871) *Petrosavia*: Nuovo genere di piante parasite della famiglia delle Melanthaceae. Nuovo Giorn Bot Ital 3:7–11

Beccari O (1878) as "1877". Descrizione di una nuova e singolare pianta parassita. Malesia 1:238–240

Bello MA, Bruneau A, Forest F, Hawkins JA (2009) Elusive relationships within order Fabales: phylogenetic analyses using *matK* and *rbcL* sequence data. Syst Bot 34:102–114

Benson-Evans K (1952) Distribution of sex in *Cryptothallus*. Nature 169:39
Benson-Evans K (1960) Some aspects of spore formation and germination in *Cryptothallus mirabilis* V. Malmb. Trans Brit Bryol Soc 3:729–735
Benson-Evans K (1961) Environmental factors and bryophytes. Nature 191:255–260
Bentham G (1883) Ordo CLXVIII Burmanniaceae. In: Bentham G, Hooker JD (eds) Genera plantarum, vol 3. Reeve & Co., London, pp 455–460
Bidartondo MI (2005) The evolutionary ecology of mycoheterotrophy. New Phytol 167:335–352
Bidartondo MI, Bruns TD (2001) Extreme specificity in epiparasitic Monotropoideae (Ericaceae): widespread phylogenetic and geographic structure. Mol Ecol 10:2285–2295
Bidartondo MI, Bruns TD (2002) Fine-level mycorrhizal specificity in the Monotropoideae (Ericaceae): specificity for fungal species groups. Mol Ecol 11: 557–569
Bidartondo MI, Burghardt B, Gebauer G, Bruns TD, Read D (2004) Changing partners in the dark: isotopic and molecular evidence of ectomycorrhizal liaisons between forest orchids and trees. Proc R Soc Lond B 271:1799–1806
Bidartondo MI, Duckett JG (2010) Conservative ecological and evolutionary patterns in liverwort-fungal symbioses. Proc R Soc Lond B 277:485–492
Bidartondo MI, Bruns TD, Weiß M, Sérgio C, Read DJ (2003) Specialized cheating of the ectomycorrhizal symbiosis by an epiparasitic liverwort. Proc R Soc Lond B 270:835–842
Bidartondo MI, Redecker D, Hijiri I, Wiemken A, Bruns TD, Dominguez LS, Sérsic A, Leake JR, Read DJ (2002) Epiparasitic plants specialized on arbuscular mycorrhizal fungi. Nature 419:389–392
Bierhorst DW (1969) On *Stromatopteris* and its ill-defined organs. Am J Bot 56:160–174
Bierhorst DW (1971) Morphology of vascular plants. Macmillan, New York
Bierhorst DW (1977) The systematic position of *Psilotum* and *Tmesipteris*. Brittonia 29:3–13
Blumenfeld H (1935) Beiträge zur Physiologie des Wurzelpilzes von *Limodorum abortivum* (L.) Sw. Inaugural Dissertation, Philosophische Fakultät der Universität Basel, Riga, Lettland
Bobrov AVFC, Melikian AP, Yembaturova EY (1999) Seed morphology, anatomy and ultrastructure of *Phyllocladus* L. C. & A. Rich. ex Mirb. (Phyllocladaceae (Pilg.) Bessey) in connection with the generic system and phylogeny. Ann Bot 83:601–618
Born MG, Maas PJM, Dressler RL, Westra LYT (1999) A revision of the saprophytic orchid genera *Wullschlaegelia* and *Uleiorchis*. Bot Jahrb Syst 121:45–74
Bougoure J, Brundrett M, Brown A, Grierson P (2008) Habitat characteristics of the rare underground orchid *Rhizanthella gardneri*. Aust J Bot 56:501–511
Bougoure J, Brundrett M, Grierson P (2010) Carbon and nitrogen supply to the underground orchid, *Rhizanthella gardneri*. New Phytol 186:947–956
Bougoure J, Dearnaley JD (2005) The fungal endophytes of *Dipodium variegatum* (Orchidaceae). Aust Mycol 24:15–19
Bougoure J, Ludwig M, Brundrett M, Grierson P (2009) Identity and specificity of the fungi forming mycorrhizas with the rare mycoheterotrophic orchid *Rhizanthella gardneri*. Mycol Res 113:1097–1106
Brade AC (1943) Saprófitas do Itatiaia. Arq Serv Flor 2:45–49
Bremer K, Friis EM, Bremer B (2004) Molecular phylogenetic dating of asteroid flowering plants shows early Cretaceous diversification. Syst Biol 53:496–505
Brownlie G (1969) Flore de la Nouvelle-Calédonie et Dépendances 3. Ptéridophytes. Museum National d'Histoire Naturelle, Paris
Brownsey PJ, Lovis JD (1987) Chromosome numbers for the New Zealand species of *Psilotum* and *Tmesipteris*, and the phylogenetic relationships of the Psilotales. New Zeal J Bot 25:439–454
Brummitt RK (1992) Vascular plant families and genera. Royal Botanic Gardens, Kew
Burgeff H (1932) Saprophytismus und Symbiose: Studien an tropischen Orchideen. Gustav Fischer, Jena
Caddick LR, Rudall PJ, Wilkin P (2000a) Floral morphology and development in Dioscoreales. Feddes Repert Spec Nov Regni Veg Beih 111:189–230
Caddick LR, Rudall PJ, Wilkin P, Chase MW (2000b) Yams and their allies: systematics of Dioscoreales. In: Wilson KJ, Morrison DA (eds) Monocots: systematics and evolution. CSIRO Publishing, Melbourne, pp 475–487
Caddick LR, Rudall PJ, Wilkin P, Hedderson TA, Chase MW (2002a) Phylogenetics of Dioscoreales based on combined analyses of morphological and molecular data. Bot J Linn Soc 138:123–144
Caddick LR, Wilkin P, Rudall PJ, Hedderson TA, Chase MW (2002b) Yams reclassified: a recircumscription of Dioscoreaceae and Dioscoreales. Taxon 51:103–114
Cameron DD, Bolin JF (2010) Isotopic evidence of partial mycoheterotrophy in the Gentianaceae: *Bartonia virginica* and *Obolaria virginica* as case studies. Am J Bot 97:1272–1277
Cameron DD, Preiss K, Gebauer G, Read DJ (2009) The chlorophyll-containing orchid *Corallorhiza trifida* derives little carbon through photosynthesis. New Phytol 183:358–364
Cameron KM (2003) Vanilloideae. In: Pridgeon AM, Cribb PJ, Chase MW, Rasmussen FN (eds) Genera Orchidacearum, vol 3. University Press, Oxford
Cameron KM, Chase MW, Rudall P (2003) Recircumscription of the monocotyledonous family Petrosaviaceae to include *Japonolirion*. Brittonia 55:214–225
Camp WH (1940) Aphyllous forms in *Pyrola*. Bull Torrey Bot Club 67:453–465
Campbell EO (1968) An investigation of *Thismia rodwayi* F. Muell. and its associated fungus. Trans Proc Roy Soc New Zeal 3:209–219
Carpenter RJ (1988) Early tertiary *Tmesipteris* (Psilotaceae) macrofossil from Tasmania. Aust Syst Bot 1:171–176

Castroviejo S (1998) Psilotaceae. In: Castroviejo S, Laínz M, López-González G, Montserrat P, Muñoz-Garmendia F, Paiva J, Villar LL (eds) Flora iberica, vol 1. Real Jardín Botánico, Madrid, pp 29–30

Cha JY, Igarashi T (1996) *Armillaria jezoensis*, a new symbiont of *Galeola septentrionalis* (Orchidaceae) in Hokkaido. Mycoscience 37:21–24

Chantanaorrapint S, Chantanaorrapint A (2009) *Thismia clavigera* (Thismiaceae), a new record for Thailand. Thai For Bull 37:27–31

Chase MW (2001) The origin and biogeography of Orchidaceae. In: Pridgeon AM, Cribb PJ, Chase MW, Rasmussen F (eds) Genera Orchidacearum, vol. 2. Orchidoideae (Part 1). University Press, Oxford, pp 1–5

Chase MW (2004) Monocot relationships: an overview. Am J Bot 91:1645–1655

Chase MW (2005) Classification of Orchidaceae in the age of DNA data. Curtis's Bot Mag 22:2–7

Chase MW, Fay MF, Devey DS, Maurin O, Ronsted N, Davies TJ, Pillon Y, Petersen G, Seberg O, Tamura MN, Asmussen CB, Hilu K, Borsch T, Davis JI, Stevenson DW, Pires JC, Givnish TJ, Sytsma KJ, McPherson MA, Graham SW, Rai HS (2006) Multigene analyses of monocot relationships: a summary. Aliso 22:63–75

Chase MW, Soltis DE, Soltis PS, Rudall PJ, Fay MF, Hahn WH, Sullivan S, Joseph J, Molvray M, Kores PJ, Givnish TJ, Sytsma KJ, Pires JC (2000) Higher-level systematics of the monocotyledons: an assessment of current knowledge and a new classification. In: Wilson KL, Morrison DA (eds) Monocots: systematics and evolution. CSIRO, Collingwood, pp 3–16

Chase MW, Soltis DE, Olmstead RG, Morgan D, Les DH, Mishler BD, Duvall MR, Price RA, Hills HG, Qiu Y-L, Kron KA, Rettig JH, Conti E, Palmer JD, Manhart JR, Sytsma KJ, Michaels HJ, Kress WJ, Karol KG, Clark WD, Hedrén M, Gaut BS, Jansen RK, Kim K-J, Wimpee CF, Smith JF, Furnier GR, Strauss SH, Xiang Q-Y, Plunkett GM, Soltis PS, Swensen SM, Williams SE, Gadek PA, Quinn CJ, Equiarte LE, Golenberg E, Learn GH, Graham SW, Barrett SCH, Dayanandan S, Albert VA (1993) Phylogenetics of seed plants: an analysis of nucleotide sequences from the plastid gene *rbcL*. Ann Missouri Bot Gard 80:528–580

Chase MW, Stevenson DW, Wilkin P, Rudall PJ (1995) Monocot systematics: a combined analysis. In: Rudall PJ, Cribb PJ, Cutler DF, Humphries CJ (eds) Monocotyledons: systematics and evolution. Royal Botanic Gardens, Kew, pp 685–730

Cheek M (2003a) A new species of *Afrothismia* (Burmanniaceae) from Kenya. Kew Bull 58:951–955

Cheek M (2003b) Kupeaeae, a new tribe of Triuridaceae from Africa. Kew Bull 58:939–949

Cheek M (2006) African saprophytes: new discoveries. In: Ghazanfar SA, Beentje HJ (eds) Taxonomy and ecology of African plants, their conservation and sustainable use. Royal Botanic Gardens, Kew, pp 693–697

Cheek M (2007) *Afrothismia amietii* (Burmanniaceae), a new species from Cameroon. Kew Bull 61:605–607

Cheek M (2009) Burmanniaceae. In: Timberlake JR, Martins ES (eds) Flora Zambesiaca, Vol. 12, part 2. Royal Botanic Gardens, Kew, pp 141–143

Cheek M, Jannerup P (2005) A new species of *Afrothismia* (Burmanniaceae) from Tanzania. Kew Bull 60:593–596

Cheek M, Williams SA (1999) A review of African saprophytic flowering plants. In: Timberlake J, Kativu S (eds) African plants: biodiversity, taxonomy and uses. Proceedings of the 1997 AETFAT congress, Harare, Zimbabwe. Royal Botanic Gardens, Kew, pp 39–49

Cheek M, Cable S, Hepper FN, Ndam N, Watts J (2006) Mapping plant biodiversity on Mount Cameroon. In: van der Maesen LJG, van der Burgt XM, van Medenbach de Rooy JM (eds) The biodiversity of African plants: proceedings, XIVth AETFAT Congress. Kluwer, Dordrecht, pp 110–120

Cheek M, Williams SA, Etuge M (2003) *Kupea martinetugei*, a new genus and species of Triuridaceae from western Cameroon. Kew Bull 58:225–228

Chen S, Luo Y (2002) *Pachystoma nutans*, an undescribed orchid from Myanmar. Acta Phytotax Sin 40:173–175

Chen X, Tamura MN (2000) *Petrosavia*. In: Wu ZY, Raven PH (eds) Flora of China, Vol. 24 (Flagellariaceae through Marantaceae). Science Press, Beijing, and Missouri Botanical Garden Press, St. Louis, pp 77

Chinnock RJ (1998a) Psilotaceae. In: Flora of Australia, Vol. 48. CSIRO Publishing, Collingwood, pp 47–53

Chinnock RJ (1998b) Ophioglossaceae. In: Flora of Australia, Vol. 48. CSIRO Publishing, Collingwood, pp 99–109

Christenhusz MJM, Zhang X-C, Schneider H (2011a) A linear sequence of extant lycophytes and ferns. Phytotaxa 19:7–54

Christenhusz MJM, Reveal JL, Farjon A, Gardner MF, Mill RR, Chase MW (2011b) A new classification and linear sequence of extant gymnosperms. Phytotaxa 19:55–70

Copeland HF (1934) The structure of the flower of *Newberrya*. Madroño 2:137–142

Copeland HF (1935) On the genus *Pityopus*. Madroño 3:154–168

Copeland HF (1937) The reproductive structures of *Pleuricospora*. Madroño 4:1–16

Copeland HF (1938) The structure of *Allotropa*. Madroño 4:137–153

Copeland HF (1939) The structure of *Monotropsis* and the classification of the Monotropoideae. Madroño 5:105–119

Copeland HF (1941) Further studies on Monotropoideae. Madroño 6:97–119

Courty P-E, Walder F, Boller T, Ineichen K, Wiemken A, Rousteau A, Selosse M-A (2011) C and N metabolism in mycorrhizal networks and mycoheterotrophic plants of tropical forests: a stable isotope analysis. Plant Physiol 156:952–961

Cowley EJ (1988) Burmanniaceae. In: Polhill RM (ed) Flora of tropical East Africa. Royal Botanic Gardens, Kew, pp 1–9

Crandall-Stotler BJ, Forrest LL, Stotler RE (2005) Evolutionary trends in the simple thalloid liverworts (Marchantiophyta, Jungermanniopsida subclass Metzgeriidae). Taxon 54:299–316
Crandall-Stotler BJ, Stotler RE, Long DG (2009) Morphology and classification of the marchantiophyta. In: Goffinet B, Shaw AJ (eds) Bryophyte biology, 2nd edn. Cambridge University Press, Cambridge, pp 1–54
Cribb P (1985) The saprophytic genus *Corsia* in the Solomon Islands. Kew Mag 2:320–323
Cribb P, Fischer E, Killmann D (2010) A revision of *Gastrodia* (Orchidaceae: Epidendroideae, Gastrodieae) in tropical Africa. Kew Bull 65:315–321
Cribb P, Wilkin P, Clements M (1995) Corsiaceae: a new family for the Falkland Islands. Kew Bull 50:171–172
Cronquist A (1970) The evolution and classification of flowering plants. Nelson, London
Cronquist A (1981) An integrated system of classification of flowering plants. Columbia University Press, New York
Cronquist A (1988) The evolution and classification of flowering plants, 2nd edn. The New York Botanical Garden, New York
Crum HA, Bruce J (1996) A new species of *Cryptothallus* from Costa Rica. Bryologist 99:433–438
Cullings K (2000) Reassessment of phylogenetic relationships of some members of the Monotropoideae based on partial 28S ribosomal RNA gene sequencing. Can J Bot 78:1–2
Cullings KW, Szaro TM, Bruns TD (1996) Evolution of extreme specialization within a lineage of ectomycorrhizal epiparasites. Nature 379:63–66
Dahlgren RMT, Clifford HT, Yeo PF (1985) The families of the monocotyledons. Springer, Berlin
Dauby G, Parmentier I, Stévart T (2007) *Afrothismia gabonensis* sp. nov. (Burmanniaceae) from Gabon. Nord J Bot 25:268–271
Davies T, Savolainen V, Chase MW, Goldblatt P, Barraclough TG (2005) Environment, area, and diversification in the species-rich flowering plant family Iridaceae. Am Nat 166:418–425
Davis DF, Holderman DF (1980) The white redwoods: ghosts of the forest. Naturegraph Publishers, Happy Camp
Davis EC, Shaw AJ (2008) Biogeographic and phylogenetic patterns in diversity of liverwort-associated endophytes. Am J Bot 95:914–924
Davis JI, Stevenson DW, Petersen G, Seberg O, Campbell LM, Freudenstein JV, Goldman DH, Hardy CR, Michelangeli FA, Simmons MP, Specht CD, Vergara-Silva F, Gandolfo MA (2004) A phylogeny of the monocots, as inferred from *rbcL* and *atpA* sequence variation, and a comparison of methods for calculating jackknife and bootstrap values. Syst Bot 29:467–510
de Laubenfels DJ (1959) Parasitic conifer found in New Caledonia. Science 130:97
de Laubenfels DJ (1972) Flore de la Nouvelle Caledonia et Dependances. No. 4. gymnospermes. Muséum National d'Histoire Naturelle, Paris
de Vega C, Arista M, Ortiz PL, Talavera S (2010) Anatomical relations among endophytic holoparasitic angiosperms, autotrophic host plants and mycorrhizal fungi: a novel tripartite interaction. Am J Bot 97:730–737
Dearnaley JD (2006) The fungal endophytes of *Erythrorchis cassythoides*—is this orchid saprophytic or parasitic? Aust Mycol 25:51–57
Dearnaley JD, Bougoure J (2010) Isotopic and molecular evidence for saprotrophic Marasmiaceae mycobionts in rhizomes of *Gastrodia sesamoides*. Fungal Ecol 3:288–294
Dearnaley JDW, Le Brocque AF (2006) Molecular identification of the primary root fungal endophytes of *Dipodium hamiltonianum* (Orchidaceae). Aust J Bot 54:487–491
Denis M (1919) Sur quelques thalles d'*Aneura* depourvus de chlorpophylle. Compt Rend Hebd Séances Acad Sci 168:64–66
Diels L (1902) Gleicheniaceae. In: Engler A, Prantl K (eds) Die Natürlichen Pflanzenfamilien. W. Engelmann, Leipzig, pp 350–356
Dimitri MJ (1972) Una nueva especie del genero *Arachnitis* Phil. (Corsiaceae). Revista Fac Agron Univ Nac La Plata 48:37–45
Dixon KW (2003) *Rhizanthella gardneri*. Orchidaceae. Curtis's Bot Mag 20:94–100
Dowie NJ, Hemenway JJ, Trowbridge SM, Miller SL (2011) Mycobiont overlap between two mycoheterotrophic genera of Monotropoideae (*Pterospora andromedea* and *Sarcodes sanguinea*) found in the Greater Yellowstone Ecosystem. Symbiosis 54:29–36
Duckett JG, Burch J, Fletcher PW, Matcham HW, Read DJ, Russell AJ, Pressel S (2004) *In vitro* cultivation of bryophytes: a review of practicalities, problems, progress and promise. J Bryol 26:3–20
Du Puy D, Cribb P (2007) The gemus *Cymbidium*. Royal Botanic Gardens, Kew
Earle CJ (1997) Gymnosperm database. Available at http://www.conifers.org. Accessed Aug 2011
Eckenwalder JE (2009) Conifers of the world. Timber Press, Portland
Engler A (1888) Burmanniaceae. In: Engler A, Prantl K (eds) Die Natürlichen Pflanzenfamilien, 1st edn. W. Engelmann, Leipzig, pp 44–51
Engler A (1897) Burmanniaceae. In: Engler A, Prantl K (eds) Die Natürlichen Pflanzenfamilien, Nachträge zum Teil II-IV. W. Engelmann, Leipzig, p 96
Engler A (1905) *Thismia winkleri* Engl., eine neue afrikanische Burmanniacee. Bot Jahrb Syst 38:89–91
Eriksen B, Persson C (2007) Polygalaceae. In: Kubitzki K (ed) The families and genera of flowering plants, vol 9. Springer, Berlin, pp 345–363
Eriksson O, Kainulainen K (2011) The evolutionary ecology of dust seeds. Perspect Plant Ecol Evol Systemat 13:73–87
Ernst A, Bernard C (1912) Beiträge zur Kenntnis der Saprophyten Javas. IX. Entwicklungsgeschichte des Embryosackes und des Embryos von *Burmannia candida* Engl. und *B. championii* Thw. Ann Jard Bot Buitenzorg 25 sér 2:161–188

Fay MF, Chase MW, Ronsted N, Devey DS, Pillon Y, Pires JC, Petersen G, Seberg O, Davis J (2006) Phylogenetics of Liliales: summarized evidence from combined analyses of five plastid and one mitochondrial loci. Aliso 22:559–564

Fay MF, Rudall PJ, Sullivan S, Stobart KL, de Bruijn AY, Reeves G, Qamaruz-Zaman F, Hong W-P, Joseph J, Hahn WJ, Conran JG, Chase MW (2000) Phylogenetic studies of Asparagales based on four plastid DNA regions. In: Wilson KL, Morrison DA (eds) Monocots: systematics and evolution. CSIRO, Collingwood, pp 360–371

Feild TS, Brodribb TJ (2005) A unique mode of parasitism in the conifer coral tree *Parasitaxus ustus* (Podocarpaceae). Plant Cell Environ 28:1316–1325

Figdor W (1897) Über *Cotylanthera* Bl. Ein Beitrag zur Kenntnis tropischer Saprophyten. Ann Jard Bot Buitenzorg 14:213–239

Fineran BA, Ingerfeld M (1985) Cross-linked mitochondrial aggregates in the primitive vascular plant *Tmesipteris* (Psilotopsida). Eur J Cell Biol 38:185–194

Fitch WH (1924) Illustrations of the British flora. Reeve & Co., London

Flora of North America Editorial Committee (2009) Flora of North America, Vol. 8. Magnoliophyta: Paeoniaceae to Ericaceae. Oxford University Press, New York

Forest F, Chase MW, Persson C, Crane PR, Hawkins JA (2007) The role of biotic and abiotic factors in evolution of ant dispersal in the milkwort family (Polygalaceae). Evolution 61:1675–1694

Forrest LL, Davis EC, Long DG, Crandall-Stotler BJ, Clark A, Hollingsworth ML (2006) Unraveling the evolutionary history of the liverworts (Marchantiophyta): multiple taxa, genomes and analyses. Bryologist 109:303–334

Franke T (2002) The myco-heterotrophic *Voyria flavescens* (Gentianaceae) and its associated fungus. Mycol Prog 1:367–376

Franke T (2004) *Afrothismia saingei* (Burmanniaceae, Thismieae), a new myco-heterotrophic plant from Cameroon. Syst Geogr Pl 74:27–33

Franke T (2007) Miscellaneous contributions to the taxonomy and mycorrhiza of AMF-exploiting myco-heterotrophic plants. PhD thesis. Ludwig-Maximilians-Universität, Münich

Franke T, Beenken L, Hahn C (2000) *Triuridopsis intermedia* spec. nov. (Triuridaceae), a new myco-heterotrophic plant from Bolivia. Pl Syst Evol 225:141–144

Franke T, Beenken L, Döring M, Kocyan A, Agerer R (2006) Arbuscular mycorrhizal fungi of the *Glomus*-group A lineage (Glomerales; Glomeromycota) detected in myco-heterotrophic plants from tropical Africa. Mycol Prog 5:24–31

Franke T, Sainge MN, Agerer R (2004) A new species of *Afrothismia* (Burmanniaceae; tribe Thismieae) from the western foothills of Mount Cameroon. Blumea 49:451–456

Freudenstein JV (1997) A monograph of *Corallorhiza* (Orchidaceae). Harv Pap Bot 10:5–51

Freudenstein JV (1999) Relationships and character transformation in Pyroloideae (Ericaceae) based on ITS sequences, morphology, and development. Syst Bot 24:398–408

Freudenstein JV, Barrett CF (2010) Mycoheterotrophy and diversity in Orchidaceae. In: Seberg O, Petersen G, Barfod AS, Davis JI (eds) Diversity, phylogeny, and evolution in the monocotyledons. Aarhus University Press, Arhus, pp 25–37

Furman TE, Trappe JM (1971) Phylogeny and ecology of mycotrophic achlorophyllous angiosperms. Quat Rev Biol 46:219–225

Furness CA, Rudall PJ (2006) Comparative structure and development of pollen and tapetum in Pandanales. Int J Plant Sci 167:341–348

Furness CA, Rudall P, Eastman A (2002) Contribution of pollen and tepatal characters to the systematics of Triuridaceae. Pl Syst Evol 235:209–218

Fuse S, Tamura MN (2000) A phylogenetic analysis of the plastid *matK* gene with emphasis on Melanthiaceae *sensu lato*. Plant Biol 2:415–427

Gandolfo MA, Nixon KC, Crepet WL (2002) Triuridaceae fossil flowers from the Upper Cretaceous of New Jersey. Am J Bot 89:1940–1957

Gandolfo MA, Nixon KC, Crepet WL, Stevenson DW (1998) Oldest known fossils of monocotyledons. Nature 394:532–533

Gebauer G, Meyer M (2003) 15N and 13C natural abundance of autotrophic and myco- heterotrophic orchids provides insight into nitrogen and carbon gain from fungal association. New Phytol 160:209–223

Giesen H (1938) Triuridaceae. In: Engler A (ed) Das Pflanzenreich. IV.18. W. Engelmann, Leipzig, pp 1–84

Girlanda M, Segreto R, Cafasso D, Liebel HT, Rodda M, Ercole E, Cozzolino S, Gebauer G, Perotto S (2011) Photosynthetic Mediterranean meadow orchids feature partial mycoheterotrophy and specific mycorrhizal associations. Am J Bot 98:1148–1163

Girlanda M, Selosse M-A, Cafasso D, Brilli F, Delfine S, Fabbian R, Ghignone S, Pinelli P, Segreto R, Loreto F, Cozzolino S, Perotto S (2006) Inefficient photosynthesis in the Mediterranean orchid *Limodorum abortivum* is mirrored by specific association to ectomycorrhizal Russulaceae. Mol Ecol 15:491–504

Givnish TJ, Pires JC, Graham SW, McPherson MA, Prince LM, Patterson TB, Rai HS, Roalson EH, Evans TM, Hahn WJ, Millam KC, Meerow AW, Molvray M, Kores PJ, O'Brien HE, Hall JC, Kress WJ, Sytsma KJ (2006) Phylogenetic relationships of monocots based on the highly informative plastid gene *ndhF*: evidence for widespread concerted convergence. Aliso 22:28–51

Goldblatt P, Manning JC (2008) The Iris family. Natural history and classification. Timber Press, London

Goldblatt P, Manning JC (2010) *Geosiris albiflora* (Geosiridoideae), a new species from the Comoro Archipelago. Bothalia 40:169–171

Goldblatt P, Rodriguez A, Powell MP, Davies TJ, Manning JC, van der Bank M, Savolainen V (2008) Iridaceae 'Out of Australasia'? Phylogeny, biogeography, and divergence time based on plastid DNA sequences. Syst Bot 33:495–508

Govaerts R, Pfahl J, Campacci MA, Holland Baptista D, Tigges H, Shaw J, Cribb P, George A, Kreuz K, Wood J (2011) World Checklist of Orchidaceae. The Board of Trustees of the Royal Botanic Gardens, Kew. Published on the Internet http://www.kew.org/wcsp/. Accessed Sep 2011

Graham SW, Zgurski JM, McPherson MA, Cherniawsky DM, Saarela JM, Horne EFC, Smith SY, Wong WA, O'Brien HE, Biron VL, Pires JC, Olmstead RG, Chase MW, Rai HS (2006) Robust inference of monocot deep phylogeny using an expanded multigene plastid data set. Aliso 22:3–21

Groenendijk JP, van Dulmen ATJ, Bouman F (1997) The "forest floor" saprophytes *Voyria spruceana* and *V. aphylla* (Gentianaceae) growing as epiphytes in Colombian Amazonia. Ecotropica 3:129–131

Groom P (1895) On *Thismia aseroe* Beccari and its mycorrhiza. Ann Bot 9:327–361

Haig D (2008) Homologous versus anthetic alternation of generations and the origin of sporophytes. Bot Rev 74:395–418

Hamada M (1939) Studien über die Mykorrhiza von *Galeola septentrionalis* Reichb. f.—ein neuer Fall der Mykorrhiza-Bildung durch intraradicale Rhizomorpha. Jap J Bot 10:151–211

Hasebe M, Wolf PG, Pryer KM, Ueda K, Ito M, Sano R, Gastony GJ, Yokoyama J, Manhart JR, Murakami M, Crane EH, Haufler CH, Hauk WD (1995) Fern phylogeny based on *rbcL* nucleotide sequences. Am Fern J 85:134–181

Hashimoto T (1990) A taxonomic review of the Japanese *Lecanorchis* (Orchidaceae). Ann Tsukuba Bot Gard 9:1–40

Hatusima S (1976) Two new species of Burmanniaceae from Japan. J Geobot 24:2–6

Hazard C, Lilleskov EA, Horton TR (2011) Is rarity of pinedrops (*Pterospora andromedea*) in eastern North America lined to rarity of its unique fungal symbiont. Mycorrhiza 22:393–402

He-Nygrén X, Juslén A, Ahonen I, Glenny D, Piippo S (2006) Illuminating the evolutionary history of liverworts (Marchantiophyta)—towards a natural classification. Cladistics 22:1–31

Heinrichs J, Hentschel J, Wilson R, Feldberg K, Schneider H (2007) Evolution of leafy liverworts (Jungermannniidae, Marchantiophyta): estimating divergence times from chloroplast DNA sequences using penalized likelihood with integrated fossil evidence. Taxon 56:31–44

Henderson A, Stevenson DW (2004) Burmanniaceae. In: Smith N, Mori SA, Henderson A, Stevenson DW, Heald SV (eds) Flowering plants of the Neotropics. University Press, Princeton, pp 421–423

Hentrich H, Kaiser R, Gottsberger G (2010) The reproductive biology of *Voyria* (Gentianaceae) species in French Guiana. Taxon 59:867–880

Holub JL (1983) Validation of generic names in Lycopodiaceae: with a description of a new genus *Pseudolycopodiella*. Folia Geobot Phytotax 18: 439–442

Hooker JD (1886–1887) Icones Plantarum, vol VI. Williams and Norgate, London

Hsieh C, Hsieh TH, Lai I (1995) *Epirixanthes elongata* Bl.—a new record to the Flora of Taiwan. Taiwania 40:381–384

Hsu TW, Chang HM, Su SH, Chung SW, Wang WC (2005) *Burmannia cryptopetala* Makino (Burmanniaceae), a newly recorded plant in Taiwan. Taiwania 50:302–306

Hutchinson J (1934) The families of flowering plants. Monocotyledons. Clarendon, Oxford

Hutchinson J (1959) The families of flowering plants. Vol. 2. Ed. 2. Monocotyledons. Clarendon, Oxford

Hynson NA, Preiss K, Gebauer G, Bruns TD (2009) Isotopic evidence of full and partial myco-heterotrophy in the plant tribe Pyroleae (Ericaceae). New Phytol 182:719–726

Hynson NA, Bruns TD (2009) Evidence of a myco-heterotroph in the plant family Ericaceae that lacks mycorrhizal specificity. Proc R Soc Lond B 276:4053–4059

Ibisch PL, Neinhuis C, Rojas NP (1996) On the biology, biogeography, and taxonomy of *Arachnitis* Phil. nom. cons. (Corsiaceae) in respect to a new record from Bolivia. Willdenowia 26:321–332

Imhof S (1997) Root anatomy and mycotrophy of the achlorophyllous *Voyria tenella* Hook. (Gentianaceae). Bot Acta 110:298–305

Imhof S (1998) Subterranean structures and mycotrophy of the achlorophyllous *Triuris hyalina* (Triuridaceae). Can J Bot 76:2011–2019

Imhof S (1999a) Anatomy and mycotrophy of the achlorophyllous *Afrothismia winkleri*. New Phytol 144:533–540

Imhof S (1999b) Root morphology, anatomy and mycotrophy of the achlorophyllous *Voyria aphylla* (Jacq.) Pers. (Gentianaceae). Mycorrhiza 9:33–39

Imhof S (1999c) Subterranean structures and mycorrhiza of the achlorophyllous *Burmannia tenella* (Burmanniaceae). Can J Bot 77:637–643

Imhof S (2001) Subterranean structures and mycotrophy of the achlorophyllous *Dictyostega orobanchoides* (Hook.) Miers (Burmanniaceae). Rev Biol Trop 49:239–247

Imhof S (2003) A dorsiventral mycorrhizal root in the achlorophyllous *Sciaphila polygyna* (Triuridaceae). Mycorrhiza 13:327–332

Imhof S (2004) Morphology and development of the subterranean organs of the achlorophyllous *Sciaphila polygyna* (Triuridaceae). Bot J Linn Soc 146: 295–301

Imhof S (2006) Two distinct fungi colonize roots and rhizomes of the myco-heterotrophic *Afrothismia gesnerioides* (Burmanniaceae). Can J Bot 84:852–861

Imhof S (2007) Specialized mycorrhizal colonization pattern in achlorophyllous *Epirixanthes* spp. (Polygalaceae). Plant Biol 9:786–792

Imhof S (2009) Arbuscular, ecto-related, orchid mycorrhizas—three independent structural lineages towards

mycoheterotrophy: implications for classification? Mycorrhiza 19:357–363
Imhof S, Weber HC (1997) Root anatomy and mycotrophy (AM) of the achlorophyllous *Voyria truncata* (Standley) Standley & Steyermark (Gentianaceae). Bot Acta 110:127–134
IPNI (2011) The International Plant Names Index. Published on the Internet http://www.ipni.org. Accessed Oct 2011
Janse JM (1897) Les endophytes radicaux de quelques plantes Javanaises. Ann Jard Bot Buitenzorg 14:3–201
Janssen T, Bremer K (2004) The age of major monocot groups inferred from 800+ *rbcL* sequences. Bot J Linn Soc 146:385–398
Jensen SR, Schripsema J (2002) Chemotaxonomy and pharmacology of Gentianaceae. In: Struwe L, Albert VA (eds) Gentianaceae—systematics and natural history. Cambridge University Press, Cambridge, pp 573–631
Jolles D (2007) Phylogenetics and biogeography of the *Pyrola picta* species complex (Pyroleae: Monotropoideae: Ericaceae). M.Sc. Thesis. The Ohio State University, Colombus
Jones RL (2005) Plant life of Kentucky: an illustrated guide to the vascular flora. University of Kentucky Press, Lexington
Jones DL, Gray B (2008) *Corsia dispar* D. L. Jones & B. Gray (Corsiaceae), a new species from Australia, and a new combination of *Corsia* for a New Guinea taxon. Austrobaileya 7:717–722
Jonker FP (1938) A monograph of the Burmanniaceae. Meded Bot Mus Herb Rijksuniv Utrecht 51:1–279
Jonker FP (1939) Les Géosiridaceae, une nouvelle famille de Madagascar. Recueil Trav Bot Néerl 36:473–479
Judd WS, Campbell CS, Kellogg EA, Stevens PF (1999) Plant systematics: a phylogenetic approach. Sinauer Associates, Sunderland
Julou T, Burghardt B, Gebauer C, Berveiller D, Damesin C, Selosse MA (2005) Mixotrophy in orchids: insights from a comparative study of green individuals and nonphotosynthetic individuals of *Cephalanthera damasonium*. New Phytol 166:639–653
Karsten H (1858) Über die Stellung einiger Familien parasitischer Pflanzen im natürlichen system. Nov Actorum Acad Caes Leop Carol Nat Cur 26:885–927
Kato M (1983) The classification of major groups of pteridophytes. J Math Sci Univ Tokyo 13:263–283
Kato M (1990) Ophioglossaceae: a hypothetical archetype for the angiosperm carpel. Bot J Linn Soc 102:303–311
Kato M (1996) Plant-pollinator interactions in the understory of a lowland mixed Dipterocarp forest in Sarawak. Am J Bot 83:732–743
Keng H (1974) Rediscovery of *Cheilotheca malayana* and the identity of *Cheilotheca*, *Andresia*, and *Monotropastrum* (Ericaceae-Monotropoideae). Reinwardtia 9:77–83
Kennedy AH, Taylor DL, Watson LE (2011) Mycorrhizal specificity in the fully mycoheterotrophic *Hexalectris* Raf. (Orchidaceae: Epidendroideae). Mol Ecol 20:1303–1316
Kennedy AH, Watson LE (2010) Species delimitations and phylogenetic relationships within the fully mycoheterotrophic *Hexalectris* (Orchidaceae). Syst Bot 35:64–76
Kenrick P, Crane PR (1997) The origin and early diversification of land plants: a cladistic study. Smithsonian Institution, Washington, DC
Khandelwal S (1990) Chromosome evaluation in genus *Ophioglossum* L. Bot J Linn Soc 102:205–217
Kissling J (2012) Taxonomy of *Exochaenium* and *Lagenias*: two resurrected genera of tribe Exaceae (Gentianaceae). Syst Bot 37:238–253
Kissling J, Yuan Y, Küpfer P, Mansion G (2009) The polyphyletic genus *Sebaea* (Gentianaceae): a step forward in understanding the morphological and karyological evolution of the Exaceae. Mol Phylogenet Evol 53:734–748
Klackenberg J (1985) The genus *Exacum* (Gentianaceae). Opera Bot 84:1–144
Klackenberg J (1990) Famille 168—Gentianacées. In: Morat P (ed) Flore de Madagascar et des Comores. Museum National d' Histoire Naturelle, Paris, pp 108–117
Klackenberg J (2002) Tribe Exaceae. In: Struwe L, Albert VA (eds) Gentianaceae—systematics and natural history. Cambridge University Press, Cambridge, pp 66–108
Klackenberg J (2006) *Cotylanthera* transferred to *Exacum* (Gentianaceae). Bot Jahrb Syst 126:477–481
Klooster MR, Culley TM (2009) Comparative analysis of the reproductive ecology of *Monotropa* and *Monotropsis*: two mycoheterotrophic genera in the Monotropoideae (Ericaceae). Am J Bot 96: 1337–1347
Köpke E, Musselman LJ, de Laubenfels DJ (1981) Studies on the anatomy of *Parasitaxus ustus* and its root connections. Phytomorphology 31:85–89
Kottke I, Nebel M (2005) The evolution of mycorrhiza-like associations in liverworts: an update. New Phytol 167:330–334
Kovács GM, Balázs T, Pénzes Z (2007) Molecular study of arbuscular mycorrhizal fungi colonizing the sporophyte of the eusporangiate rattlesnake fern (*Botrychium virginianum*, Ophioglossaceae). Mycorrhiza 17:597–605
Kretzer AM, Bidartondo MI, Szaro TM, Grubisha L, Bruns TD (2000) Regional specialization of *Sarcodes sanguinea* on a single fungal symbiont from the *Rhizopogon ellenae* species complex. Am J Bot 87:1778–1783
Kron KA, Judd WS, Stevens PF, Crayn DM, Anderberg AA, Gadek PA, Quinn CJ, Luteyn JL (2002) Phylogenetic classification of Ericaceae: molecular and morphological evidence. Bot Rev 68:335–423
Kusano S (1911) *Gastrodia elata* and its symbiotic association with *Armillaria mellea*. J Coll Agr Imp Univ Tokyo 4:1–65
Lan J, Xu JT, Li JS (1994) Study on symbiotic relation between *Gastrodia elata* and *Armillariella mellea* by autoradiography. Acta Mycol Sin 13:219–222
Leake JR (1994) The biology of myco-heterotrophic ('saprophytic') plants. New Phytol 127:171–216

Leake JR (2005) Plants parasitic on fungi: unearthing the fungi in myco-heterotrophs and debunking the 'saprophytic' plant myth. Mycologist 19:113–122

León B (2006) Burmanniaceae endémicas del Perú. Rev Peru Biol 13:S738

Lewis DQ (2002) Burmanniaceae. In: Flora of North America Editorial Committee (eds) Flora of North America North of Mexico, Vol. 26. Magnoliophyta: Liliidae: Liliales and Orchidales. Flora of North America Association, New York and Oxford, pp 486–489

Lichthardt J, Mancuso M (1991) Report on the conservation status of *Allotropa virgata* on the Nez Perce National Forest I. Field survey and first- and second-year monitoring results. Idaho Department of Fish and Game, Boise

Liebel HT, Bidartondo MI, Preis K, Segreto R, Stöckel M, Rodda M, Gebauer G (2010) C and N stable isotope signatures reveal constraints to nutritional modes in orchids from the Mediterranean and Macaronesia. Am J Bot 97:1–10

Liebel HT, Gebauer G (2011) Stable isotope signatures confirm carbon and nitrogen gain through ectomycorrhizas in the ghost orchid *Epipogium aphyllum* Swartz. Plant Biol 13:270–275

Ligrone R, Pocock K, Duckett JG (1993) A comparative ultrastructural analysis of endophytic basidiomycetes in the parasitic achlorophyllous hepatics *Cryptothallus mirabilis* and the closely allied photosynthetic species *Aneura pinguis* (Metzgeriales). Can J Bot 71:666–679

Lindley J (1846) The vegetable kingdom. Bradbury & Evans, London

Liu Z-W, Zhou J, Liu E-D, Peng H (2010) A molecular phylogeny and a new classification of *Pyrola* (Pyroleae, Ericaceae). Taxon 59:1690–1700

Maas H, Maas PJM (1987) A new *Thismia* (Burmanniaceae) from French Guiana. Brittonia 39:376–378

Maas PJM, Ruyters P (1986) *Voyria* and *Voyriella* (saprophytic Gentianaceae) In: Organization for Flora Neotropica (eds) Flora Neotropica, Vol. 41. New York Botanical Garden, New York, pp 1–93

Maas PJM, Maas H (1989) Burmanniaceae. In: Flora of the Venezuelan Guayana—VII Contributions to the flora of the Cerro Aracamuni, Venezuela, eds. J. A. Steyermark and B. K. Holst, pp. 956–958. Ann Missouri Bot Gard 76:945–992

Maas PJM, Rübsamen T (1986) Triuridaceae. In: Organization for Flora Neotropica (eds) Flora Neotropica, Vol. 40. New York Botanical Garden, New York, pp 1–55

Maas PJM, Maas-van de Kamer H, van Benthem J, Snelders HCM, Rübsamen T (1986) Burmanniaceae. In: Organization for Flora Neotropica (eds) Flora Neotropica, Vol. 42. New York Botanical Garden, New York, pp 1–189

Maas PJM, Maas-van de Kamer H (2010) Thismiaceae. In: Sosef MSM, Florence J, Bourobou H, Ngok Banak L (eds) Flore du Gabon, vol 41. Margraf Publishers, Weikersheim, pp 60–62

Maas-van de Kamer H (1995) Triuridiflorae—Gardner's Delight? In: Rudall PJ, Cribb PJ, Cutler DF, Humphries CJ (eds) Monocotyledons: systematics and evolution. Royal Botanic Gardens, Kew, pp 287–301

Maas-van de Kamer H (1998) Burmanniaceae. In: Kubitzki K (ed) The families and genera of flowering plants, Volume 3, Lilianae (except Orchidaceae). Springer, Berlin, pp 154–164

Maas-van de Kamer H (2003) *Afrothismia gesnerioides*, another new species of *Afrothismia* (Burmanniaceae) from tropical Africa. Blumea 48:475–478

Maas-van de Kamer H, Maas PJM (1994) *Triuridopsis*, a new monotypic genus in Triuridaceae. Pl Syst Evol 192:257–262

Maas-van de Kamer H, Weustenfeld T (1998) Triuridaceae. In: Kubitzki K (ed) The families and genera of flowering plants, Volume 3, Lilianae (except Orchidaceae). Springer, Berlin, pp 452–458

Mabberley DJ (1997) The plant-book. A portable dictionary of the vascular plants, 2nd edn. University Press, Cambridge

Magallón S, Castillo A (2009) Angiosperm diversification through time. Am J Bot 96:349–365

Maguire B, Boom BM (1989) Gentianaceae, part 3. In: Maguire et al. (eds.) The botany of the Guyana Highland—Part XIII. Mem. New York Bot Gard 51:2–56

von Malmborg S (1933) *Cryptothallus* nov. gen. Ein saprophytisches Lebermoos. Ann Bryol 6:122–123

Mandal S, Das PC, Joshi PC (1992) Naturally occuring xanthones from terrestrial flora. J Ind Chem Soc 69:611–636

Márquez-Gúzman J, Vázquez-Santana S, Engleman EM, Martínez-Mena A, Martínez E (1993) Pollen development and fertilization in *Lacandonia schismatica* (Lacandoniaceae). Ann Missouri Bot Gard 80:891–897

Martínez E, Ramos CH (1989) Lacandoniaceae (Triuridales): una nueva familia de Mexico. Ann Missouri Bot Gard 76:128–135

Martos F, Dulormne M, Pailler T, Bonfante P, Faccio A, Fournel J, Dubois M-P, Selosse M-A (2009) Independent recruitment of saprotrophic fungi as mycorrhizal partners by tropical achlorophyllous orchids. New Phytol 184:668–681

McLintock HA (1966) An encyclopaedia of New Zealand. R. E. Owen, Wellington

McKendrick SL, Leake JR, Read DJ (2000) Symbiotic germination and development of myco-heterotrophic plants in nature: transfer of carbon from ectomycorrhizal *Salix repens* and *Betula pendula* to the orchid *Corallorhiza trifida* through shared hyphal connections. New Phytol 145:539–548

McKendrick SL, Leake JR, Taylor DL, Read DJ (2002) Symbiotic germination and development of the myco-heterotrophic orchid *Neottia nidus-avis* in nature and its requirement for locally distributed *Sebacina* spp. New Phytol 154:233–247

McLennan EI (1958) *Thismia rodwayi* F. Muell. and its endophyte. Aust J Bot 6:25–37

Meerendonk JPM van der (1984) Triuridaceae. In: van Steenis CGGJ (ed) Flora Malesiana, ser. 1, vol. 10. Kluwer Academic, Dordrecht, pp. 109–121

Melo A, Alves M (2012) The discovery of *Lacandonia* (Triuridaceae) in Brazil. Phytotaxa 40:21–25
Merckx V, Bakker FT, Huysmans S, Smets E (2009) Biass and conflict in phylogenetic inference of myco-heterotrophic plants: a case study in Thismiaceae. Cladistics 25:64–77
Merckx V, Bidartondo MI (2008) Breakdown and delayed cospeciation in the arbuscular mycorrhizal mutualism. Proc R Soc Lond B 275:1029–1035
Merckx V, Chatrou LW, Lemaire B, Sainge MN, Huysmans S, Smets E (2008) Diversification of myco-heterotrophic angiosperms: evidence from Burmanniaceae. BMC Evol Biol 8:178
Merckx V, Huysmans H, Smets E (2010a) Cretaceous origins of myco-heterotrophy in Dioscoreales. In: Seberg O, Petersen G, Barfod AS, Davis JI (eds) Diversity, phylogeny, and evolution in the monocotyledons. Aarhus University Press, Arhus, pp 39–53
Merckx VSFT, Janssens SB, Hynson NA, Specht CD, Bruns TD, Smets E (2012) Mycoheterotrophic interactions are not limited to a narrow phylogenetic range of arbuscular mycorrhizal fungi. Mol Ecol 21:1524–1532
Merckx V, Schols P, Maas-van de Kamer H, Maas P, Huysmans S, Smets E (2006) Phylogeny and evolution of Burmanniaceae (Dioscoreales) based on nuclear and mitochondrial data. Am J Bot 93:1684–1698
Merckx V, Stöckel M, Fleischmann A, Bruns TD, Gebauer G (2010b) 15N and 13C natural abundance of two mycoheterotrophic and a putative partially mycoheterotrophic species associated with arbuscular mycorrhizal fungi. New Phytol 188:590–596
Miers J (1841) On some new Brazilian plants allied to the natural order Burmanniaceae. Trans Linn Soc Lond 18:535–556
Miers J (1847) On a new genus of plants of the family Burmanniaceae. Proc Linn Soc Lond 1:328–329
Miers J (1866) On *Myostoma*, a new genus of the Burmanniaceae. Trans Linn Soc Lond 25:461–476
Mill RR (2003) Towards a biogeography of the Podocarpaceae. In: Mill RR, Banks L, Brickell C (eds) Proceedings of the fourth international conifer conference: conifers for the future? Acta Hort 615. International Society for Horticultural Science, Leuven, pp 137–147
Minoletti ML (1986) *Arachnitis uniflora* Phil. una curiosa monocotiledonea de la flora chilena. Bol Soc Biol Concepción Chile 57:7–20
Molina J, Struwe L (2009) Utility of secondary structure in phylogenetic reconstructions using nrDNA ITS sequences—an example from Potalieae (Gentianaceae: Asteridae). Syst Bot 34:414–428
Momose K, Yumoto T, Nagamitsu T, Kato M, Nagamasu H, Sakai S, Harrison RD, Itioka T, Hamid AA, Inoue T (1998) Pollination biology in a lowland Dipterocarp forest in Sarawak. Malaysia. I. Characteristics of the plant-pollinator community in a lowland Dipterocarp forest. Am J Bot 85:1477–1501
Moore LB, Edgar E (1970) Flora of New Zealand. Volume II. Indigenous Tracheophyta—Monocotyledons except Graminae. First electronic edition, Landcare Research, June 2004. Available at http://FloraSeries.LandcareResearch.co.nz. Accessed 24 May 2011
Motomura H, Selosse M, Martos F, Kagawa A, Yukawa T (2010) Mycoheterotrophy evolved from mixotrophic ancestors: evidence in *Cymbidium* (Orchidaceae). Ann Bot 106:573–581
Nakai T (1930) Notulae ad plantas Japoniae & Koreae. Bot Mag Tokyo 44:7–40
Neinhuis C, Ibisch PL (1998) Corsiaceae. In: Kubitzki K (ed) The families and genera of vascular plants, Volume 3, Lilianae (except Orchidaceae). Springer, Berlin, pp 198–201
Nemomissa S (2002) Flora of tropical East Africa. Gentianaceae. A. A. Balkema, Rotterdam
Neyland R (2002) A phylogeny inferred from large ribosomal subunit 26S rDNA gene sequences suggests that Burmanniales are polyphyletic. Aust Syst Bot 15:1–10
Neyland R, Hennigan M (2003) A phylogenetic analysis of large-subunit (26S) ribosome DNA sequences suggests that the Corsiaceae are polyphyletic. New Zeal J Bot 41:1–11
Nixon KC, Crepet WL (1993) Late Cretaceous fossil flowers of Ericalean affinity. Am J Bot 80:616–623
Oehler E (1927) Entwicklungsgeschichtlich-zytologische Untersuchungen an einige saprophytischen Gentianaceen. Planta 3:641–733
Ogura-Tsujita Y, Gebauer G, Hashimoto T, Umata H, Yukawa T (2009) Evidence for novel and specialized mycorrhizal parasitism: the orchid *Gastrodia confusa* gains carbon from saprotrophic *Mycena*. Proc R Soc Lond B 22:761–767
Ogura-Tsujita Y, Yukawa T (2008) High mycorrhizal specificity in a widespread mycoheterotrophic plant, *Eulophia zollingeri* (Orchidaceae). Am J Bot 95:93–97
Ohashi H (2000) *Petrosavia* (Petrosaviaceae) in Taiwan and Hainan. Taiwania 45:263–269
Ohwi J (1965) Flora of Japan. Smithsonian Institution, Washington
Oliver FW (1890) On *Sarcodes sanguinea* Torr. Ann Bot Lond 4:303–326
Øllgaard B (1987) A revised classification of the Lycopodiaceae s. lat. Opera Bot 92:153–178
Øllgaard B (2012) New combinations in Neotropical Lycopodiaceae. Phytotaxa 57:10–22
Paduano C, Rodda M, Ercole E, Girlanda M, Perotto S (2011) Pectin localization in the Mediterranean orchid *Limodorum abortivum* reveals modulation of the plant interface in response to different mycorrhizal fungi. Mycorrhiza 21:97–104
Page CN (1990) Podocarpaceae. In: Kubitzki K (ed) Families and genera of vascular plants, vol 1. Springer, Berlin, pp 332–346
Pedersen HA, Watthana S, Roy M, Suddee S, Selosse M (2009) *Cephalanthera exigua* rediscovered: new insights in the taxonomy, habitat requirements and breeding system of a rare mycoheterotrophic orchid. Nord J Bot 27:460–468

Pendry CA (2010) *Epirixanthes compressa* Pendry, a new mycoheterotrophic species of Polygalaceae from Thailand. Thai For Bull 38:184–186

Perrier de la Bathie H (1946) Triuridacées, Famille 27. In: Humbert H (ed) Flore de Madagascar et des Comores. Imprimerie Officielle, Tananarive

Peterson PM (1972) *Cryptothallus mirabilis* Malb. found on Disko Island, West Greenland (69°15' N., 53°34' W.). Lindbergia 1:189–190

Pfeiffer NE (1914) Morphology of *Thismia americana*. Bot Gaz 57:122–135

Pires JC, Maureira IJ, Givish TJ, Sytsma KJ, Petersen G, Seberg O, Asmussen C, Davis JI, Stevenson DW, Rudall PJ, Fay MF, Chase MW (2006) Phylogeny, genome size, and chromosome evolution of the Asparagales. Aliso 22:287–304

Preiss K, Adam IK, Gebauer G (2010) Irradiance governs exploitation of fungi: fine-tuning of carbon gain by two partially myco-heterotrophic orchids. Proc R Soc Lond B 277:1333–1336

Pressel S, Bidartondo MI, Ligrone R, Duckett JG (2010) Fungal symbioses in bryophytes: new insights in the Twenty First Century. Phytotaxa 9:238–253

Preußing M, Olsson S, Schäfer-Verwimp A, Wickett NJ, Wicke S, Quandt D, Nebel M (2010) New insights in the evolution of the liverwort family Aneuraceae (Metzgeriales, Marchantiophyta), with emphasis on the genus *Lobatiriccardia*. Taxon 59:1424–1440

Ramirez SR, Gravendeel B, Singer RB, Marshall CR, Pierce NE (2007) Dating the origin of the Orchidaceae from a fossil orchid with its pollinator. Nature 448:1042–1045

Rasmussen HN (1995) Terrestrial orchids from seed to mycotrophic plant. Cambridge University Press, Cambridge

Rasmussen HN (2002) Recent developments in the study of orchid mycorrhiza. Plant Soil 244:149–163

Raubeson LA, Jansen RK (1992) Chloroplast DNA evidence on the ancient evolutionary split in vascular land plants. Science 225:1697–1699

Raynal-Roques A (1967a) Étude critique des genres *Voyria* et *Leiphaimos* (Gentianaceae) et revision des *Voyria* d'Afrique. Adansonia 7:53–71

Raynal-Roques A (1967b) Sur un *Sebaea* Africain saprophyte (Gentianaceae). Adansonia 7:207–219

Remizowa MV, Sokoloff DD, Rudall PJ (2006a) Patterns of floral structure and orientation in *Japonolirion*, *Narthecium*, and *Tofieldia*. Aliso 22:159–171

Remizowa M, Sokoloff D, Rudall PJ (2006b) Evolution of the monocot gynoecium: evidence from comparative morphology and development in *Tofieldia*, *Japonolirion*, *Petrosavia* and *Narthecium*. Pl Syst Evol 258:183–209

Roberts N, Wapstra M, Duncan F, Woolley A, Morley J, Fitzgerald N (2003) Shedding some light on *Thismia rodwayi* F. Muell. (Fairy Lanterns) in Tasmania: distribution, habitat and conservation status. Papers Proc Roy Soc Tasmania 137:55–66

Rodriguez S, Marston A, Wolfender J-L, Hostettmann K (1998) Iridoids and secoiridoids in the Gentianaceae. Curr Org Chem 2:627–648

Rothmaler W (1964) Pteridophyten-Studien I. Repert Spec Nov Regni Veg 54:55–82

Rothwell GW, Stockey RA (1989) Fossil Ophioglossales of the Paleocene of western North America. Am J Bot 76:637–644

Roy M, Watthana S, Stier A, Richard F, Vessabutr S, Selosse M-A (2009a) Two mycoheterotrophic orchids from Thailand tropical dipterocarpacean forests associate with a broad diversity of ectomycorrhizal fungi. BMC Biol 7:51

Roy M, Yagame T, Yamato M, Isawe K, Heinz C, Faccio A, Bonfante P, Selosse M-A (2009b) Ectomycorrhizal *Inocybe* species associate with the mycoheterotrophic orchid *Epipogium aphyllum* but not its asexual propagules. Ann Bot 104:595–610

Rübsamen T (1980) Beiträge zur Mikromorphologie der Testa und zur Embryologie amerikanischer Burmanniaceen und Triuridaceen 1:1–355; 2:356–528. Thesis. Ruhr-Universität, Bochum

Rübsamen T (1986) Morphologische, embryologische und systematische Untersuchungen an Burmanniaceae und Corsiaceae (Mit Ausblick auf die Orchidaceae-Apostasioideae). Diss Bot 902:1–310. J. Cramer, Berlin

Rudall PJ (2002) Homologies of inferior ovaries and septal nectaries in monocotyledons. Int J Plant Sci 163:261–276

Rudall PJ (2003) Monocot pseudanthia revisited: floral anatomy and systematics of the mycoheterotrophic family Triuridaceae. Int J Plant Sci 164:S307–S320

Rudall PJ (2008) Fascicles and filamentous structures: comparative ontogeny of morphological novelties in Triuridaceae. Int J Plant Sci 169:1023–1037

Rudall PJ, Bateman RM (2006) Morphological phylogenetic analysis of Pandanales: testing contrasting hypotheses of floral evolution. Syst Bot 31:223–238

Rudall P, Eastman A (2002) The questionable affinities of *Corsia* (Corsiaceae): evidence from floral anatomy and pollen morphology. Bot J Linn Soc 138:315–324

Rudall PJ, Strange A, Cunniff J, Cheek M (2007) Floral anatomy of *Kupea martinetugei* (Triuridaceae). Kew Bull 62:282–292

Sainge MN, Franke T (2005) A new species of *Afrothismia* (Burmanniaceae) from Cameroon. Nordic J Bot 23: 299–303

Sainge MN, Franke T, Agerer R (2005) A new species of *Afrothismia* (Burmanniaceae, tribe Thismieae) from Korup National Park, Cameroon. Willdenowia 35:287–291

Salmia A (1989) General morphology and anatomy of chlorophyll-free and green forms of *Epipactis helleborine* (Orchidaceae). Ann Bot Fenn 26:95–105

Sasidharan N, Sujanapal P (2000) Rediscovery of *Haplothismia exannulata* Airy Shaw (Burmanniaceae) from its type locality. Rheedea 10:131–134

Schlechter R (1906) Burmanniaceae africanae. Bot Jarhb Syst 38:137–143

Schlechter R (1921) Die Thismieae. Notizbl Bot Gart Berlin-Dahlem 8:31–45

Schuster RM (1992) The Hepaticae and Anthocerotae of North America, vol V. Field Museum of Natural History, Chicago

Selosse MA, Faccio A, Scappaticci G, Bonfante P (2004) Chlorophyllous and achlorophyllous specimens of *Epipactis microphylla* (Neottieae, Orchidaceae) are associated with ectomycorrhizal septomycetes, including truffles. Microb Ecol 47:416–426

Selosse MA, Roy M (2009) Green plants that feed on fungi: facts and questions about mixotrophy. Trends Plant Sci 14:64–70

Selosse M-A, Weiss M, Jany J-L, Tillier A (2002) Communities and populations of sebacinoid basidiomycetes associated with the achlorophyllous orchid *Neottia nidus-avis* (L.) L.C.M. Rich. and neighbouring tree ectomycorrhizae. Mol Ecol 11:1831–1844

Sérgio C, Draper D, Garcia C (2005) Modeling the distribution of *Cryptothallus mirabilis* Malmb. (Aneuraceae, Hepaticopsida) in the Iberian Peninsula. J Hattori Bot Lab 97:309–316

Shieh WC, DeVol CE (1994) Ophioglossaceae. In: Hsieh CF, Huang TC, Keng H, Shieh WC, Tsai JL, Hu JM, Shen CF, Yang KC, Yang SY (eds) Flora of Taiwan (Pteridophyta and Gymnospermae), vol 1. Sandos Chromagraph Printing, Taipei, pp 63–67

Shukun C, Haiying M, Parnell JAN (2008) Polygalaceae In: Wu ZY, Raven PH, Hong DY (eds) Flora of China, Vol. 11 (Oxalidaceae through Aceraceae). Science Press, Beijing, and Missouri Botanical Garden Press, St. Louis, pp 139–159

Sinclair WT, Mill RR, Gardner MF, Woltz P, Jaffré T, Preston J, Hollingsworth ML, Ponge A, Möller M (2002) Evolutionary relationships of the New Caledonian heterotrophic conifer, *Parasitaxus usta* (Podocarpaceae), inferred from chloroplast *trnL-F* intron/spacer and nuclear rDNA ITS2 sequences. Plant Syst Evol 233:79–104

Smith JJ (1909) Burmaniaceae, Corsiaceae. Nova Guinea 8:193–197

Smith SE, Read DJ (2008) Mycorrhizal symbiosis, 3rd edn. Academic, London

Soltis DE, Soltis PS, Chase MW, Mort ME, Albach TD, Zanis M, Savolainen V, Hahn WH, Hoot SB, Fay MF, Axtell M, Swensen SM, Prince LM, Kress WJ, Nixon KC, Farris JS (2000) Angiosperm phylogeny inferred from 18S rDNA, *rbcL*, and *atpB* sequences. Bot J Linn Soc 133:381–461

Soltis DE, Soltis PS, Endress PK, Chase MW (2005) Phylogeny and evolution of angiosperms. Sinauer Associates, Sunderland

Spitmann A (1975) Entwicklungsgeschichtliche Untersuchungen an *Burmannia stuebelii* Hieron. et Schltr. Unveröff. Staatsexamensarb. Univ. Bochum

Sporne KR (1962) The morphology of pteridophytes. Hutchinson & Co., London

Stevens PF (2001) and onward. Angiosperm Phylogeny Website. Version 9, June 2008 [and more or less continuously updated since]. Available at http://www.mobot.org/MOBOT/Research/APweb. Accessed Jan-Dec 2011

Stevenson DW, Loconte H (1996) Ordinal and familial relationships of Pteridophyta genera. In: Camus JM, Gibby M, Johns RJ (eds) Pteridology in perspective. Royal Botanic Gardens Kew, Kew, pp 435–467

Stöckel M, Meyer C, Gebauer G (2011) The degree of mycoheterotrophic carbon gain in green, variegated and vegetative albino individuals of *Cephalanthera damasonium* is related to leaf chlorophyll concentrations. New Phytol 189:790–796

Stone BC (1980) Rediscovery of *Thismia clavigera* (Becc.) F.v.M. (Burmanniaceae). Blumea 26:419–425

Struwe L, Albert VA (2002) Gentianaceae: systematics and evolution. Cambridge University Press, Cambridge

Struwe L, Albert VA, Bremer B (1994) Cladistics and family level classification of the Gentianales. Cladistics 10:175–205

Struwe L, Kadereit JW, Klackenberg J, Nilsson S, Thiv M, Von Hagen KB, Albert VA (2002) Systematics, chracter evolution, and biogeography of Gentianaceae, including a new tribal and subtribal classification. In: Struwe L, Albert VA (eds) Gentianaceae—systematics and natural history. Cambridge University Press, Cambridge, pp 21–301

Summerhayes VS (1955) A revision of the genus *Brachycorythis*. Kew Bull 10:221–264

Sun B-Y, Kim MH, Kim CH, Park C-W (2001) *Mankyua* (Ophioglossaceae), a new fern genus from Cheju Island, Korea. Taxon 50:1019–1024

Takahashi H, Nishio E, Hayashi H (1993) Pollination biology of the saprophytic species *Petrosavia sakuraii* (Makino) van Steenis in Central Japan. J Pl Res 106:213–217

Takhtajan AL (1996) Validation of some previously described families of flowering plants. Bot Zhurn 81:85–86

Takhtajan AL (1997) Diversity and classification of flowering plants. Columbia University Press, New York

Tamura MN (1998) Nartheciaceae. In: Kubitzki K (ed) The families and genera of flowering plants, Volume 3, Lilianae (except Orchidaceae). Springer, Berlin, pp 381–392

Tamura MN, Yamashita J, Fuse S, Haraguchi M (2004) Molecular phylogeny of monocotyledons inferred from combined analysis of plastid *matK* and *rbcL* sequences. J Pl Res 117:109–120

Tanaka H (1978) Pollination biology of *Monotropastrum globosum*. Jap J Bot 53:201–202

Taylor DL, Bruns TD (1997) Independent, specialized invasions of ectomycorrhizal mutualism by two nonphotosynthetic orchids. Proc Natl Acad Sci USA 94:4510–4515

Taylor DL, Bruns TD (1999) Population, habitat and genetic correlates of mycorrhizal specialization in the 'cheating' orchids *Corallorhiza maculata* and *C. mertensiana*. Mol Ecol 8:1719–1732

Taylor DL, Bruns TD, Hodges SA (2004) Evidence for mycorrhizal races in a cheating orchid. Proc R Soc Lond B 271:35–43

Taylor DL, Bruns TD, Szaro TM, Hodges SA (2003) Divergence in mycorrhizal specialization within *Hexalectris spicata* (Orchidaceae), a nonphotosynthetic desert orchid. Am J Bot 90:1168–1179

Tedersoo L, Pellet P, Kõljalg U, Selosse M-A (2007) Parallel evolutionary paths to mycoheterotrophy in

understorey Ericaceae and Orchidaceae: ecological evidence for mixotrophy in Pyroleae. Oecologia 151:206–217
Terashita T (1996) Biological species of *Armillaria* symbiotic with *Galeola septentrionalis*. Nippon Kingakukai Kaiho 37:45–49
Terashita T, Kawakami Y (1991) An endomycorrhizal fungus of *Burmannia liukiuensis*. Trans Mycol Soc Jap 32:207–215
Thiele KR, Jordan P (2002) *Thismia clavarioides* (Thismiaceae), a new species of fairy lantern from New South Wales. Telopea 9:765–771
Thiv M, Struwe L, Albert VA, Kadereit JW (2002) Tribe Saccifolieae. In: Struwe L, Albert VA (eds) Gentianaceae—systematics and natural history. Cambridge University Press, Cambridge, pp 55–66
Thomas D, Chuyong G (2006) The establishment of long-term forest monitoring plots in Southeast Cameroon. Available at http://carpe.umd.edu/resources/Documents/SI_VegMonitoring_SECameroon_TechnicalRpt_Nov 2006.pdf. Accessed Feb 2011
Thorne RF (1972) Major disjunctions in the geographic ranges of seed plants. Quart Rev Biol 47:365–411
Thorne RF (1992) Classification and geography of the flowering plants. Bot Rev 58:225–348
Thulin M (2001) *Exacum* (Gentianaceae) on the Arabian peninsula and Socotra. Nord J Bot 21:243–247
Tobe H (2008) Embryology of *Japonolirion* (Petrosaviaceae, Petrosaviales): a comparison with other monocots. J Pl Res 121:407–416
Tobe H, Takahashi M (2009) Embryology of *Petrosavia* (Petrosaviaceae, Petrosaviales): evidence for the distinctness of the family from other monocots. J Pl Res 122:597–610
Tomimatsu H, Hoya A, Takahashi H, Ohara M (2004) Genetic diversity and multilocus genetic structure in the relictual endemic herb *Japonolirion osense* (Petrosaviaceae). J Pl Res 117:13–18
Tomlinson PB, Braggins JE, Rattenbury JA (1997) Contrasted pollen capture mechanisms in Phyllocladaceae and certain Podocarpaceae (Coniferales). Am J Bot 84:214–223
TROPICOS (2011) Missouri Botanical Garden. Available at http://www.tropicos.org. Accessed Sep 2011
Tryon RM, Tryon AF (1982) Ferns and allied plants, with special reference to tropical America. Springer, Berlin
Tsukaya H, Nakajima M, Okada H (2011) *Kalimantanorchis*: a new genus of mycotrophic orchid from West Kalimantan. Borneo Syst Bot 36:49–52
Tsukaya H, Yokoyama J, Imaichi R, Ohba H (2008) Taxonomic status of *Monotropastrum humile*, with special reference to *M. humile* var. *glaberrimum* (Ericaceae, Monotropoideae). J Plant Res 121:271–278
Umata H (1995) Seed germination of *Galeola altissima*, an achlorophyilous orchid, with aphyllophorales fungi. Mycoscience 36:369–372
Umata H (1997a) Formation of endomycorrhizas by an achlorophyllous orchid, *Erythrorchis ochobiensis*, and *Auricularia polytricha*. Mycoscience 38:335–339
Umata H (1997b) In vitro germination of *Erythrorchis ochobiensis* (Orchidaceae) in the presence of *Lyophyllum shimeji*, an ectomycorrhizal fungus. Mycoscience 38:355–357
Umata H (1998a) A new biological function of Shiitake mushroom, *Lentinula edodes*, in a myco-heterotrophic orchid, *Erythrorchis ochobiensis*. Mycoscience 39:85–88
Umata H (1998b) In vitro symbiotic associations of an achlorophyllous orchid, *Erythrorchis ochobiensis*, with orchid and non-orchid fungi. Mem Fac Agr Kagoshima Univ 34:97–107
Uphof JCT (1929) Beiträge zur Kenntnis der Burmanniacee *Apteria aphylla* (Nutt.). Barnh Österr Bot Z 78:71–80
Ushimaru A, Imamura A (2002) Large variation in flower size of the myco-heterotrophic plant. Monotropastrum globosum: effect of floral display on female reproductive success. Plant Species Biol 17:147–153
Van der Pijl L (1934) Die Mycorrhiza von *Burmannia* und *Epirrhizanthes* und die Fortpflanzung ihres Endophyten. Recueil Trav Bot Néerl 31:761–779
Van Royen P (1972) Sertulum Papuanum 17. Corsiaceae of New Guinea and surrounding areas. Webbia 27:223–255
Vanderpoorten A, Goffinet B (2009) Introduction to Bryophytes. University Press, Cambridge
Vergara-Silva F, Espinosa-Matiás S, Ambrose BA, Vázquez-Santana A, Martinez-Mena J, Márquez-Guzmán E, Meyerowitz EM, Alvarez-Buylla E (2003) Inside-out flowers characteristic of *Lacandonia schismatica* (Lacandoniaceae: Triuridales) evolved at least before the divergence from its sister taxon, *Triuris brevistylis*. Int J Plant Sci 164:345–357
Vogel S (1962) Duftdrüsen im Dienste der Bestäubung. Akad Wiss Abh Math Naturwiss Kl 10:601–763
Vogel S (1978) Pilzmückenblumen als Pilzmimeten. Flora 167:329–398
Vogel S (1990) The role of scent glands in pollination: on the structure and function of osmophores. Amerind, New Delhi
Voirin B, Jay M (1997) Sur la présence d'amentoflavone chez *Tmesipteris tannensis*. Phytochemistry 16: 2043–2044
Vollesen K (1982) A new species of *Seychellaria* (Triuridaceae) from Tanzania. Kew Bull 36:733–736
Wachowiak W, Baczkiewicz A, Chudzinska E, Buczkowska K (2007) Cryptic speciation in liverworts—a case study in the *Aneura pinguis* complex. Bot J Linn Soc 155:273–282
Wagner WH (1993) Schizaeaceae. In: Morin N (ed) Flora of North America North of Mexico, Volume 2: Pteridophytes and Gymnosperms. Oxford University Press, New York, pp 112–113
Wagner WH, Beitel JM (1993) Lycopodiaceae. In: Morin N (ed) Flora of North America North of Mexico, Volume 2: Pteridophytes and Gymnosperms. Oxford University Press, New York, pp 18–37
Wagner WH, Wagner FS (1993) Ophioglossaceae. In: Morin N (ed) Flora of North America North of Mexico, Volume 2: Pteridophytes and Gymnosperms. Oxford University Press, New York, pp 85–106

Wagstaff SJ (2004) Evolution and biogeography of the austral genus *Phyllocladus* (Podocarpaceae). J Biogeogr 31:1569–1577

Wallace GD (1975) Studies of the Monotropoideae (Ericaceae): taxonomy and distribution. Wasmann J Biol 33:1–88

Wallace GD (1987) Transfer of *Eremotropa sciaphila* to *Monotropastrum* (Ericaceae: Monotropoideae). Taxon 36:128–130

Wang B, Qiu YL (2006) Phylogenetic distribution and evolution of mycorrhizas in land plants. Mycorrhiza 16:299–363

Wapstra M, French B, Davies N, O'Reilly-Wapstra J, Peters D (2005) A bright light on the dark forest floor: observations on fairy lanterns *Thismia rodwayi* F. Muell. (Burmanniaceae) in Tasmanian forests. Tasmanian Nat 127:2–18

Warcup JH (1985) *Rhizanthella gardneri* (Orchidaceae), its *Rhizoctonia* endophyte and close associations with *Melaleuca uncinata* (Myrtaceae) in Western Australia. New Phytol 99:273–280

Warcup JH (1991) The *Rhizoctonia* endophytes of *Rhizanthella* (Orchidaceae). Mycol Res 95:656–659

Warming ES (1901) Sur quelques Burmanniacées receuillies au Brésil par le Dr. A. Glaziou. Overs. Kongel. Danske Vidensk. Selsk. Forh. Meddlemers Arbeider 173–188

White F (1986) La végétation de l'Afrique—Mémoire accompagnant la carte de végétation de l'Afrique. Unesco/AETFAT/UNSO. ORSTOM et l'Organisation des Nations Unies, Paris

Wickett NJ, Goffinet B (2008) Origin and relationships of the myco-heterotrophic liverwort *Cryptothallus mirabilis* Malmb. (Metzgeriales, Marchantiophyta). Bot J Linn Soc 156:1–12

Wickett NJ, Fan Y, Lewis PO, Goffinet B (2008a) Distribution and evolution of pseudogenes, gene losses, and a gene rearrangement in the plastid genome of the nonphotosynthetic liverwort, *Aneura mirabilis* (Metzgeriales, Jungermanniopsida). J Mol Evol 67:111–122

Wickett NJ, Zhang Y, Hansen SK, Roper JM, Kuehl JV, Plock SA, Wolf PG, dePamphilis CW, Boore JL, Goffinet B (2008b) Functional gene losses occur with minimal size reduction in the plastid genome of the parasitic liverwort *Aneura mirabilis*. Mol Biol Evol 25:393–401

Wikström N, Kenrick P (1997) Phylogeny of Lycopodiaceae (Lycopsida) and the relationships of *Phylloglossum drummondii* Kunze based on *rbcL* sequences. Int J Plant Sci 158:862–871

Wikström N, Kenrick P, Chase M (1999) Epiphytism and terrestrialization in tropical *Huperzia* (Lycopodiaceae). Plant Syst Evol 218:221–243

Wikström N, Kenrick P, Vogel JC (2002) Schizaeaceae: a phylogenetic approach. Rev Palaeobot Palynol 119:35–50

Wikström N, Savolainen V, Chase MW (2001) Evolution of the angiosperms: calibrating the family tree. Proc Roy Soc Lond B 268:2211–2220

Winther JL, Friedman WE (2007) Arbuscular mycorrhizal symbionts in *Botrychium* (Ophioglossaceae). Am J Bot 94:1248–1255

Winther JL, Friedman WE (2008) Arbuscular mycorrhizal associations in Lycopodiaceae. New Phytol 177: 790–801

Winther JL, Friedman WE (2009) Phylogenetic affinity of arbuscular mycorrhizal symbionts in *Psilotum nudum*. J Plant Res 122:485–496

Wirz H (1910) Beiträge Entwicklungsgeschichte von *Sciaphila* spec. und von *Epirrhizanthes elongata* Bl. Flora 101:395–446

Woltz P, Stockey RA, Gondran M, Cherrier J-F, Bernard J (1994) Interspecific parasitism in the gymnosperms—unpublished data on two endemic New Caledonian Podocarpaceae using scanning electron microscopy. Acta Bot Gallica 141:731–746

Wood CE (1983) The genera of Burmanniaceae in the Southeastern United States. J Arnold Arbor 64: 293–307

Wood JJ, Beaman TE, Lamb A, Lun CC, Beaman JH (2011) The Orchids of Mount Kinabalu, vol 1 & 2. Natural History Publications, Kota Kinabalu

Woodward CL, Berry PE, Maas-van de Kamer H, Swing K (2007) *Tiputinia foetida*, a new mycoheterotrophic genus of Thismiaceae from Amazonian Ecuador, and a likely case of deceit pollination. Taxon 56:157–162

World Checklist of Selected Plant Families (2011) The Board of Trustees of the Royal Botanic Gardens, Kew. Published on the Internet http://www.kew.org/wcsp/. Accessed July-Sep 2011

Xu JT, Fan L (2001) Cytodifferentiation of the seeds (protocorms) and vegetative propagation corms colonized by mycorrhizal fungi. Acta Bot Sin 43:1003–1010

Yagame T, Yamato M, Suzuki A, Iwase K (2007) Developmental processes of achlorophyllous orchid. Epipogium roseum: from seed germination to flowering under symbiotic cultivation with mycorrhizal fungus. J Plant Res 120:229–236

Yagame T, Yamato M, Suzuki A, Iwase K (2008) Ceratobasidiaceae mycorrhizal fungi isolated from nonphotosynthetic orchid *Chamaegastrodia sikokiana*. Mycorrhiza 18:97–101

Yahara T, Tsukaya H (2008) *Oxygyne yamashitae*, a new species of Thismiaceae from Yaku Island, Japan. Acta Phytotax Geobot 59:97–104

Yamato M (2001) Identification of a mycorrhizal fungus in the roots of achlorophyllous *Sciaphila tosaensis* Makino (Triuridaceae). Mycorrhiza 11:83–88

Yamato M, Yagame T, Iwase K (2011a) Arbuscular mycorrhizal fungi in roots of non-photosynthetic plants *Sciaphila japonica* and *Sciaphila tosaensis* (Triuridaceae). Mycoscience 52:217–223

Yamato M, Yagame T, Shimonura N, Iwase K, Takahashi H, Ogura-Tsujita Y, Yukawa T (2011b) Specific arbuscular mycorrhizal fungi associated with non-photosynthetic *Petrosavia sakuraii* (Petrosaviaceae). Mycorrhiza 21:631–639

Yamato M, Yagame T, Suzuki A, Iwase K (2005) Isolation and identification of mycorrhizal fungi associating with an achlorophyllous plant, *Epipogium roseum* (Orchidaceae). Mycoscience 46:73–77

Yang HB, Fang RC, Jin CL (1999) Ericaceae. In: Fang RC (ed) Flora Reipublicae Popularis Sinicae 57. Science Press, Beijing, pp 1–213

Yang S-Z, Saunders RM, Hsu C-J (2002) *Thismia taiwanensis* sp. nov. (Burmanniaceae tribe Thismieae): first record of the tribe in China. Syst Bot 27:485–488

Yokoyama J, Fukuda T, Tsukaya H (2005) Molecular identification of the mycorrhizal fungi of the epiparasitic plant *Monotropastrum humile* var. *glaberrimum* (Ericaceae). J Plant Res 118:53–56

Yokoyama J, Koizumi Y, Yokota M, Tsukaya H (2008) Phylogenetic position of *Oxygyne shinzatoi* (Burmanniaceae) inferred from 18S rDNA sequences. J Pl Res 121:27–32

Yuan YM, Wohlhauser S, Möller M, Klackenberg J, Callmander MW, Küpfer P (2005) Phylogeny and biogeography of *Exacum* (Gentianaceae): a disjunctive distribution in the Indian Ocean basin resulted from long distance dispersal and extensive radiation. Syst Biol 54:21–34

Yuan YM, Wohlhauser S, Moller M, Chassot P, Mansion G, Grant J, Küpfer P, Klackenberg J (2003) Monophyly and relationships of the tribe Exaceae (Gentianaceae) inferred from nuclear ribosomal and chloroplast DNA sequences. Mol Phylogenet Evol 28:500–517

Yukawa T (1999) *Cremastra aphylla* (Orchidaceae), a new mycoparasitic species from Japan. Ann Tsukuba Bot Gard 18:59–63

Zhang DX, Saunders RMK (1999) *Burmannia larseniana* (Burmanniaceae): a new species from Thailand. Nordic J Bot 19:241–244

Zhang DX, Saunders RMK (2000) Reproductive biology of a mycoheterotrophic species, *Burmannia wallichii* (Burmanniaceae). Bot J Linn Soc 132:359–367

Zhang DX, Saunders RMK, Hu CM (1999) *Corsiopsis chinensis* gen. et sp. nov. (Corsiaceae): first record of the family in Asia. Syst Bot 24:311–314

Zimmer K, Hynson NA, Gebauer G, Allen EB, Allen MF, Read DJ (2007) Wide geographical and ecological distribution of nitrogen and carbon gains from fungi in pyroloids and monotropoids (Ericaceae) and in orchids. New Phytol 175:166–175

3 Biogeography and Conservation

Vincent S.F.T. Merckx, Erik F. Smets, and Chelsea D. Specht

3.1 Introduction

There is hardly a region in the world where mycoheterotrophy does not occur. As far as we know, all orchid species are dependent on fungi during germination and early development and are therefore classified as initial mycoheterotrophs (Chap. 1). Orchidaceae have a worldwide distribution and occur in almost every terrestrial ecosystem apart from deserts and permafrosts. The distribution of fully mycoheterotrophic species, however, is much more restricted and often shows intriguing patterns. Fully mycoheterotrophic flowering plants almost exclusively inhabit closed-canopy forests, and the majority of species occur in the tropical regions of the world. Many families and genera of mycoheterotrophic plants have remarkably widespread distributions that cross major dispersal barriers, and several botanists have suggested that these widespread lineages must be ancient (Engler 1888; Malme 1896; Jonker 1938; Rübsamen 1986; Leake 1994). In addition, the distribution ranges of mycoheterotrophic plants often show wide gaps ("disjunctions"). The mycoheterotrophic genus *Thismia* (Thismiaceae) even holds the status of what Robert Thorne considered one of the strangest distribution patterns in flowering plants (Thorne 1972), with two allegedly related species: one in northern USA and the sister species in Australia and New Zealand.

At local scale, the distribution of mycoheterotrophs is often highly patterned. Mycoheterotrophs can be extremely scarce, and some genera and species are known from only a few localities. It is not always clear whether this reflects true rarity or the plant's ability to remain unnoticed by collectors. Even in well-collected areas, the ephemeral nature of mycoheterotrophs makes their discovery challenging. In addition, some tropical species seem to have intriguingly fragmented distribution ranges, which, in some cases, appears to be linked with postulated glacial rainforest refugia (Cheek and Williams 1999; Franke 2004). Despite this rarity, certain localities and habitats seem to be more suited for mycoheterotroph survival: both in tropical and temperate zones, there is a tendency of unrelated species to grow together (e.g., Wallace 1975; Maas and Rübsamen 1986).

In general terms, the study of biogeography includes two distinct components. First, the distribution of species is determined by historical

V.S.F.T. Merckx (✉)
Naturalis Biodiversity Center, Leiden University, P.O. Box 9514, 2300 RA Leiden, The Netherlands
e-mail: vincent.merckx@naturalis.nl

E.F. Smets
Naturalis Biodiversity Center, Leiden University, P.O. Box 9514, 2300 RA Leiden, The Netherlands

Laboratory of Plant Systematics, KU Leuven, 3001 Leuven, Belgium

C.D. Specht
Department of Plant and Microbial Biology, University of California Berkeley, Berkeley, CA 94720, USA

V.S.F.T. Merckx (ed.), *Mycoheterotrophy: The Biology of Plants Living on Fungi*, DOI 10.1007/978-1-4614-5209-6_3, © Springer Science+Business Media New York 2013 factors. Historical biogeography attempts to infer how evolutionary, geological, and climatic events have shaped the distribution of a particular lineage. Molecular phylogenetic hypotheses and divergence time estimates are particularly useful to test hypotheses in this context (de Queiroz 2005; Renner 2005). Unfortunately, the lack of a rigid phylogenetic framework still prevents inference of the historical biogeography of many mycoheterotrophic lineages. Nevertheless, the limited data that is available now allows us to evaluate the origin of the distribution patterns or disjunctions at least for a few mycoheterotrophic clades. In addition to historical events, the distribution of a species is also determined by ecological aspects. Mycoheterotrophic species seem to show very specific preferences toward certain (micro)habitats, a matter we discuss in detail below. In this context, the use of molecular techniques for the identification of root-inhabiting fungi has brought a new and interesting ecological association to the forefront of mycoheterotrophic biogeography. As many mycoheterotrophs show specialization toward narrow clades of fungi, it is often hypothesized that the distribution of their host fungi is a major determinant for the distribution of the plants. The study of the biogeography of fungi is still in its infancy, and to date, only a few studies have tested this hypothesis in detail. However, the development of next-generation sequencing techniques provides promising prospects to resolve such questions in the upcoming years.

Understanding distribution patterns is also essential for effective species conservation (Whittaker et al. 2005). The rapid and extensive destruction of habitats, particularly in the tropics, has become a serious threat to native biotas. Mycoheterotrophic plants are often restricted to areas that experience exceptional loss of habitat, and due to their localized distributions, many species are extremely vulnerable to extinction. Some species may already be on the brink of extinction, and only drastic conservation efforts can prevent their disappearance.

Here we provide a detailed overview of the global distribution of mycoheterotrophic plants. This overview is considerably biased toward angiosperm mycoheterotrophs, although we also discuss in short the distribution patterns observed in nonangiosperm lineages where mycoheterotrophy is mostly restricted to the gametophytes of lycophytes, ferns, and gymnosperms. The occurrence of mycoheterotrophy in nonangiosperm gametophytes remains poorly studied, thus preventing an in-depth discussion of their distribution and habitat preferences. The taxonomy and species numbers of the lineages discussed here are based on the information provided in Chap. 2. It is important to note that the majority of the species discussed here are only *putative* full mycoheterotrophs. The assumption that a species fully relies on mycoheterotrophy is, in most cases, based on the fact that leaves are absent and chlorophyll is lacking. However, the precise characterization of a species' trophic strategy requires careful investigation, which has been carried out for only very few species. Also, new species and localities are constantly discovered. Thus, our observations, particularly about number of species, are subject to change. Nevertheless, it is unlikely that undiscovered species and localities represent significant dissimilarities to the general patterns described here.

3.2 Distribution of Nonangiosperm Mycoheterotrophs

3.2.1 Liverworts

The only non-seed plant species with a completely mycoheterotrophic life cycle is the liverwort *Aneura mirabilis*. The species has been recorded from England, Germany, France, Portugal, Russia, Scandinavia, and Greenland. A liverwort species supposedly related to *Aneura mirabilis* is described from Costa Rica, but it remains unknown whether this species is a full mycoheterotroph (Crum and Bruce 1996; Wickett and Goffinet 2008).

3.2.2 Clubmosses

The genera *Huperzia* and *Lycopodiella*, with respectively fully mycoheterotrophic and putatively partially mycoheterotrophic gametophytes, have a true cosmopolitan distribution, although the diversity of *Lycopodiella* is highest in the New World and Australasia. Species of *Lycopodium* occur in temperate and tropical zones as well, but in the tropics, *Lycopodium* is restricted to montane regions.

3.2.3 Ferns

The gametophytes of *Botrychium* and *Ophioglossum* species (Ophioglossaceae) are achlorophyllous and thus putatively mycoheterotrophic. Both genera have a cosmopolitan distribution. The gametophytes of the only other Ophioglossaceae species, *Helminthostachys zeylanica* and *Mankyua chejuense*, have not been studied in detail. The former occurs in India and Sri Lanka, Southeast Asia, Japan, and Australasia, and the latter genus is endemic to Cheju Island (Korea).

Species of the two Psilotaceae genera *Psilotum* and *Tmesipteris* have subterranean mycoheterotrophic gametophytes. *Psilotum* has a pantropical distribution, with one species extending into temperate areas. *Tmesipteris* occurs in Southeast Asia, Australasia, Pacific, and New Zealand.

The gametophyte of *Stromatopteris moniliformis* (Gleicheniaceae), which is endemic to New Caledonia, is achlorophyllous as well and thus presumably mycoheterotrophic. Mycoheterotrophic gametophytes are also observed in species of the genus *Actinostachys* (Schizaeaceae), which are native to tropical America, Southeast Asia, and temperate regions in the southern hemisphere.

3.2.4 Gymnosperms

The remarkable *Parasitaxus usta* (Podocarpaceae) is the only achlorophyllous gymnosperm, but its mycoheterotrophic status remains doubtful. It is only found growing near or on the New Caledonian sickle pine (*Falcatifolium taxoides*), which occurs in dense evergreen rainforest on New Caledonia and the adjacent Île des Pins (Feild and Brodribb 2005; Eckenwalder 2009).

3.3 Distribution of Fully Mycoheterotrophic Angiosperms

3.3.1 Tropical Regions

The majority of fully mycoheterotrophic angiosperms—ca. 75% of all species—occur in the tropical zone ("tropics") between the Tropic of Cancer and the Tropic of Capricorn. This zone is dominated by tropical rainforest, the preferred habitat of mycoheterotrophs (Fig. 3.1). Tropical rainforest is the most diverse of all plant communities and occurs in areas with a warm wet climate without pronounced cold or dry seasons.

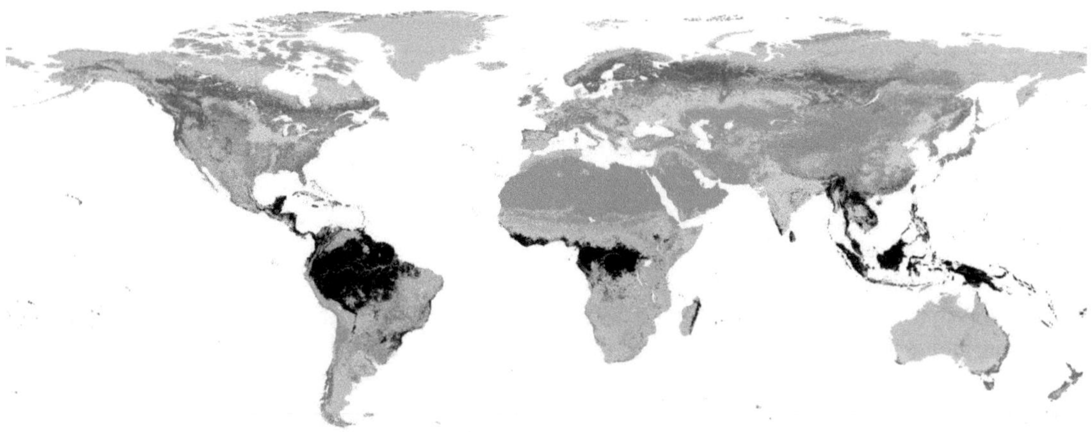

Fig. 3.1 Global distribution of tropical rainforest (*black*) in 2003. Map based on information obtained from NASA Earth Observations (2011)

Table 3.1 Approximate numbers of fully mycoheterotrophic angiosperm species, genera, and families for the main tropical regions discussed here

	Central and South America	Africa and Madagascar	Southwest India and Sri Lanka	Southeast Asia	Australasia	Pacific Islands
Families	7	6	4	9	7	3
Genera	24	19	10	29	17	9
Species	84	52	17	169	78	17

Numbers of species and genera are based on data from Chap. 2. Not all species included in this overview grow in tropical rainforest. Ericaceous mycoheterotrophs in tropical America and Asia, for example, are mainly restricted to coniferous and mixed forests at high elevations

Tropical rainforest is dominated by broad-leaved evergreen trees, and its multistoried canopy prevents light from penetrating to the forest floor.

The vast majority of tropical tree species form arbuscular mycorrhizas (AM), and AM interactions are the most common mycorrhizal associations in tropical rainforests (Smith and Read 2008). Most tropical mycoheterotrophic plants are associated with arbuscular mycorrhizal fungi, although mycoheterotrophic interactions involving ectomycorrhizal fungi (Orchidaceae and Ericaceae) and saprotrophic fungi (Orchidaceae) are also common. Ectomycorrhizal associations were long considered to be rare in the tropics, but we now know that at least three lineages of tropical trees form ectomycorrhizal associations (Alexander and Lee 2005). The most notable exception to the AM dominance in tropical forest is found in Southeast Asia. Rainforests in Southeast Asia are among the most diverse plant communities in the world and are characterized by high abundance of trees in the family Dipterocarpaceae, which are known to form associations with ectomycorrhizal fungi (Lee 1990).

While tropical rainforests on different continents are physiognomically very similar, their floras have little in common. This reflects their distinct geological and evolutionary histories (Morley 2000). In our overview, we adopt six major rainforest regions, loosely based on Primack and Corlett (2005): (1) Central and South America, (2) Africa and Madagascar, (3) India and Sri Lanka, (4) Southeast Asia, (5) Australasia, and (6) the Pacific Islands. Biologically, these regions can be subdivided into smaller areas, and we distinguish several smaller rainforest blocks in our discussion of each major region.

Mycoheterotrophs are represented in all these tropical rainforest regions, but their diversity is not equally distributed among the regions (Table 3.1). Southeast Asia contains by far the largest number of fully mycoheterotrophic species. On higher taxonomic level, the differences in diversity between the regions are less pronounced (Table 3.1), although there is little floristic overlap between the regions and many species and genera are endemic to a particular region.

3.3.2 Central and South America

Half of the world's current tropical rainforests are in South and Central America (Morley 2000; Primack and Corlett 2005), together referred to as the "Neotropics" or New World tropics (as opposed to the Paleotropics or Old World tropics). Centered on the Amazon River basin in northern and central Brazil and extending to the western foothills of the Andes and up into southern Mexico along the isthmus of Panama, the neotropical forests form the largest block of continuous rainforest on Earth (Fig. 3.2). The adjacent basin of the Orinoco River drains eastern Colombia and Venezuela and contains a large rainforest that extends into French Guiana, Surinam, and Guyana ("the Guianas"), sharing many characteristics with the Amazon River basin. A distinct patch of rainforest, the Brazilian Atlantic Forest (Mata Atlântica), runs along the coast of southeast Brazil from Recife in the north to Sao Paulo in the south. Due to excessive logging, less than 5% of the original Mata Atlântica remains; however, this forest still retains a high level of taxonomic diversity and endemism.

Fig. 3.2 Occurrence of rainforest in the Neotropics (*black*)

In Central America, rainforests extend from the Pacific coast of northwest South America to southernmost Mexico. Many larger islands of the West Indies were covered by rainforest as well, but very little of that forest now remains.

In comparison with Old World (paleotropical) tropical forests, neotropical rainforests are characterized by a high species diversity of trees belonging to the families Vochysiaceae, Bignoniaceae, Lecythidaceae, and Chrysobalanaceae. Another distinct feature of neotropical rainforests is the prominent presence of Bromeliaceae, which are the preeminent group of neotropical epiphytes (Primack and Corlett 2005). With an estimated 90,000 species of higher plants, the neotropical region is more species rich than the Paleotropics (Prance 1994). In terms of fully mycoheterotrophic plant species, however, the Neotropics contain only about a quarter of the number of species found in the Paleotropics (Table 3.1), with the flora of tropical America including ca. 80 species of fully mycoheterotrophs. The paucity of fully mycoheterotrophic species of orchids in the Neotropics is the main reason for this discrepancy. Despite the enormous diversity of tropical New World orchids, only seven species are achlorophyllous versus 17 species in tropical Africa and ca. 135 in tropical Asia.

With the exception of mycoheterotrophic Iridaceae (Madagascar), Petrosaviaceae (Southeast Asia), and Polygalaceae (Southeast Asia and Australasia), all angiosperm families with fully mycoheterotrophic species are present in the Neotropics. Of the 24 neotropical genera of mycoheterotrophic flowering plants, 17 are endemic: *Apteria*, *Dictyostega*, *Hexapterella*, *Marthella*, and *Miersiella* (Burmanniaceae); *Arachnitis* (Corsiaceae); *Lacandonia*, *Peltophyllum*, *Soridium*, *Triuris*, and *Triuridopsis* (Triuridaceae); *Degranvillea*, *Pogoniopsis*, *Uleiorchis*, and *Wullschlaegelia* (Orchidaceae); *Tiputinia* (Thismiaceae); and *Voyriella* (Gentianaceae). The Neotropics are also an important center of diversity for *Voyria* (Gentianaceae), *Gymnosiphon* (Burmanniaceae), and *Thismia* (Thismiaceae).

Although some neotropical mycoheterotrophic plant species are endemic to a restricted region, many species are widely distributed. *Apteria aphylla* (Burmanniaceae), for example, occurs from southern USA and the West Indies in the north to Paraguay and southern Brazil in the south (Maas et al. 1986). A similar distribution is observed in *Voyria aphylla* (Gentianaceae) (Maas and Ruyters 1986).

3.3.2.1 Central America

Many South American mycoheterotrophs reach their northern distribution limit in the rainforests of Central America. Central American forests also contain a few endemic mycoheterotrophic species, including *Lacandonia schismatica* (Triuridaceae), *Gymnosiphon panamensis* (Burmanniaceae), and *Voyria kupperi* (Gentianaceae). The only mycoheterotrophic orchids that grow in the rainforests of Central America are species of *Wullschlaegelia* (Orchidaceae). In terms of species richness, the floras of Panama and Costa Rica contain perhaps slightly more mycoheterotrophic species than other countries in Central America consistent with their increased species richness relative to surrounding regions.

Central America is also an important center of diversity for the orchid genera *Hexalectris* and *Corallorhiza*, which reach their southernmost distributions here. However, these mycoheterotrophs do not grow in rainforests but are found in a variety of habitats ranging from coniferous forest to mixed scrub forests and desert canyons (Salazar and Freudenstein 1998; Kennedy and Watson 2010). Central American coniferous forests at higher elevations are also home to *Hypopitys monotropa* and *Monotropa uniflora* (Ericaceae). The latter species reaches its southernmost distribution in the montane forests of Colombia (Wallace 1975). The distribution range of these species does not overlap with any rainforest species.

3.3.2.2 West Indies

The West Indies lack many groups of mycoheterotrophs that are found on continental South and Central America (e.g., Triuridaceae, Thismiaceae, *Voyriella*, and most Burmanniaceae genera). But species of *Apteria*, *Corallorhiza*, *Gymnosiphon*, *Voyria*, and *Wullschlaegelia* that are found in nearby continental forests have been recorded on many islands of the West Indies. Trinidad harbors a number of additional continental species (*Dictyostega orobanchoides*, *Hexapterella gentianoides*, and *Gymnosiphon divaricatus*) most likely as a result of its proximity to mainland South America. The rare monotypic genus *Marthella* (Burmanniaceae) is endemic to Trinidad.

3.3.2.3 Guianas and Amazonia

In South America, the Guianas are an important center of diversity for many mycoheterotrophic genera, in particular for *Voyria* and *Voyriella* (Gentianaceae), *Gymnosiphon* (Burmanniaceae), and *Sciaphila* (Triuridaceae). *Degranvillea dermaptera* (Orchidaceae) and *Thismia saülensis* (Thismiaceae) are endemic to this region. The adjacent Amazon rainforest is home to a high diversity of mycoheterotrophs, and several species are endemic to this region: *Sciaphila rubra*, *S. oligantha*, *S. corymbosa*, *Triuridopsis peruviana* (Triuridaceae), and the enigmatic species *Tiputinia foetida* and *Thismia melanomitra* (Thismiaceae) from Amazonian Ecuador. Many other mycoheterotrophs from the Amazon rainforest also occur in adjacent regions, particularly in the Guianas, Venezuela, and Colombia. There is little doubt that the forests of the Amazon Basin harbor many species that have yet to be discovered.

3.3.2.4 Southern Neotropics

Further to the south, many tropical mycoheterotrophic plants reach the edge of their distribution ranges in northern Paraguay and Bolivia. Contrarily, *Arachnitis uniflora* (Corsiaceae)

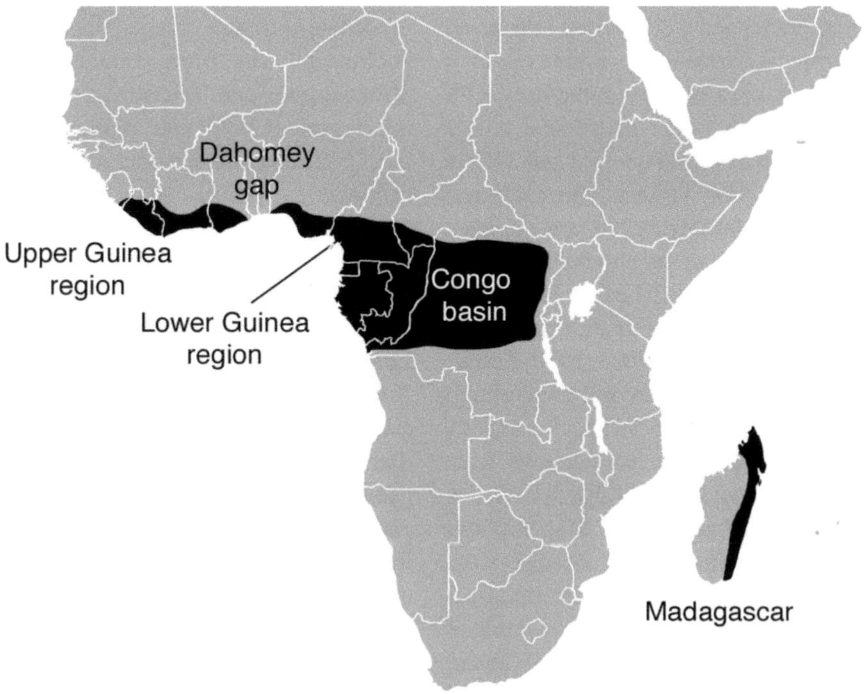

Fig. 3.3 Distribution of tropical rainforest (*black*) in Africa and Madagascar

reaches its most northern distribution in high-altitude rainforests of Bolivia that resemble the habitat of this species' southern Patagonian populations in Argentina and Chile (Ibisch et al. 1996).

3.3.2.5 Atlantic Forest

The Atlantic Forest (Mata Atlântica) on the Atlantic coast of Brazil is another important region for mycoheterotrophic plant diversity and has a high level of endemism: many species of *Thismia* (Thismiaceae) are endemic to this region (Maas et al. 1986), and the Mata Atlântica harbors the only known populations of *Lacandonia brasiliana*, *Peltophyllum caudatum*, *Triuris alata* (Triuridaceae), *Voyria obconica* (Gentianaceae), and *Pogoniopsis nidus-avis* (Orchidaceae). The Atlantic Forest is severely affected by habitat destruction, and many of these species are threatened with extinction.

3.3.3 Africa and Madagascar

Africa contains the second largest block of tropical rainforest, centered on the Congo River basin and continuing on the coast of the Gulf of Guinea to Sierra Leone but with a gap in the vicinity of Togo (the Dahomey Gap) (Fig. 3.3) (Morley 2000; White 2001). Tropical rainforest in East Africa is mostly restricted to small "islands," mainly centered on mountains. While these East African patches of rainforest cover only a small area, they contain a high percentage of endemic species due to their prolonged isolation from the forests of West and Central Africa. In Madagascar, tropical rainforest is mostly restricted to a 120 km-wide band along the eastern coast (Mittermeier et al. 1999), but due to intensive human activity, only very little of this forest remains today.

African rainforests are generally dryer, lower, and more open than rainforests found elsewhere (Morley 2000). They are relatively poor in plant species when compared with neotropical and Asian rainforests, a feature that has been attributed to the degrading effect of significant climatic changes during the Cenozoic and, more recently, to Pleistocene climatic fluctuations, although other factors such as human impact may have contributed as well (Plana 2004). African rainforests are particularly poor in palms and Lauraceae as well as in epiphytes and woody vines in general. The flora of tropical Africa is also notably poor in orchids, and it has been estimated that only 15% of the world's orchids occur in tropical Africa compared to 41% in tropical America and 34% in tropical Asia and New Guinea (Primack and Corlett 2005). Tropical Africa has always been considered relatively poor in number of mycoheterotrophic plant species (Leake 1994), but recent new discoveries started to alter this image (Cheek 2003a, b; Cheek et al. 2003, 2008; Franke 2004; Sainge et al. 2005; Cheek and Vanderburgt 2010; Cribb et al. 2010). Over 50 species of fully mycoheterotrophic plants are now known from Africa (including Madagascar) (Table 3.1). When compared to the total number of higher plants from tropical Africa (ca. 45,000; Beentje et al. 1994), the percentage of mycoheterotrophic species in the flora of Africa is not lower than that of the Neotropics, where ca. 80 species of mycoheterotrophs are recorded for an estimated 90,000 higher plants (Prance 1994). The continuous description of new taxa also illustrates the fact that certain areas in Africa remain undercollected and poorly characterized and that new discoveries, particularly from rainforests in Central Africa, are anticipated.

The mycoheterotrophic species from Africa and Madagascar belong to 19 genera, of which *Afrothismia* (Thismiaceae), *Auxopus*, *Brachycorythis* (Orchidaceae), *Kupea*, *Kihansia*, *Seychellaria* (Triuridaceae), and *Geosiris* (Iridaceae) are endemic. In general, the distribution of mycoheterotrophs in tropical Africa seems to be more patterned than in the Neotropics, as many species have very restricted distribution ranges. Exceptional widespread species include *Exochaenium oliganthum* (Gentianaceae), *Didymoplexis africana*, *Eulophia galeoloides*, and *Epipogium roseum* (Orchidaceae).

3.3.3.1 West and Central Africa

The main African rainforest block is centered in the Congo Basin and extends from the East African Albertine Rift mountains to the Atlantic Ocean in the West (Plana 2004). This rainforest region is known as the Lower Guinea Region. The Lower Guinea Region has a high diversity of mycoheterotrophic plants, and many mycoheterotrophs in Burmanniaceae, Orchidaceae, Thismiaceae, and Triuridaceae are endemic to this region (Cheek and Williams 1999; Franke 2007). These endemics include *Afrothismia* spp., *Oxygyne triandra* (Thismiaceae), *Auxopus letouzeyi*, *Gastrodia africana* (Orchidaceae), *Kupea martinetugei*, and *Sciaphila ledermannii* (Triuridaceae). The latter species is also found on the islands in the Gulf of Guinea along with *Epipogium roseum* (Orchidaceae) (Daniel 2010). A smaller western block of West African rainforest is found in the Upper Guinea Region, from Sierra Leone to Ghana (Plana 2004). This area is generally less diverse in mycoheterotrophs compared to the Lower Guinea Region. *Afrothismia* is remarkably absent from this region. Other mycoheterotrophs are generally shared with the Lower Guinea Region, including *Campylosiphon congestus*, *Gymnosiphon longistylus*, *G. bekensis* (Burmanniaceae), *Auxopus macranthus*, *A. kamerunensis*, *Epipogium roseum* (Orchidaceae), *Exochaenium oliganthum*, and *Voyria primuloides* (Gentianaceae). The species *Sciaphila africana* and *Gymnosiphon samoritoureanus* are endemic to the Upper Guinea Region.

3.3.3.2 East Africa

Rainforests in East Africa consist of small patches of forest on the East African mountains between about 1,200 and 2,500 m altitude. These "islands" of rainforests are generally surrounded by dry woodland (Primack and Corlett 2005). Although the total area of these rainforest patches is small, those on older mountains have potentially provided a stable habitat for a long period of time and have been isolated from the forests of West and Central Africa for millions of years. As a result, many of the plant and animal species are endemic to these mountains. Recent discoveries have stressed the importance of the coastal East African forests for mycoheterotrophic plant diversity (Cheek 2006). *Gymnosiphon usambaricus* (Burmanniaceae), *Afrothismia baerae*, *A. mhoroana* (Thismiaceae), *Seychellaria africana*, *Kihansia jonii*, and *Kupea lovettii* (Triuridaceae) are endemic to the region, and most of these species are known from very few restricted localities.

Kihansia jonii and *Kupea lovettii*, for example, have only been recorded at the Kihansi River Gorge in Tanzania (Cheek 2003b).

3.3.3.3 Madagascar

In Madagascar, mycoheterotrophs are restricted to the humid lowland forests along the eastern coastal strip and the subhumid forests above 600–800 m elevation. The flora of Madagascar largely evolved in isolation, and this explains the high level of endemism of mycoheterotrophic plants. Indeed, all fully mycoheterotrophic plant species that occur in Madagascar (and the adjacent Comores) are endemics. The Malagasy flora also includes one endemic mycoheterotrophic plant genus: *Geosiris* (Iridaceae), with one species known from Madagascar and the Comores and another species endemic to the island of Mayotte (Goldblatt and Manning 2010). Other mycoheterotrophic species are *Gymnosiphon danguyanus*, *G. marieae* (Burmanniaceae), *Seychellaria madagascariensis* (Triuridaceae), *Auxopus madagascariensis, Galeola humblotii*, and *Gastrodia madagascariensis* (Orchidaceae), although the generic identity of the latter has been disputed and this species may belong to *Didymoplexis* (Cribb et al. 2010). Another *Gastrodia* species (*G. similis*) is known from a single collection on La Réunion. With only eight species, Madagascar is rather poor in mycoheterotrophic plants, and most genera that occur on the east coast of Africa are absent from Madagascar (e.g., *Afrothismia, Kupea, Kihansia*). Another notable absence is that of the orchid *Epipogium roseum*, which has a widespread distribution in continental Africa, India and Sri Lanka, Southeast Asia, Australasia, and the Pacific Islands. In contrast, the genus *Seychellaria*, recorded from Madagascar, the Seychelles, and Tanzania, provides an interesting biogeographic link between Madagascar and continental Africa, perhaps indicating its potential for dispersal between Madagascar and the mainland. On the other hand, *Galeola*, for which one species is known from Madagascar and the Comores, is absent from continental Africa but is widespread in Southeast Asia. A biogeographic link between Madagascar and India/Southeast Asia is observed in several groups of organisms, including lizards (Macey et al. 2000), birds (Cooper et al. 2001), and amphibians (Bossuyt and Milinkovitch 2001), and is indicative of the geological past in which Madagascar separated from the India-Seychelles landmass several tens of millions of years after its separation from Africa (Storey et al. 1995).

3.3.4 India and Sri Lanka

India has a long but very narrow strip of rainforest running parallel to the west coast along the crest of the Western Ghats (Fig. 3.4). The southwest of the island of Sri Lanka supports rainforest as well. Southern India and Sri Lanka are separated by the relatively shallow Palk Strait, which allowed for intensive biotic interchange during the Pleistocene ice ages. As a result, Sri Lanka and the Western Ghats have very similar biota, although for some groups of fauna, differences are somewhat more pronounced than expected (e.g., Bossuyt et al. 2004). The rainforest of India and Sri Lanka has a vivid geological history and probably underwent more changes than any other rainforest region. The Indian Plate began to drift away from eastern Gondwana in the Early Cretaceous. While the Indian Plate drifted close to the African plate, it received diverse Late Cretaceous elements of the African tropical flora and lost temperate elements of the eastern Gondwanan gymnosperm flora (Morley 2000). The dramatic latitudinal and climatic changes that affected the Indian Plate during the Late Cretaceous and Tertiary, as it traveled from Gondwana and then collided with Asia, caused massive extinctions in its biota. Today, the Western Ghats–Sri Lanka rainforest covers only a very small area and is considered as a biodiversity hotspot under severe threat: 45.6% of the plant species that occur here are endemics (Myers et al. 2000).

This region contains ca. 17 fully mycoheterotrophic species within families Burmanniaceae, Orchidaceae, Thismiaceae, and Triuridaceae. Almost half of the mycoheterotrophic species from the Western Ghats–Sri Lanka hotspot are found nowhere else, while the

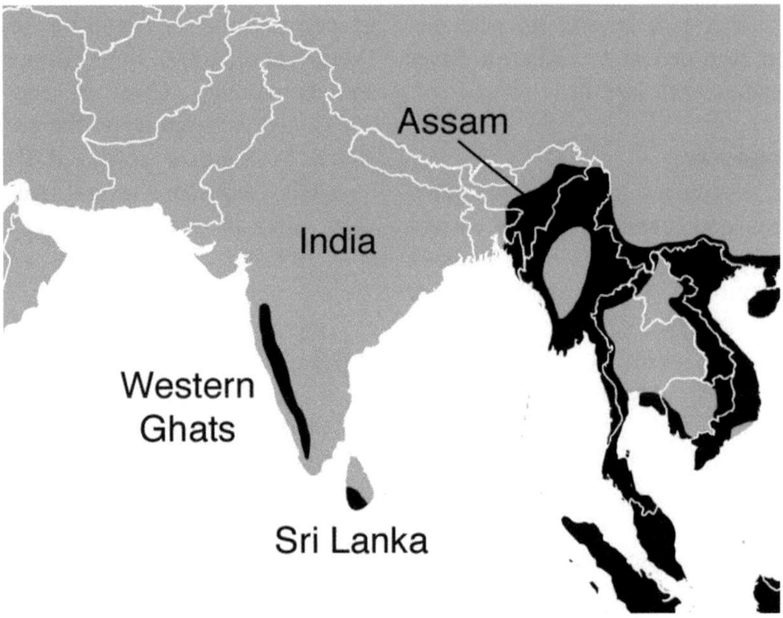

Fig. 3.4 Distribution of tropical rainforest (*black*) in India and Sri Lanka

widespread genera *Galeola* (Orchidaceae) and *Gymnosiphon* (Burmanniaceae) are remarkably absent from the southern India flora. In addition, apart from the single species in the endemic genus *Hyalisma* (Triuridaceae) and the widespread species *Aphyllorchis montana*, *Epipogium roseum* (Orchidaceae), and *Burmannia championii* (Burmanniaceae) (Mesta et al. 2011), the Western Ghats and Sri Lanka have no mycoheterotrophic species in common, indicating a high degree of local endemism in this region. For example, the enigmatic *Haplothismia exannulata* (Thismiaceae) is endemic to the Western Ghats and is only known from two restricted populations (Sasidharan and Sujanapal 2000). The rainforest of South India is also a habitat for three mycoheterotrophic *Burmannia* species: *B. candelabrum*, *B. indica*, and *B. stricta*. However, the latter two species are endemic to Southwest India. The Western Ghats also contains at least five mycoheterotrophic orchid species, of which two are endemic to the region (*Didymoplexis seidenfadenii* and *Gastrodia silentvalleyana*). In Sri Lanka, many Southeast Asian mycoheterotrophic species reach their westernmost distributions (e.g., *Sciaphila secundiflora*, *S. tenella*, *Cyrtosia javanica*, and *Eulophia zollingeri*). The island contains at least two endemic mycoheterotrophic species: *Thismia gardneriana* (Thismiaceae) and *Gastrodia zeylanica* (Orchidaceae).

The flora of India contains many other mycoheterotrophic species than those mentioned above, but all of these occur in the far northeastern state of Assam (e.g., *Burmannia nepalensis*, *Sciaphila khasiana*, *Chamaegastrodia shikokiana*, *Cymbidium macrorhizon*, *Eulophia zollingeri*, *Erythrorchis altissima*, *Galeola falconeri*, *Odontochilus asraoa*, *Aphyllorchis* spp., *Yoania* spp.). Biogeographically, these species have more in common with South China and Southeast Asia than with the Western Ghats–Sri Lanka rainforests.

3.3.5 Southeast Asia

Southeast Asia has one of the most complex geological histories in the world. The region has developed by the interaction of the Pacific, India–Australia, Eurasia, and several smaller tectonic plates, and as a result of this complicated past, several distinct centers of biological diversity can be identified within a small geographic range (Hall 1998; Sodhi et al. 2004). Until recently, rainforest covered most of the Malay Peninsula,

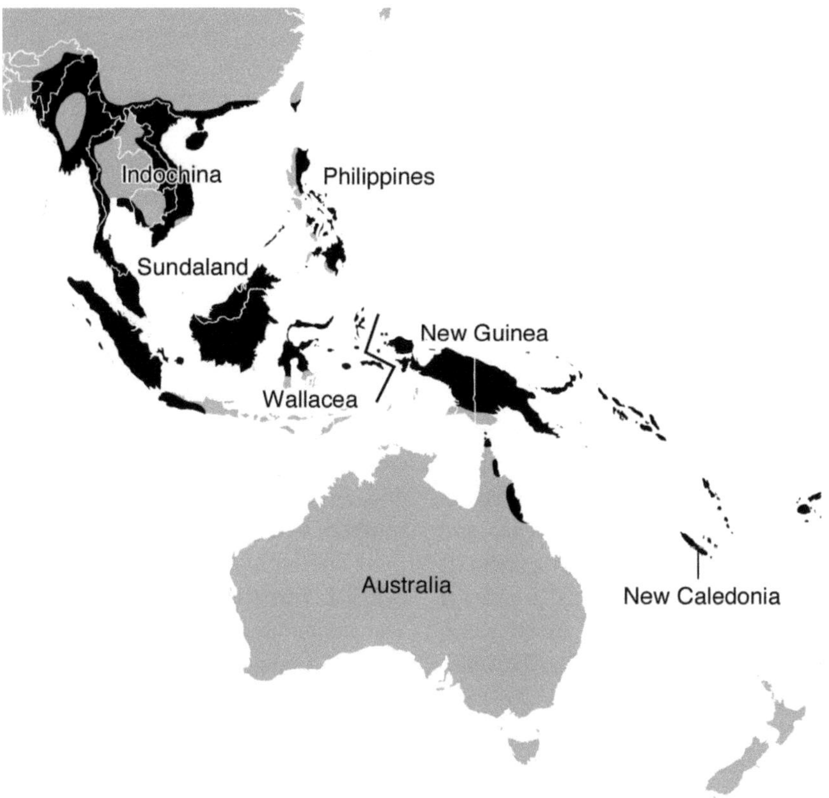

Fig. 3.5 Distribution of tropical rainforest (*black*) in tropical Asia and tropical Australasia. The barrier between Southeast Asia and Australasia is shown with a black line

Borneo, Sumatra, and Java (Fig. 3.5). This region is often referred to as "Sundaland," after the surrounding Sunda continental shelf. Lowland rainforest in this area has suffered considerably from human activity, particularly on Java. North of Sundaland rainforest extends into mainland Asia, including most of Cambodia, Laos, and Vietnam, and much of Thailand and Myanmar. However, due to the rain shadow of several long north–south mountain chains, most of the interior of Thailand and Myanmar is too dry to support rainforest. Farther north, rainforest covers South China and the southern tip of Taiwan, although most has now been cleared. The rainforests of Southeast Asia extend westward through Myanmar into Northeast India. Eastward rainforest covers Sulawesi and many of the smaller islands between Borneo and New Guinea ("Wallacea") (Fig. 3.5). As a whole, the tropical forests of Southeast Asia include four biodiversity hotspots (Indo-Burma, Sundaland, the Philippines, Wallacea) containing high concentrations of endemic species and undergoing immense and rapid habitat loss (Myers et al. 2000; Sodhi et al. 2004).

Southeast Asian rainforests can be characterized as "dipterocarp forests," with a canopy dominated by large trees of the family Dipterocarpaceae. Dipterocarps are particularly important forest elements of the Malay Peninsula, Borneo, Java, Sumatra, and the wetter parts of the Philippines. The dipterocarp forests of Southeast Asia are extremely rich in mycoheterotrophic plants. Except for Iridaceae mycoheterotrophs, all angiosperm families with fully mycoheterotrophic species are represented in Southeast Asia. Orchids are particularly species rich, with over 100 fully mycoheterotrophic species. The region is a center of diversity for mycoheterotrophic orchids of the genera *Aphyllorchis*, *Gastrodia*, *Didymoplexis*, *Didymoplexiella*, *Cyrtosia*, *Lecanorchis*, and *Galeola*. It is the only part of the world where fully mycoheterotrophic orchids of the genera

Cystorchis, *Platanthera*, *Silvorchis*, *Kalimantanorchis*, and *Tropidia* can be found. Other mycoheterotrophic orchids from Southeast Asia belong to *Erytrorchis*, *Pseudovanilla*, *Chamaegastrodia*, *Odontochilus*, *Cephalanthera*, *Epipogium*, *Eulophia*, and *Stereosandra*. The poorly known ericaceous genus *Cheilotheca*, which contains two mycoheterotrophic species, is also endemic to Southeast Asia: *Cheilotheca* species occur in pine and oak forest at high elevations, and therefore, their distribution does not overlap with other tropical mycoheterotrophs.

The region is also a major center of diversity for arbuscular mycorrhizal mycoheterotrophs in Burmanniaceae, *Petrosavia* (Petrosaviaceae), *Epirixanthes* (Polygalaceae), Thismiaceae, and Triuridaceae. The distribution of many genera of mycoheterotrophic plants from tropical Southeast Asia extends into the subtropics, mostly into subtropical China and Japan (e.g., *Thismia*, *Sciaphila*, *Gymnosiphon*, *Andruris*, *Exacum*, *Petrosavia*, *Cyrtosia*, *Erythrorchis*, *Lecanorchis*, *Aphyllorchis*, *Gastrodia*, *Odontochilus*, and *Didymoplexiella*). But many mycoheterotrophs from Southeast Asia are also found in Australasian rainforest, and a few species extend their distribution into Oceania. Particularly widespread species include *Aphyllorchis montana*, *Didymoplexis pallens*, *Stereosandra javanica*, *Epipogium roseum*, and *Eulophia zollingeri*.

3.3.5.1 Indochina

The rainforests of mainland Southeast Asia ("Indochina" or "Indo-Burma") are located in Cambodia, Laos, Vietnam, Myanmar, Singapore, Thailand, and South China (southeastern Tibet, southern Yunnan, Guangxi, southwestern Guangdong, Taiwan, and Hainan), although at higher latitudes, these rainforests are gradually replaced by monsoon and subtropical forests (Zhu 1997; Morley 2000). The area is rich in fully mycoheterotrophic plants, and the mycoheterotrophic flora consists of a mix of genera that are widespread in tropical Southeast Asia with genera that have a more temperate distribution. The majority of fully mycoheterotrophic species found here are orchids, and the region is an important center of diversity for the genera *Cyrtosia*, *Galeola*, *Lecanorchis*, *Chamaegastrodia*, and *Odontochilus*. Several species in these genera are endemic to the region. The Old World species of mycoheterotrophic *Cephalanthera* orchids are restricted to mainland Southeast Asia. Apart from Orchidaceae, mainland Southeast Asia contains full mycoheterotrophs that belong to *Burmannia*, *Gymnosiphon* (Burmanniaceae), *Petrosavia* (Petrosaviaceae), *Thismia* (Thismiaceae), *Sciaphila* (Triuridaceae), *Exacum* (Gentianaceae), and *Epirixanthes* (Polygalaceae). *Corsiopsis* (Corsiaceae) is endemic to the region; the only described species, *C. chinensis*, is known from a single collection from a subtropical forest in Guangdong, southern China (Zhang et al. 1999).

3.3.5.2 Sundaland

The major areas of rainforest of the Sundaland region are located in Borneo, Java, Sumatra, and Peninsular Malaysia. Sixty percent of the 25,000 plant species recorded from this region are endemic, but with natural habitat disappearing at an alarming rate, Sundaland is one of the most threatened rainforest areas of the world (Myers et al. 2000). This area is home to the most species-rich plant communities on Earth and is likewise one of the richest regions for fully mycoheterotrophic plants. The diversity of mycoheterotrophic orchids is particularly remarkable, with a high number of species of *Aphyllorchis* and *Gastrodia*. It is also a center of diversity for *Didymoplexiella* and *Didymoplexis*, and the mycoheterotrophic orchid genera *Kalimantanorchis*, *Tropidia*, and *Silvorchis* are endemic to Sundaland. Orchid genera *Cyrtosia*, *Erytrorchis*, *Galeola*, *Pseudovanilla*, *Cystorchis*, *Epipogium*, *Eulophia*, and *Stereosandra* are represented in this region by mycoheterotrophic species as well.

Sundaland rainforests are also rich in arbuscular mycorrhizal mycoheterotrophs. These belong to *Burmannia* and *Gymnosiphon* (Burmanniaceae), *Andruris*, *Sciaphila* (Triuridaceae), *Epirixanthes* (Polygalaceae), *Exacum* (Gentianaceae), *Petrosavia* (Petrosaviaceae), and *Thismia* (Thismiaceae). Several of these species are endemic to Sundaland, or even to a particular island within the Sundaland region. Of the three major islands, Borneo has by far the highest diversity of fully mycoheterotrophic species. The flora of Borneo contains at least 62 mycoheterotrophic species, which is more than the number of species known from all of tropical Africa. Endemism on Borneo is high and particularly pronounced in Orchidaceae and Thismiaceae: at least 16 species of mycoheterotrophic orchids are endemic to Borneo (*Aphyllorchis kemulensis*, *A. siantanensis*, *A. spiculaea*, *Cystorchis saprophytica*, *Didymoplexis latilabris*, *Didymoplexiella borneensis*, *D. cinnabarina*, *D. forcipata*, *D. kinabaluensis*, *Gastrodia grandilabris*, *G. sahabensis*, *G. spathulatha*, *Platanthera saprophytica*, *Kalimantanorchis nagamasui*, *Tropidia saprophytica*, and *T. connata*). All seven *Thismia* species recorded on the island are also endemic. The island of Java is also rich in mycoheterotrophs. The orchid genus *Silvorchis* is found nowhere else, and several *Didymoplexis* and *Gastrodia* species are endemic to Java as well.

3.3.5.3 Philippines

Until a few centuries ago, at least 95% of the Philippines was covered by tropical rainforest (Heaney and Regalado 1998). Today, only 3% of that rainforest remains, mainly in montane areas, and the Philippines has the questionable honor of being the second "hottest" diversity hotspot on Earth (Myers et al. 2000). The flora of the Philippines has fewer plant species than Sundaland and Indochina (which cover larger areas) but includes ca. 24 species of fully mycoheterotrophic angiosperms. Six species of *Sciaphila* (Triuridaceae) are native to the Philippines, but only a single mycoheterotrophic *Burmannia* species (*B. nepalensis*) and remarkably no *Gymnosiphon* species have been recorded for the Burmanniaceae. *Epirixanthes* (Polygalaceae), found throughout Australasia and Sundaland, is also absent. Mycoheterotrophic orchids from the Philippines are of the genera *Cyrtosia*, *Erythrorchis*, *Galeola*, *Lecanorchis*, *Pseudovanilla*, *Cystorchis*, *Aphyllorchis*, *Didymoplexis*, *Epipogium*, *Eulophia*, *Gastrodia*, and *Stereosandra*. Endemic mycoheterotrophs are *Aphyllorchis halconensis*, *Didymoplexis philippinensis*, *Pseudovanilla philippinensis* (Orchidaceae), *Exacum loheri* (Gentianaceae), and *Thismia gigantea* (Thismiaceae).

3.3.5.4 Wallacea

Wallacea includes the island of Sulawesi, the Maluku Islands, and the Lesser Sunda Islands. The area is one of the most geologically complex regions in the world. The islands originated from land fragments that rifted from Gondwana at different geological time periods, and they were never physically connected to Southeast Asia ("Wallace's Line") (Audley-Charles 1983). Due to their prolonged isolation, each island evolved highly endemic faunas, but the proximity of Sundaland caused a large influx of tropical Southeast Asian plants, which started during the mid-Miocene. The flora of Wallacea contains an estimated 10,000 species, of which ca. 1,500 are endemic (Myers et al. 2000). Ca. 21 species of mycoheterotrophic plants are known from Wallacea. Most of them have distribution ranges that include other parts of Southeast Asia, Australasia, or both: *Burmannia lutescens*, *B. championii*, *Gymnosiphon aphyllus*, *G. papuanus* (Burmanniaceae), *Sciaphila arfakiana*, *S. corniculata*, *S. densiflora*, *S. tenella* (Triuridaceae), *Petrosavia stellaris* (Petrosaviaceae), *Cyrtosia javanica*, *Cystorchis aphylla*, *Didymoplexis micradenia*, *Epipogium roseum*, and *Eulophia zollingeri* (Orchidaceae). *Gymnosiphon minahassae*, *Pseudovanilla ternatensis*, *Galeola nudifolia*, *Aphyllorchis acuminata*, *A. angustipetala*, *A. gracilis*, and *Gastrodia celebica* are endemic to the region. Remarkably, in Wallacea there are no records of *Thismia*, a genus that is otherwise widespread in Southeast Asia and Australasia.

3.3.6 Australasia

Most of the island of New Guinea is covered by what is now the third largest and most intact rainforest area of the world (Primack and Corlett 2005). Australia also supports a small area of rainforest in the northeast, along the coast between Cooktown and Townsville (Fig. 3.5). The geological history of New Guinea and Australia differs from that of Southeast Asia: New Guinea and

Australia are located on the Australian plate, while the rest of the Southeast Asian tropical forest is located on the Asian plate. The mid-Miocene collision of the Australian and Asian plate caused a large influx of Asian rainforest plants into New Guinea. The absence of a dry land connection prevented a similar influx of vertebrates. As a result, the composition of the flora of New Guinea is relatively similar to that of Southeast Asian rainforests, while their vertebrate faunas are very different. The rainforests of Australia were much less affected by the post-Miocene influx of Asian plants. The flora of Australian rainforests contains more Gondwanian and Australian components as well as more early diverging angiosperm families, such as Winteraceae, Eupomatiaceae, Monimiaceae, Lauraceae, and Cunoniaceae (Primack and Corlett 2005). Rainforests in New Guinea do contain Dipterocarpaceae trees, but they are much less dominant than in Southeast Asian rainforests.

Australasian rainforests harbor a high diversity of mycoheterotrophic plants, although the total number of mycoheterotrophic species is considerably lower than the number of species recorded from Southeast Asia (Table 3.1). Australasian mycoheterotrophs are part of seven different flowering plant families: Burmanniaceae, Thismiaceae, Triuridaceae, Corsiaceae, Polygalaceae, Orchidaceae, and Gentianaceae. Ca. 30 species belong to Orchidaceae—which is lower than the number of mycoheterotrophic orchids from Southeast Asia, but still high compared to the number of species occurring in Africa and the Neotropics. Reflecting their general floristic overlap, Australasian rainforests share many genera and species of mycoheterotrophs from rainforests in Southeast Asia and Oceania. A notable exception is the genus *Corsia* (Corsiaceae), of which all 25 species are endemic to Australasia. Conversely, Southeast Asian mycoheterotrophic species of *Petrosavia* (Petrosaviaceae), *Cyrtosia*, *Erythrorchis*, *Cystorchis*, *Odontochilus*, *Platanthera*, *Cephalanthera*, *Didymoplexiella*, *Tropidia*, and *Yoania* (Orchidaceae) do not occur in Australasia.

For mycoheterotrophic Orchidaceae, the region is an important center of diversity for *Pseudovanilla*, *Lecanorchis*, and *Aphyllorchis*. The orchid genera *Galeola*, *Didymoplexis*, *Epipogium*, *Eulophia*, *Gastrodia*, and *Stereosandra* are also represented by fully mycoheterotrophic species. Arbuscular mycorrhizal mycoheterotrophs in Australasia belong to the genera *Burmannia*, *Gymnosiphon* (Burmanniaceae), *Thismia* (Thismiaceae), *Sciaphila*, *Andruris* (Triuridaceae), *Epirixanthes* (Polygalaceae), *Exacum* (Gentianaceae), and *Corsia* (Corsiaceae).

3.3.6.1 New Guinea

By far the largest part of Australasian rainforest is found on New Guinea, and this is where most Australasian mycoheterotrophs occur. The world's second largest island is home to arbuscular mycorrhizal species of *Burmannia*, *Epirixanthes*, *Exacum*, *Gymnosiphon*, *Andruris*, *Sciaphila*, and *Thismia*. The island is also the center of diversity of the little-known genus *Corsia* (Corsiaceae), where 23 of the 25 described species occur as endemics. Mycoheterotrophic orchids belong to the genera *Galeola*, *Lecanorchis*, *Pseudovanilla*, *Aphyllorchis*, *Didymoplexis*, *Epipogium*, *Eulophia*, and *Gastrodia*. In addition to the 23 species of *Corsia*, island endemics include *Aphyllorchis elata*, *A. exilis*, *A. torricellensis*, *Didymoplexis torricellensis*, *Gastrodia crassisepala*, *G. papuana*, *Lecanorchis bicarinata*, *L. ciliolata*, *L. neglecta*, *Pseudovanilla gracilis*, *P. vanilloides* (Orchidaceae), *Burmannia micropetala*, *Thismia appendiculata*, *Andruris wariana*, *Sciaphila quadribullifera*, *S. papillosa*, *Gymnosiphon affinis*, *G. oliganthus*, and *G. pauciflorus*.

3.3.6.2 Australia

Tropical rainforest once covered Australia, but as a result of the continued northward drift of Australia, coupled with global climatic cooling, the climate of Australia became much drier in the Middle and Late Miocene and Late Pliocene, resulting in a withdrawal of tropical rainforests from all but the northeast coast region of Queensland. This tiny tip of tropical rainforest contains ca. 15 species of mycoheterotrophic angiosperms in the genera *Aphyllorchis*, *Andruris*,

Corsia, *Didymoplexis*, *Epipogium*, *Eulophia*, *Gastrodia*, *Pseudovanilla*, and *Thismia*. Endemic mycoheterotrophs are *Thismia yorkensis*, *Corsia dispar*, *Andruris australasica*, *Aphyllorchis anomala*, *A. queenslandica*, *Didymoplexis pachystomoides*, *Gastrodia crebrifolia*, *G. queenslandica*, and *G. urceolata*.

3.3.6.3 New Caledonia and Vanuatu

The Southwest Pacific Island group of New Caledonia is also part of Australasia, and the region is recognized as a global biodiversity "hotspot" (Myers et al. 2000), with a very high degree of plant endemism (Jaffré et al. 1998). New Caledonia formed part of the eastern margin of Gondwana, until it became separated by the Tasman Sea about 90 million years ago (Ma) (Wilford and Brown 1994). Thus, the flora of New Caledonia results from a Gondwanan origin and mainly evolved in isolation (Jaffré 1992), although volcanic islands between Australia and the New Caledonian region could have facilitated plant migration during the Neogene (Wilford and Brown 1994), and island chains are also implicated along the Norfolk and Reinga Ridges toward New Zealand during this time (Herzer et al. 1997). In contrast to the continental origin of New Caledonia, the Vanuatu archipelago is oceanic in origin, and its development results from tectonic events ranging from 11.2 to 2.0 Ma (Kroenke and Rodda 1984). The flora of New Caledonia includes relatively few mycoheterotrophic plants, and none of them are endemic to the region except for the sole putative mycoheterotrophic gymnosperm *Parasitaxus usta*. New Caledonian mycoheterotrophic flowering plants include three Triuridaceae species (*Sciaphila densiflora*, *S. corallophyton*, and *S. corniculata*) and three orchids (*Didymoplexis micradenia*, *Dipodium squamatum*, and *Epipogium roseum*). Burmanniaceae are absent from the flora of New Caledonia, but the mycoheterotroph *Burmannia lutescens* occurs on the neighboring Vanuatu archipelago (Yohan Pillon, pers. comm.), which represents the easternmost distribution record for the family. Other Vanuatu mycoheterotrophs are *Sciaphila arfakiana*, *S. aneitensis* (Triuridaceae), *Didymoplexis micradenia*, *D. pallens*, *Dipodium squamatum*, *Epipogium roseum*, and *Gastrodia cunninghamii* (Orchidaceae) (van de Meerendonk 1984; Govaerts et al. 2011).

3.3.7 Pacific Islands

The islands in the Pacific Ocean are composed mostly of volcanic emergents and coral atolls that arose from the sea in geologically recent times, many of them in the Pleistocene. They were created either by hotspot volcanism or as island arcs pushed upward by the collision and subduction of tectonic plates. The islands range from tiny islets, sea stacks, and coral atolls to large mountainous regions containing complex ecosystems. Only these larger islands harbor tropical rainforests that are potential habitats for mycoheterotrophs. The flora of the islands is comprised entirely from long-distance dispersal events and is generally characterized by a low diversity but high endemism (Carlquist 1967). For mycoheterotrophic plants, only a few widespread species in Burmanniaceae, Triuridaceae, and Orchidaceae have been able to colonize some islands. Burmanniaceae are only present on the Caroline Islands, where *Burmannia ledermannii*, *Gymnosiphon aphyllus*, *G. papuanus*, and *G. okamotoi* occur. The latter is endemic to the islands. For Triuridaceae, *Sciaphila arfakiana* and *S. consimilis* have been recorded on Fiji (van de Meerendonk 1984). The latter species is widespread in Southeast Asia and is also found on the Caroline Islands, together with *S. corallophyton* and *S. multiflora*. Triuridaceae reach their easternmost distribution on Futuna, where the widespread *Sciaphila aneitensis* occurs. In Orchidaceae, the widespread species *Didymoplexis micradenia* occurs on the Marianas, Niue, Samoa, Tonga, Wallis–Futuna, and Caroline Islands, *D. pallens* is known from Niue, *Stereosandra javanica* from Samoa, and *Epipogium roseum* from Fiji. None of these species are endemic to the region. In contrast, *Pseudovanilla ponapensis* is only known from the Caroline Islands, and *Pseudovanilla anomala* is endemic to Fiji, but it is unclear whether these species are (partial) mycoheterotrophs. There are no records of mycoheterotrophic plants from French Polynesia and Hawaii. Whether this is the result of their isolated position or the absence of certain habitat characteristics necessary for the establishment of mycoheterotrophic plants remains unknown.

3.3.8 Temperate Regions

Similarly to the tropics, mycoheterotrophs in temperate regions are mainly restricted to forests. In temperate regions, forest coverage mainly consists of coniferous, deciduous, or mixed forests, and all these forest types provide habitats for mycoheterotrophic plants. Temperate forests predominantly occur in the northern hemisphere, although the southern hemisphere has small pockets of temperate forest as well. Temperate forests of the northern hemisphere are located in eastern North America, western North America, Europe (including southwestern Asia), and eastern Asia (Fig. 3.6). Many flora and fauna elements have a disjunct distribution across these forest regions, which is illustrated by the similarities of biotas between eastern North America and temperate Asia (Wen 1999; Donoghue et al. 2001). This current disjunct distribution has been attributed to the historical existence of a widespread evergreen vegetation type ("boreotropical" forest) earlier during the Tertiary, when warm and wet climatic conditions prevailed over northern latitudes (Milne and Abbott 2002). In response to climatic cooling from the start of the Oligocene, deciduous elements moved southward forming a northern hemisphere flora (the "mixed mesophytic forest") that contained a mix of deciduous and evergreen trees with increasing numbers of associated understory herbs (Tiffney 1985; Milne and Abbott 2002). During the Late Miocene to Pliocene, global temperatures dropped and the flora retreated, leading to increased extinction particularly in western North America. Further depauperation of the flora occurred during the Quaternary glaciations, which had a strong impact on the European flora (Milne and Abbott 2002; Donoghue and Smith 2004).

In the southern hemisphere, temperate forests are restricted to small patches in southern South America, South Africa, Australia, and New Zealand. These forests are thought to have originated from southern temperate Gondwanan floristic elements. The sequential breakup of Gondwana started 165 Ma ago and resulted in a successive division of the ancestral biota. However, there has been a documented exchange of plants between the major landmasses following their breakup; for example, trans-Tasman dispersal between Australia and New Zealand appears to be quite common (Hill 2004; Sanmartin and Ronquist 2004).

The diversity of mycoheterotrophic plants in temperate regions is lower than in the tropics, with temperate mycoheterotrophs mainly belonging to Ericaceae and Orchidaceae. In contrast to the dominance of arbuscular mycorrhizal mycoheterotrophs in the tropics, mycoheterotrophic interactions in temperate regions generally rely on ectomycorrhizal fungi. Nevertheless, many predominantly tropical arbuscular mycorrhizal mycoheterotrophic families extend their distribution into subtropical and even temperate regions, particularly in Asia (e.g., *Thismia*, *Petrosavia*, *Sciaphila*).

3.3.9 North America

North America contains a large variety of different forest types, providing habitat for a very distinct group of mycoheterotrophs. Of a total of 30 fully mycoheterotrophic species in North America, only the two widespread species *Hypopitys monotropa* and *Monotropa uniflora* (Ericaceae) occur outside North America, indicating a high level of continental endemism. The distribution of the partial mycoheterotroph *Corallorhiza trifida* (Orchidaceae) extends only into temperate Eurasia. North American forests are particularly rich in ericaceous mycoheterotrophs: with the exception of *Cheilotheca* and *Monotropastrum*, all genera of ericaceous mycoheterotrophs occur here. Many Ericaceae genera are endemic to North America (*Monotropsis*, *Pterospora*), most occurring only in western North America (*Pityopus*, *Allotropa*, *Hemitomes*,

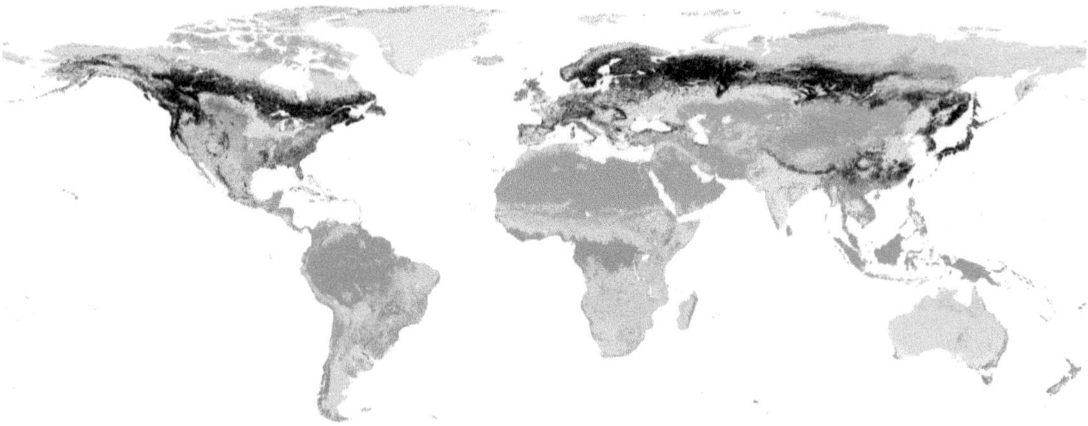

Fig. 3.6 Distribution of coniferous and mixed forest (*black*) in the temperate parts of the northern hemisphere. Map based on information obtained from NASA Earth Observations (2011)

Pleuricospora, *Sarcodes*). The only fully mycoheterotrophic species in *Pyrola* (Ericaceae), *P. aphylla*, is also endemic to North America.

North American mycoheterotrophic orchids include *Cephalanthera austiniae*, all *Corallorhiza* species, all *Hexalectris* species, and the partial mycoheterotroph *Liparis liliifolia*. As far as is known, these mycoheterotrophs are associated with ectomycorrhizal fungi. In contrast, the North American partially mycoheterotrophic Gentianaceae species of *Bartonia* and *Obolaria* are associated with arbuscular mycorrhizal fungi (Cameron and Bolin 2010). The full mycoheterotroph *Thismia americana* (Thismiaceae), which has been recorded from a prairie near Chicago, was probably also associated with arbuscular mycorrhizal fungi (see further). This species is probably extinct.

3.3.10 Europe

The temperate forests of Europe are less diverse in plant species than those of North America. This difference has been explained by greater survival of plant species in North America during the Quaternary glaciation, although some evidence indicates that floristic differentiation between Europe and North America started in the late Tertiary (Davis 1983). The diversity of mycoheterotrophic plants in Europe is likewise significantly lower than that of North America and temperate Asia. The coniferous, mixed, and deciduous forests of Europe harbor only seven mycoheterotrophic species, the ericaceous *Hypopitys monotropa* and six orchid species: *Neottia nidus-avis*, *Limodorum abortivum*, *L. rubriflorum*, *L. trabutianum*, *Corallorhiza trifida*, and *Epipogium aphyllum*. The distribution range of some of these species extends into North Africa and temperate Asia.

3.3.11 Temperate Asia

In temperate Asia, mycoheterotrophic flowering plants occur in coniferous, mixed, and deciduous forests. Toward the south, there is a transition from temperate through subtropical forests into the tropical rainforest of Indochina, allowing for influx of tropical mycoheterotrophic groups. Typical temperate elements are the genera *Hypopitys* and *Monotropa* (Ericaceae), of which both monotypic species are present in temperate Asia. Other ericaceous mycoheterotrophs *Monotropastrum humile* and *M. sciaphilum* occur in the subtropical zones of Asia. There is a relatively large diversity of fully mycoheterotrophic orchids in temperate and subtropical Asia. With about 13 species, *Neottia* is particularly species rich. Five species of *Aphyllorchis*, a genus with a center of diversity in tropical Asia, are found in South and Central China and Tibet. All tree species of *Epipogium* are found in temperate and subtropical Asia as well. *Yoania japonica* and *Y. prainii* are also present in the subtropical zone of Asia, as well as *Risleya atropurpurea*. In addition, the distribution of a few *Burmannia* (Burmanniaceae) species extends into Central China, Assam, and Nepal (e.g., *B. nepalensis*, *B. wallichii*, *B. itoana*, *B. cryptopetala*, *B. chinensis*).

3.3.12 Japan

The flora of Japan includes a large number of mycoheterotrophic plants. The wet subtropical evergreen forest in southern Japan is a habitat for many species of arbuscular mycoheterotrophs that belong to genera with a mainly tropical distribution: *Burmannia*, *Thismia*, *Petrosavia*, *Andruris*, and *Sciaphila*. Three species of *Oxygyne* (Thismiaceae) have been discovered in these forests as well, which is remarkable since the only other species of the genus is recorded from Cameroon. With ca. 25 species, the Japanese flora is also rich in fully mycoheterotrophic Orchidaceae. Genera with a mainly tropical distribution, as well as genera from temperate Asia, are represented in Japan. Japanese mycoheterotrophic orchids include *Cyrtosia septentrionalis*, *Erythrorchis altissima*, *Lecanorchis* spp., *Chamaegastrodia shikokiana*, *Odontochilus poilanei*, *Aphyllorchis montana*, *Cymbidium macrorhizon*, *Didymoplexiella siamensis*, *Epipogium japonicum*, *E. roseum*, *Eulophia zollingeri*, *Gastrodia* spp., *Neottia* spp., *Stereosandra javanica*, and *Yoania* spp. Lastly, three ericaceous mycoheterotrophs *Hypopitys monotropa*, *Monotropa uniflora*, and *Monotropastrum humile* occur in Japan as well.

3.3.13 South Chile and Argentina

Temperate forests in southern South America are located on the Pacific coast of southern Chile, on the west-facing slopes of the southern Chilean coast range, and the Andes Mountains in both Chile and Argentina down to the southern tip of South America (Patagonia). These wet *Nothofagus* forests are home to *Arachnitis uniflora* (Corsiaceae), an arbuscular mycorrhizal mycoheterotroph that also occurs in tropical Bolivia and on the Falkland Islands (Ibisch et al. 1996).

3.3.14 South Africa

In South Africa, forests have a patchy distribution and occur in frost-free areas with more than 725 mm rainfall during the wet season. Indigenous forest is the smallest biome in South Africa (Eeley et al. 1999). Only two mycoheterotrophic orchid species have been recorded from the South African flora. *Didymoplexis verrucosa* is endemic to KwaZulu-Natal. *Gastrodia sesamoides*, which occurs natively in South and East Australia, was introduced to South Africa and is naturalized in the Cape Province near Kirstenbosch (Linder and Kurzweil 1999; Linder et al. 2005; Cribb et al. 2010).

3.3.15 Temperate Australia, Tasmania, and New Zealand

The forests of temperate Australia and Tasmania comprise ca. eight species of mycoheterotrophic orchids, and all of them are endemic to this region. *Cryptostylis hunteriana*, *Burnettia cuneata*, and *Dipodium roseum* are endemic to southeast Australia, and the latter two also occur in Tasmania. *Erythrorchis cassythoides* and *Dipodium variegatum* are endemic to East Australia, and *Gastrodia lacista* is endemic to western Australia. Temperate Australia is also the only place in the world where the enigmatic underground orchid, *Rhizanthella*, is found. Three species of *Rhizanthella* are known: *R. gardneri* from southwest Australia and *R. slateri* and *R. omissa* from southeast Australia. In southwest Australia, *Rhizanthella* is found growing in much drier habitats than the eastern Australian species. The temperate rainforests and wet sclerophyll forests in New South Wales, Victoria, and

Tasmania are also home to the arbuscular mycorrhizal mycoheterotroph *Thismia rodwayi* (Thismiaceae). Another *Thismia* species, *T. clavarioides*, is only known from Morton National Park in New South Wales.

In New Zealand, mycoheterotrophs only occur on the North Island, mostly in coniferous podocarp forests and broadleaf evergreen forests. Apart from *Thismia rodwayi* (Thismiaceae) and *Danhatchia australis* (Orchidaceae), which also occurs in Australia, all New Zealand mycoheterotrophs are endemic and belong to Orchidaceae: *Corybas cryptanthus*, *Gastrodia cunninghamii*, and *G. minor* (Moore and Edgar 1970).

3.4 Biogeographic Patterns

3.4.1 General Patterns of Diversity

Most fully mycoheterotrophic angiosperms occur in the tropics, and—in terms of the number of species—Southeast Asia is the most important region. Southeast Asia contains more than twice the number of mycoheterotrophic species found in the Neotropics (Table 3.1). This difference is remarkable but is almost solely the result of the high number of mycoheterotrophic orchid species in Southeast Asia (Fig. 3.7). Indeed, the flora of the Neotropics includes only seven species of mycoheterotrophic orchids. With 17 species, the diversity of orchid species in Africa and Madagascar is only slightly higher. In tropical Asia, the mycoheterotrophic orchid diversity peaks, with 27 species in Australasia and ca. 100 species in Southeast Asia. Most of these species belong to species-poor genera, although a few genera, such as *Lecanorchis*, *Aphyllorchis*, and *Gastrodia*, are particularly species rich. Mycoheterotrophic orchids associate with saprotrophic or ectomycorrhizal fungi, while tropical tree species are generally associated with arbuscular mycorrhizal fungi (Smith and Read 2008). However, in Southeast Asia—and to a lesser extent also in Australasia—tropical forests are dominated by Dipterocarpaceae trees which are known to form associations with ectomycorrhizal fungi (Lee 1990). These mycorrhizal networks

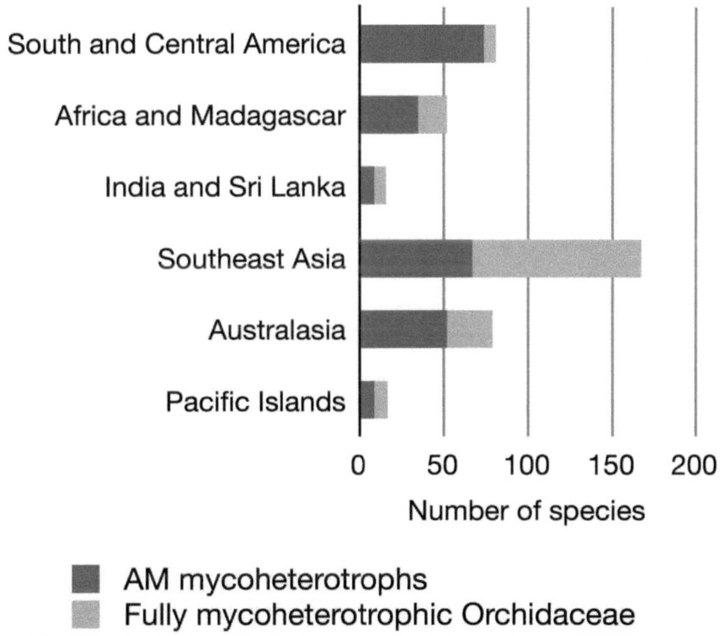

Fig. 3.7 Comparison of the species richness between major rainforest regions for arbuscular mycorrhizal (AM) mycoheterotrophs and mycoheterotrophic Orchidaceae, which associate with ectomycorrhizal and saprotrophic fungi

are potential "hosts" for ectomycorrhizal mycoheterotrophic orchids: indeed, mycoheterotrophic orchids from a dipterocarp forest in Thailand have been found to obtain carbon from dipterocarps through shared ectomycorrhizal fungi (Roy et al. 2009). Thus, it is possible that the dominance of ectomycorrhizal trees in Asian rainforests explains the high number of mycoheterotrophic orchids in these forests.

Differences between major rainforest regions in the number of arbuscular mycorrhizal mycoheterotrophs are less pronounced. The Neotropics have the most species, but in comparison with floras of Southeast Asia and Australasia, which have in general less plant species but also cover a smaller area, species diversity of AM mycoheterotrophs is more pronounced in tropical Asia (Fig. 3.7). However, the Neotropics are home to 19 genera of AM mycoheterotrophs, Africa and Madagascar have 12 genera of AM mycoheterotrophs, and Southeast Asia and Australasia only have nine and eight genera, respectively, or ten when considered as one area. Generic diversity in the Neotropics is particularly high in Burmanniaceae and Triuridaceae (Table 3.2). It has been hypothesized that Burmanniaceae originated in South America (or western Gondwana) and only reached the Old World during the Eocene (Merckx et al. 2008). The early diversification of Burmanniaceae in the Neotropics may explain the comparatively high number of genera in this region.

In the temperate zone, East Asia and Japan are the most species-rich regions for mycoheterotrophic plants. Their floras contain a mix of temperate and subtropical elements and are particularly rich in fully mycoheterotrophic orchids. Influx from the tropics probably also explains the high number of arbuscular mycorrhizal mycoheterotrophs in these regions (e.g., species of *Burmannia*, *Thismia*, *Sciaphila*, *Petrosavia*). The number of mycoheterotrophic species in North America is only slightly lower than in Asia and Japan. The diversity of Ericaceae in North America is remarkable, and the region is also relatively rich in mycoheterotrophic orchids (e.g., *Corallorhiza*, *Hexalectris*). The diversity of mycoheterotrophs in Europe is relatively low. Besides a few mycoheterotrophic orchids, European forests only contain the most widespread ericaceous mycoheterotroph *Hypopitys monotropa* and no arbuscular mycorrhizal species. Perhaps this is due to the lack of an historical connection with tropical floras, restricting all present-day mycoheterotrophic plants to be relictual elements of taxa with an ancient boreotropical distribution.

In the southern hemisphere, the much higher proportion of sea to land creates conditions favoring temperate rainforest on west-facing coasts (Beard 1990). The identity of mycoheterotrophic plants in these forests is very different from that of their northern hemisphere counterparts, and the species diversity is lower. Most mycoheterotrophs in the forests of the southern hemisphere belong to groups with mainly tropical distributions, indicating independent tropical origins of the southern hemisphere mycoheterotrophs. Southern hemisphere mycoheterotrophs include both species that are living on arbuscular mycorrhizal fungi (*Thismia*, *Arachnitis*) and orchids linked with ectomycorrhizal or saprotrophic fungi, but mycoheterotrophic Ericaceae are absent from the southern hemisphere.

3.4.2 Widespread Distributions

A few tropical mycoheterotrophic families are particularly widespread. Burmanniaceae and Triuridaceae can be found in all rainforest regions of the world (Table 3.2). Thismiaceae have a similar distribution but are absent from Madagascar and the Pacific Island region. Mycoheterotrophic Gentianaceae and Orchidaceae are also found in nearly all tropical regions, but their mycoheterotrophic species have evolved independently in different lineages. These lineages show little overlap in their distribution ranges, and it is likely that mycoheterotrophy has evolved on different continents independently. At the genus level, *Sciaphila* (Triuridaceae) has the most widespread distribution, and it is only absent from East Africa, Madagascar, and Southwest India. However, in these regions, Triuridaceae are represented by *Seychellaria* and *Hyalisma*, which may be

Table 3.2 List of families with fully mycoheterotrophic species present in each rainforest region

Central and South America	Africa and Madagascar	India and Sri Lanka	Southeast Asia	Australasia	Pacific Islands
Burmanniaceae (8/24)	Burmanniaceae (3/10)	Burmanniaceae (1/4)	Burmanniaceae (2/15)	Burmanniaceae (2/8)	Burmanniaceae (2/4)
Corsiaceae (1/1)	Gentianaceae (2/2)	Orchidaceae (4/7)	Corsiaceae (1/1)	Corsiaceae (1/26)	Orchidaceae (5/8)
Ericaceae (1/2)	Iridaceae (1/2)	Thismiaceae (2/2)	Ericaceae (1/2)	Gentianaceae (1/1)	Triuridaceae (2/5)
Gentianaceae (2/19)	Orchidaceae (7/17)	Triuridaceae (2/3)	Gentianaceae (1/4)	Orchidaceae (9/28)	
Orchidaceae (4/7)	Thismiaceae (2/13)		Orchidaceae (21/99)	Polygalaceae (1/2)	
Thismiaceae (2/14)	Triuridaceae (4/8)		Petrosaviaceae (1/3)	Thismiaceae (1/3)	
Triuridaceae (6/17)			Polygalaceae (1/5)	Triuridaceae (2/12)	
			Thismiaceae (1/24)		
			Triuridaceae (2/15)		

For each family the number of present genera/species is indicated between brackets congeneric with *Sciaphila*. *Gymnosiphon* (Burmanniaceae) is extremely widespread in the tropics as well, yet has not been recorded in India, Sri Lanka, and Australia. The distribution of *Burmannia* (Burmanniaceae) is also pantropical, although the genus is represented in East Africa and Madagascar by chlorophyllous species only. There is only one fully mycoheterotrophic *Burmannia* species in the Neotropics, which limits the South American distribution range of *Burmannia* mycoheterotrophs. In Orchidaceae, there are no tropical mycoheterotrophic genera that occur both in the Old and the New World, although a few Old World genera are extremely widespread (e.g., *Epipogium*, *Didymoplexis*, *Eulophia*).

In the temperate zones, Monotropoideae (Ericaceae) are extremely widespread, spanning the entire northern temperate region and extending into the tropics both in South America and Southeast Asia. *Hypopitys* is the most widespread genus of this group, and its distribution is almost identical to the entire Monotropoideae. *Corallorhiza* (Orchidaceae) is also widespread and occurs both in the temperate zone of the Old and the New World. In the Old World, however, the genus is represented only by *C. trifida*, which retains chlorophyll but obtains most of its carbon from fungi. Mycoheterotrophic *Neottia* species (Orchidaceae) also have a widespread distribution, ranging from Europe and North Africa into Asia and Japan.

3.4.3 Widespread Species

While many species of mycoheterotrophic plants have restricted distributions (see "Rarity"), some species have remarkable widespread distribution ranges. In the Neotropics, *Apteria aphylla* (Burmanniaceae) and *Voyria aphylla* (Gentianaceae) are particularly widespread, and their distribution almost completely overlaps with that of the entire neotropical rainforest biome. However, the Old World tropics are home to species with an even wider distribution especially given the challenges to dispersal across Old World tropical forests. The orchid species *Epipogium roseum* has the most widespread distribution of all tropical mycoheterotrophic plants. The species occurs throughout tropical Africa, India and Sri Lanka, South China, Japan, Southeast Asia, Australasia, and the Pacific Islands. Remarkably, *Epipogium roseum* has not been reported from Madagascar. Another widespread mycoheterotrophic orchid is *Didymoplexis pallens*, which occurs from Afghanistan to India, Southeast Asia, Australasia, and the Pacific Islands. The distribution of *Eulophia zollingeri* is only slightly more restricted and covers Sri Lanka, South China, Southeast Asia, Japan, and Australasia.

Hypopitys monotropa (Ericaceae) and the partial mycoheterotroph *Corallorhiza trifida* have boreotropical distributions that cover almost the entire northern temperate region. The distribution of *Monotropa uniflora* is slightly more restricted as this species is absent from Europe and most of temperate Asia. The orchid *Epipogium aphyllum* is extremely widespread in the temperate regions of Europe, Asia, and Japan but is not known from North America.

While widespread species in the Neotropics are members of lineages that are associated with arbuscular mycorrhizal fungi (e.g., *Apteria aphylla*, *Voyria aphylla*), widespread species in the Paleotropics are generally orchids that are associated with saprotrophic fungi (e.g., *Epipogium roseum*, *Eulophia zollingeri*). Widespread northern temperate species are all associated with ectomycorrhizal fungi (e.g., *Hypopitys monotropa*, *Monotropa uniflora*, *Corallorhiza trifida*, *Epipogium aphyllum*).

3.4.4 Disjunct Distribution Patterns

The distributions of a few groups of mycoheterotrophic plants show remarkable disjunctions (i.e., distributions that are geographically separated). On familial level, the distribution of Corsiaceae is intriguing: *Arachnitis* grows in Argentina, Chile, Bolivia, and the Falkland Islands, *Corsia* has its center of diversity in New Guinea and the Solomon Islands, and *Corsiopsis* was collected once in southern China (Fig. 3.8a). A Southern American–Australasian–Chinese disjunction is unusual in flowering plants, although similar patterns are observed at both familial

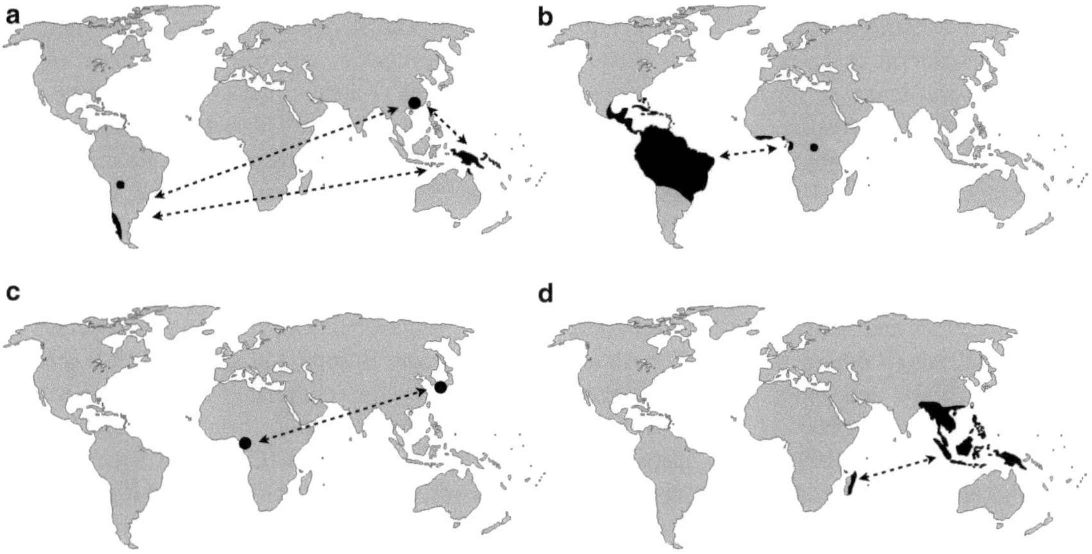

Fig. 3.8 Examples of disjunct distribution patterns in mycoheterotrophic plants. (**a**) The distribution of Corsiaceae. The *large dot* represents the single locality of *Corsiopsis*. Note the disjunct distribution of *Arachnitis* in South America. (**b**) The trans-Atlantic distribution of *Voyria* (Gentianaceae). (**c**) Species of *Oxygyne* (Burmanniaceae) have only been found in Cameroon and Japan. (**d**) *Galeola* (Orchidaceae) is widespread in Southeast Asia and Australasia, but one species is endemic to Madagascar and the Comores. See text for further discussion

(Stylidiaceae; Good 1974) and generic (*Coriaria*; Coriariaceae) level (van Balgooy 1966; Zhang et al. 1999). The South American–Australasian disjunction (*Arachnitis–Corsia*) in the family has been hypothesized to result from a Gondwanan tectonic link through Antarctica (Zhang et al. 1999). However, there is doubt that all genera of Corsiaceae are closely related (Neyland and Hennigan 2003; Petersen et al. 2006), and molecular phylogenetic studies are needed to resolve this issue.

At the generic level, disjunct distributions are quite common in full mycoheterotrophs. An interesting disjunct pattern is observed in *Voyria* (Gentianaceae), which comprises 18 species from tropical South and Central America and one species from tropical West Africa (Fig. 3.8b). It has been hypothesized that this pattern is either the result of a long-distance dispersal event or the relictual distribution of a previously continuous, boreotropical distribution across the North Atlantic (Albert and Struwe 1997). Due to the estimated origin of the Gentianaceae (ca. 50 Ma; Yuan et al. 2003), it is unlikely that this distribution is caused by ancient continental drift (vicariance). The small dustlike seeds of *Voyria* make a dispersal event plausible, although Albert and Struwe (1997) note that the seeds of the African *Voyria* species and its presumed neotropical sister species are not among the most highly modified for wind dispersal compared to other species in the genus. Also, a scenario of migration through Laurasia during the Eocene (e.g., Davis et al. 2002) cannot be excluded. Disjunct relationships between Africa and the Neotropics are not uncommon. Very similar South American–West African disjunctions are present in the plant families Rapateaceae and Bromeliaceae, which have been explained as the result of long-distance dispersal events rather than continental drift (Givnish et al. 2004; Renner 2004).

In Thismiaceae, some notable disjunct distribution patterns are found in *Thismia* and *Oxygyne*. *Thismia* is widespread in tropical South America, Southeast Asia, and Australasia but is absent from Africa. The occurrence of *Thismia* in the Chicago area, many thousands of kilometers from other *Thismia* populations, still remains unexplained (see below). The distribution of *Oxygyne*, with records only from Cameroon and Japan, also represents an extreme case of a geographic disjunction unique in flowering plants (Fig. 3.8c).

However, this distribution pattern has to be interpreted with caution as the relationships between the African and Japanese specimens of *Oxygyne* are in need of close investigation.

In Triuridaceae, *Seychellaria* occurs in Madagascar, the Seychelles, the Comores, and Eastern Africa (Tanzania). Vollesen (1982) suggested that this distribution pattern results from the breakup of Gondwana during the Cretaceous. But Madagascar and Africa have been separated since ca. 120 Ma (Rabinowitz et al. 1983; Ali and Aitchison 2008), and most African–Malagasy plant disjunctions have been explained by dispersal (Yoder and Nowak 2006). Thus, dispersal by wind or water (Renner 2004; Ali and Huber 2010), perhaps aided by the occurrence of land bridges (McCall 1997), may offer a more plausible explanation for the current distribution of *Seychellaria.* Other Triuridaceae genera have disjunct distribution ranges as well, although the pattern may be influenced by collecting bias: *Kupea* occurs both in Cameroon and Tanzania, but the genus was only recently discovered, and it is possible that populations exist connecting these two regions. The South American genus *Triuris* has a patchy distribution with large gaps between collection sites, but this also may reflect a sampling artifact rather than its actual distribution.

Geographic disjunctions are also observed in several mycoheterotrophic lineages within Orchidaceae. Species of *Galeola*, which are all putative full mycoheterotrophs, occur in tropical forests in Southeast Asia and New Guinea. However, one species, *G. humboltii*, is endemic to Madagascar and the Comoros (Fig. 3.8d). The achlorophyllous species of *Cephalanthera* occur as disjunct between North America and Southeast Asia, but the distribution of achlorophyllous *Cephalanthera* species overlaps with that of the achlorophyllous species and also covers intervening areas. It is not yet clear whether full mycoheterotrophy arose more than once in the genus. In this case, the apparent disjunct distribution may in fact be indicative of multiple origins of mycoheterotrophy in a particular lineage.

In Australia, *Rhizanthella gardneri* from western Australia is separated from its relatives *R. omissa* and *R. slateri* in southeastern Australia by 3,500 km of desert. Dixon (2003) suggests that *Rhizanthella* may have been present in the paratropical forests that once covered the Australian continent. These forests disappeared when the climate of Australia became much drier in the Middle and Late Miocene and Later Pliocene, possibly separating *Rhizanthella* populations to their current distribution.

Lastly, in liverworts, the mycoheterotrophic species *Aneura mirabilis* (Aneuraceae) occurs in northwestern Europe and Greenland. A species thought to be closely related to *Aneura mirabilis* has been reported from Costa Rica providing an interesting disjunction (Crum and Bruce 1996). However, it remains to be confirmed if the species from Costa Rica is the closest relative of *Aneura mirabilis* or if this disjunct distribution is also the result of convergent evolution of mycoheterotrophy in *Aneura.*

At the species level, it is mainly tropical mycoheterotrophic species that have disjunct distributions. In many cases (e.g., tropical species in Thismiaceae and Triuridaceae), these may represent sampling artifacts because the species are known from only very few collections. A few "real" disjunctions are notable. *Thismia rodwayi* (Thismiaceae) occurs on Tasmania, mainland Australia (Victoria, New South Wales, and Queensland), and the North Island of New Zealand. This distribution may be the result of long-distance seed dispersal but can also be interpreted as a relict of a widespread ancestor (e.g., Heads 2009). It is doubtful whether there still is genetic exchange between the populations on different landmasses. Another disjunction can be found in the distribution of *Arachnitis uniflora* (Corsiaceae), where Bolivian populations are separated from populations in southern Argentina, southern Chile, and the Falkland Islands by a 2,000–km-broad-belt of vegetation types that are clearly unsuitable habitat for *Arachnitis*: the Atacama desert, high mountain deserts and grasslands of the Andes, dry forests, the shrubby "monte" vegetation, and the Patagonian grasslands (Ibisch et al. 1996). While recent long-distance dispersal remains a possible explanation for this pattern, Ibisch et al. (1996) hypothesized that the current gap was bridged along the Andes by islands of montane forests during glacial times (18,000–19,000 years ago (ya)) allowing *Arachnitis* to migrate into Bolivian tropical montane forests when the postglacial era started (10,000–11,000 ya) (Ibisch et al. 1996).

Hypopitys monotropa and *Monotropa uniflora* (Ericaceae) have remarkable transoceanic distribution ranges as well: the two species are found in temperate zones on both sides of the Pacific Ocean. *Monotropa uniflora* grows in large parts of North America extending into montane regions of Central and South America. It also occurs in Asia, where it is found in Japan, southern China, India to Nepal, and Bhutan (Wallace 1975). *Hypopitys monotropa* has an even wider distribution that mainly overlaps with that of *M. uniflora* but also includes temperate Europe. Thorne (1972) categorized this distribution as "circumboreal." Although there are no divergence time estimates for these species, migration through Beringia, a land bridge that connected North America and temperate Asia at various times during the Pleistocene ice ages, seems the most likely explanation for this distribution pattern (Donoghue and Smith 2004).

The partial mycoheterotroph *Corallorhiza trifida* (Orchidaceae) also has a circumboreal distribution, while all other *Corallorhiza* species are restricted to the New World. *Corallorhiza trifida* is not the earliest diverging species in the genus (Freudenstein and Senyo 2008; Barrett and Freudenstein 2008) and possibly dispersed through Beringia, similar to the dispersal hypothesized for both *Monotropa* and *Hypopitys* (see above).

3.5 Biogeographic History

3.5.1 Distributions of Families and Genera: Vicariance Versus Long-Distance Dispersal

In his treatment of the Burmanniaceae, Adolf Engler noted the transoceanic distribution of several genera and assumed that these genera must be extremely old: "Aus dem Vorkommen der Gattungen *Burmannia, Gymnosiphon, Dictyostega* und *Thismia* in der alten und neuen Welt ergiebt sich, dass die Familie der Burmanniaceae sehr hohen Alters sein muss und dass höchstwahrscheinlich in der Tertiärperiode ihre Verbreitung sich bis nach den Polen hin erstreckt hat" (Engler 1888, p. 46). Written at a time when the Earth's geography was assumed to have been stable, ancient dispersal across the poles seemed the only possible explanation for the widespread distribution of these genera. In 1915, the German meteorologist Alfred Wegener published his theory of continental drift (Wegener 1915), although, due to the lack of convincing mechanisms that might have caused the movement and splitting of huge landmasses, his theory only became widely accepted in the 1960s after the discovery of plate tectonics. From that time, the fragmentation of Gondwana became an appealing explanation for the widespread and disjunct distribution patterns observed in many groups of plants (Raven and Axelrod 1974) and has been used to explain the disjunct distribution of Corsiaceae (Zhang et al. 1999) and *Seychellaria* (Vollesen 1982). However, the application of molecular dating of lineage divergences has suggested that most cases of transoceanic distributions are unlikely to be the result of tectonic vicariance simply because the lineages are too young to have been dispersed through ancient land connections (Renner 2004; de Queiroz 2005).

The lack of fossil data and hypotheses of phylogenetic relationships have prevented detailed studies on the biogeographic histories of most mycoheterotrophic groups to date. There is only one series of fossils that may be assigned to an extant mycoheterotrophic lineage. These fossils are from the Late Cretaceous (about 90 Ma) and were found in New Jersey (Gandolfo et al. 1998). A phylogenetic analysis based on morphological characters placed the two fossil genera *Mabelia* and *Nuhliantha* within extant Triuridaceae (Gandolfo et al. 2002). However, it remains questionable whether these plants were in fact mycoheterotrophic because the fossilized material only consists of flowers (Gandolfo et al. 2002). Moreover, the fossilized pollen lacks distinctive features of pollen of extant members of the Triuridaceae (Furness et al. 2002). If the fossils represent genuine remnants of Triuridaceae, they

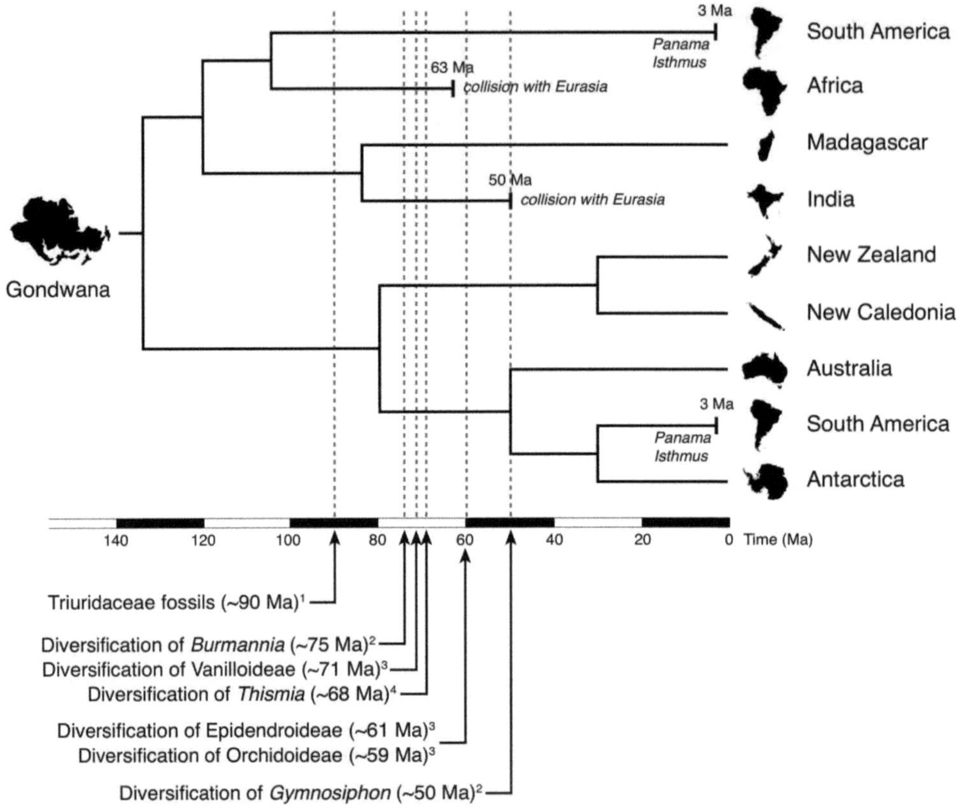

Fig. 3.9 Geological area cladogram showing the progressive breakup of Gondwana. *Vertical lines* indicate collisions between landmasses. South America is shown twice to illustrate its separation from both Africa and Antarctica. Representations of landmasses are not drawn to scale. Cladogram modified from Sanmartin and Ronquist (2004) and Cox and Moore (2010). [1]Age estimate from Gandolfo et al. (2002). Fossils were found in North America (Laurasia). [2]Mean molecular clock estimate taken from Merckx et al. (2008). [3]Molecular clock estimates of crown node ages of Orchidaceae subfamilies taken from Ramirez et al. (2007). [4]Mean molecular clock estimate from Merckx et al. (2010a). In this study, *Thismia* is paraphyletic and includes *Haplothismia* and *Tiputinia*

show that the family occurred at higher latitudes than today, at least during the Late Cretaceous when the climate was warmer than present. Also, if major Triuridaceae lineages had diverged by the Late Cretaceous, the breakup of Gondwana may have played a significant role in the distribution of the family (Fig. 3.9). Nevertheless, the occurrence of *Sciaphila* species on isolated oceanic islands (e.g., Bonin, see Fosberg and Sachet 1980; Fiji, see van de Meerendonk 1984) indicates that long-distance dispersal is certainly an important factor in the widespread distribution of this genus and probably for Triuridaceae in general.

In the absence of fossil data, divergence time estimates for biogeographic reconstructions can be obtained through broadscale molecular clock analyses which allow for inclusion of fossil data from distant groups or by using secondary calibration points. These strategies have been used to investigate the biogeographic history of Burmanniaceae (Merckx et al. 2008). The obtained age estimates suggest that Burmanniaceae genera are relatively ancient and diverged from each other during the Cretaceous probably in South America. However, the obtained hypothesis suggests that the widespread genera *Burmannia* and *Gymnosiphon* reached their transoceanic distribution range during the Eocene, when the breakup of Gondwana was well under way (Fig. 3.9). This demonstrates that both genera obtained their current pantropical distribution by dispersal, probably aided by various land bridges. In particular, dispersal of *Burmannia* and *Gymnosiphon* out of South America may have been possible by "boreotropical" migration routes: during the Eocene, global temperatures peaked, and tropical vegetation occurred at high latitudes. This allowed for migration of tropical flora between the Neotropics and the Paleotropics over the "North Atlantic Land Bridge" (Tiffney 1985; Davis et al. 2002). The disjunct distribution of *Voyria* (Gentianaceae) between the Neotropics and Africa has also been explained as a relict of an ancient boreotropical distribution (Albert and Struwe 1997; see above). Similarly, molecular clock analyses suggest a Cretaceous origin for Thismiaceae, but diversification occurred well after the breakup of Gondwana, rendering a "Gondwanian aborigine" explanation for its widespread distribution unlikely (Merckx and Bidartondo 2008; Merckx et al. 2010a; Fig. 3.9).

In Orchidaceae, mycoheterotrophy evolved multiple times independently in subfamilies Vanilloideae, Orchidoideae, and Epidendroideae, all of which diversified after the start of the breakup of Gondwana (Ramirez et al. 2007; Fig. 3.9; Chap. 5). Thus, any mycoheterotrophic genus in Orchidaceae with a distribution that covers multiple Gondwanan fragments must have acquired their distribution through dispersal rather than vicariance. The widespread genus *Gastrodia*, for instance, with a distribution that covers tropical Africa, Madagascar, Southeast Asia, Australasia, and the Pacific Islands, belongs to the Epidendroideae. *Gastrodia* therefore diverged and diversified after the breakup of Africa, Madagascar, and India (Fig. 3.9) and must have migrated by long-distance dispersal. Similar observations can be made for widespread genera *Didymoplexis*, *Galeola*, *Epipogium*, *Eulophia*, and *Gastrodia*. In addition, distributions that include oceanic islands must have been the result of long-distance dispersal (e.g., *Stereosandra*).

In conclusion, widespread lineages of mycoheterotrophic plants are generally too young to have acquired their distribution by tectonic vicariance. In some cases, temporary land bridges may have aided migration between fragments of Gondwana. However, we hypothesize that transoceanic dispersal has been the most important factor in the widespread distributions of mycoheterotrophic plants. Most mycoheterotrophic plants produce large amounts of small dustlike seeds which promotes dispersal and increases the likelihood of reaching suitable microsites for recruitment (Eriksson and Kainulainen 2011).

3.5.2 Species Distributions

Only few phylogeographic studies on fully mycoheterotrophic plants have been carried out. Recent phylogeographic studies on *Hypopitys monotropa* in North America (Beatty and Provan 2011a) and Europe (Beatty and Provan 2011b) provide the first detailed hypotheses about the glacial history of a mycoheterotrophic plant species (see also Chap. 6). Population genetic data and ecological niche modeling suggest that the current east–west disjunct distribution of *H. monotropa* in North America results from the existence of separate eastern and western refugia during the last glaciation. In Europe, *H. monotropa* probably recolonized northern Europe from refugia in the Balkans and southern Europe, a scenario that is suggested for many European taxa (Provan and Bennett 2008). The high species diversity of mycoheterotrophic plants in tropical postulated glacial refugia (see further) suggests that glacial and interglacial cycles that characterized the Quaternary period (ca. 2.6 Ma to present) have had a significant effect on the distributions of tropical mycoheterotrophic species as well.

3.6 Habitat Characteristics

3.6.1 Light

Most mycoheterotrophic plants grow in forests with a dense overstory that produces deep shade. Probably due to the lack of light, these microhabitats often have little or no herbaceous ground flora (Leake 1994; Cheek and Williams 1999). Since mycoheterotrophs do not require light to grow, the occurrence of most of these plants to deeply shaded sites lacking herbaceous autotrophs suggests their mode of life evolved to provide escape from competitive exclusion in the shaded conditions of forest understory habitats (Bidartondo et al. 2004). Indeed, in partially mycoheterotrophic orchids, it has been shown that low light levels result in strong mycoheterotrophy, while higher irradiances successively drive the orchids toward autotrophy (Preiss et al. 2010).

3.6.2 Water

Besides their preference for low-light habitats, the availability of water seems an important feature for mycoheterotrophic plant habitats. Bakshi (1959) noted that *Pterospora* is particularly associated with moist soils. Summerhayes (1951) assumed that a moist but well-drained soil is thought to favor the fungal symbiont of *Epipogium aphyllum.* Mycoheterotrophic plants from tropical rainforests seem to prefer local habitats with very moist soils as well (Maas et al. 1986; Cheek and Williams 1999). Moreover, some species are often found near perennial or seasonable streams (Maas et al. 1986; Cheek and Williams 1999; Taylor and Roberts 2011). In Africa, areas with a high diversity of mycoheterotrophic plants seem to overlap with postulated rainforest refugia (Cheek and Ndam 1996). The occurrence of these refugia is strongly correlated with rainfall as well (Linder 2001), indicating minimal seasonality and a stable, moist environment year-round. In general, mycoheterotrophs seem to prefer habitats with high rainfall and short dry seasons. For example, on Mount Kupe, one of the richest sites for mycoheterotrophic plants in Africa, annual rainfall exceeds 4,000 mm with only 3 months with less than 100 mm precipitation (Hofer et al. 2000). Similarly, Reserva Ducke in Brazil receives over 2,000 mm of rainfall and has a dry season of 3 months (Iriondo and Latrubesse 1994; ter Steege et al. 2003). There is some evidence that mycoheterotrophic plants are sensitive to desiccation. In dry summers, the flowering stems of *Monotropa uniflora*, *Hypopitys monotropa*, and *Neottia nidus-avis* suffer severely from drought and often fail to expand before drying out (Snetselaar and Whitney 1990; Leake 1994). Klooster and Culley (2009) observed that a drought in the summer and fall of 2007 likely contributed to the overall decline in reproductive effort and output of *Monotropa* with some populations experiencing 100% floral abortion.

3.6.3 Soil Types

Both in temperate and tropical zones, mycoheterotrophic plants occur on a variety of soil types, including organic soils, clay, loam, and white sand (Wallace 1975; Maas et al. 1986; Maas and Ruyters 1986). They generally prefer acidic soils. Reported soil pH values for mycoheterotrophic habitats in temperate regions range from 5.2 to 6.2 (Wallace 1975; Gebauer and Meyer 2003; Bougoure et al. 2008), but Merckx et al. (2010b) reported a soil pH of 3.8 at a tropical rainforest site in French Guiana where *Dictyostega* and *Voyria* species were found. Mycoheterotrophs are often found in places with a thick layer of decaying leaf litter (Graham 1953; Paul 1964; Van Royen 1972; Richards 1976; Maas and Ruyters 1986). In some cases, their flowers fail to emerge through this layer, for example, *Monotropsis* (Copeland 1939), *Thismia rodwayi* (Campbell 1968), *Epipogium* and *Neottia* spp. (Davies et al. 1988). *Rhizanthella* species remain entirely underground when flowering and fruiting (George 1980).

The roots of some species, for example, *Voyria rosea* and *Campylosiphon purpurascens*, can penetrate the soil to a depth of 20 cm and more (Maas et al. 1986; VSFTM pers. oberv.). Contrarily, the roots of a few species (e.g., *Voyria aphylla*, *V. flavescens*, *V. primuloides*, *Afrothismia foertheriana*) are entirely located in the uppermost soil strata, which consists of lose decaying leaf litter (Franke 2002; Franke et al. 2004; VSFTM and CDS pers. observ.).

3.6.4 Elevation

Mycoheterotrophic plants occur from sea level to montane forests generally up to 2,000 m. In North America, Monotropoideae can occur up to

4,000 m (Wallace 1975). In the Neotropics, *Apteria*, *Dictyostega*, and *Gymnosiphon* species are sometimes found above 2,000 m (Maas et al. 1986). In East Africa, mycoheterotrophs often occur at high elevations; *Epipogium roseum* is recorded from Mount Kilimanjaro at 2,500 m (Cheek and Williams 1999). In New Guinea, species of *Corsia* generally inhabit forests between 900 and 1,500 m, extending into the upper part of lowland forest and into montane forest (Van Royen 1972).

3.6.5 Plant Communities

Mycoheterotrophic plants occur in a variety of forest communities. In temperate zones, Monotropoideae occur in forests dominated by conifers (pines and cypress), beech, and oak (Wallace 1975; Ogura-Tsujita et al. 2009). Similarly, fully mycoheterotrophic orchids are generally found in forests dominated by pines, beech, and oak trees (Gebauer and Meyer 2003; Bidartondo et al. 2004; Zimmer et al. 2008; Taylor and Roberts 2011). These forests tend to have higher pH soils than other temperate forest habitats.

Information on plant communities is purely anecdotal for species that occur in the tropics. In the Neotropics, species of *Gymnosiphon* (Burmanniaceae) are often found in *Mora* forest (Maas et al. 1986). In West Africa, *Afrothismia* (Thismiaceae) occurs in proximity of a wide variety of rainforest trees, including species of *Cola*, *Diospyros*, and *Tabernaemontana* (Franke 2004; Sainge et al. 2005; Dauby et al. 2007). Tropical mycoheterotrophic orchid species of *Aphyllorchis* and *Cephalanthera* have been found in Fagaceae and Dipterocarpaceae forest in Thailand (Roy et al. 2009). In Southeast Asia, mycoheterotrophic plants are sometimes found in bamboo thickets, for example, *Epirixanthes* (Chen et al. 2008), *Thismia* (Chantanaorrapint 2008), and *Gastrodia confusa* (Ogura-Tsujita et al. 2009).

Arbuscular mycorrhizal mycoheterotrophs are sometimes reported from forest communities dominated by ectomycorrhizal trees. In New Guinea, species of *Corsia* (Corsiaceae), for example, are mostly found in beech and oak forests (Van Royen 1972). In Southeast Asia, species of *Thismia* are often collected from forest dominated by Dipterocarpaceae or Fagaceae trees (Jarvie 1996; Yang et al. 2002; Tsukaya and Okada 2005). In Japan, *Oxygyne shinzatoi* (Thismiaceae) has been found in *Castanopsis* forest (Yokoyama et al. 2008). However, in Argentina, *Arachnitis uniflora* (Corsiaceae) was found to link with the same arbuscular mycorrhizal fungi as *Osmorhiza chilensis* (Apiaceae), *Austrocedrus chilensis* (Cupressaceae), and *Nothofagus dombeyi* (Nothofagaceae) (Bidartondo et al. 2002). The latter species had been previously reported as ectomycorrhizal (Fontenla et al. 1998).

A very strong preference toward a particular plant community is observed for the underground orchid *Rhizanthella gardneri* (Orchidaceae), which only occurs in *Melaleuca uncinata* thickets in West Australia (Bougoure et al. 2008). In Tasmania, *Thismia rodwayi* (Thismiaceae) is always found in wet forest dominated by *Eucalyptus* trees (Wapstra et al. 2005).

3.6.6 Exceptional Habitats

There are plenty of exceptions to the general habitat preferences listed above. In the Neotropics, *Apteria aphylla* (Burmanniaceae) and several *Voyria* species (Gentianaceae) are known to occur in wet grasslands and savannas (Maas et al. 1986). *Arachnitis* (Corsiaceae) has been recorded on the treeless East Falkland Island, "growing in sand amongst rocks on an eroded sandstone ridge" (Cribb et al. 1995). In Africa, mycoheterotrophic *Brachycorythis* orchids occur in woodland and wooded grassland (Cheek and Williams 1999). Monotropoideae are often found in open vegetations, such as dune slacks (Wallace 1975; Leake 1994). The enigmatic species *Thismia americana* (Thismiaceae) was discovered in the margin of a grass field (Pfeiffer 1914), and the western underground orchid *Rhizanthella gardneri* grows in shrublands, in habitats of low-nutrient availability and high light levels (Bougoure et al. 2008). *Hexalectris spicata* (Orchidaceae) occurs in diverse habitats: from swamps to oak canyons rising out of the desert (Luer 1975). And while mycoheterotrophic species are usually terrestrial,

Voyria spruceana and *V. aphylla* (Gentianaceae) have been found growing as epiphytes up to 30 m high in trees in Colombia (Groenendijk et al. 1997). *Burmannia kalbreyeri* (Burmanniaceae) is also known to grow epiphytically, but this species retains chlorophyll and has well-developed leaves (Maas et al. 1986). Lastly, several species of *Sciaphila* (Triuridaceae) are often found growing on termite nests (van de Meerendonk 1984; Maas and Ruyters 1986), and Franke (2002) reported a specimen of *Voyria flavescens* (Gentianaceae) growing top of a termite mound of *Embiratermes neothenicus* (Isoptera, Nasutitermitinae).

3.6.7 Population Size

Even within suitable habitats, populations of mycoheterotrophic plants are mostly thinly scattered. Population sizes are usually small, with less than 15 individuals per population, but plants can only be detected when aboveground parts are present (flowering and fruiting stages), so the actual population size is difficult to determine. In some cases, populations of over hundreds of flowering specimens have been reported (e.g., *Gymnosiphon* Cheek and Williams 1999; *Burmannia* Fensham 1993).

3.6.8 Co-occurrence

Many authors have noted that different species of mycoheterotrophs (often from different families), both in temperate and tropical zones, have the tendency to grow together (e.g., van der Pijl 1934; Jonker 1938; Van Royen 1972; van de Meerendonk 1984; Maas and Rübsamen 1986; Cheek and Williams 1999). Indeed, after a mycoheterotrophic plant is spotted in a forest, closer inspection of the area will often reveal other species growing close by (Maas et al. 1986; Cheek and Williams 1999). There is no explanation for this phenomenon. A few authors have suggested that co-occurring mycoheterotrophs are possibly sharing the same mycorrhizal fungus (Cheek and Williams 1999; Cheek 2003b) but no evidence has been found to support this. In contrast, Merckx et al. (2010b) sampled specimens of *Voyria aphylla* (Gentianaceae) and *Dictyostega orobanchoides* (Burmanniaceae) co-occurring at a rainforest plot in French Guiana and found that they were associated with nonoverlapping AMF lineages. A specimen of *Campylosiphon purpurascens* (Burmanniaceae) that occurred at this spot was found to utilize distinct fungal lineages as well (VM, unpublished data). Similarly, Courty et al. (2011) identified distinct lineages of arbuscular mycorrhizal fungi in roots of *Voyria aphylla* (Gentianaceae), *Apteria aphylla*, and *Gymnosiphon* sp. (Burmanniaceae) growing at a site in Guadeloupe. Interestingly, many patches of rainforest in French Guiana that are rich in mycoheterotrophic Burmanniaceae and Gentianaceae are also inhabited by the mycoheterotrophic orchid *Wullschlaegelia calcarata* (Orchidaceae) (VSFTM pers. oberv.), yet *Wullschlaegelia* is associated with saprotrophic fungi (Martos et al. 2009), while Burmanniaceae and Gentianaceae exploit arbuscular mycorrhizal mycorrhiza. In the temperate forests of Northwest America, different species of ericaceous mycoheterotrophs are often found growing together (Wallace 1975) but are each specialized on different mycorrhizal fungi (Bidartondo and Bruns 2001, 2002). Similarly, Taylor and Bruns (1999) investigated the mycorrhizal associations of the mycoheterotrophic orchids *Corallorhiza maculata* and *C. mertensiana* over a wide geographic range and found that they never shared fungal species, even when growing intermixed. In these cases, co-occurrence of mycoheterotrophs cannot be explained by specialization on the same "host" fungus. The pattern may reflect access to different resources by different fungal taxa and thus would result from competition rather than convergence. Plant sister species that occur in sympatry but grow with different fungal lineages may result from a speciation process driven by mycorrhizal specialization (Chap. 5). Other factors that may cause co-occurrence of mycoheterotrophs include similar preferences of mycoheterotrophs and/or their associated mycorrhizal fungi toward certain microhabitats or similar dispersal biases.

3.7 Biodiversity Hotspots

A few localities are notorious for their high number of mycoheterotrophic species. For example, Mount Kupe in Southwest Cameroon is often cited as the richest site for mycoheterotrophic plants in Africa (Cheek and Cable 1997; Franke 2004). Two decades of intensive surveys on the slopes of this mountain have revealed twelve different mycoheterotrophic species, including two *Afrothismia* species yet to be described (Franke 2007; Moses N. Sainge, pers. comm.). Therefore, Mount Kupe is home to about one-fourth of all mycoheterotrophic species known from continental tropical Africa. The adjacent Mount Cameroon has a comparable diversity of mycoheterotrophic plants, although some species may be extinct due to extensive habitat destruction (Schlechter 1906; Cheek and Williams 1999; Franke et al. 2004). Mount Kupe and Mount Cameroon are part of the Lower Guinea rainforest region, which is a center of diversity and endemism for mycoheterotrophic plants (see above) and for flowering plants in general (Linder 2001; Plana 2004). The stunning diversity of plants in this region can be explained by modern patterns in rainfall seasonality, while the high level of endemism is probably related to paleoclimatic fluctuations, and the area has likely served as a rainforest refugium during glacial maxima (Linder 2001; Plana 2004). In Africa, centers of diversity of mycoheterotrophic plants all occur in areas that have been thought to be glacial refugia. Since many mycoheterotrophic plants are vulnerable to disturbance, have very narrow habitat preferences, and seem to have very slow dispersal rates, it has been proposed that mycoheterotrophic plants may be suitable indicators of Pleistocene rainforest refugia (Cheek and Ndam 1996; Cheek and Williams 1999).

Another mycoheterotrophic plant "hotspot" is the Reserva Ducke near Manaus (Brazil). In an area of only 100 km^2, no less than 22 mycoheterotrophic species have been found, or over 25% of the total species diversity of mycoheterotrophic plants in the Neotropics. Two species of Triuridaceae (*Sciaphila oligantha* and *S. rubra*) are endemic to the reserve (Ribeiro et al. 1999; Maas and Rübsamen 1986; Maas and Maas 2005). The Reserva Ducke is characterized by an extremely high diversity of flowering plants (Ribeiro et al. 1999). Patterns of tree diversity in the Amazon rainforest have been linked to soil gradients and the current climate, and particularly rainfall seasonality (ter Steege et al. 2003, 2006), although recent research has indicated that paleoclimate probably had a much greater effect on current patterns of tree diversity than current climate (Hoorn et al. 2010; ter Steege et al. 2010). Indeed, the high diversity of plant species found in Reserva Ducke has been attributed to its position at the contact point among several Tertiary and Quaternary refugia, indicating that its diversity may have been enhanced by the coalescence of the distributions of former allopatric species in this area (Oliveira and Daly 1999; Oliveira and Mori 1999).

The most species-rich locality of mycoheterotrophic plants in Southeast Asia is Mount Kinabalu in Malaysian Borneo. Since botanical explorations on the slopes of the 4,095 m high mountain started in the second half of the nineteenth century, 29 species of fully mycoheterotrophic plants have been recorded. These include 16 species of orchids, of which *Didymoplexiella kinabaluensis* and *Gastrodia spathulata* (formerly *Neoclemensia spathulata*) have not been found elsewhere (Wood et al. 2011). Burmanniaceae and Thismiaceae are both represented with two fully mycoheterotrophic species, Triuridaceae with five species, Polygalaceae with three fully mycoheterotrophic species, and Petrosaviaceae with a single species (Beaman and Beaman 1998; Beaman and Anderson 2004). Mount Kinabalu is ca. 1.5 million years old and therefore comparatively young in geological terms. The flora is exceptionally diverse with a high percentage of endemicity (Wong and Phillipps 1996; Wood et al. 2011). The high diversity of plant species on Mount Kinabalu has been attributed by the occurrence of a wide range of soil types and climatic conditions. The endemic species of Mount Kinabalu may be relicts of ancient, more widespread distribution ranges, or recent species that result from rapid adaptive radiation, catastrophic selection and drift, and dispersal of propagules from distant and neighboring mountain systems (Wong and Phillipps 1996).

Other areas with high numbers of mycoheterotrophic species include Saül (French Guiana) (Clarke and Funk 2005) and Mabura (Guyana) (Ek and ter Steege 1997). In temperate Northwest America, mainly in the coastal areas of northern California, Oregon, and Washington, there are many small "hotspots" where distribution ranges of several Monotropoideae species overlap (Wallace 1975).

Molecular clock evidence indicates that Burmanniaceae and Thismiaceae have Cretaceous origins and major diversification began shortly after the mass extinctions at the K/T boundary, suggesting that glacial periods may have had a significant influence on their current distributions (Merckx and Bidartondo 2008; Merckx et al. 2008). In addition, these analyses indicate that speciation events within mycoheterotrophic Dioscoreales lineages predate the Pleistocene glaciations. Based on this evidence, it seems that during glacial maxima rainforest refugia have acted as "museums" of ancient lineages of mycoheterotrophic Dioscoreales rather than as speciation "engines" (see Plana 2004). However, mycoheterotrophic species of other families that are endemic to a particular hotspot may result from recent speciation events. For example, several species of orchids endemic to Mount Kinabalu have evolved from vertical altitudinal radiation of lowland congeners or dispersal from neighboring mountain systems most likely after the uplift of the mountain (1.5 million years ago) and can be considered as "neoendemics" (Barkman and Simpson 2001; Wood et al. 2011). The endemic mycoheterotrophic orchids of Mount Kinabalu may therefore be recent species as well. Thus, hotspots of mycoheterotrophic plant diversity may be both "museums" and "cradles" of diversity.

3.8 Endemism and Rarity

Mycoheterotrophic plants are seldom found in high abundances. That does not mean that all species are rare. Some species have wide distribution ranges and occur in a variety of habitat types (e.g., *Hypopitys monotropa*, *Epipogium roseum*). But even widespread mycoheterotrophs may still be extremely rare on a local scale. *Epipogium aphyllum*, for example, has a widespread distribution in temperate Eurasia, but on a local scale, it is often considered to be extremely rare. In Britain, it has been frequently described as the rarest orchid and even as Britain's rarest plant (Taylor and Roberts 2011). In addition, many mycoheterotrophs seem to have very limited distribution ranges and/or have extremely low abundances (e.g., known from very few populations). For example, most neotropical *Thismia* species are only known from one or two collections (Maas et al. 1986). A large number of African mycoheterotrophs are also known from only one or two localities (e.g., species in the genera *Afrothismia*, *Kupea*, *Kihansia*, *Oxygyne*, *Auxopus*, and *Gastrodia*). Similarly, the enigmatic underground orchid (*Rhizanthella gardneri*) is known from only a handful of sites in Southwest Australia (Bougoure et al. 2008). Numerous other mycoheterotrophic species from all continents share this apparent rarity.

It is important to note that our knowledge about the occurrence of fully mycoheterotrophic plants may be considerably biased by the plants' ability to remain unnoticed by collectors. Many species, particularly those from tropical rainforests, are only known from remote areas where botanical inventories have yet to be carried out. Moreover, mycoheterotrophs can only be spotted when they are flowering or fruiting, mostly for a short period of time only and often in the wet season, when few botanists are eager or able to enter the forest. The rest of the year, they remain underground hiding from discovery, and they may not even flower each year. We lack detailed information on the phenology of most mycoheterotrophs, but, for example, in the ghost orchid (*Epipogium aphyllum*), it has been observed that populations can disappear for up to 30 years between successive flowering episodes at the same site (Leake 1994). This suggests that some mycoheterotrophs may have very cryptic flowering cycles. Finally, mycoheterotrophs are generally very difficult to spot due to their small size and hyaline coloration. They often fail to protrude above the dense leaf litter and remain covered by fallen leaves,

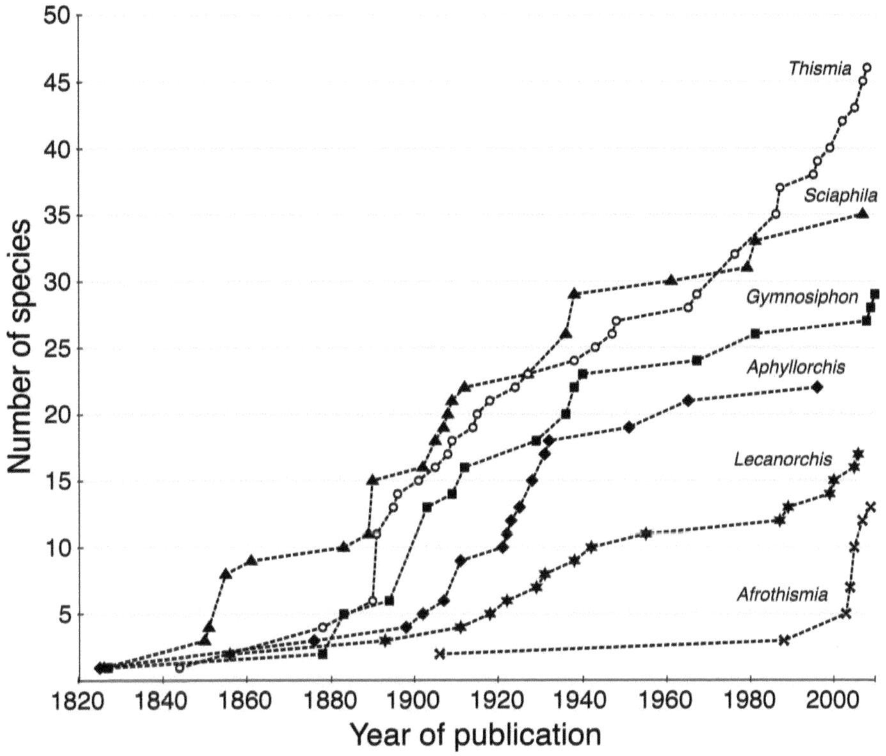

Fig. 3.10 Number of species in a selection of mycoheterotrophic genera that occur in the tropics (*Thismia*, *Sciaphila*, *Gymnosiphon*, *Aphyllorchis*, *Lecanorchis*, and *Afrothismia*) and their year of publication. Data based on the World Checklist of Selected Plant Families (2010)

even when flowering. It is little wonder that mycoheterotrophs are often spotted by mushroom hunters or by a botanist during a sanitary break; some species may be more abundant than we presume because we just fail to find them even when actively looking for them. The fact that new species are constantly being described and thus escaped discovery for a long time illustrates the secret nature of mycoheterotrophs (Fig. 3.10).

Some new species were spotted only after extensive long-term monitoring. Notorious is the discovery of two new *Afrothismia* (Thismiaceae) species in Korup Forest Dynamic Plot in Cameroon (Sainge and Franke 2005; Sainge et al. 2005). This 50-ha plot was established in 1994 and is constantly monitored, yet two *Afrothismia* species escaped discovery for almost a decade, despite the fact that a path through the plot was also going through one of the *Afrothismia* populations (Franke 2007). The spectacular species *Tiputinia foetida* (Thismiaceae), with a flower of 5 cm in diameter, was discovered in 2005 in a biological station in Ecuador growing within a meter of the path linking the station's dining hall to the laboratory (Woodward et al. 2007).

The influence of collection effort has been addressed for the rare species *Thismia rodwayi* (Roberts et al. 2003; Wapstra et al. 2005). From its discovery in 1890 until 2002, there were only five records of *T. rodwayi* in Tasmania (Roberts et al. 2003). Since the discovery of two specimens at a new site in Tasmania, subsequent searches on this and other sites with similar habitat characteristics revealed a total of 110 *T. rodwayi* flowers (Roberts et al. 2003), and *T. rodwayi* is now known from 26 sites from 7 disparate locations in Tasmania (Wapstra et al. 2005) (Fig. 3.11). Although these numbers certainly do not upgrade *T. rodwayi* to a common species, it can be concluded that it is at least more abundant

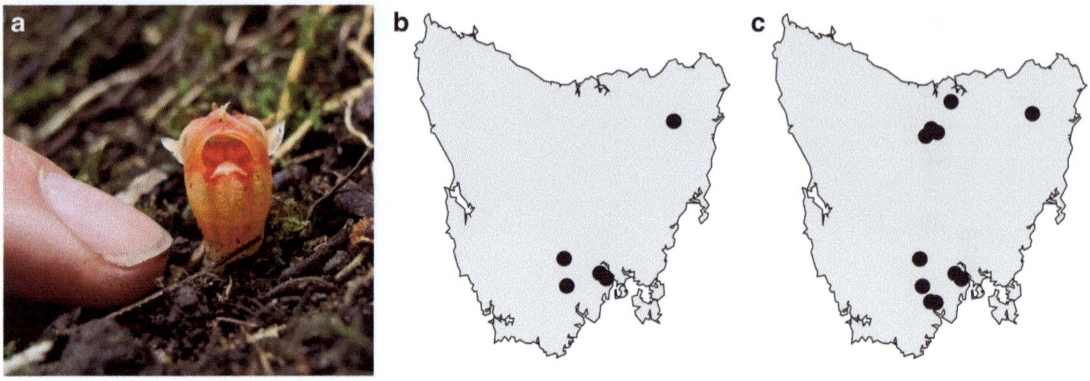

Fig. 3.11 Distribution of *Thismia rodwayi* (Thismiaceae) in Tasmania. (**a**) A flower of *T. rodwayi*. (**b**) Known distribution of *Thismia rodwayi* based on records before 2002. (**c**) Distribution of *T. rodwayi* based on all records up to 2005. Maps adapted from Roberts et al. (2003) and Wapstra et al. (2005)

than presumed by previous collections. As standard biological inventories failed to encounter *T. rodwayi* (Roberts et al. 2003), another conclusion that can be drawn from this study is that this species, and other inconspicuous mycoheterotrophs, can only be reliably recorded by targeted surveys. Because very few botanists search tropical rainforests specifically for mycoheterotrophic plants, the majority of collections result from chance encounters, hence explaining the lack of collections for so many known mycoheterotrophic species. The few intensive searches for mycoheterotrophic plants that have been carried out lead, in many cases, to the discovery of unexpected mycoheterotrophic plant diversity or even to the discovery of undescribed taxa (e.g., Franke 2007).

Despite the probable impact of collection effort, there is no doubt that the paucity of records for many mycoheterotrophic species is the result of "real" rarity and high local endemism (Kruckeberg and Rabinowitz 1985). This is obvious for species that are part of well-known floras (e.g., *Epipogium aphyllum* in Britain, *Thismia americana* in North America). In lesser-known tropical regions, differences in rarity among species are becoming more obvious with targeted collecting. Comparing herbarium records between species with similar habit, it becomes clear that some species are encountered more frequently than others (Fig. 3.12). In the Neotropics, for example, there is a very pronounced difference in collection frequency between mycoheterotrophic Burmanniaceae and Gentianaceae on the one hand and Thismiaceae and Triuridaceae on the other. This can only be explained by differences in distribution and abundance. According to Harper (1981), rare plant species can be classified based on space, group, or time relatedness. Space-limited species may be locally abundant but only occur in a limited number of sites. Their distributions may be restricted due to high niche specificity or dispersal barriers. These species are often local endemics. Group-dependent species occupy a specialized niche with a limited distribution, associated with certain ecotypes often at ecological frontiers for species. Rarity in time-dependent species results from fluctuations in population numbers following adverse sporadic or cyclical events, such as drought or fire (Swarts and Dixon 2009). Most rare mycoheterotrophic species seem to belong to the first and second categories, although time-dependent rarity may occur as well.

The influence of abiotic factors on the distribution of mycoheterotrophic plants remains to be determined. The fact that (rare) mycoheterotrophs of different species are often found growing at the same site, but in association with different fungi, possibly indicates that distribution of mycoheterotrophic species, or their associated fungi, is restricted by adaptations to similar

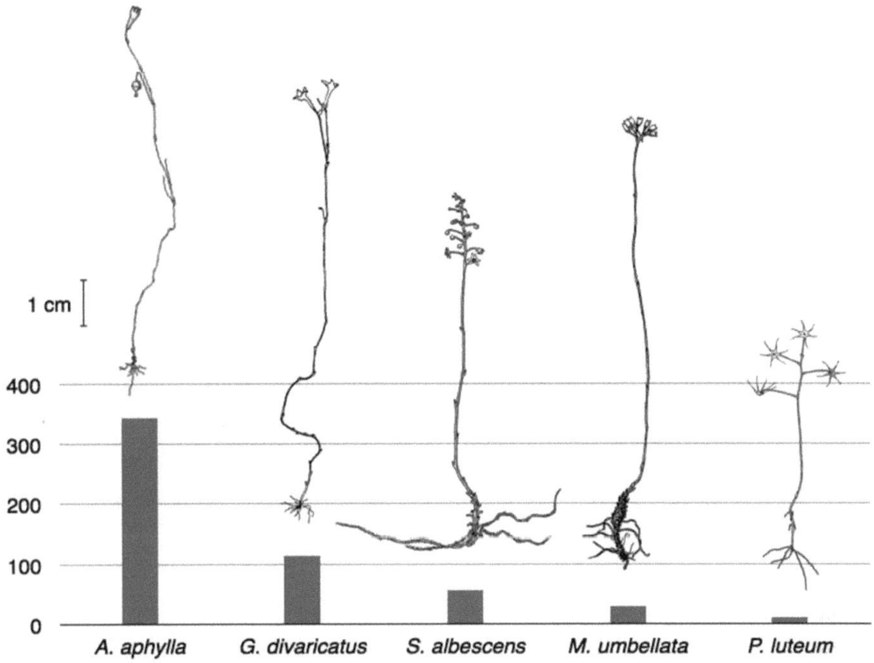

Fig. 3.12 The number of herbarium records cited in the Flora Neotropica for a selection of neotropical Burmanniaceae (*Apteria aphylla*, *Gymnosiphon divaricatus*, *Miersiella umbellata*) and Triuridaceae (*Sciaphila albescens*, *Peltophyllum luteum*). Data obtained from Maas et al. (1986) and Maas and Rübsamen (1986). For comparison, the habit of each species drawn at the same scale is shown above

microhabitats. These microhabitats can be characterized by certain abiotic factors such as soil type, humidity, and water availability. Alternatively, this pattern of co-occurrence may be explained by similar dispersal and colonization patterns.

Next to abiotic factors, biotic dependency may play an important role in the distribution range and abundance of mycoheterotrophic plants (Fig. 3.13). As fully mycoheterotrophic plants are completely dependent on their (often very specific) associated fungi, the availability of these fungi will be essential for the plant's establishment. Mycoheterotrophs that are specialized on rare fungi may thus be a priori severely limited in their distribution range. The range of the fungi may be restricted to certain abiotic habitat requirements and dispersal barriers (see further). In addition, some mycorrhizal fungi may also show specificity toward particular autotrophic plants. This can create a very complex tripartite biotic dependency for a mycoheterotrophic plant. For example, *Rhizanthella gardneri* is associated with a specific fungus, which is possibly only able to form ectomycorrhizas with *Melaleuca uncinata* (Bougoure et al. 2009). Furthermore, it is possible that the host fungus is not always able to serve as a suitable host for a mycoheterotrophic plant. Theoretically, a mycoheterotrophic plant will only successfully grow and reproduce if its associated fungus is able to provide enough carbon from coassociated trees or, in case of saprotrophic fungi, from dead material. For obligate mycorrhizal fungi, this may be influenced by the size of their network and the number, age, and identity of its associated autotrophic plants. Competition between different fungus species may influence their ability to allocate nutrients as well (Bever et al. 2009).

Specificity to aboveground biotic interactions may also induce rarity (Swarts and Dixon 2009). Specialized pollination systems, in a manner similar to the level of specialization in mycorrhizal associations, may play a role in causing rarity in

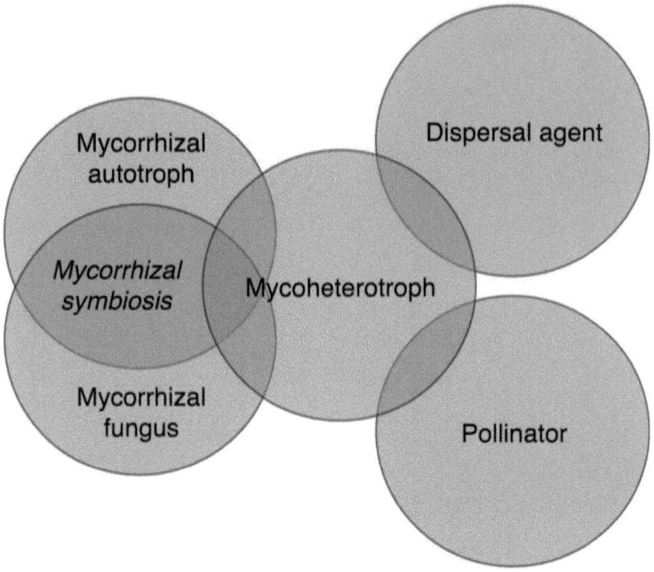

Fig. 3.13 Primary biotic agents that may limit the abundance and distribution of mycoheterotrophic plants. Overlap indicates potential biological dependency of the mycoheterotroph on that factor. See text for further discussion. Figure based on Swarts and Dixon (2009)

mycoheterotrophs. It has been hypothesized that mycoheterotrophic plants that show high specificity in their mycorrhizal interactions probably have generalist pollination syndromes and/or exhibit autogamous self-pollination due to the evolutionary instability inherent to specializing on two lineages (Bidartondo 2005). We know little about the pollination strategy of mycoheterotrophic plants, but support for Bidartondo's hypothesis concerning absence of specialized pollination syndromes has been found for mycoheterotrophic orchids (Dressler 1981; Benzing and Atwood 1984; Arditti 1992; Molvray et al. 2000) and mycoheterotrophic plants in general (Leake 1994). However, detailed studies on the reproductive biology of *Hypopitys*, *Monotropa*, and *Monotropsis* (Klooster and Culley 2009) and *Voyria* (Hentrich et al. 2010) revealed the presence of outcrossing and specialization in pollination interactions (see Chap. 7). Therefore, it is possible that, in some cases, both pollinator and fungal specificity affects the distribution mycoheterotrophic plants (Waterman and Bidartondo 2008). Comparative population approaches (Thompson 2005), where both mycorrhizas and reproductive traits for multiple populations of a single mycoheterotrophic species are examined, are needed to explore the effect of both mutualisms on the distribution of mycoheterotrophs. A recent study has addressed this interaction for achlorophyllous orchids of the tribe Coryciinae, which show specificity both in pollinators and associated fungi (Waterman et al. 2011). Interestingly, when orchids were subjected to transplant experiments, it was found that effective pollination does not occur outside native regions, whereas effective fungi can be recruited. This strongly suggests that pollination specificity has more influence on the local distribution of the species than does mycorrhizal specificity. However, it can be argued that mycorrhizal selection pressure is less influential for initially mycoheterotrophic plants than for full mycoheterotrophs that completely rely on the allocation of adequate nutrients from the fungi for their entire life cycle.

Finally, specialization toward seed dispersal agents may limit plant distributions as well. This factor seems to have little impact on mycoheterotrophs, which produce large amounts of dust seeds that are presumably dispersed by wind (Leake 1994; Eriksson and Kainulainen 2011). However, the dust seeds of orchids generally disperse only over short distances, and long-distance dispersal seems to be rare (Diez 2007; Jacquemyn et al. 2007). In *Voyria*, there is evidence of endozoochory (Hentrich et al. 2010), and seed dispersal of the rare underground orchid *Rhizanthella gardneri* is thought the carried out by a native mammal that is now extirpated from all known localities of the orchid (Dixon 1991). Therefore, limitations to seed dispersal cannot be ruled out as a determining factor in the distribution of mycoheterotrophic plants.

3.9 Extinct Species

According to the IUCN Red List, a taxon is extinct when there is no reasonable doubt that the last individual has died. A taxon is presumed extinct when exhaustive surveys in known and/or expected habitat, at appropriate times (diurnal, seasonal, annual), throughout its historical range have failed to record an individual (IUCN 2010). Since many mycoheterotrophic species, particularly those occurring in tropical rainforests, grow in inaccessible areas and are extremely difficult to spot (see above), it is impossible to declare any mycoheterotrophic species as extinct with confidence. Even when the type locality is destroyed and a species has not been seen for many decades, it is still possible that other populations escaped discovery. Sometimes species have been rediscovered after a notably long hiatus. *Haplothismia exannulata* (Thismiaceae) was rediscovered at its type locality in India in 2000, 49 years after its discovery and only a few years after being declared "extinct" (Sasidharan and Sujanapal 2000). The second collection of *Thismia clavigera* (Thismiaceae) was made 115 years after the first and over 1,000 km from the type locality (Stone 1980).

Despite these rediscoveries, many other rare species have not been collected for a remarkably long period of time. For some, it remains plausible they have escaped extinction. For example, *Marthella trinitatis* (Burmanniaceae) is only known from Mount Tucuche on Trinidad and was last collected in 1898 (Maas et al. 1986). But large patches of undisturbed forests are still present on Mt. Tucuche, retaining the possibility of a rediscovery of this species (Paul Maas, pers. comm.). In other cases, however, chances for survival of the species seem grim because the type locality and surrounding habitat has been destroyed. One of the most famous, now destroyed, localities is the "Alto Macahé" near Nova Friburgo (Rio de Janeiro), which is part of the coastal rainforest of southeast Brazil. In the nineteenth century, John Miers and Auguste Glaziou collected many remarkable mycoheterotrophic plants at this location. As a result, Alto Macahé is the type locality of *Peltophyllum caudatum* (Triuridaceae), *Thismia fungiformis*, *T. caudata*, *T. macahensis*, *T. janeirensis*, and *T. glaziovii* (Thismiaceae). Of these species, only *Thismia janeirensis* and *T. glaziovii* were later collected at another location. All other species have not been recorded since the type collection, and because 95% of the original Mata Atlântica rainforest has been replaced by farmland (Prance et al. 2000; Murray-Smith et al. 2009), little hope remains that these species escaped extinction (Maas et al. 1986). A similar fate was suffered by the endemics of Mount Cameroon, where most of the forest has been replaced by farmland, thereby destroying the type localities of *Oxygyne triandra*, *Afrothismia pachyantha*, *A. winkleri* (Thismiaceae), and *Burmannia densiflora* (Burmanniaceae) (Schlechter 1906, 1921). The latter two species were later found at other nearby locations, but *Oxygyne triandra* and *Afrothismia pachyantha* have not been collected for more than 100 years and may be extinct.

Arguably, the most mysterious of all mycoheterotrophic plant discoveries is that of *Thismia americana* (Thismiaceae). This tiny plant was discovered in August 1912 by Norma E. Pfeiffer in a low prairie near Chicago, Illinois (USA) (Pfeiffer 1914). *Thismia americana* was observed at this locality for several subsequent summers and was probably last seen in 1916. The type locality of *Thismia americana* has been replaced by an industrial complex, and numerous attempts to relocate this enigmatic species have been unsuccessful. Therefore, the species is currently listed as "possibly extinct" (Lewis 2002). *Thismia* species are generally found in the leaf litter of moist tropical rainforests in South and Central America, Southeast Asia, and Australasia, although some species occur in subtropical and temperate rainforests in Japan, Australia, and northern New Zealand. While this widespread distribution indicates that there is considerable variety in ecological preferences among *Thismia* species, the occurrence of a *Thismia* species in a prairie in temperate North America, more than 3,500 km from the nearest *Thismia* site (southern Costa Rica), is truly remarkable. The average temperature in the Chicago area lowers to −5°C during winter, by far the lowest temperature for any Thismiaceae site. This led Pfeiffer (1914) to the suggestion that the plant was perennial and that the underground parts of the plant were able to hibernate. Based on morphological similarities, it has been suggested that the closest known relative of *T. americana* is *T. rodwayi* from Australia and New Zealand (Jonker 1938; Maas et al. 1986; but see Thiele and Jordan 2002), forming one of the "most anomalous disjunctions known in flowering plants" (Thorne 1972, p. 407). Was this *Thismia* population the result of a human introduction, a recent long-distance dispersal, or the last remnant of an ancient boreotropical *Thismia* distribution? Unless the plant is rediscovered, this mystery will remain unsolved. Many people assume that the species is still present in the area. The only certainty is that if *T. americana* still exists, it is extremely difficult to find. In a letter to Prof. Warren H. Wagner in 1956, Pfeiffer recalled that it took her 3 h to relocate the plants when she returned to the exact same spot shortly after her first discovery.

3.10 Distribution of Host Fungi

The distribution of mycoheterotrophic plants is strongly geographically patterned both on global and local scales. This may be the result of constraints imposed by the physical environment or the biogeographic history of the plant lineages (see above). However, as many mycoheterotrophic plants show specificity toward narrow lineages of fungi (see Chaps. 5 and 7), an obvious question emerges: is the distribution of mycoheterotrophic plants also limited by the distribution of their associated fungi?

On a broad geographic scale, a correlation between the distributions of mycoheterotrophs and their host fungi seems apparent. Arbuscular mycorrhizal mycoheterotrophs are almost exclusively found in tropical forests, where the native trees form arbuscular mycorrhizal associations. In contrast, ectomycorrhizal mycoheterotrophic orchids and Ericaceae species are mainly restricted to temperate forests where ectomycorrhizal associations are predominantly formed by the native tree species. Forests of Southeast Asia, with their pronounced diversity of mycoheterotrophic orchids, are dominated by large trees of the family Dipterocarpaceae which are known to form associations with ectomycorrhizal fungi (Lee 1990; Moyersoen 2006). Indeed, species of *Aphyllorchis* and *Cephalanthera* mycoheterotrophs have been shown to use ectomycorrhizal fungi to obtain carbon from dipterocarps (Roy et al. 2009), and it is likely that many other mycoheterotrophic orchids from tropical Asia rely on similar associations.

A few ericaceous mycoheterotrophs do occur at tropical latitudes: *Cheilotheca* is restricted to Southeast Asia, and the distribution of the mostly temperate *Monotropa uniflora* reaches as far south as Colombia (Wallace 1975). However, tropical ericaceous mycoheterotrophs grow in pine, beech, and oak forests at high elevations where the presence of ectomycorrhizal fungi allows the establishment of these species. Interestingly, there are several reports of arbuscular mycorrhizal mycoheterotrophs growing in forests dominated by ectomycorrhizal trees (see above). In these cases, the arbuscular mycorrhizal "host" is likely associated with arbuscular mycorrhizal understory plants.

On a finer spatial scale, the relation between mycoheterotroph and fungus distributions is a much more complicated issue. Why are some mycoheterotrophs rare, even when their habitat is relatively common (e.g., *Rhizanthella*)? Are these mycoheterotrophs associated with "rare" fungi, and is their distribution limited by that of the fungus? In contrast to the obvious distribution limitations of specialized holoparasitic plants imposed by their host plant range, the relationship between the distribution range of mycoheterotrophic plants and their associated fungi is far less obvious. A major obstacle to assess this question is our lack of detailed knowledge of the distribution ranges of fungi. Historically, the assumption that fungi and other eukaryotic microorganisms have "global" geographic ranges was widely accepted. This would imply that the distributions of fungi are independent of biogeography and if conditions are right, the appropriate fungi will appear (de Candolle 1820). Similar environments, tropical rainforests, for example, would thus harbor similar fungi, and the potential distribution of mycoheterotrophic plants would be constrained by the physical environment rather than the distribution of their associated fungi. Recently, the assumption that "every fungus is everywhere" has been challenged by molecular studies of historical biogeography, ecology, and population genetics of fungi (Taylor et al. 2006; Lumbsch et al. 2008; Öpik et al. 2010; Peay et al. 2010a). These studies show that although some fungi are capable of long-distance dispersal (Moncalvo and Buchanan 2008), the actual distributions of most species reflect the same major dispersal barriers (e.g., oceans and mountains) that drive vicariance events in other organisms (James et al. 1999; Matheny et al. 2009). Geographic patterning is also evident at a more local scale (Collier and Bidartondo 2009; Peay et al. 2010b). This implies that the actual distribution of fungi is not necessarily equal to the potential distribution and that mycoheterotrophs that are specialized on particular fungi may fail to invade new areas solely because their obligate host fungi are not present. However, evidence that availability of fungi poses a real limitation to the distribution ranges of some mycoheterotrophic plants is not eminent. One speculative example for this phenomenon may be presented by the Hawaiian Islands. Despite the presence of seemingly suitable habitats, mycoheterotrophic plants are absent from the flora of Hawaii. In addition, only three species of orchids are native to the Hawaiian Islands, a surprisingly small number for a tropical region (Ziegler 2002). This is remarkable, because the small dust seeds of orchids seem ideal for long-distance dispersal by air, and both mycoheterotrophic and green orchids are found on remote Pacific Islands like Vanuatu, Fiji, and Samoa. Carlquist (1980) hypothesized that orchid seeds may not be resistant to freezing temperatures of the upper air layers or that the pollination requirements of potential colonizing species are not met. Another possible explanation for the scarcity of orchids—and the absence of mycoheterotrophic plants in general—in Hawaii may be the absence of suitable fungi necessary for their establishment. Similarly, a limited number of suitable ectomycorrhizal plants may explain the absence neottioid orchids in the Macaronesian region (Liebel et al. 2010).

On a smaller geographic scale, there is evidence that the rarity of the full mycoheterotroph *Pterospora andromedea* (Ericaceae) in eastern North America is influenced by the distribution and rarity of its fungal symbiont. A recent study by Hazard et al. (2011) showed that *P. andromedea* from the east coast of North America consists of a single haplotype that grows only with a single narrow lineage of *Rhizopogon* fungi. This fungal lineage appears to be rare in eastern white pine forests, and this may be a contributing factor to the rarity of eastern *P. andromedea* plants. In contrast, in western North America, five haplotypes of *P. andromedea* have been identified, and these haplotypes show preference for either *Rhizopogon salebrosus* or *R. arctostaphyli*, even when they co-occur (Bidartondo and Bruns 2002). Kjøller and Bruns (2003) found that these fungus species are common in the soil spore bank of the Sierra National Forest in western North America, where *P. andromedea* is common as well. Similarly, the initial mycoheterotroph *Caladenia huegelii*, a rare terrestrial orchid from Australia, partners with a specific Sebacinales fungus, which is ecologically efficacious only under a highly limited range of habitat and environmental conditions. Thus, rarity of this orchid species is potentially caused by a high degree of mycorrhizal specialization (Swarts et al. 2010). A similar study on species of *Drakaea* orchids from western Australia showed that all species show high mycorrhizal specificity and germinate only in a particular microhabitat. However, within this microhabitat, rare and common species exhibit no difference in germination rates, and both germinate in suitable habitat not currently occupied. The extreme rarity of some *Drakaea* species is therefore attributed to their highly specific pollination systems rather than their mycorrhizal specificity (Phillips et al. 2011).

A putative case of plant rarity induced by the rarity of a specific fungal symbiont is the distribution of the western underground orchid (*Rhizanthella gardneri*), which is only known from five sites in Southwest Australia (Bougoure et al. 2008). *Rhizanthella gardneri* is found in *Melaleuca uncinata* shrublands, in habitats of low-nutrient availability and high light levels. However, comparison between *Rhizanthella* sites revealed that *R. gardneri* can tolerate a range of habitat conditions and may be more widespread than previously thought, given that there are extensive areas of *Melaleuca* thickets with similar habitat characteristics across Southwest Australia (Bougoure et al. 2008). Underground orchids are extremely difficult to find, and thus, additional populations may remain to be discovered. However, *Rhizanthella* orchids seem to be linked with *Melaleuca uncinata* plants by a specific *Ceratobasidium* fungus (Bougoure et al. 2009). A possible explanation for the scarcity of *Rhizanthella* populations compared to the wide range of sites that appear to be suitable habitat may be found in the availability of the required fungus. Similarly, the rarity of *Petrosavia sakuraii* in the understory of Japanese cypress plantations—a common vegetation type in Japan—may be the result of a preference of the mycoheterotroph toward specific and range-restricted fungi (Yamato et al. 2011).

These observations suggest that mycorrhizal specificity may play an important role in both the global and local distribution of mycoheterotrophic plants, in addition to other biogeographic and ecological factors. However, it must be noted that for mycoheterotrophic plants, host specificity does not necessarily lead to restricted geographic ranges. The mycoheterotrophic orchid *Eulophia zollingeri* has an extremely wide distribution in tropical and subtropical Asia, yet individuals from seven populations in Japan, Myanmar, and Taiwan were found to associate with the same narrow lineage of fungi related to *Psathyrella candolleana* (Coprinaceae). Based on this observation, the authors conclude that a mycoheterotrophic plant can achieve a wide distribution even with a high mycorrhizal specificity, so long as the fungal partner is widely distributed (Ogura-Tsujita and Yukawa 2008).

3.11 Threats

3.11.1 Value of Diversity

Mycoheterotrophic plants do not have a direct economical value. They are not useful for consumption or for pharmaceutical purposes with the only exception being *Gastrodia elata* (Orchidaceae), used in Chinese traditional medicine (Xu and Guo 2000). Mycoheterotrophs are not essential ecological components of forest habitats; however, their presence in forest ecosystems may offer an indirect economical value through recreational services for humans. While few people will actually visit forests only to see mycoheterotrophic plants, most will be intrigued when they encounter one due to their exceptional and often mysterious habit. Rare plants hardly ever reach the status as a "flagship" species, such as the Galápagos tortoise or the giant panda, but they certainly add an unusual biological aspect to an ecosystem that can be appreciated by visitors. Apart from this value, mycoheterotrophic plants present a unique model system to study mycorrhizal mutualisms and ecological symbioses in general. Their inability to fix carbon through photosynthesis clearly shows the potential of mycorrhizal fungi to transport carbon, and perhaps other nutrients, between plants. This has led to the discovery of partial mycoheterotrophic plants. In addition, mycoheterotrophic plants offer excellent research opportunities to study the evolution of multipartite symbioses.

Like all organisms, mycoheterotrophic plants carry a certain "existence value" (Primack 2008). Existence value can be defined as the amount that people are willing to pay to prevent species from going extinct, habitats from being destroyed, and genetic variation from being lost (Martín-Lopez et al. 2007). As a unique part of their ecosystems (deciduous forests, rainforests, savannas, etc.), their presence contributes to the existence value that is attached to these ecosystems. In addition, their rarity is sometimes used in favor of protection of specific areas. An example is *Kupea martinetugei* (Triuridaceae), which is often used to support conservation efforts for the threatened rainforest on Mount Kupe in Cameroon (Cheek et al. 2004; Baird 2006). The genus *Kihansia* (Triuridaceae) is one of the endemics of the Kihansi River Gorge in Tanzania, and its presence draws attention to the necessity of conservation efforts for protection of this severely threatened habitat (Davis and Mvungi 2004). Also, the charismatic underground orchids of Australia are often featured to highlight conservation necessity (Swarts and Dixon 2009). And *Thismia americana* still helps to protect native prairie in the Calumet area near Chicago, USA, even though it was last seen almost 100 years ago (Chew 2004).

Lastly, fully mycoheterotrophic plants offer serious potential for horticulture. "Exceptional" plants, such as rare orchids and carnivorous plants, have always generated interest from plant enthusiasts. Indeed, commercial growing of rare or bizarre plants is a profitable business. If the difficulties of culturing fully mycoheterotrophic plants can be overcome, these "ghost plants" will find their way into the greenhouses of orchid collectors but may also attract the attention of the average consumer looking for an extraordinary plant.

3.11.2 Habitat Destruction

The major threat for the existence of mycoheterotrophic plants is habitat destruction. This is the inevitably result of the expansion of human populations and human activities. Habitat destruction is the primary cause of the loss of biodiversity in terrestrial ecosystems (Pimm and Raven 2000). It is likely that most of the original habitat of some tropical mycoheterotrophic species has already been destroyed. The distribution of many mycoheterotrophic plants overlaps with the biodiversity "hotspots" assigned by Myers et al. (2000). These areas comprise exceptional concentrations of endemic species and are undergoing exceptional loss of habitat. Particularly strongly affected are the Philippines and Indo-Burma, where less than 5% of the primary vegetation is retained (Myers et al. 2000). For mycoheterotrophic plants, endemism is particularly pronounced in the biodiversity hotspots of Sundaland, the Western Ghats and Sri Lanka, Madagascar, the Eastern Arc and Coastal Forests of Tanzania/Kenya, West Africa, and the Atlantic Forest in Brazil. These areas rank among the "hottest" hotspots with the highest number of endemics and the most severe habitat loss (Myers et al. 2000). The effect of habitat loss on mycoheterotrophic plant diversity is probably best illustrated by Brazil's Atlantic Forest, where only 5% of the original vegetation remains (Murray-Smith et al. 2009; Fig. 3.14). Several Thismiaceae and Triuridaceae species are only known from this biodiversity hotspot. Because most of the historical localities of these species are now destroyed, little hope remains for their survival (Maas et al. 1986; Maas and Rübsamen 1986).

Apart from habitat destruction, habitat fragmentation is an important factor that contributes to the loss of biodiversity in ecosystems. Habitat fragments are not only isolated from one another by a highly modified or degraded landscape, but edges of each fragment experience an altered set of environmental conditions, referred to as "edge effects" (Primack 2008). It remains unclear how this affects diversity of mycoheterotrophic plants, but effects may be profound. It can be assumed that various edge effects, such as a lower canopy density, lower relative humidity, and lower soil-moisture content (Laurance et al. 2002), will have significant influence on the occurrence of mycoheterotrophic plants. Their occurrence may be even more influenced by the availability of their host fungi. Little is known about the effect of fragmentation, isolation, and concurrent edge effects on the belowground diversity. A few studies have shown that ectomycorrhizal fungal species richness is significantly reduced on smaller and more isolated habitat fragments in temperate (Peay et al. 2007, 2010b) and tropical forests (Tedersoo et al. 2007). Similar observations have been made for

Fig. 3.14 The original distribution of the Atlantic Forest in Brazil in 1500 AD and its distribution in the 1990s. Adapted from Morellato and Haddad (2000)

saprotrophic fungi (Penttilä et al. 2006). For ectomycorrhizal fungi, species composition may be very different at edges (Dickie and Reich 2005). All these factors may influence host availability and thus successful establishment for mycoheterotrophic plants. Mycoheterotrophic plants that are specialized on particular "rare" fungi may thus be in jeopardy if habitat size proves to be a strong determinant of fungal richness.

Even without destruction or fragmentation, ecosystems can suffer from human activities causing pollution. Pollution that impacts plant and fungal diversity is commonly caused by pesticides, sewage, fertilizers from agricultural fields, industrial chemicals and wastes, emissions from factories and automobiles, and sediment deposits from eroded hillsides (Relyea 2005). Pollution can significantly alter plant species richness, and there is no doubt that its negative impact will affect mycoheterotrophic plants as well (Brandle et al. 2001; Stevens et al. 2004).

3.11.3 Global Climate Change

The emission of greenhouse gasses has been steadily increasing over the past 100 years. There is convincing scientific evidence that the increased levels of greenhouse gasses produced by human activity have affected the world's climate and ecosystems already and that these effects will increase in the future (Primack 2008). Global surface temperatures have increased by 0.6°C during the last century and are likely to increase by 2–4°C by 2100 (Solomon et al. 2007). This global warming will affect a complete set of climate characteristics, leading to a global climate change. The effect of global climate change on rainforests remains poorly known. Different rainforest regions may experience different effects. It is expected that there is severe risk of forest retreat, especially in eastern Amazonia, Central America, and parts of Africa, but there are also indications for potential of expansion in other regions, for example, around the Congo Basin (Malhi et al. 2009; Zelazowski et al. 2011).

Global climate change may be especially harmful for montane forests and their associated mycoheterotrophs. The vegetation zonation on tropical mountains is strongly controlled by temperature. A 3°C warming would result in temperature zones moving 500 m vertically up the mountain, permitting lowland plants to migrate upward and eliminating the species in the highest zones (Pounds et al. 1999; Foster 2001). In the northern temperate zone, it is expected that global warming-induced changes in the ratio of extinctions and colonizations at the northern and southern boundaries of species ranges will result in northward range shifts for many species. However, given the current landscape fragmentation of forest habitats and the short time period involved, colonization will be very difficult for most forest plant species, and their survival will depend on their environmental tolerance. Plants with low environmental tolerance may be replaced by mobile generalist species invading from the south (McKinney and Lockwood 1999; Honnay et al. 2002). The fragmented distribution patterns of many mycoheterotrophic plant species suggests slow dispersal rates and/or high ecological specificity. Therefore, we hypothesize that temperate mycoheterotrophic plants are particularly vulnerable to global climate change. At least for one mycoheterotrophic species, *Hypopitys monotropa* in Europe, the impact of global climate change has been estimated using species distribution modeling based on projected future climate (Beatty and Provan 2011b). The results indicate that climate change will have substantial impact on the distribution range of the species, with a loss of southern and central European populations, and a potential northward expansion. Finally, there is no doubt that global climate change will affect diversity on all levels, including diversity of mycorrhizal fungi (Bellgard and Williams 2011). Changes in community structures of mycorrhizal fungi may have an indirect effect on the occurrence of mycoheterotrophic plants as well.

3.11.4 Disturbance

It is unclear whether (limited) human disturbance, such as selective logging, is a threat for mycoheterotrophs. It is often noted that mycoheterotrophic species prefer areas that have been devoid of disturbance in recent history (Cheek and Williams 1999; Taylor and Roberts 2011). This seems to suggest that mycoheterotrophic plants are sensible to disturbance or that recolonization of areas that have been prone to disturbance is slow. However, little research has been done on this issue. Bergman et al. (2006) studied the distribution of the mycoheterotrophic orchid *Wullschlaegelia calcarata* in a rainforest in Puerto Rico and concluded that its occurrence is correlated with land use history as the orchid was most abundant in areas which had been minimally impacted by human activity. Fensham (1993) reported that a fully mycoheterotrophic *Burmannia* species on Bathurst Island, Australia, prefers monsoon forest sites that are cleared of leaf litter by seasonal flooding, but their occurrence is negatively influenced by disturbance by pigs. However, in Tasmania, *Thismia rodwayi* often occurs at sites that have been subject to intensive and relatively recent forest activities, including clear-cutting and regeneration burns (Roberts et al. 2003). Similarly, there are many records of mycoheterotrophs that have been collected from secondary forests (Maas and Rübsamen 1986; Maas and Ruyters 1986; Lok et al. 2009; Cheek and Vanderburgt 2010; Klooster and Culley 2010; Pendry 2010). In Japan, the rare mycoheterotroph *Petrosavia sakuraii* occurs in *Chamaecyparis obtusa* plantations (Yamato et al. 2011). *Eulophia zollingeri* is frequently found in the introduced *Calliandra calothyrsus* forest in Java (Comber 1990). Therefore, at least some species are able to withstand moderate disturbance or have the ability to migrate into areas that have been affected by human activities. A key factor in the recolonization of secondary forests may be the presence of preferred host fungi, although other factors such as the presence of specific pollinators, seed dispersal barriers, or abiotic requirements may be of importance as well.

Herbivory may also have a negative impact on reproductive success of mycoheterotrophs (Klooster and Culley 2009; Taylor and Roberts 2011). Thus, introduction of herbivores into mycoheterotrophic plant habitats could pose potential threat for the local distribution of mycoheterotrophs. Finally, rare mycoheterotrophic plants may suffer from overzealous botanists, who collect specimens and trample populations during searches (Taylor and Roberts 2011).

3.11.5 Vulnerable Taxa

Only a dozen of mycoheterotrophic plant species are featured in the IUCN Red List (2010). Those that are listed are all categorized as "threatened" and range from "vulnerable" to "critically endangered." Unfortunately, this low number illustrates the incomplete state of the list rather than the absence of vulnerability of mycoheterotrophic species. Due to the continuing loss of rainforest area, it can be assumed that many rainforest mycoheterotrophs are at risk of extinction. In particular, species that are known from only a few localities, close to human settlements or outside protected areas, and species with a very limited distribution ranges and small population sizes are threatened. More intensive studies on the occurrence and distribution of mycoheterotrophic plants, particularly in the tropics, are urgently needed to assess the conservation status of these species. At the current rate of habitat destruction in the tropics, it is likely that some mycoheterotrophic plant species become extinct before they are even discovered.

3.12 Conservation

3.12.1 Habitat Conservation

Protecting populations is the key to preserving species. The best and most straightforward strategy to protect mycoheterotrophic plants is to protect the habitats in which they occur. Large parks are probably the only way to preserve complete rainforest ecosystems, but smaller rainforest reserves can also play a valuable role in the protection of rainforest species. Besides legal protection, investments in management are essential to preserve the biodiversity within protected areas.

3.12.2 Inventory and Monitoring

Our knowledge about the abundance and ecology of most species of mycoheterotrophs remains very poor. Only through careful inventories and observations in the field can the true status of a species and its habitat be determined. In addition, because flowering times of mycoheterotrophic plants can be cryptic, the conservation status of a species can only be established by studying populations over time. Detailed inventory studies and long-term ecological monitoring often lead to the discovery of new populations and new species, stressing the importance of these types of studies as well as our limited knowledge about mycoheterotrophic plant distributions (e.g., Roberts et al. 2003; Franke 2007; Cheek and Vanderburgt 2010).

3.12.3 Establishing New Populations

Conservation of rare and endangered species can be supported by establishing new populations. It is unclear whether this strategy is applicable to rare mycoheterotrophic plants. Seed germination of fully mycoheterotrophic orchids and Ericaceae has been achieved by burying seed packages near ectomycorrhizal trees (McKendrick et al. 2000b; Bidartondo and Bruns 2001; Bidartondo 2005), showing the possibility of introducing (or reintroducing) mycoheterotrophs into existing suitable habitats. However, given the paucity of knowledge on ecological requirements of mycoheterotrophs (mycorrhiza, pollinators, etc.), only crude assessments can be made to ensure a site is suitable for introduction of a species. Given the high ecological specificity of many mycoheterotrophic species, establishing new populations may be extremely difficult. Translocation of mycoheterotrophic plants from natural populations into new sites either as propagated seeds or as adult plants is likely to lead to failure due to the breakage of vital mycorrhizal connections, but if attempted, one should take care to fully characterize the existing mycoheterotrophic flora of the transplant area in order to avoid any negative impacts on existing mycoheterotroph populations.

3.12.4 Ex Situ Conservation

The best strategy for the conservation of biodiversity is the preservation of existing ecosystems. However, if the last remaining populations of a rare and endangered species are too small to maintain the species, if they are declining despite conservation efforts, or if the remaining individuals are found outside protected areas, then in situ—or on-site—preservation may not be effective. It is likely that the only way species in such circumstances can be prevented from going extinct is to maintain individuals in artificial conditions under human supervision (Primack 2008). The worlds' botanic gardens are a safe house for many plant species: about 80,000 species of plants are currently being cultivated (Guerrant et al. 2004), and several of these species are extinct in the wild. Botanic gardens thus play a key role in plant conservation, and this aspect should become even more important in the future. However, due to the complexity of their mode of life, cultivation of mycoheterotrophic plants remains problematic and, in most cases, currently impossible. There are several reports of successful germination and subsequent development of mycoheterotrophic orchids. Umata (1995, 1999) reported successful in vitro germination and formation of lateral roots of *Erythrorchis altissima. Gastrodia elata* and *Epipogium roseum* have been in vitro germinated and grown up to the formation of flowers (Xu and Mu 1990; Xu and Guo 2000; Yagame et al. 2007). In these experiments, the orchids germinated and grew in media that were inoculated with saprotrophic fungi, presumably closely related or identical to those with which the plants grow in nature. Asymbiotic germination to and cultivation to flowering stage have been reported for *Cymbidium macrorhizon* (as *Cymbidium nipponicum*) and *Didymoplexis pallens* (Mizuno et al. 1991; Irawati 2002). Successful long-term cultivation of mycoheterotrophs that are associated with fungi that are mycorrhizal with surrounding trees has not been achieved to date. Bruns and Read (2000) were able to germinate seeds of *Sarcodes* and *Pterospora* (Ericaceae) in vitro with ectomycorrhizal *Rhizopogon* fungi, which were isolated from mature plants in nature, but further development of seedlings could not be accomplished. Adult plants of *Thismia rodwayi, Afrothismia winkleri*, and *A. foertheriana* (Thismiaceae) have been grown for considerable amounts of time after transplantation from the field to the laboratory, but it remains unclear whether an effective tripartite symbiosis (mycoheterotroph–fungus–autotroph) was achieved under these conditions (Wood 2010; Franke, pers. comm.; VM pers. obs.). Establishment of such tripartite symbioses have been successfully accomplished in microcosm studies linking naturally germinated seedlings of the partial mycoheterotroph *Corallorhiza trifida* and seedlings of *Betula* and *Salix* trees by a shared mycorrhizal fungus (McKendrick et al. 2000a). Similar microcosm experiments successfully linked *Aneura mirabilis* and *Betula* seedlings by a shared mycorrhizal fungus (Read et al. 2000). Nevertheless, more research is urgently needed to investigate ex situ cultivation possibilities for mycoheterotrophs.

Ex situ seedling culture and subsequent reintroduction into appropriate habitats have been successfully achieved for orchids (e.g., McKendrick 1995; Yam et al. 2010) and may similarly be used for conservation purposes of mycoheterotrophic plants. However, there is a potential danger to this approach. It is observed that some orchids are able to germinate and develop ex situ with a wider range of mycorrhizal fungi than those found in natural populations (Masuhara and Katsuya 1994; Perkins and McGee 1995). Similar observations have been made for the germination of mycoheterotrophic Ericaceae (Bruns and Read 2000; Bidartondo 2005). Introducing these plants into natural habitats will also lead to the introduction of an "alien" fungus with potential ecological harm (Zettler et al. 2005).

3.13 Conclusions

Mycoheterotrophic plants are found in all forest biomes of the world and show a pronounced preference for damp habitats with primary forest and a closed canopy cover. The diversity of mycoheterotrophic flowering plants peaks in the tropics and is particularly high in Southeast Asia. Many families, genera, and species of mycoheterotrophic plants have a widespread distribution that covers multiple continents. Evidence from molecular clock analyses suggests that most widespread lineages are too young to have acquired their distribution before the breakup of Gondwana. While some transoceanic distribution may be explained by migration over temporary land bridges, oceanic dispersal probably played the most important role in the acquisition of widespread distributions. Due to their minute habit and ephemeral occurrence, mycoheterotrophic plants are not often found and collected. For many species, the paucity of records is also the result of extreme rarity and high local endemism. The rarity of mycoheterotrophic plants is probably related to particular habitat preferences. High specificity toward particular lineages of fungi may significantly limit the dispersal potential of mycoheterotrophs, although specificity toward other biotic (pollinators, dispersal agents, plant communities) and abiotic factors (humidity, soil composition) may be influential as well. The continuous destruction of their habitats and the increase of global temperatures threaten many species with extinction, and a few species may already have disappeared. The protection of their habitats is the best and currently the only way for the effective conservation of mycoheterotrophic plants.

References

Albert VA, Struwe L (1997) Phylogeny and classification of *Voyria* (saprophytic Gentianaceae). Brittonia 49:466–479

Alexander IJ, Lee SS (2005) Mycorrhizas and ecosystem processes in tropical rain forest: implications for diversity. In: Burslem DFRP, Pinard MA, Hartley SE (eds) Biotic interactions in the tropics: their role in the maintenance of species diversity. Cambridge University Press, New York, pp 165–203

Ali JR, Aitchison JC (2008) Gondwana to Asia: plate tectonics, paleogeography and the biological connectivity of the Indian sub-continent from the Middle Jurassic through latest Eocene (166–35 Ma). Earth Sci Rev 88:145–166

Ali JR, Huber M (2010) Mammalian biodiversity on Madagascar controlled by ocean currents. Nature 463:653–656

Arditti J (1992) Fundamentals of orchid biology. Wiley, New York

Audley-Charles MG (1983) Reconstruction of eastern Gondwanaland. Nature 306:48–50

Baird D (2006) Newsletter of the UK Darwin Initiative, Issue6. http://darwin.defra.gov.uk/newsletter/DARWIN_NEWS_6.pdf

Bakshi TS (1959) Ecology and morphology of *Pterospora andromedea*. Bot Gaz 120:203–217

Barkman TJ, Simpson BB (2001) Origin of high-elevation *Dendrochilum* species (Orchidaceae) endemic to Mount Kinabalu, Sabah. Malaysia Syst Bot 26:658–669

Barrett CF, Freudenstein JV (2008) Molecular evolution of *rbcL* in the mycoheterotrophic coralroot orchids (*Corallorhiza* Gagnebin, Orchidaceae). Mol Phylogenet Evol 47:665–679

Beaman JH, Anderson C (2004) The plants of Mount Kinabalu. 5: Dicotyledon families Magnoliaceae to Winteraceae. Natural History Publications (Borneo), Kota Kinabalu

Beaman JH, Beaman RS (1998) The plants of Mount Kinabalu: 3 gymnosperms and non-orchid monocotyledons. Natural History Publications (Borneo), Kota Kinabalu

Beard JS (1990) Temperate forests of the southern hemisphere. Vegetatio 89:7–10

Beatty GE, Provan J (2011a) Phylogeographic analysis of North American populations of the parasitic herbaceous plant *Monotropa hypopitys* L. reveals a complex history of range expansion from multiple late glacial refugia. J Biogeogr 38:1585–1599

Beatty GE, Provan J (2011b) Comparative phylogeography of two related plant species with overlapping ranges in Europe, and the potential effects of climate change on their intraspecific genetic diversity. BMC Evol Biol 11:29

Beentje HJ, Adams B, Davis SD, Hamilton AC (1994) Regional overview: Africa. In: Davis SD, Heywood VH, Hamilton AC (eds) Centres of plant diversity. IUCN Publication Unit, Cambridge, pp 101–148

Bellgard S, Williams S (2011) Response of mycorrhizal diversity to current climatic changes. Diversity 3:8–90

Benzing DH, Atwood JT (1984) Orchidaceae: ancestral habitats and current status in forest canopies. Syst Bot 9:155–1665

Bergman E, Ackerman J, Thompson J, Zimmerman J (2006) Land-use history affects the distribution of the saprophytic orchid *Wullschlaegelia calcarata* in Puerto Rico's tabonuco forest. Biotropica 38:492–499

Bever JD, Richardson SC, Lawrence BM, Holmes J, Watson M (2009) Preferential allocation to beneficial symbiont with spatial structure maintains mycorrhizal mutualism. Ecol Lett 12:13–21

Bidartondo MI (2005) The evolutionary ecology of mycoheterotrophy. New Phytol 167:335–352

Bidartondo MI, Bruns TD (2001) Extreme specificity in epiparasitic Monotropoideae (Ericaceae): widespread phylogenetic and geographic structure. Mol Ecol 10:2285–2295

Bidartondo MI, Bruns TD (2002) Fine-level mycorrhizal specificity in the Monotropoideae (Ericaceae): specificity for fungal species groups. Mol Ecol 11:557–569

Bidartondo MI, Burghardt B, Gebauer G, Bruns TD, Read DJ (2004) Changing partners in the dark: isotopic and molecular evidence of ectomycorrhizal liaisons between forest orchids and trees. Proc R Soc Lond B 271:1799–1806

Bidartondo MI, Redecker D, Hijri I, Wiemken A, Bruns TD, Domìnguez L, Sérsic A, Leake JR, Read DJ (2002) Epiparasitic plants specialized on arbuscular mycorrhizal fungi. Nature 419:389–392

Bougoure JJ, Brundrett M, Brown A, Grierson PF (2008) Habitat characteristics of the rare underground orchid. Aust J Bot 56:501–511

Bougoure JJ, Ludwig M, Brundrett M, Grierson PF (2009) Identity and specificity of the fungi forming mycorrhizas with rare, mycoheterotrophic *Rhizanthella gardneri* (Orchidaceae). Mycol Res 113: 1097–1106

Bossuyt F, Meegaskumbura M, Beenaerts N, Gower DJ, Pethiyagoda R, Roelants K, Mannaert A, Wilkinson M, Bahir MM, Manamendra-Arachchi K, Ng PKL, Schneider CJ, Oommen OV, Milinkovitch MC (2004) Local endemism within the western Ghats-Sri Lanka biodiversity hotspot. Science 306:479–481

Bossuyt F, Milinkovitch MC (2001) Amphibians as indicators of Early Tertiary "Out-of-India" dispersal of vertebrates. Science 292:93–95

Brandle M, Amarell U, Auge H, Klotz S, Brandl R (2001) Plant and insect diversity along a pollution gradient: understanding species richness across trophic levels. Biodivers Conserv 10:1497–1511

Bruns TD, Read DJ (2000) In vitro germination of nonphotosynthetic, mycoheterotrophic plants stimulated by fungi isolated from the adult plants. New Phytol 148:335–342

Cameron DD, Bolin JF (2010) Isotopic evidence of partial mycoheterotrophy in the Gentianaceae: *Bartonia virginica* and *Obolaria virginica* as case studies. Am J Bot 97:1272–1277

Campbell EO (1968) An investigation of *Thismia rodwayii* F. Muell and its associated fungus. Trans R Soc N Z Bot 3:209–219

Carlquist S (1967) The biota of long-distance dispersal. V. Plant dispersal to the Pacific Islands. Bull Torr Bot Club 64:129–162

Carlquist S (1980) Hawaii. A natural history, 2nd edn. SB Printers, Honolulu

Chantanaorrapint S (2008) *Thismia angustimitra* (Thismiaceae), a new species from Thailand. Blumea 53:524–526

Cheek M (2003a) A new species of *Afrothismia* (Burmanniaceae) from Kenya. Kew Bull 58:951–955

Cheek M (2003b) Kupeaeae, a new tribe of Triuridaceae from Africa. Kew Bull 58:939–949

Cheek M (2006) African saprophytes: new discoveries. In: Ghanzafar SA, Beentje HJ (eds) Taxonomy and ecology of African plants, their conservation and sustainable use. Royal Botanic Gardens, Kew, pp 693–697

Cheek M, Cable S (1997) Plant inventory for conservation management: the Kew Earthwatch programme in Western Cameroon, 1993–1996. In: Doolan S (ed) African rainforest and the conservation of biodiversity, Proceedings of the Limbe Conference, 17–24 January 1997. Limbe Botanic Garden, Limbe, Cameroon, pp 29–38

Cheek M, Ndam N (1996) Saprophytic flowering plants of Mount Cameroon. In: van der Maesen LJG, van der Burgt XM, van Medenbach de Rooy JM (eds) The biodiversity of African plants. Kluwer Academic Publishers, Dordrecht, pp 612–617

Cheek M, Pollard BJ, Darbyshire I, Onana J-M, Wild C (2004) The plants of Kupe, Mwanenguba and the Bakossi Mountains, Cameroon: a conservation checklist. Kew Publishing, Kew

Cheek M, Vanderburgt X (2010) *Gymnosiphon samoritoureanus* (Burmanniaceae) a new species from Guinea, with new records of other achlorophyllous heteromycotrophs. Kew Bull 65:83–88

Cheek M, Williams S (1999) A review of African saprophytic flowering plants. In: Timberlake J, Kativu S (eds) African plants: biodiversity, taxonomy. Royal Botanic Gardens, Kew, pp 39–49

Cheek M, Williams S, Brown A (2008) *Gymnosiphon marieae* sp. nov. (Burmanniaceae) from Madagascar, a species with tepal-mediated stigmatic extension. Nord J Bot 26:230–234

Cheek M, Williams S, Etuge M (2003) *Kupea martinetugei*, a new genus and species of Triuridaceae from western Cameroon. Kew Bull 58:225–228

Chen S, Ma H, Parnell JAN (2008) Polygalaceae. Missouri Botanical Garden, St. Louis, MO & Harvard University Herbaria, Cambridge, MA. Published on the Internet http://www.efloras.org. Accessed June 2011

Chew R (2004) *Thismia americana*. A mystery that still haunts—and helps—the Calumet region. *Chicago Wilderness Magazine*. Published online http://www.chicagowildernessmag.org/issues/summer2004/thismia.html

Clarke HD, Funk VA (2005) Using checklists and collections data to investigate plant diversity: II. An analysis of five florulas from northeastern South America. Proc Acad Nat Sci Phil 154:29–37

Collier F, Bidartondo MI (2009) Waiting for fungi: the ectomycorrhizal invasion of lowland heathlands. J Ecol 97:950–963

Comber JB (1990) Orchids of java. Royal Botanic Gardens, Kew

Cooper A, Lalueza-Fox C, Anderson S, Rambaut A, Austin J, Ward R (2001) Complete mitochondrial genome sequences of two extinct moas clarify ratite evolution. Nature 409:704–707

Copeland HF (1939) The structure of *Monotropsis* and the classification of the Monotropoideae. Madroño 5:105–119

Courty P-E, Walder F, Boller T, Ineichen K, Wiemken A, Rousteau A, Selosse M-A (2011) C and N metabolism in mycorrhizal networks and mycoheterotrophic plants of tropical forests: a stable isotope analysis. Plant Physiol 156:952–961

Cox CB, Moore PD (2010) Biogeography. An ecological and evolutionary approach, 8th edn. Wiley, Hoboken

Cribb P, Fischer E, Killmann D (2010) A revision of *Gastrodia* (Orchidaceae: Epidendroideae, Gastrodieae) in tropical Africa. Kew Bull 65:315–321

Cribb P, Wilkin P, Clements M (1995) Corsiaceae: a new family for the Falkland Islands. Kew Bull 50: 171–172

Crum H, Bruce J (1996) A new species of *Cryptothallus* from Costa Rica. Bryologist 99:433–438

Daniel DF (2010) *Sciaphila ledermannii* (Triuridaceae), a biogeographically significant holosaprophyte newly reported from Príncipe in the Gulf of Guinea. Proc Cal Acad Sci 15:617–622

Dauby G, Parmentier I, Stévart T (2007) *Afrothismia gabonensis* sp. nov. (Burmanniaceae) from Gabon. Nord J Bot 25:268–271

Davies P, Davies J, Huxley A (1988) Wild orchids of Britain and Europe. The Hogarth Press, London

Davis CC, Bell CD, Mathews S, Donoghue MJ (2002) Laurasian migration explains Gondwanan disjunctions: evidence from Malpighiaceae. Proc Natl Acad Sci USA 99:6833–6837

Davis AP, Mvungi EF (2004) Two new and endangered species of *Coffea* (Rubiaceae) from the Eastern Arc Mountains (Tanzania) and notes on associated conservation issues. Bot J Linn Soc 146:237–245

Davis MB (1983) Quaternary history of deciduous forests of eastern North America and Europe. Ann Missouri Bot Gard 70:550–563

de Candolle A (1820) Essai élementaire de géographie botanique. In: Lomolino MV, Sax DF, Brown JH (eds) Foundations of biogeography. University of Chicago Press, Chicago, pp 28–48

de Queiroz A (2005) The resurrection of oceanic dispersal in historical biogeography. Trends Ecol Evol 20:68–73

Dickie IA, Reich PB (2005) Ectomycorrhizal fungal communities at forest edges. J Ecol 93:244–255

Diez JM (2007) Hierarchical patterns of symbiotic orchid germination linked to adult proximity and environmental gradients. J Ecol 95:159–170

Dixon KW (1991) Seeder/clonal concepts in Western Australian orchids. In: Wells TCE, Willems JH (eds) Population ecology of terrestrial orchids. SPB Academic Publishing, The Hague, pp 111–124

Dixon KW (2003) *Rhizanthella gardneri*. Orchidaceae. Curtis's Bot Mag 20:94–100

Donoghue MJ, Bell CD, Li J (2001) Phylogenetic patterns in northern hemisphere plant geography. Int J Plant Sci 162:S41–S52

Donoghue M, Smith S (2004) Patterns in the assembly of temperate forests around the Northern Hemisphere. Philos Trans R Soc Lond B 359:1633–1644

Dressler RL (1981) The orchids. Harvard University Press, London

Eckenwalder JE (2009) Conifers of the world. Timber Press, Portland

Eeley HAC, Lawes MJ, Piper SE (1999) The influence of climate change on the distribution of indigenous forest in KwaZulu-Natal, South Africa. J Biogeogr 26:595–617

Ek RC, ter Steege H (1997) The flora of the Mabura Hill area, Guyana. In: Ek RC (ed) Botanical diversity in the tropical rainforest of Guyana. Tropenbos-Guyana Series. Koeltz Scientific Books, Koenigstein, pp 1–85

Engler HGA (1888) Burmanniaceae. In: Engler HGA, Prantl K (eds) Die Natürlichen Pflanzenfamilien, 1st edn. W. Engelmann, Leipzig, pp 44–51

Eriksson O, Kainulainen K (2011) The evolutionary ecology of dust seeds. Perspect Plant Ecol Evol Syst 13: 73–87

Feild TS, Brodribb TJ (2005) A unique mode of parasitism in the conifer coral tree *Parasitaxus ustus* (Podocarpaceae). Plant Cell Environ 28:1316–1325

Fensham RJ (1993) The impact on pig rooting on populations of *Burmannia* sp., a rare rainforest herb on Bathurst Island. Proc Roy Soc Queensland 103:5–12

Fontenla S, Godoy R, Rosso P, Havrylenko M (1998) Root associations in *Austrocedrus* forests and seasonal dynamics of arbuscular mycorrhizas. Mycorrhiza 8: 29–33

Fosberg F, Sachet M (1980) A new *Sciaphila* (Triuridaceae) from the Palau Islands. Pac Sci 34:25–26

Foster P (2001) The potential negative impacts of global climatic change on tropical montane cloud forests. Earth Sci Rev 55:73–106

Franke T (2002) The myco-heterotrophic *Voyria flavescens* (Gentianaceae) and its associated fungus. Mycol Prog 1:367–376

Franke T (2004) *Afrothismia saingei* (Burmanniaceae, Thismieae), a new myco-heterotrophic plant from Cameroon. Syst Geogr Pl 74:27–33

Franke T (2007) Miscellaneous contributions to the taxonomy and mycorrhiza of AMF- exploiting mycoheterotrophic plants. PhD. thesis. München: Fakultät für Biologie der Ludwig-Maximilians-Universität

Franke T, Sainge MN, Agerer R (2004) A new species of *Afrothismia* (Burmanniaceae; tribe: Thismieae) from the western foothills of Mount Cameroon. Blumea 49:451–456

Freudenstein JV, Senyo DM (2008) Relationships and evolution of *matK* in a group of leafless orchids (*Corallorhiza* and Corallorhizinae; Orchidaceae: Epidendroideae). Am J Bot 94:498–505

Furness CA, Rudall P, Eastman A (2002) Contribution of pollen and tapetal characters to the systematics of Triuridaceae. Plant Syst Evol 235:209–218

Gandolfo MA, Nixon KC, Crepet WL (2002) Triuridaceae fossil flowers from the Upper Cretaceous of New Jersey. Am J Bot 89:1940–1957

Gandolfo MA, Nixon KC, Crepet WL, Sevenson DW (1998) Oldest known fossils of monocotyledons. Nature 394:532–533

Gebauer G, Meyer M (2003) 15N and 13C natural abundance of autotrophic and myco- heterotrophic orchids provides insight into nitrogen and carbon gain from fungal association. New Phytol 160:209–223

George AS (1980) *Rhizanthella gardneri* R. S. Rogers the underground orchid of Western Australia. Am Orchid Soc Bull 49:631–646

Givnish TJ, Millam KC, Evans TM, Hall JC, Pires JC, Berry PE, Sytsma KJ (2004) Ancient vicariance or recent long-distance dispersal? Inferences about phylogeny and South American-African disjunctions in Rapateaceae and Bromeliaceae based on ndhF sequence data. Int J Plant Sci 165:S35–S54

Goldblatt P, Manning JC (2010) *Geosiris albiflora* (Geosiridoideae), a new species from the Comoro Archipelago. Bothalia 40:169–171

Good R (1974) The geography of flowering plants, 4th edn. Longman, London

Govaerts R, Pfahl J, Campacci MA, Holland Baptista D, Tigges H, Shaw J, Cribb P, George A, Kreuz K, Wood J (2011) World Checklist of Orchidaceae. The Board of Trustees of the Royal Botanic Gardens, Kew. Published on the Internet; http://www.kew.org/wcsp/. Accessed 26 May 2011

Graham R (1953) Epipogium aphyllum Sw. in Buckinghamshire. Watsonia 3:33

Guerrant EO, Havens K, Maunder M (2004) Ex situ conservation of species survival in the wild. Island Press, Washington, DC

Groenendijk JP, van Dulmen ATJ, Bouman F (1997) The "forest floor" saprophytes *Voyria spruceana* and *V. aphylla* (Gentianaceae) growing as epiphytes in Colombian Amazonia. Ecotropica 3:129–131

Hall R (1998) The plate tectonics of Cenozoic SE Asia and the distribution of land and sea. In: Hall R, Holloway JD (eds) Biogeography and geological evolution of SE Asia. Backhuys Publishers, Leiden, pp 99–131

Harper JL (1981) The meaning of rarity. In: Synge H (ed) The biological aspects of rare plant conservation. Wiley, New York, pp 189–203

Hazard C, Lilleskov EA, Horton TR (2011) Is rarity of pinedrops (*Pterospora andromedea*) in eastern North America lined to rarity of its unique fungal symbiont? Mycorrhiza 22:393–402

Heads M (2009) Inferring biogeographic history from molecular phylogenies. Biol J Linn Soc 98:757–774

Heaney LR, Regalado JG (1998) Vanishing treasures of the Philippine rain forest. University of Chicago Press, Chicago

Hentrich H, Kaiser R, Gottsberger G (2010) The reproductive biology of *Voyria* (Gentianaceae) species in French Guiana. Taxon 59:867–880

Herzer RH, Chaproniere GCH, Edwards AR, Hollis CJ, Pelletier B, Raine JI, Scott GH, Stag-Poole V, Strong CP, Symonds P, Wilson GJ, Zhu H (1997) Seismic stratigraphy and structural history of the Reinga Basin and its margins, southern Norfolk Ridge system. New Zeal J Geol Geophys 40:425–451

Hill R (2004) Origins of the southeastern Australian vegetation. Philos Trans R Soc B 359:1537–1549

Hofer U, Bersier LF, Bocard D (2000) Ecotones and gradient as determinants of herpetofaunal community structure in the primary forest of Mount Kupe. Cameroon J Trop Ecol 16:517–533

Honnay O, Verheyen K, Butaye J, Jacquemyn H, Bossuyt B, Hermy M (2002) Possible effects of habitat fragmentation and climate change on the range of forest species. Ecol Lett 5:525–530

Hoorn C, Wesselingh F, ter Steege H, Bermudez M, Mora A, Sevink J, Sanmartin I, Sanchez-Meseguer A, Anderson C, Figueiredo J, Jamaerillo C, Riff D, Negri F, Hooghiemstra H, Lundberg J, Stadler T, Särkinen T, Antonelli A (2010) Amazonia through time: Andean uplift, climate change, landscape evolution, and biodiversity. Science 330:927–931

Ibisch PL, Neinhuis C, Rojas NP (1996) On the biology, biogeography, and taxonomy of *Arachnitis* Phil nom. cons. (Corsiaceae) in respect to a new record from Bolivia. Willdenowia 26:321–332

Irawati HJ (2002) Seed germination of the mycotrophic orchid, *Didymoplexis pallens*. In Proceedings of the 17th World Orchid Conference. Natural History Publications, Kota Kinabalu, p 299–301

Iriondo M, Latrubesse EM (1994) A probable scenario for a dry climate in Central Amazonia during the Late Quaternary. Quat Int 21:121–128

IUCN (2010) IUCN Red List of Threatened Species. Version 2010.3. http://www.iucnredlist.org. Downloaded on 02 Sep 2010

Jacquemyn H, Brys R, Vandepitte K, Honnay O, Roldán-Ruiz I, Wiegand T (2007) A spatially explicit analysis of seedling recruitment in the terrestrial orchid *Orchis purpurea*. New Phytol 176:448–459

Jaffré T (1992) Floristic and ecological diversity of the vegetation on ultramafic rocks in New Caledonia. In: Baker AJM, Proctor J, Reeves RD (eds) The vegetation of ultramafic (serpentine) soils. Intercept Ltd, Andover, pp 101–107

Jaffré T, Bouchet P, Veillon J-M (1998) Threatened plants of New Caledonia: is the system of protected areas adequate? Biodivers Conserv 7:109–135

James TY, Porter D, Hamrick JL, Vilgalys R (1999) Evidence for limited intercontinental gene flow in the cosmopolitan mushroom, *Schizophyllum commune*. Evolution 53:1665–1677

Jarvie JK (1996) *Thismia lauriana* (Burmanniaceae), a new species from Central Kalimantan. Blumea 41:257–259

Jonker FP (1938) A monograph of the Burmanniaceae. Meeded Bot Mus Herb Rijks Univ Utrecht 51:1–279

Kennedy AH, Watson LE (2010) Species delimitations and phylogenetic relationships within the fully mycoheterotrophic *Hexalectris* (Orchidaceae). Syst Bot 35:64–76

Kjøller R, Bruns TD (2003) *Rhizopogon* spore bank communities within and among California pine forests. Mycologia 95:603–616

Klooster M, Culley T (2009) Comparative analysis of the reproductive ecology of *Monotropa* and *Monotropsis*: two mycoheterotrophic genera in the Monotropoideae. Am J Bot 96:1337–1347

Klooster M, Culley T (2010) Population genetic structure of the mycoheterotroph *Monotropa hypopitys* L. (Ericaceae) and differentiation between red and yellow color forms. Int J Plant Sci 171:167–174

Kroenke LW, Rodda P (1984) Cenozoic tectonic development of the Southwest Pacific. United Nations Economic and Social Commission, Committee for Co-ordination of Joint Prospecting for Mineral Resources in South Pacific Offshore Areas (CCOP/SOPAC)

Kruckeberg AR, Rabinowitz D (1985) Biological aspects of endemism in higher-plants. Annu Rev Ecol Syst 16:447–479

Laurance WF, Lovejoy TE, Vasconcelos HL, Bruna EM, Didham RK, Stouffer PC, Gascon C, Bierregaard RO, Laurance SG, Sampaio E (2002) Ecosystem decay of Amazonian forest fragments: a 22-year investigation. Conserv Biol 16:605–618

Leake JR (1994) The biology of myco-heterotrophic ('saprophytic') plants. New Phytol 127:171–216

Lee SS (1990) The mycorrhizal association of the Dipterocarpaceae in the tropical rain forests of Malaysia. Ambio 19:383–385

Lewis DQ (2002) Burmanniaceae Blume—Burmannia family. In: Argus GW, Gandhi K, Goldblatt P, Hess WJ, Kiger RW, Strother JL, Utech FH, Zarucchi JL (eds) Flora of North America, vol. 26: Magnoliophyta: Liliidae: Liliales and Orchidales. Oxford University Press, New York, pp 486–489

Liebel HT, Bidartondo MI, Preiss K, Segreto R, Stöckel M, Rodda M, Gebauer G (2010) C and N stable isotope signatures reveal constraints to nutritional modes in orchids from the Mediterranean and Macaronesia. Am J Bot 97:903–912

Linder HP (2001) Plant diversity and endemism in sub-Saharan tropical Africa. J Biogeogr 28:169–182

Linder HP, Kurzweil H (1999) Orchids of southern Africa. Balkema, Rotterdam

Linder HP, Kurzweil H, Johnson DD (2005) The Southern African orchid flora: composition, sources and endemism. J Biogeogr 32:29–47

Lok AFSL, Ang WF, Tan HTW (2009) The status of *Gastrodia javanica* (Bl.) Lindl. in Singapore. Nat Sing 2:415–419

Luer CA (1975) The native orchids of the United States and Canada excluding Florida. New York Botanical Garden, New York

Lumbsch HT, Buchanan PK, May TW, Mueller GM (2008) Phylogeography and biogeography of fungi. Mycol Res 112:423–424

Maas H, Maas PJM (2005) Flora da Reserva Ducke, Amazonas, Brasil: Burmanniaceae. Rodriguésia 56: 125–130

Maas PJM, Maas-van de Kamer H, van Benthem J, Snelders HCM, Rübsamen T (1986) Burmanniaceae. Fl Neotrop 42:1–189

Maas PJM, Rübsamen T (1986) Triuridaceae. Fl Neotrop 40:1–55

Maas PJM, Ruyters P (1986) *Voyria* and *Voyriella* (saprophytic Gentianaceae). Fl Neotrop 41:1–93

Macey JR, Schulte JA II, Larson A, Ananjeva NB, Wang Y, Pethiyagoda R, Rastegar-Pouyani N, Papenfuss TJ (2000) Evaluating trans-Tethys migration: an example using acrodont lizard phylogenetics. Syst Biol 49:233–256

Malhi Y, Aragao LEOC, Galbraith D, Huntingford C, Fisher R, Zelazowski P, Sitch S, McSweeney C, Meir P (2009) Exploring the likelihood and mechanism of a climate-change-induced dieback of the Amazon rainforest. Proc Natl Acad Sci USA 106: 20610–20615

Malme GOA (1896) Die Burmannien der ersten Regnel'schen Expedition. Bih Kongl Svenska Vetensk Akad Handl 22:1–32

Martín-Lopez B, Montes C, Benayas J (2007) The non-economic motives behind the willingness to pay for biodiversity conservation. Biol Conserv 139:67–82

Martos F, Dulormne M, Pailler T, Bonfante P, Faccio A, Fournel J, Dubois M-P, Selosse M-A (2009) Independent recruitment of saprotrophic fungi as mycorrhizal partners by tropical achlorophyllous orchids. New Phytol 184:668–681

Masuhara G, Katsuya K (1994) *In situ* and *in vitro* specificity between *Rhizoctonia* spp. and *Spiranthes sinensis* (Persoon.) Ames. var. amoena (M. Beiberstein) Hara (Orchidaceae). New Phytol 127:711–718

Matheny PB, Aime MC, Bougher NL, Buyck B, Desjardin DE, Horak E, Kropp BR, Lodge DJ, Soytong K, Trappe JM, Hibbett DS (2009) Out of the palaeotropics? Historical biogeography and diversification of the cosmopolitan ectomycorrhizal mushroom family Inocybaceae. J Biogeogr 36:577–592

McCall RA (1997) Implications of recent geological investigations of the Mozambique Channel for the mammalian colonization of Madagascar. Proc R Soc Lond B 264:663–665

McKendrick SL (1995) The effects of herbivory and vegetation on laboratory-raised *Dactylorhiza praetermissa* (Orchidaceae) planted into grassland in southern England. Biol Conserv 73:215–220

McKendrick SL, Leake JR, Read DJ (2000a) Symbiotic germination and development of myco-heterotrophic plants in nature: transfer of carbon from ecto-mycorrhizal *Salix repens* and *Betula pendula* to the orchid *Corallorhiza trifida* through shared hyphal connections. New Phytol 145:539–548

McKendrick SL, Leake JR, Taylor DL, Read DJ (2000b) Symbiotic germination and development of mycoheterotrophic plants in nature: ontogeny of *Corallorhiza trifida* and characterization of its mycorrhizal fungi. New Phytol 145:523–537

McKinney ML, Lockwood JL (1999) Biotic homogenization: a few winners replacing many losers in the next mass extinction. Trends Ecol Evol 14:450–453

Merckx V, Bidartondo MI (2008) Breakdown and delayed cospeciation in the arbuscular mycorrhizal mutualism. Proc R Soc Lond B 275:1029–1035

Merckx V, Chatrou LW, Lemaire B, Sainge MN, Huysmans S, Smets EF (2008) Diversification of mycoheterotrophic angiosperms: evidence from Burmanniaceae. BMC Evol Biol 8:178–194

Merckx V, Huysmans S, Smets E (2010a) Cretaceous origins of myco-heterotrophic lineages in Dioscoreales. In: Seberg O, Petersen G, Barfod AS, Davis J (eds) Diversity, phylogeny, and evolution in the monocotyledons. Århus University Press, Århus, pp 39–53

Merckx V, Stöckel M, Fleischmann A, Bruns TD, Gebauer G (2010b) 15N and 13C natural abundance of two mycoheterotrophic and a putative partially mycoheterotrophic species associated with arbuscular mycorrhizal fungi. New Phytol 188:590–596

Mesta DK, Hegde HV, Upadhya V, Kholkute SD (2011) *Burmannia championii* Thwaites (Dioscoreales: Burmanniaceae), a new addition to the flora of Karnataka. J Threat Taxa 3:1465–1468

Milne RI, Abbott RJ (2002) The origin and evolution of tertiary relict floras. Adv Bot Res 38:281–314

Mittermeier RA, Myers N, Gil PR, Mittermeier CG (1999) Hotspots: earth's biologically richest and most endangered terrestrial ecoregions. Cemex, Conservation International and Agrupacion Sierra Madre, Monterrey

Mizuno N, Hiyama I, Higuchi H (1991) Aseptic culture, in vitro flowering, and in vitro fruiting of a mycoparasitic orchid, *Cymbidium nipponicum*. In Proceedings of NIOC'91. NIOC, Nagoya, p 141–143

Molvray M, Kores PJ, Chase MW (2000) Polyphyly of mycoheterotrophic orchids and functional influences on floral and molecular characters. In: Wilson KL, Morrison DA (eds) Monocots: systematics and evolution. CSIRO, Melbourne, pp 441–447

Moncalvo JM, Buchanan PK (2008) Molecular evidence for long distance dispersal across the southern hemisphere in the *Ganoderma applanatum-australe* species complex (Basidiomycota). Mycol Res 112:425–436

Moore LB, Edgar E (1970) Flora of New Zealand. Volume II. Indigenous Tracheophyta—Monocotyledons except Graminae. First electronic edition, Landcare Research, June 2004. Transcr. A.D. Wilton, and I.M.L. Andres. http://FloraSeries.LandcareResearch.co.nz. Accessed 10 May 2011

Morellato LPC, Haddad CFB (2000) Introduction: the Brazilian Atlantic Forest. Biotropica 32:786–792

Morley RJ (2000) Origin and evolution of tropical rain forests. Wiley, New York

Moyersoen B (2006) *Pakaraimaea dipterocarpacea* is ectomycorrhizal, indicating an ancient Gondwanaland origin for the ectomycorrhizal habit in Dipterocarpaceae. New Phytol 172:753–762

Murray-Smith C, Brummit NA, Oliveira-Filho AT, Bachman S, Moat J, Lughada EM, Lucas EJ (2009) Plant diversity hotspots in the Atlantic coastal forests of Brazil. Conserv Biol 23:151–163

Myers N, Mittelmeier RA, Mittelmeier CG, da Fonseca GAB, Kent J (2000) Biodiversity hotspots for conservation priorities. Nature 403:853–858

NASA Earth Observations (2011) http://neo.sci.gsfc.nasa.gov/. Accessed May 2011

Neyland R, Hennigan M (2003) A phylogenetic analysis of large-subunit (26 S) ribosome DNA sequences suggests that the Corsiaceae are polyphyletic. New Zeal J Bot 41:1–11

Ogura-Tsujita Y, Gebauer G, Hashimoto T, Umata H, Yukawa T (2009) Evidence for novel and specialised mycorrhizal parasitism: the orchid *Gastrodia confusa* gains carbon from saprotrophic *Mycena*. Proc R Soc Lond B 267:761–767

Ogura-Tsujita Y, Yukawa T (2008) High mycorrhizal specificity in a widespread mycoheterotrophic plant *Eulophia zollingeri* (Orchidaceae). Am J Bot 95:93–97

Oliveira AA, Daly DC (1999) Geographic distribution of tree species occurring in the region of Manaus, Brazil: implications for regional diversity and conservation. Biodivers Conserv 8:1245–1259

Oliveira AA, Mori SA (1999) A central Amazonian terra firme forest. I. High tree species richness on poor soils. Biodivers Conserv 8:1219–1244

Öpik M, Vanatoa A, Vanatoa E, Moora M, Davison J, Kalwij JM, Reier U, Zobel M (2010) The online database MaarjAM reveals global and ecosystemic distribution patterns in arbuscular mycorrhizal fungi (Glomeromycota). New Phytol 188:223–241

Paul VN (1964) Epipogium aphyllum. Reading Nat 16:29–30

Peay KG, Bidartondo MI, Arnold A (2010a) Not every fungus is everywhere: scaling to the biogeography of fungal–plant interactions across roots, shoots and ecosystems. New Phytol 185:878–882

Peay KG, Bruns TD, Kennedy PG, Bergemann SE, Garbelotto M (2007) A strong species-area relationship for eukaryotic soil microbes: island size matters for ectomycorrhizal fungi. Ecol Lett 10:470–480

Peay KG, Garbelotto M, Bruns TD (2010b) Evidence of dispersal limitation in soil microorganisms: isolation reduces species richness on mycorrhizal tree islands. Ecology 91:3631–3640

Pendry CA (2010) *Epirixanthes compressa* Pendry, a new mycoheterotrophic species of Polygalaceae from Thailand. Thai For Bull 38:184–186

Penttilä R, Lindgren M, Miettinen O, Rita H, Hanski I (2006) Consequences of forest fragmentation for polyporous fungi at two spatial scales. Oikos 114: 225–240

Perkins AJ, McGee PA (1995) Distribution of the orchid mycorrhizal fungus *Rhizoctonia solani*, in relation to its host, *Pterostylis acuminata*, in the field. Aust J Bot 43:565–575

Petersen G, Seberg O, Davis JI, Goldman DH, Stevenson DW, Campbell LM, Michelangeli FA, Specht CD, Chase CW, Fay MF, Pires JC, Freudenstein JV, Hardy CR, Simmons MP (2006) Mitochondrial data in monocot phylogenetics. In: Columbus JT, Friar EA, Porter JM, Prince LM, Simpson MG (eds) Monocots: comparative biology and evolution (excluding Poales). Rancho Santa Ana Botanic Garden, Claremont, pp 52–62

Pfeiffer NE (1914) Morphology of *Thismia americana*. Bot Gaz 57:122–135

Phillips RD, Barrett MD, Dixon KW, Hopper SD (2011) Do mycorrhizal symbioses cause rarity in orchids? J Ecol 99:858–869

Pimm SL, Raven P (2000) Extinction by numbers. Nature 403:843–845

Plana V (2004) Mechanisms and tempo of evolution in the African Guineo-Congolian rainforest. Philos Trans R Soc Lond B 359:1585–1594

Pounds JA, Fogden MPL, Campbell JH (1999) Biological response to climate change on a tropical mountain. Nature 398:611–615

Prance GT (1994) A comparison of the efficacy of higher taxa and species numbers in the assessment of biodiversity in the Neotropics. Philos Trans R Soc Lond B 345:89–99

Prance GT, Beentje H, Dransfield J, Johns R (2000) The tropical flora remains undercollected. Ann Missouri Bot Gard 87:67–71

Preiss K, Adam IK, Gebauer G (2010) Irradiance governs exploitation of fungi: fine-tuning of carbon gain by two partially mycoheterotrophic orchids. Proc R Soc Lond B 277:1333–1336

Primack RB (2008) A primer of conservation biology, 4th edn. Sinauer Associates, Sunderland

Primack RB, Corlett RT (2005) Tropical rainforests: an ecological and biogeographical comparison. Blackwell Publishing, Malden

Provan J, Bennett K (2008) Phylogeographic insights into cryptic glacial refugia. Trends Ecol Evol 23:564–571

Rabinowitz PD, Coffin MF, Falvey D (1983) The separation of Madagascar and Africa. Science 220:67–69

Ramirez SR, Gravendeel B, Singer RB, Marshall CR, Pierce NE (2007) Dating the origin of the Orchidaceae from a fossil orchid with its pollinator. Nature 448:1042–1045

Raven PH, Axelrod DI (1974) Angiosperm biogeography and past continental movements. Ann Missouri Bot Gard 61:539–673

Read DJ, Duckett JG, Francis R, Ligrone R, Russell A (2000) Symbiotic fungal associations in 'lower' land plants. Philos Trans R Soc Lond B 355:815–831

Relyea RA (2005) The impact of insecticides and herbicides on the biodiversity and productivity of aquatic communities. Ecol App 15:618–627

Renner S (2004) Plant dispersal across the tropical Atlantic by wind and sea currents. Int J Plant Sci 165:S23–S33

Renner SS (2005) Relaxed molecular clocks for dating historical plant dispersal events. Trends Plant Sci 10:550–558

Ribeiro JELS, Hopkins MJG, Vicentini A, Sothers CA, Costa MAS, Brito JM, Souza MAD, Martins LH, Lohmann LG, Assunção PA, Pereira EC, Silva CF, Mesquita MR, Procópio LC (1999) Flora da Reserva Ducke: Guia de identificação das plantas vasculares de uma floresta de terrafirme na Amazônia Central. Instituto Nacional de Pesquisas da Amazônia, Manaus

Richards PW (1976) The tropical rain forest. University Press, Cambridge

Roberts N, Wapstra M, Duncan F, Woolley A, Morley J, Fitzgerald N (2003) Shedding some light on *Thismia rodwayi* F. Muell. (fairy lanterns) in Tasmania: distribution, habitat and conservation status. Pap Proc Roy Soc Tasmania 137:55–66

Roy M, Whatthana S, Richard F, Vessabutr S, Selosse M-A (2009) Mycoheterotrophic orchids from Thailand tropical dipterocarpacean forests associate with a broad diversity of ectomycorrhizal fungi. BMC Biol 7:51

Rübsamen T (1986) Morphologische, embryologische und systematische Untersuchungen an Burmanniaceae und Corsiaceae (Mit Ausblick auf die Orchidaceae-Apostasioideae). Diss Bot 902:1–310

Sainge MN, Franke T (2005) A new species of *Afrothismia* (Burmanniaceae) from Cameroon. Nordic J Bot 23:299–303

Sainge MN, Franke T, Agerer R (2005) A new species of *Afrothismia* (Burmanniaceae, tribe Thismieae) from Korup National Park, Cameroon. Willdenowia 35: 287–291

Salazar GA, Freudenstein JV (1998) Identity and typification of *Corallorhiza punctata* and *C. bulbosa* (Orchidaceae). Taxon 47:51–54

Sanmartin I, Ronquist F (2004) Southern hemisphere biogeography inferred by event-based models: plant versus animal patterns. Syst Biol 53:216–243

Sasidharan N, Sujanapal P (2000) Rediscovery of *Haplothismia exannulata* Airy Shaw (Burmanniaceae) from its type locality. Rheedea 10:131–134

Schlechter R (1906) Burmanniacae africanae. Bot Jahrb Syst 38:137–143

Schlechter R (1921) Die Thismieae. Notizbl Bot Gart Berlin-Dahlem 8:31–45

Smith SE, Read DJ (2008) Mycorrhizal symbiosis, 3rd edn. Academic, London

Snetselaar KM, Whitney KD (1990) Fungal calcium oxalate in mycorrhizae of *Monotropa uniflora*. Can J Bot 68:533–543

Sodhi NS, Koh LP, Brook BW, Ng PKL (2004) Southeast Asia biodiversity: an impending disaster. Trends Ecol Evol 19:654–660

Solomon S, Qin D, Manning M, Chen Z, Marquis M, Averyt KB, Tignor M, Miller HL (2007) Contribution of Working Group I to the Fourth Assessment Report of the Intergovernmental Panel on Climate Change. University Press, Cambridge

Stevens CJ, Dise NB, Mountford JO, Gowing DJ (2004) Impact of nitrogen deposition on species richness of grasslands. Science 303:1876–1879

Stone B (1980) Rediscovery of *Thismia clavigera* (Becc.) F.v.M. (Burmanniaceae). Blumea 26:419–425

Storey M, Mahoney JJ, Saunders AD, Duncan RA, Kelley SP, Coffin MF (1995) Timing of hot spot-related volcanism and the breakup of Madagascar and India. Science 267:852–855

Summerhayes VS (1951) Wild orchids of Britain. Collins, London

Swarts ND, Dixon KW (2009) Terrestrial orchid conservation in the age of extinction. Ann Bot 104: 543–556

Swarts ND, Sinclair EA, Francis A, Dixon KW (2010) Ecological specialization in mycorrhizal symbiosis leads to rarity in an endangered orchid. Mol Ecol 19:3226–3242

Taylor DL, Bruns TD (1999) Population, habitat and genetic correlates of mycorrhizal specialization in the 'cheating' orchids *Corallorhiza maculata* and *C. mertensiana*. Mol Ecol 8:1719–1732

Taylor L, Roberts D (2011) Biological flora of the British Isles: *Epipogium aphyllum* Sw. J Ecol 99:878–890

Taylor JW, Turner E, Townsend JP, Dettman JR, Jacobson D (2006) Eukaryotic microbes, species recognition and the geographic limitation of species: examples from the kingdom fungi. Philos Trans R Soc Lond B 361:1947–1963

Tedersoo L, Suvi T, Beaver K, Koljalg U (2007) Ectomycorrhizal fungi of the Seychelles: diversity patterns and host shifts from the native *Vateriopsis seychellarum* (Dipterocarpaceae) and *Intsia bijuga* (Caesalpiniaceae) to the introduced *Eucalyptus robusta* (Myrtaceae), but not *Pinus caribea* (Pinaceae). New Phytol 175:321–333

ter Steege H, Pitman NC, Phillips OL, Chave J, Sabatier D, Duque A, Molino JF, Prévost MF, Spichiger R, Castellanos H, von Hildebrand P, Vásquez R (2006) Continental-scale patterns of canopy tree composition and function across Amazonia. Nature 443:444–447

ter Steege H, Amazon Tree Diversity Network, RAINFOR (Amazon Forest Inventory Network) (2010) Contributions of current and historical processes to patterns of tree diversity and composition of the Amazon. In: Hoorn C, Wesselingh F (eds) Amazonia: landscape and species evolution. Wiley, Oxford, p 349–359

ter Steege H, Pitman N, Sabatier D (2003) A spatial model of tree α-diversity and tree density for the Amazon. Biodivers Conserv 12:2255–2277

Thiele KR, Jordan P (2002) *Thismia clavarioides* (Thismiaceae), a new species of fairy lantern from New South Wales. Telopea 9:765–771

Thompson JN (2005) The geographic mosaic of coevolution. University of Chicago Press, Chicago

Thorne RF (1972) Major disjunctions in the geographic ranges of seed plants. Quart Rev Biol 47:365–411

Tiffney BH (1985) The Eocene North Atlantic land bridge: its importance in Tertiary and modern phytogeography of the Northern Hemisphere. J Arnold Arbor 66:243–273

Tsukaya H, Okada H (2005) *Thismia mullerensis* (Burmanniaceae), a new species from Muller Range, Central Kalimantan. Acta Phytotax Geobot 56:129–133

Umata H (1995) Seed germination of *Galeola altissima*, an achlorophyllous orchid, with aphyllophorales fungi. Mycoscience 36:369–372

Umata H (1999) Germination and growth of *Erythrorchis ochobiensis* (Orchidaceae) accelerated by monokaryons and dikaryons of *Lenzites betulinus* and *Trametes hirsute*. Mycoscience 40:367–371

van Balgooy MMJ (1966) *Coriaria* L. In: van Steenis CGGJ, van Balgooy MMJ (eds) Pacific plant areas, vol 2. Rijksherbarium/Hortus Botanicus, Leiden, pp 122–123

van de Meerendonk JPM (1984) Triuridaceae. Fl Males Ser 1(10):109–121

van der Pijl L (1934) Die Mykorrhiza von *Burmannia* und *Epirrhizanthes* und die Fortflanzung ihres Endophyten. Rec Trav Bot Ned 31:701–779

Van Royen P (1972) Sertulum Papuanum 17. Corsiaceae of New Guinea and surrounding areas. Webbia 27:223–255

Vollesen K (1982) A new species of *Seychellaria* (Triuridaceae). Kew Bull 36:733–736

Wallace GD (1975) Studies of the Monotropoideae (Ericaceae): taxonomy and distribution. Wasmann J Biol 33:1–88

Wapstra M, French B, Davies N, O'Reilly-Wapstra J, Peters D (2005) A bright light on the dark forest floor: observations on fairy lanterns *Thismia rodwayi* F. Muell. (Burmanniaceae) in Tasmanian forests. Tasmanian Nat 127:2–18

Waterman RJ, Bidartondo MI (2008) Deception above, deception below: linking pollination and mycorrhizal biology of orchids. J Exp Bot 59:1085–1096

Waterman RJ, Bidartondo MI, Stofberg J, Combs J, Gebauer G, Savolainen V, Barraclough TG, Pauw A (2011) The effects of above- and belowground mutualisms on orchid speciation and coexistence. Am Nat 177:E54–E68

Wegener A (1915) Die Entstehung der Kontinente und Ozeane. Vieweg, Braunschweig

Wen J (1999) Evolution of eastern Asian and eastern North American disjunct distributions in flowering plants. Annu Rev Ecol Syst 30:421–455

White LJT (2001) The African rain forest: climate and vegetation. In: Weber W, White LJT, Vedder A, Naughton-Trevers L (eds) African rain forest ecology: an interdisciplinary perspective. Yale University Press, New Haven, pp 3–29

Whittaker RJ, Araulo MB, Paul J, Ladle RJ, Watson JEM, Willis KJ (2005) Conservation biogeography: assessment and prospects. Divers Distr 11:3–23

Wickett NJ, Goffinet B (2008) Origin and relationships of the myco-heterotrophic liverwort *Cryptothallus mirabilis* Malmb. (Metzgeriales, Marchantiophyta). Bot J Linn Soc 156:1–12

Wilford GE, Brown PJ (1994) Maps of late Mesozoic-Cenozoic Gondwana break-up: some palaeogeographical implications. In: Hill RS (ed) History of the Australian vegetation: cretaceous to recent. Cambridge University Press, Cambridge, pp 5–13

Wong KM, Phillipps A (1996) Kinabalu, summit of Borneo. The Sabah Society, Kota Kinabalu

Wood J (2010) Initial observations of seed and fruit development in *Thismia rodwayi* (Fairy Lanterns). Tasm Nat 132:78–80

Wood JJ, Beaman TE, Lamb A, Chew C, Beaman JH (2011) The orchids of Mount Kinabalu, vol 1. Natural History Publications, Kota Kinabalu

Woodward CL, Berry PE, Maas-van de Kamer H, Swing K (2007) *Tiputinia foetida*, a new myco-heterotrophic genus of Thismiaceae from Amazonian Ecuador, and a likely case of deceit pollination. Taxon 56:157–162

Xu JT, Guo SX (2000) Retrospect on the research of the cultivation of *Gastrodia elata*, a rare traditional Chinese medicine. Chinese Med J 113:686–692

Xu JT, Mu C (1990) The relation between growth of *Gastrodia elata* protocorms and fungi. Acta Bot Sin 32:26–31

Yagame T, Yamato M, Mii M, Suzuki A, Iwase K (2007) Developmental processes of achlorophyllous orchid, *Epipogium roseum*: from seed germination to flowering under symbiotic cultivation with mycorrhizal fungus. J Plant Res 120:229–236

Yam T, Chua J, Tay F, Ang P (2010) Conservation of the native orchids through seedling culture and reintroduction—a Singapore experience. Bot Rev 76:263–274

Yamato M, Yagame T, Shimomura N, Iwase K, Takahashi H, Ogura-Tsujita Y, Yukawa T (2011) Specific arbuscular mycorrhizal fungi associated with non-photosynthetic *Petrosavia sakuraii* (Petrosaviaceae). Mycorrhiza 21:631–639

Yang SZ, Saunders RMK, Hsu CJ (2002) *Thismia taiwanensis* sp. nov. (Burmanniaceae tribe Thismieae): first record of the tribe in China. Syst Bot 27: 485–488

Yoder A, Nowak M (2006) Has vicariance or dispersal been the predominant biogeographic force in Madagascar? Only time will tell. Annu Rev Ecol Evol Syst 37:405–431

Yokoyama J, Koizumi Y, Yokota M, Tsukaya H (2008) Phylogenetic position of *Oxygyne shinzatoi* (Burmanniaceae) inferred from 18S rDNA sequences. J Plant Res 121:27–32

Yuan Y-M, Wohlhauser S, Möller M, Chassot P, Mansion G, Grant J, Küpfer P, Klackenberg J (2003) Monophyly and relationships of the tribe Exaceae (Gentianaceae) inferred from nuclear ribosomal and chloroplast DNA sequences. Mol Phylogenet Evol 28:500–517

Zhang D, Saunders RM, Hu C-M (1999) *Corsiopsis chinensis* gen. et sp. nov. (Corsiaceae): first record of the family in Asia. Syst Bot 24:311–314

Zhu H (1997) Ecological and biogeographical studies on the tropical rain forest of south Yunnan, SW China with a special reference to its relation with rain forests of tropical Asia. J Biogeogr 24:647–662

Zelazowski P, Mahli Y, Huntingford C, Sitch S, Fisher J (2011) Changes in the potential distribution of humid tropical forests on a warmer planet. Philos Trans R Soc Lond A 369:137–160

Zettler LW, Perlman S, Dennis DJ, Hopkins SF, Poulter SB (2005) Symbiotic germination of a federally endangered Hawaiian endemic, *Platanthera holochila* (Orchidaceae), using a mycobiont from Florida: a conservation dilemma. Selbyana 26:269–276

Ziegler AC (2002) Hawaiian natural history, ecology, and evolution. University of Hawaii Press, Honolulu

Zimmer K, Meyer C, Gebauer G (2008) The ectomycorrhizal specialist orchid *Corallorhiza trifida* is a partial myco-heterotroph. New Phytol 178:395–400

Subterranean Morphology and Mycorrhizal Structures

4

Stephan Imhof, Hugues B. Massicotte, Lewis H. Melville, and R. Larry Peterson

4.1 Introduction

For most, if not all plants, subterranean parts are less known than their aerial counterparts, due in part to the difficulty in extracting a complete root system (see Kutschera and Lichtenegger 1982, 1992; Kutschera et al. 2009) and the lack of morphological information in floras and taxonomic descriptions because many herbarium specimens do not include underground parts such as roots and rhizomes. Likewise, information from the fossil record is biased towards aerial structures (Peterson 1992) although there have been discoveries of fossils showing fungal associations with underground organs (e.g., Kidston and Lang 1921, Remy et al. 1994; Taylor et al. 1995; LePage et al. 1997; Stockey et al. 2001). To date, fossils of root-fungal associations of mycoheterotrophic plants are unknown.

S. Imhof(✉)
Spezielle Botanik und Mykologie, Fachbereich Biologie, Philipps-Universität, 35032 Marburg, Germany
e-mail: imhof@uni-marburg.de

H.B. Massicotte
Ecosystem Science and Management Program, University of Northern British Columbia, 3333 University Way, Prince George, BC, Canada

L.H. Melville • R.L. Peterson
Department of Molecular and Cellular Biology, University of Guelph, Guelph, ON, Canada

In autotrophic plants, many scientific questions can be dealt with using generalized concepts of root structure and function (e.g., Kutschera and Lichtenegger 1992; Polomski and Kuhn 1998; Gregory 2006). However, this certainly does not hold for mycoheterotrophic (MH) plants. The structure of roots, rhizomes, or subterranean scale leaves of MH plants intimately linked to the association with soil fungi is of critical ecological relevance because these plants essentially depend upon fungi for their carbon and perhaps other nutrient needs. Hence, the subterranean organs of MH plants often show remarkable morphological and anatomical adaptations to meet their specific requirements. This chapter, therefore, addresses the importance of morphology and anatomy to complement modern methods for understanding the fungal colonization patterns in MH plants and their relationships to function.

In the following, we summarize the current knowledge of structural aspects of the underground parts (for a peculiar exception, see *Afrothismia*) of MH plants ranging from bryophytes to angiosperms, the latter in systematical order following the Angiosperm Phylogeny Group (APG 2009), which has been regularly updated by Stevens (2001 onwards). We are aware of the gradual differences between species in terms of mycorrhizal dependence, however, due to space limitations, we focus on the visibly achlorophyllous species, and only include the partially mycoheterotrophs where they add to the common picture.

V.S.F.T. Merckx (ed.), *Mycoheterotrophy: The Biology of Plants Living on Fungi*,
DOI 10.1007/978-1-4614-5209-6_4, © Springer Science+Business Media New York 2013

The final section interprets the available information in terms of detecting phylogenetic trends of MH plants, in order to understand their evolutionary history, a subject that is receiving considerable attention in the mycorrhizal literature (see Brundrett 2002).

4.2 Nonvascular Plants

4.2.1 *Aneura*

Aneura mirabilis (Aneuraceae/Hepaticae) was described as *Cryptothallus mirabilis* by von Malmborg (1933, 1934), although it was noted earlier around 1914 (Schiffner 1934) and suggested to be either an *Aneura* or *Riccardia* species (e.g., Denis 1919; Schiffner 1934). Recently, *Cryptothallus* was formally transferred to *Aneura* by Wickett and Goffinet (2008) based on molecular and morphological characteristics. This decision is supported by the observation that the endophyte in *Aneura mirabilis* belongs to the same genus (*Tulasnella*) as that in *Aneura pinguis* (Bidartondo et al. 2003), and the mycorrhizal pattern in both species is very similar (Ligrone et al. 1993).

Aneura mirabilis mostly occurs in maritime climates (e.g., Sjörs 1949; Williams 1950; Petersen 1972; Wiehle et al. 1989; Sergio and Seneca 1997; Sergio and Garcia 1999; Boudier et al. 1999) in cool, humid, mostly peaty environments with large mats of bryophytes (Wiehle et al. 1989). Only a part of the seta and the sporangium is elevated above the surrounding mats consisting of several moss species (von Malmborg 1933; Wiehle et al. 1989; Sergio and Garcia 1999; Boudier et al. 1999). The whitish, vermiform to lobular-coralloid, brittle gametophytes are only a few centimeters in length and remain embedded within the mosses or litter. Male and female gametophytes differ in lobe shape (Williams 1950; Benson-Evans 1952; Wiehle et al. 1989).

The first structural work on the mycorrhiza in *A. mirabilis* by Denis (1919) revealed intracellular fungal colonization with hyphal coils in the ventral (lower) part of the thallus, although he considered the specimen as an albino of another chlorophyllous *Aneura* species. Von Malmborg (1933) observed hyphae growing through the seta into the sporangium and assumed that the fungus is distributed together with the spores. Williams (1950), unable to confirm this statement of von Malmborg (1933), published the first detailed investigations and provided drawings of the full life cycle, including the pattern of mycorrhizal colonization. The thallus lobes bearing antheridia or archegonia are devoid of hyphae; starch is deposited in the upper part of the thallus and around the gametangia. In an ultrastructural comparison of green hepatics and *Aneura mirabilis* (still called *Cryptothallus*), Pocock and Duckett (1984) confirmed the concentration of fungal colonization in the lower half of the thallus, but more recently, Ligrone et al. (1993) showed that the upper parts of the thallus can also become colonized in later stages. Rhizoids are also colonized, albeit in an uncoiled manner (Duckett et al. 1990). Most likely, these straight hyphae within rhizoids represent the connection to the external substrate. The carbon of this liverwort probably comes from surrounding beech (*Fagus sylvatica*) trees (Read et al. 2000; Bidartondo et al. 2003), with which it is connected via the mutual *Tulasnella* mycorrhizal fungus, although Ligrone et al. (1993) found dissimilar dolipore structures in the endophytes of birch (*Betula* spp.) and *Cryptothallus*. The fungal coils within the thallus cells eventually degenerate to dark masses (von Malmborg 1933; Williams 1950; Pocock and Duckett 1984), interpreted as digestion, and the cells can be reinfected (Ligrone et al. 1993). Williams (1950) and Pocock and Duckett (1984) stressed the difference in fungal identity between Aneuraceae hosting basidiomycetes and resembling orchid mycorrhiza, in contrast to other liverworts hosting "phycomycetous" (today considered as Glomeromycota, Schüßler et al. 2001) endophytes forming arbuscular mycorrhiza (AM) in higher plants. This fact, together with the identification of the fungi in *A. mirabilis* as *Tulasnella* spp. (Read et al. 2000; Bidartondo et al. 2003), has led to the hypothesis of a novel acquisition of *Tulasnella* spp. as associates in Aneuraceae. By attaining an epiphytic habit during phylogeny, liverworts may have lost the original symbiotic relationship with Glomeromycota. Secondarily terrestrial Aneuraceae then could have associated with new fungal partners (Kottke and Nebel 2005; Bidartondo and Duckett 2009).

4.3 Seedless Vascular Plants

Mycoheterotrophy in the seedless vascular plants is restricted to their gametophytic phase (Read et al. 2000; Smith and Read 2008). Genera possessing acholorophyllous gametophytes (and photosynthetic sporophytes) belong to Lycopodiaceae (e.g., *Lycopodium*, *Huperzia*, Fig. 4.1a, b), Ophioglossaceae (*Ophioglosssum*, *Botrychium*, *Helminthostachys*, *Mankyua* Fig. 4.1d–f), Psilotaceae (*Psilotum*, *Tmesipteris*, Fig. 4.1g, h), some species of *Schizaea* and *Actinostachys* in the Schizaeaceae, and the monotypic species *Stromatopteris moniliformis* in the Gleicheniaceae.

4.3.1 Lycopodiaceae (Fig. 4.1a–c)

It was recognized very early that subterranean gametophytes of several *Lycopodium* species are associated with endophytic fungi (Treub 1885, 1890; Bruchmann 1885, 1910; Lang 1899; Burgeff 1938). Illustrations in Burgeff (1938) and Boullard (1979) clearly show that fungi colonize the basal region of gametophytes shortly after spore germination. Mature subterranean gametophytes show variations in form from disc-shaped with convoluted margins (*L. clavatum*, Fig. 4.1b, *L. obscurum*) to elongated, cylindrical structures (*L. complanatum* = *Diphasiastrum complanatum*, Bierhorst 1971; Gifford and Foster 1996). Gametophytes of all species have fungal colonization restricted to a zone underlying more surficial cells that give rise to antheridia and archegonia (Bierhorst 1971, Fig. 4.1c).

Although the identity of the fungus was unknown in these early studies, it was described as being aseptate and forming intracellular hyphal coils. An ultrastructural investigation of the fungal endophyte in association with achlorophyllous gametophytes of *L. clavatum* showed that complex hyphal coils and vesicles formed but arbuscules were absent (Schmid and Oberwinkler 1993a). Entrance of the fungus occurred either through rhizoids, degenerated epidermal cells, or between epidermal cells. Once within parenchyma cells of the gametophyte, host-derived plasma membrane and wall material was deposited around invading hyphae. Hyphae were multinucleate and contained bacterium-like organelles (BLOs). Hyphae became progressively more vacuolated and ultimately degenerated. The authors came to the conclusion, based on a number of unusual structural features, that this fungus-gametophyte interaction was unlike anything described in the literature and could not be attributed to a known mycorrhizal association. They therefore proposed a new term "lycopodioid mycothallus interaction" to describe the association.

More recently, based on structural features of the fungi within cells, the fungal symbionts in the gametophytes of all seedless vascular plants were suspected to be members of the Glomeromycota and to have the *Paris*-type arbuscular mycorrhiza association (Read et al. 2000). Molecular studies have confirmed this for the fungus associated with two subterranean gametophytes of *Huperzia hypogaea* collected in Ecuador: the fungus was identified as belonging to a specific clade of *Glomus*-Group A (Winther and Friedman 2008). Observations of these sectioned gametophytes confirmed earlier reports that hyphal coils are restricted to the basal region and that arbuscules are not formed.

4.3.2 Ophioglossaceae (Fig. 4.1d–f)

Gametophytes of *Ophioglossum* may be cylindrical (*O. nudicaule*, *O. vulgatum*, Boullard 1957; Gifford and Foster 1996), globose (*O. crotalophoroides*, Mesler 1976), or highly branched (*O. palmatum*, Mesler 1975). Fungal hyphae may be evenly distributed but avoiding the meristematic area and gametangia (Bierhorst 1971). Fungal colonization occurs immediately after spore germination (Campbell 1908) and gametophytes do not develop unless they are associated with the appropriate fungus. Hyphal coils, some of which

Fig. 4.1 (**a–c**) Lycopodiaceae, (**d–f**) Ophioglossaceae, (**g–j**) Psilotaceae. (**a**) *Lycopodium obscurum* sporophyte showing strobili. (**b**) *Lycopodium clavatum* mycoheterotrophic gametophytes with shoots (*arrowheads*). (**c**) Section of *L. obscurum* gametophyte showing zone of arbuscular mycorrhizal fungi (*arrowheads*) and antheridia (*arrows*).

have undergone degeneration, are illustrated in gametophyte cells of *O. pendulum* (Burgeff 1938). Mesler (1975) described the endophytic hyphae in gametophytes of *O. palmatum* as being non-septate and multi-nucleate. He also showed what he interpreted as vesicles in some gametophyte cells. Mesler (1976) gave a similar description of the fungal endophyte in *O. crotalophoroides*. Details at the ultrastructural level are lacking for gametophytes of *Ophioglossum* spp. and the identity of the fungus remains unknown.

Gametophytes of *Botrychium* also vary in their morphology from being tuber-like to disc-shaped (Bruchmann 1906; Burgeff 1943; Gifford and Foster 1996; Winther and Friedman 2007); endophytic fungi are restricted to a basal zone of parenchymatous cells (Bruchmann 1906; Bierhorst 1971). The fungus in *B. lunaria* has been described as forming aseptate intracellular coils and irregular vesicles (Bruchmann 1906). An ultrastructural study of the fungus-gametophyte interaction of this species (Schmid and Oberwinkler 1993b) has provided additional details. The intracellular hyphae contain vacuoles, endoplasmic reticulum, mitochondria, and lipid-like bodies. They are enclosed by host-derived plasma membrane and wall material that shows irregular outgrowths. Vesicles, some very irregular in shape, contain BLOs, and lipids; they can become very enlarged and then undergo degeneration. The identity of the fungal endophyte has been determined for the subterranean gametophytes of *B. crenulatum* (Fig. 4.1f) and *B. lanceolatum* based on DNA sequence data (Winther and Friedman 2007). The endophytes in both species belong to a major clade of glomalean fungi, *Glomus*-group A.

A third genus in the Ophioglossaceae, *Helminthostachys*, is monotypic (*H. zeylanica*) and native to the Indo-Malayan region (Gifford and Foster 1996). It also forms achlorophyllous mycoheterotrophic gametophytes (Lang 1902) but little is known of the fungal association.

A new genus and species (*Mankyua chejuense*) has been described from Cheju Island, Korea (Sun et al. 2001) based on differences in sporophyte morphological characters from the other genera in the family. Gametophytes have not been described but are presumed to be subterranean.

4.3.3 Psilotaceae (Fig. 4.1g–j)

The two genera, *Psilotum*, with two species and *Tmesipteris*, with ten species, have historically been of considerable interest because of the belief that they represented some of the most primitive extant seedless vascular plants (Bierhorst 1971; Gifford and Foster 1996). The lack of roots and the presence of much reduced leaf-like structures of the sporophyte strengthened this view. However, based on molecular evidence, Smith et al. (2006) include this family within the Psilotales, an order belonging to the extant ferns.

Subterranean gametophytes of *Psilotum* are highly variable cylindrical structures (Fig. 4.1h) sometimes showing repeated branching (Bierhorst 1971). Asexual reproductive propagules (gemmae) are frequently developed (Bierhorst 1971). Darnell-Smith (1917) was the first to succeed in achieving spore germination and to monitor early stages in gametophyte development. He reported that endophytic fungi appeared as dense "skeins" within interior cells of gametophytes and that hyphae entered rhizoids. Other authors have described an aseptate intracellular fungus thought to be a phycomycete in either field-collected gametophytes (Burgeff 1938; Boullard 1957) or

Fig. 4.1 (continued) (**d**) Shoot of *Botrychium virginianum* with fertile segment of leaf with sporangia (*arrowhead*). (**e**) Mycoheterotrophic gametophyte (*arrow*) of *B. virginianum* with a root (*arrowhead*) and base of a shoot (*double arrowhead*). (**f**) Intracellular hyphal coils of *Glomus*-Group A in a *Botrychium crenulatum* mycoheterotrophic gametophyte. Photo courtesy of Jennifer Winther. (**g**) Shoots of *Psilotum nudum* with synangia (*arrowheads*). (**h**) Branched mycoheterotrophic gametophyte of *P. nudum*. (**i**) Intracellular hyphal coils of an arbuscular mycorrhizal fungus in a sectioned *P. nudum* mycoheterotrophic gametophyte stained with Toluidine blue O. (**j**) Transmission electron micrograph of hyphae within a mycoheterotrophic gametophyte of *P. nudum* showing the interface consisting of host plasma membrane (perifungal membrane) (*arrowheads*) and host-derived intracellular matrix (*arrows*). (**k**) *Buxbaumia aphylla* sporophyte, 1.5 cm high gametophytes growing in greenhouse pots containing various angiosperm species (Bierhorst 1953). Aspects of the ultrastructure of the gametophyte-fungus interaction have been described from gametophytes collected from greenhouse pots (Davis 1976; Peterson et al. 1981). The fungus in these gametophytes is aseptate and forms complex coils (Fig. 4.1i) that undergo degeneration; arbuscules are not formed. Intracellular hyphae are separated from the gametophyte cell cytoplasm by host-derived plasma membrane (perifungal membrane) and interfacial matrix material (Peterson et al. 1981, Fig. 4.1j), characteristics of arbuscular mycorrhizal associations (Bonfante and Perotto 1995). To date, the fungus has not been identified but the structural characteristics are typical of a *Paris*-type arbuscular mycorrhiza.

The fungal endophyte in subterranean gametophytes of *Tmesipteris tannensis* was described by Lawson (1917) and Holloay (1921) as consisting of intracellular fungal coils (pelotons). As with *Psilotum* gametophytes, the identity of the fungus has not been determined.

4.3.4 Schizaeaceae

The gametophytes of the genus (*Schizaea*) in this leptosporangiate fern family may either be surficial and green, subterranean and achlorophyllous, or a combination of both, depending on species and habitat (Bierhorst 1968, 1971). Gametophytes of all species are associated with endophytic fungi that have been described as aseptate and frequently associated with rhizoids (Bierhorst 1971; Swatzell et al. 1996).

Gametophytes of all species in the genus *Actinostachys* are axial structures that are subterranean with fungal hyphae confined to a distinctive zone (Bierhorst 1968). The identity of the fungi associated with achlorophyllous gametophytes in these two genera is unknown.

4.3.5 Gleicheniaceae

The monotypic genus *Stromatopteris moniliformis* (subfamily Stromatopteridaeae), has axial subterranean gametophytes with coiled fungal hyphae (Bierhorst 1971), reminiscent of *Paris*-type arbuscular mycorrhizas. Although Bierhorst (1971) concluded that the fungus present in the gametophyte is the same as that in the photosynthetic sporophyte, this needs to be confirmed with molecular methods.

Experimental evidence confirming transfer of nutrients from fungi to the subterranean gametophytes of all seedless vascular plants is lacking.

4.4 Gymnosperms

4.4.1 Podocarpaceae

The New Caledonian endemic *Parasitaxus usta* (not *P. ustus*, as many authors repeated the linguistically incorrect transfer from *Podocarpus* to the feminine genus *Parasitaxus* by de Laubenfels 1972) is a succulent shrub or small tree (up to 2 m high) with wine-red scale leaves (Cherrier et al. 1992; Schneckenburger 1999), unable to photosynthesize (Feild and Brodribb 2005) and only occurring closely associated with *Falcatifolium taxoides* (also Podocarpaceae, Sinclair et al. 2002). Root graft-like subterranean connections between the two species have led to the notion of parasitism in *P. usta* (de Laubenfels 1959; Köpke et al. 1981). However, Cherrier et al. (1992) and an English version of that paper adding a SEM micrograph (Woltz et al. 1994) found an endophytic mycelium (called "ectendomycelium") in both species, together with haustorial-like connections apparent at the cellular level developing in tissues up to the cambium of *F. taxoides*. The authors assume a symbiotic association of the three partners but, based on their anatomical observations, are convinced of parasitism in this case. The latest investigation on *P. usta* confirms the intimate vascular association of both species, but results from stable carbon isotope investigations suggest that most carbon is provided by the fungus (Feild and Brodribb 2005). With respect to water physiology however, *P. usta* has higher stomatal conductance and lower water potential values relative to its host, which is typical for parasitic angiosperms (Feild and Brodribb 2005). Hence, apart from being a gymnosperm, woody, and relatively large, *Parasitaxus* is even more unique among heterotrophic organisms in possibly being a parasitic and mycoheterotrophic plant at the same time.

4.5 Monocots

4.5.1 Petrosaviaceae (*Petrosavia*)

The three species of *Petrosavia* are distributed from Japan to Java. The external morphology of the underground structures does not differ much among the species. Their subterranean rhizomes can be branched and thus, may bear several 10–15 cm high scapes with terminal racemes or corymbs of white flowers. Rhizomes measure up to 1.5 mm in diameter and are densely covered by sheathing scale leaves (Groom 1895a; Makino 1903; Stant 1970; Jessop 1979; Chen and Tamura 2000; Cameron et al. 2003). The filiform, hairless, approximately 0.5 mm thick and sparsely branched adventitious roots, are initiated from the rhizome, especially close to the base of the scape. They most likely originate from the axils of the scale leaves, as do the rhizome branches (Groom 1895a). In *Petrosavia sakuraii*, the roots predominantly grow horizontally through the substrate and can be up to 20 cm long (Watanabe 1944). This author also reports hyphae penetrating into the roots 2–5 mm proximal from the root tip.

The epidermis is either ephemeral (Groom 1895a) or partly persistent (Stant 1970). The cortex consists of a suberized exodermis, 4–6 layers of parenchyma cells, and an endodermis with particularly strong u-shaped tertiary thickenings surrounding the tetrarch central cylinder (Groom 1895a; Watanabe 1944; Stant 1970). This is similar to many mycoheterotrophic Burmanniaceae (Johow 1889; Uphof 1929) with *Dictyostega orobanchoides* as an extreme example (Imhof 2001, Fig. 4.4f). Watanabe (1944) mentions that segments of older roots lose the cortex parenchyma but remain connected to the rhizome by the central cylinder that is surrounded by the fortified endodermis. The maintenance of connectivity between roots and rhizomes bearing inflorescences is particularly important for MH plants having filiform roots, since not only water and nutrients but also carbohydrates must be transported through these comparatively long structures. A tertiary endodermis, a synapomorphy of monocots (Esau 1965), seems to be less costly than the production of layers of lignified tissue, which is the equivalent option for non-monocots in order to protect the connectivity. This economical advantage of monocots may be part of the explanation as to why monocots include disproportionately so many MH plants (Imhof 2010).

Previous investigations on *Petrosavia* (Groom 1895a; Watanabe 1944; Stant 1970) report coiled mycorrhizal hyphae within the cortex parenchyma cells. The figures and descriptions of Watanabe (1944) resemble a *Paris*-type AM but without the typical lateral arbuscules, which is similar to the mycorrhiza found in *Voyria truncata* (Gentianaceae, Imhof and Weber 1997). The advantage of the frequent feature of MH plants of having a specialized mycorrhizal colonization pattern allowing a selective digestion of hyphae while keeping the fungus alive (see further), is not apparent. Petrosaviaceae are a rather basal clade of the monocots (Cameron et al. 2003; APG 2009), which might explain its plesiomorphic, i.e., basic mycorrhizal pattern. Most recently, Yamato et al. (2011a) confirmed the structural descriptions of Watanabe (1944), and revealed this mycorrhiza as an association with a highly specific clade of *Glomus*-group A.

4.5.2 Thismiaceae (Figs. 4.2–4.4 and 4.10)

Thismiaceae are either considered to be a tribe, Thismieae, in the Burmanniaceae (e.g., Jonker 1938, Cronquist 1988) or a separate family (e.g., Agardh 1858; Thorne 1992; Takhtajan 1997; Stevens 2001 onwards). APG (2009) is still reluctant to separate them from Burmanniaceae but acknowledge the arguments for separation given by Merckx et al. (2006). We regard them as a family based on floral morphology (e.g., Maas et al. 1986; Caddick et al. 2000) and molecular evidence (Merckx et al. 2006).

Fig. 4.2 (**a–h**) *Afrothismia hydra*, (**i, j**) *Afrothismia winkleri* (Thismiaceae). (**a**) Seed (*left*, 0.6 mm long) and an early germination stage (*right*) of *A. hydra* with disrupted seed coat (sc), giving rise to a first root tubercle. (**b–d**) More tubercles develop successively at the base of the initial one and the root extensions elongate. The seed coat (sc)

4.5.2.1 *Haplothismia, Oxygyne, Tiputinia*

The extremely rare *Haplothismia exannulata* from India has vermiform to tuberous, up to 3.5 cm long, roots radiating from the shoot base (Airy Shaw 1952; Sasidharan and Sujanapal 2000). For *Oxygyne triandra* from Cameroon (probably extinct, Yokoyama et al. 2008), the subterranean organs are unknown. The Japanese species, *O. shinzatoi* and *O. yamashitae*, only known from their type localities, have vermiform roots (Yokoyama et al. 2008; Yahara and Tsukaya 2008), and the original description of *O. hyodoi*, also from Japan, states "rhizoma repens" (Abe and Akasawa 1989). *Tiputinia foetida* is represented by a single specimen from Ecuador (Woodward et al. 2007), measuring about 9 cm in length. The largest part of it is an orthotropous, vermiform, 4 mm thick root, giving rise to two subterranean shoots, with only the terminal flower being epiterrestrial. The root cortex contains "intracellular, looped, septate" hyphae (Woodward et al. 2007).

4.5.2.2 *Thismia* (Fig. 4.10g, h)

This genus is by far the largest of the family, with a worldwide, although mostly tropical, distribution. The underground structures are quite variable. Most species have horizontal runner-like, vermiform roots of 1–2 mm in diameter which bear root-borne shoots (e.g., Groom 1895b; Warming 1901; Pfeiffer 1914; Chantanaorrapint 2008; Chiang and Hsieh 2011) and give rise to additional similar roots where the shoots emerge, thus forming star-like clusters (e.g., Groom 1895b; Bernard and Ernst 1910; Pfeiffer 1914; Larsen 1965; Saunders 1996; Yang et al. 2002; Wapstra et al. 2005). This indicates the trend towards a star-like radiating root system, typical for MH plants. The runner-like parts of the roots can be short (e.g., *Thismia appendiculata*, Schlechter 1919), and in this case, the shoots emerge in nest-like tufts above the soil surface. In other species, the root system is reduced to a coralloid structure, e.g., *Thismia yorkensis* (Cribb 1995), *T. goodii* (Kiew 1999), or *T. clandestina* and *T. versteegii* (Bernard and Ernst 1911). *Thismia versteegii* shows similarities to the unique fan-shaped roots of *Thismia clavigera* (Stone 1980), which probably develop through short, dichotomously branched and congenitally merged roots. *Thismia annamensis* and *T. tentaculata* have short rhizomes bearing a dense covering of vermiform roots (Larsen and Averyanov 2007), also resulting in a star-like root system. This is morphologically similar but ontogenetically quite different from the other species mentioned above. The decision whether a condensed root system is developed by root-borne shoots or shoot-borne roots is often difficult to make and sometimes requires anatomical investigations (see Imhof 2004). Finally, some neotropical species have globose tubers (see Fig. 4.10g), from which a shoot as well as numerous filiform roots arise (e.g., *Thismia hyalina*, Miers 1866, *T. glaziovii*, Poulsen 1890a, *T. janeirensis*, Warming 1901, *T. panamensis*, Maas et al. 1986, Fig. 4.8j, *T. saülensis*, Maas and Maas 1987). Inferred from *T. luetzelburgii*, these tubers are roots, giving rise to up to four endogenous flowering shoots. The filiform roots can develop new tubers at their apices (Goebel and Süssenguth 1924).

The fungal colonization of *Thismia* spp. has been investigated quite early and in great detail. Like many other MH plants, *Thismia* also shows different fungal morphologies in distinct tissue

Fig. 4.2 (continued) is still attached. The whitish content is the fungal colonization. (**e**) The rhizome has developed into a shoot terminated by a 1 cm long flower. (**f**) *A. hydra* in its natural habitat with the filiform root elongations superficially clinging to the substrate. (**g**) *A. hydra* showing the strictly sympodial flowering mode with clusters of root tubercles at the base of each pedicel. Three basal flowers (A, B, C) have already detached, the following are in dissemination stage with placentophore developed (D), in fruit (E) and in bud (F). Note early tubercle (tb) development at the base of the flower bud. (**h**) Close up of the placentophore (pl). (**i**) Flower of *A. winkleri*, measuring 1.5 cm from the subtending scale leaf to the bending of the tube. (**j**) Root/rhizome system of *A. winkleri*

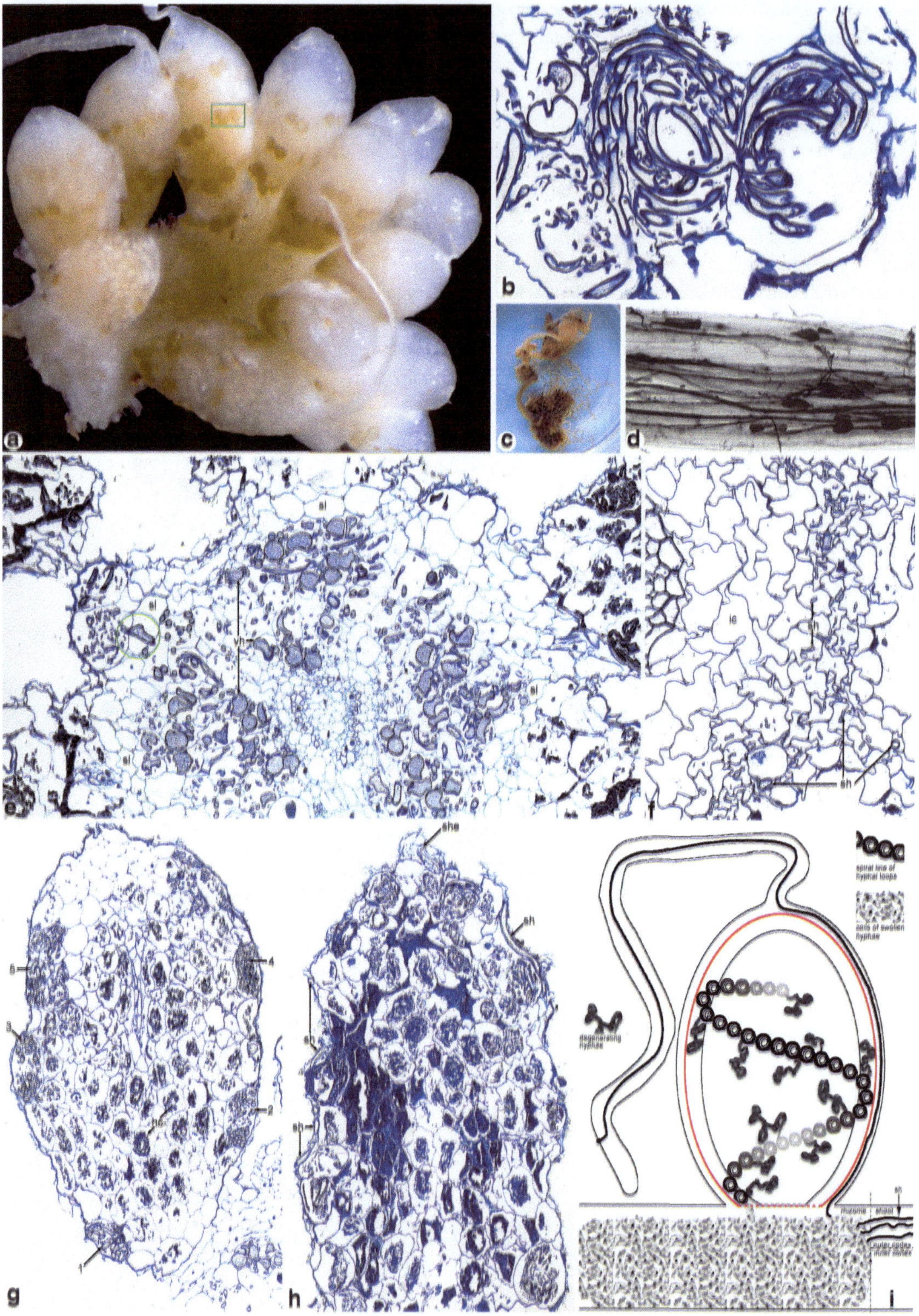

Fig. 4.3 *Afrothismia saingei* (Thismiaceae). (**a**) Rhizome tip with many root tubercles; some tubercles have been detached. The characteristic hyphal loops (*green rectangle*), developed in spiral lines within the tubercles, are

compartments, which sometimes are anatomically different. In *T. clandestina* (coralloid root system) and *T. aseroe* (vermiform roots), the outer cortex parenchyma layers contain straight hyphae with only a few coils; in the middle cortex layers, the hyphae are coiled but not digested; the inner layers show amorphous fungal material (Groom 1895b; Janse 1896; Meyer 1909; Bernard and Ernst 1911). In *T. americana*, *T. rodwayi*, and *T. javanica* (vermiform roots), straight hyphae are missing, instead, the outer cortex layer is occupied by coiled hyphae which do not degenerate. Digestion takes place in the inner cortex (Bernard and Ernst 1910; Pfeiffer 1914; Coleman 1936; McLennan 1958; Campbell 1968). Of the species having root tubers, *T. luetzelburgii* (Goebel and Süssenguth 1924) and T. *glaziovii* (Poulsen 1890a) have been investigated. They also show compartmentation of digested and undigested fungal material, whereas the digestion is more prominent in the proximal and central part of the tuber. The filiform roots connecting the mother tuber with smaller daughter tubers bear straight undigested hyphae linking the two tubers (Goebel and Süssenguth 1924). This is partly reminiscent of structures found in *Afrothismia* spp. (see next paragraph).

Due to the structural characteristics typical of a *Paris*-type AM, the fungus colonizing *Thismia* spp. almost certainly belongs to the Glomeromycota. This has recently been confirmed for *T. rodwayi* using molecular identification methods (Merckx et al. 2012).

4.5.2.3 *Afrothismia* (Figs. 4.2–4.4)

The genus *Afrothismia* from tropical Africa is characterized by its dense aggregates of small tuberous roots elongated by a filiform extension of various lengths between the species (Figs. 4.2g+j, 4.3a+c, and 4.4b). Although our chapter deals with subterranean organs, this is not entirely correct for some *Afrothismia* spp. In fact, the peculiar root/rhizome/shoot systems often grow entirely epiterrestrially (Fig. 4.2f), the filiform part of the roots clinging to rotten wood, leaf litter, or bare soil (e.g., *A. foertheriana*, Franke et al. 2004, *A. hydra*, Sainge and Franke 2005, *A. winkleri*, Imhof pers. observ., Fig. 4.2i–j). Only *A. baerae* (Cheek 2003a) and *A. gesnerioides* (Imhof pers. observ., Fig. 4.4a) are known to be rooted in the soil. The latter two species also differ by their conspicuously short filiform parts of the roots (Fig. 4.4b, Cheek 2003a; Maas-van de Kamer 2003). *Afrothismia zambesiaca*, described from a herbarium specimen collected in 1955, is only inferred to have an underground stem with bulbils (Cheek 2009). The ontogeny of *Afrothismia hydra* from seed to the open fruit has been

Fig. 4.3 (continued) visible from the outside. (**b**) Close up of the green rectangle in (**a**). Tangential section through two hypodermal cells colonized by looped hyphae that are connected to each other. (**c**) Specimen (Wilks No. 1179) of *A. saingei* under investigation from the Herbarium in Utrecht, labeled as *Afrothismia winkleri*. (**d**) Cleared preparation of a filiform root extension showing straight growing hyphae and vesicles. (**e**) Transverse section through a rhizome of *A. saingei* with coils of enlarged hyphae (vh) in the cortex. Mostly only once per tubercle these hyphae transit (*green circle*) over an inconspicuous separating layer (sl) into the hypodermis of a tubercle to start the spiral line of hyphal loops (see (**a, b**)). (**f**) Transverse section through a shoot/pedicel of *A. saingei*. The enlarged hyphae in the rhizome cortex (**e**) are continuous with the straight growing, also quite large hyphae (sh) in the outer shoot cortex, thus connecting the spatially separated tubercle clusters along the plant (**c**). The inner cortex (ic) is free of hyphae. (**g**) Longitudinal section through a young root tubercle showing the looped coils in an alternating pattern in the hypodermis as to be expected from its spiral arrangement (1–5) whereas all other cortex cells contain degenerated hyphal coils (hc–). (**h**) Longitudinal section through an old root tubercle where the digestion of hyphal coils has advanced but the epidermis now contains straight growing, nondegenerated hyphae (sh) linking those in the filiform root extension (she, see (**d**)) with the enlarged hyphae in the rhizome cortex (see (**e**)). (**i**) Schematic view of the mycorrhizal colonization pattern in *A. saingei*: Straight hyphae grow through root extension and tubercle epidermis, enter the rhizome cortex becoming enlarged and coiled, transit once per tubercle into the hypodermis of the tubercle starting a spiral line of looped interconnected hyphae around it (hl), and from there send hyphal branches into the rest of the cortex parenchyma for digestion (dh). The red line signifies an impenetrable barrier to the fungus. The spatially separated clusters of tubercles share the fungus via straight hyphae growing along the shoot axis in its outer cortex parenchyma

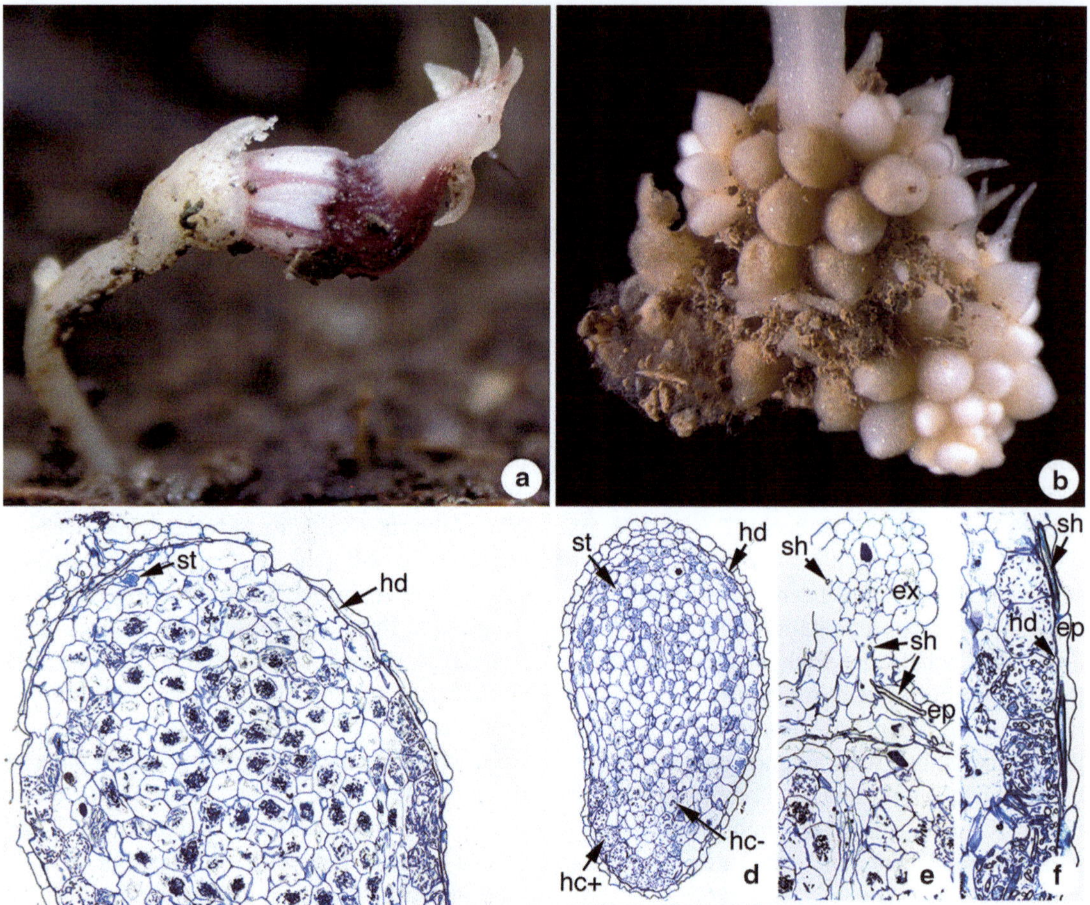

Fig. 4.4 *Afrothismia gesnerioides* (Thismiaceae). (**a**) *A. gesnerioides* emerging about 1.5 cm above the soil. In contrast to many other *Afrothismia* spp., roots and rhizome are subterranean. (**b**) Root/rhizome system of

described (Imhof and Sainge 2008), and since structurally the genus is quite consistent, this example will be detailed here to represent the whole genus. With germination, a root tubercle without a filiform extension is generated (Fig. 4.2a). Successively, more tubercles develop on a rhizome that is gradually increasing in size and the root extensions elongate (Fig. 4.2b–d). This creates a globose to ovate, coarsely echinate structure due to the characteristic roots. At some point, the rhizome proceeds to grow without root development. This axis, now more accurately called a shoot, will terminate with a flower (Fig. 4.2e), and a side shoot appears in the uppermost scale leaf of the shoot. The base of this side shoot also bears a cluster of tubercles with extensions, and this shoot will also end in a flower. This sympodial pattern is repeated several times (Fig. 4.2g). The fruit is a pyxidium, opening by means of a placentophore (Fig. 4.2h, see details in Imhof and Sainge 2008).

The fungal colonization of *Afrothismia saingei* is an extreme example of mycorrhizal complexity (Imhof 1999a, treated as *A. winkleri*[1]). Briefly, the pattern of colonization is as follows (schematic view on Fig. 4.3i). The filiform root extension bears straight hyphae, continuous with those in the epidermis of the tubercle (Fig. 4.3d+h). These hyphae never pass from the epidermis into the cortex of the tubercle but proceed around it towards the rhizome. As soon as the fungus reaches the rhizome at the tubercle base, it colonizes the rhizome cortex tissue with coiled, swollen, vesicle-like structures, but still does not show signs of degeneration (Fig. 4.3e). From there, few hyphae re-enter the tubercle from the rhizome cortex, and grow towards the subepidermal layer of the tubercle (Fig. 4.3e). Characteristic loops of hyphae are developed in the subepidermal cells (Fig. 4.3b+g), and an upward spiral line of cells containing such looped hyphae proceed around the tubercle (Fig. 4.3a). No digestion of hyphae occurs to this stage. Side branches from these hyphal loops enter the other cells of the tubercle cortex, where they degenerate to amorphous clumps (Fig. 4.3g, h). Connections to more distant tubercle clusters along the plant are provided by straight growing hyphae in the outer cortex of the shoot internodes (Fig. 4.3f, see details in Imhof 1999a). This complicated plant structure and colonization pattern represent a sophisticated and ecologically functional system. The filiform root extensions increase the surface for contact with and invasion by hyphae, the root tubercle increases the number of cells for colonization by

[1]According to Maas-van de Kamer and Maas (2010), the material under investigation in Imhof 1999a (=Wilks no. 1179, received from the herbarium of Utrecht, labeled as *A. winkleri*) turned out to be *A. saingei* (Franke 2004), synonymous to *A. gabonensis* (Dauby et al. 2008).

Fig. 4.4 (continued) *A. gesnerioides*, about 1 cm wide. (**c**) Transverse section through a rhizome (rh) and longitudinal sections of root tubercles (tb) showing starch grains (st) in the inner rhizome cortex and uncolonized root cortex, straight hyphae (sh) in the outer rhizome cortex, non-degenerated dense hyphal coils in the third cell layer of the root tubercle (hc+) and degenerated hyphal coils in the tubercle parenchyma (hc−). The hypodermis (hd) is largely collapsed. (**d**) Longitudinal section through a young tubercle of *A. gesnerioides* where the hypodermis (hd) is still visible. Fungal colonization has just started from the tubercle base in the third cell layer (hc+) and starch depositions (st) are still present in the parenchyma, which will disappear when fungal colonization proceeds. First degenerated hyphal coils (hc−) are also present in the inner parenchyma. The root extension has not yet developed. (**e**) Longitudinal section through a root tip of an older tubercle showing the root extension (ex) partly in transverse view, colonized by straight hyphae (sh) which proceed into the root epidermis (ep, see (**f**)). (**f**) Root tubercle epidermis (ep) only contains straight growing hyphae (sh) which never penetrate the hypodermis (hd, collapsed) but are continuous with those in the outer rhizome cortex. (**g**) Four neighboring tissues of the root tubercle hosting distinct morphotypes of fungal colonization: epidermis (ep) with straight hyphae (sh), hypodermis (hd) as a barrier to the fungus, the third root layer with nondegenerated dense hyphal coils (hc+) and the multilayered root parenchyma containing degenerated hyphal coils (hc−). (**h**) Transition of colonization (*green oval*) between the straight hyphae (sh) in the outer rhizome cortex and the nondegenerating hyphal coils (hc+) in the third root layer of the tubercle across a layer continuous with the otherwise impenetrable hypodermis (hd). Uncolonized parenchyma cells and the inner rhizome cortex contain starch grains (st). (**i**) Schematic view of the mycorrhizal colonization pattern in *A. gesnerioides*: straight hyphae grow through root extension, tubercle epidermis and outer rhizome cortex, transit at the base of the tubercle into its third layer to form dense coils (dc, green texture), and branches from there colonize the inner tubercle parenchyma to become digested (dh). The red marked hypodermis is impenetrable for the fungus hyphae and eventual digestion, representing the locations of the beginning and end of the mycorrhizal colonization pattern. Between these events, the different hyphal forms serve three fundamental functions: (1) transportation and distribution of carbohydrates and nutrients within the root-rhizome-complex, (2) storage, and finally (3) as a carbon source for the plant following digestion. The straight hyphae in the filiform root extension and the epidermis allow for rapid transport of nutrients and carbohydrates towards the rhizome. The swollen hyphae in the rhizome cortex store these substances, eventually for the benefit of the plant. The spiral line of hyphal loops is the geometrically and economically optimal distribution mode around the parenchyma of the tubercle. With a minimum of living hyphae, this provides short distances and limits the number of cell passages for side branches to penetrate into all parenchyma cells, necessary due to the quick degeneration process therein. The fungus in *Afrothismia gesnerioides* shows a similar colonization pattern with straight hyphae in the short root extension (Fig. 4.4e) and the root epidermis (Fig. 4.4f, g), as well as digestive tissue in the inner root parenchyma (Fig. 4.4c, see details in Imhof 2006). However, it does not develop a spiral line of hyphae around the tubercle parenchyma. Instead, dense coils of living irregular hyphae develop in the third root layer, encompassing the parenchyma in a collar-like pattern (Fig. 4.4c+f–h). Economically speaking, this pattern is less efficient than that in *A. saingei*, considering the amount of living fungal biomass necessary to supply the digesting cells with hyphal branches. Moreover, the rhizome of *A. gesnerioides* contains straight growing hyphae in its outer cortex, whereas the inner cortex cells contain starch deposits (Fig. 4.4c), as does the uncolonized tubercle cortex (Fig. 4.4d+h, schematic view on Fig. 4.4i). This means that *A. gesnerioides*, in contrast to *A. saingei*, converts the carbon delivered by the fungus into starch grains. In the case of *Afrothismia* spp. however, this appears as a unnecessary metabolic step, since the carbon source is permanently present. Therefore, although the mycorrhizal patterns in *Afrothismia* spp. are highly complex, they still show signs for an ongoing evolutionary progression of mycorrhizal structures within the genus, whereas *A. gesnerioides* can be considered to be less advanced than *A. saingei*. More of the 12 *Afrothismia* species described so far should be investigated to determine if intermediate structures exist (see 4.8 Trends, Conclusions, and Future Directions).

The fungal species associated with *Afrothismia* spp., as identified by molecular methods, all belong to *Glomus*-group A (Franke et al. 2006), and are species-specific (Merckx and Bidartondo 2008).

4.5.3 Burmanniaceae (Figs. 4.5–4.7)

Of the ten genera in this family, only *Burmannia* contains green representatives. *Burmannia tenella* is the only entirely achlorophyllous neotropical species, others occur in Africa (e.g., *B. hexapterella*) and Asia (e.g., *B. championii, B. candida*). However, many species with intermediate mycoheterotrophic status, between

Fig. 4.5 (continued) structures (a), and vesicles (v). (**f**) Transverse section through a root of *D. orobanchoides* showing the central cylinder (cc) consisting of a central tracheary element surrounded by one ring of smaller tracheary elements and the pericycle, a thickened endodermis (en), two layers of small parenchyma layers, and epidermal cells (ep). The epidermal cells contain hyphal coils (hc+) and arbuscules (a), the latter partly degraded. (**g**) SEM micrograph of a rhizome of *D. orobanchoides* imbricately covered by peltate scale leaves with fringed margins. The leaf interstitials contain fungal hyphae (hy). (**h**) Transverse section through a rhizome covered with imbricate scale leaves (le) showing fungal colonization including vesicles (v) within (ihy) and in between (ohy) the scale leaves. The rhizome axis (rh) is not colonized but contains starch grains (st). (**i**) Tangential section through a rhizome (rh) including the imbricate scale leaves (le) showing dense hyphal masses in the leaf interstitials (li). (**j**) Schematic view of the mycorrhizal colonization pattern of *D. orobanchoides*: the peltate scale leaves and their interstitials are colonized by hyphal coils and vesicles. The root is colonized only in the epidermis by hyphal coils, arbuscules, and vesicles; the arbuscules are the first to degenerate

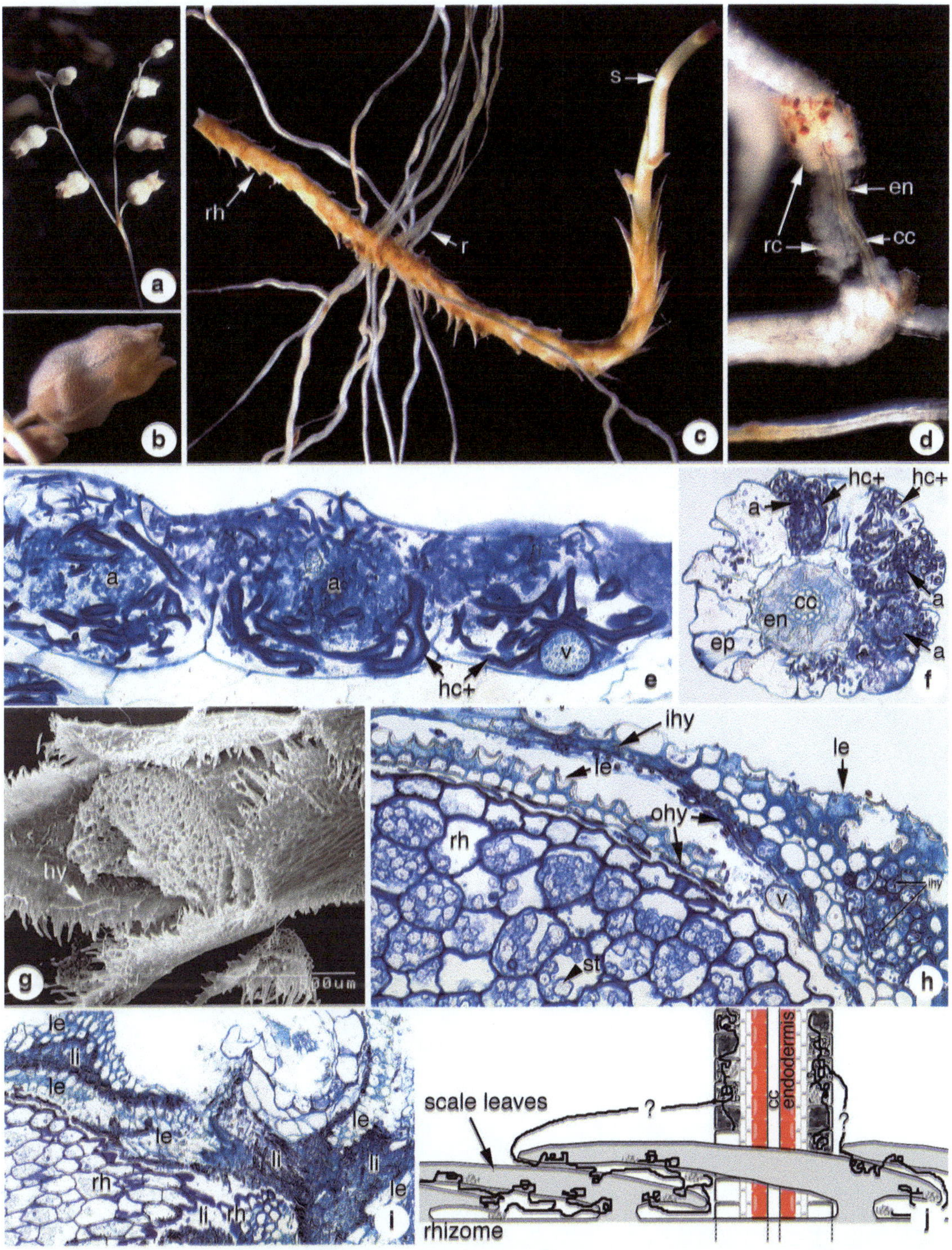

Fig. 4.5 *Dictyostega orobanchoides* (Burmanniaceae). (**a**) Inflorescence of *D. orobanchoides*, composed of a bifurcate cincinnus. (**b**) Single preserved flower of *D. orobanchoides*, 2.5 mm long. (**c**) Rhizome (rh) of *D. orobanchoides* about 1.5 mm thick with a tuft of filiform roots (r). Apically the rhizome turns into a shoot (s). (**d**) Root of *D. orobanchoides*, cortex (rc) partly detached, exposing the thickened endodermis (en) which encloses the central cylinder (cc). The whitish cell contents are fungal coils. (**e**) Longitudinal section through a root epidermis of *D. orobanchoides*. Fungal colonization consists of coiled hyphae (hc+), partly decomposed arbusculate

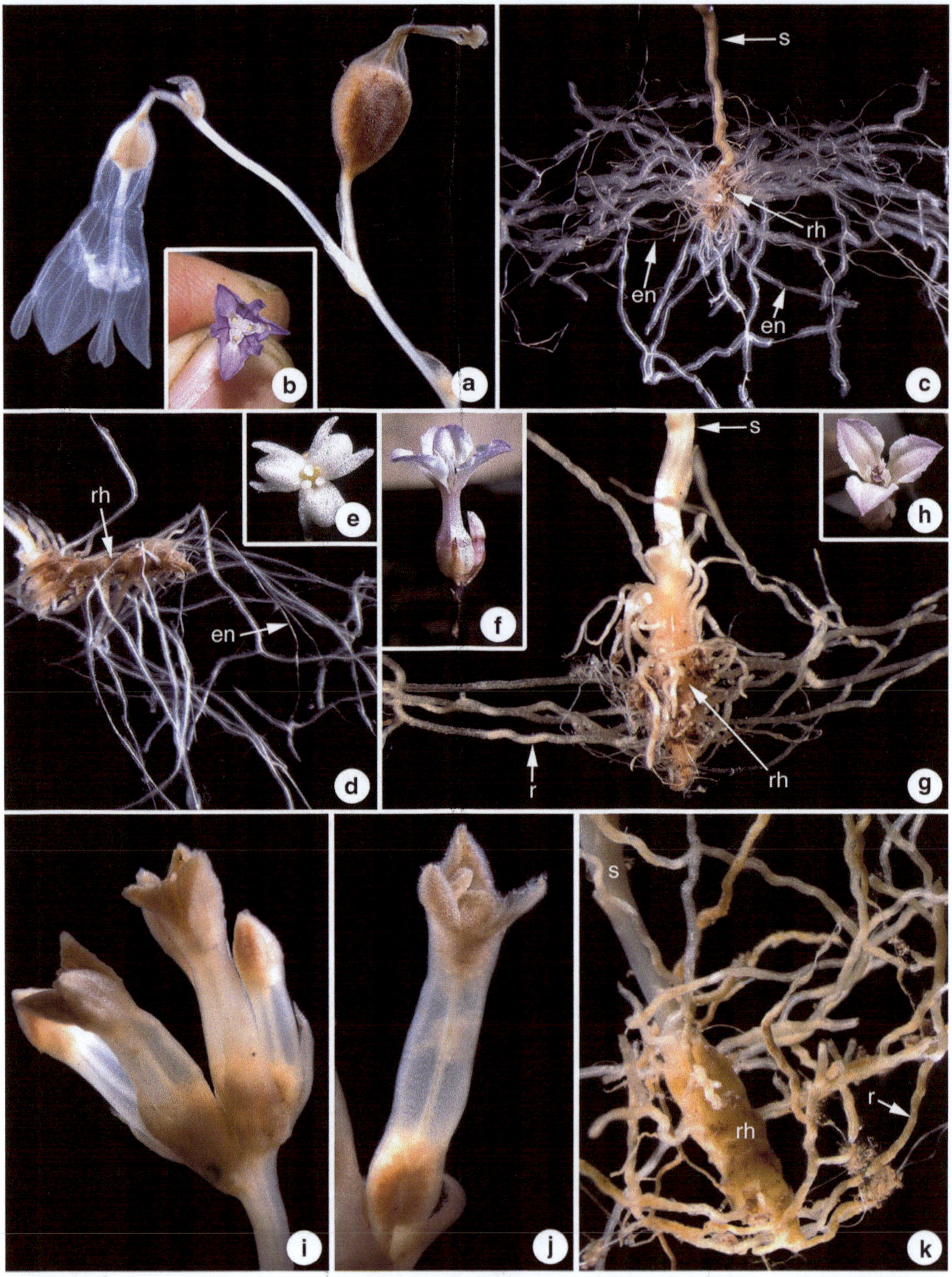

Fig. 4.6 (**a–c**) *Apteria aphylla*, (**d–e**) *Gymnosiphon divaricatus*, (**f–h**) *Hexapterella gentianoides*, (**i–k**) *Campylosiphon congestus* (Burmanniaceae). (**a**) Preserved flower (9 mm long) and fruit of *A. aphylla*. (**b**) Top view of a flower of *A. aphylla* (courtesy of H and PJM Maas). (**c**) Subterranean system of *A. aphylla*. The shoot (s) is continuous with the short (3 mm) orthotropous rhizome (rh) bearing numerous filiform roots. The root cortex parenchyma is often disrupted, leaving the thickened endodermis (en) with central cylinder enclosed as the only connection with the rhizome.

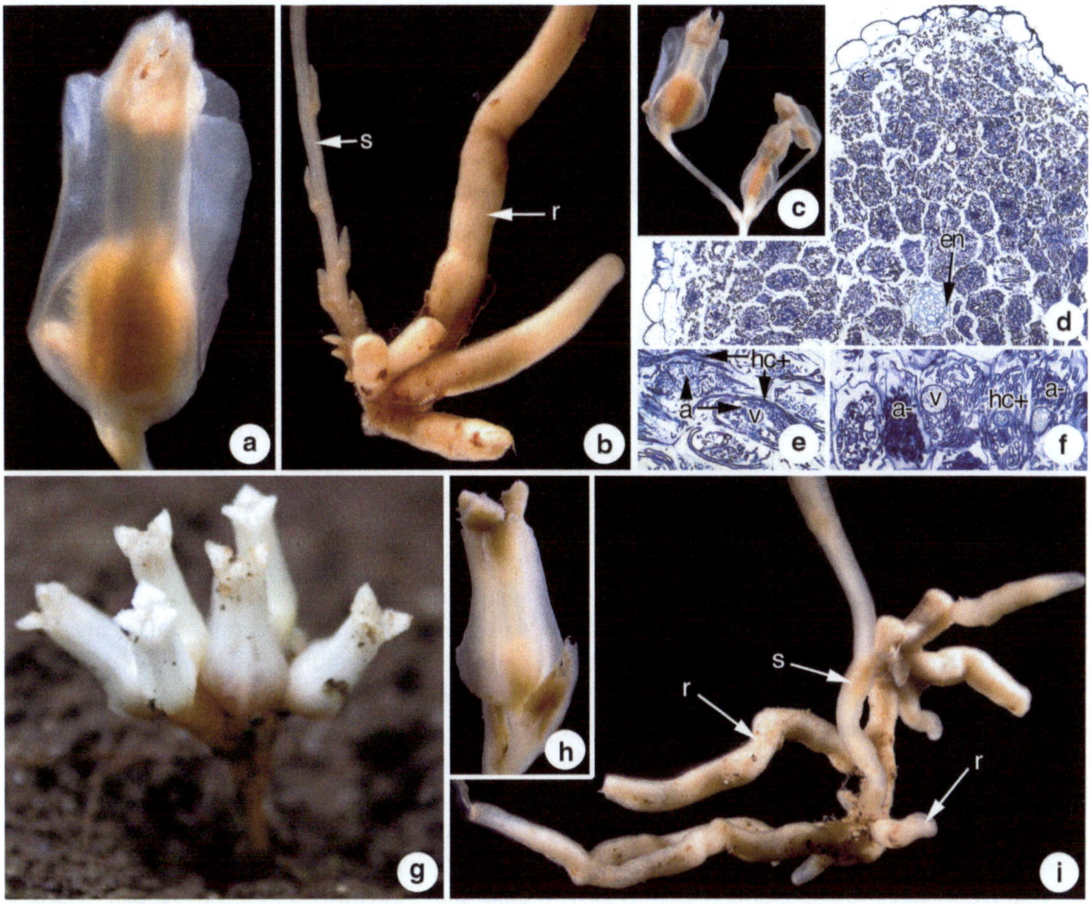

Fig. 4.7 (**a–f**) *Burmannia tenella*, (**g–i**) *Burmannia hexaptera* (Burmanniaceae). (**a**) Preserved flower (6 mm long) of *B. tenella*. (**b**) Root system of *B. tenella* with several star-like radiating vermiform roots (r), about 1.2 mm thick, at the base of the shoot (s). (**c**) Inflorescence of *B. tenella*, the bifurcate cincinnus usually consists of a few flowers. (**d**) Transverse section through a root of *B. tenella* with extensive fungal colonization of the multilayered cortex parenchyma. The central cylinder is reduced and surrounded by a tertiary endodermis (en). (**e**, **f**) The fungus in the root parenchyma cells of *B. tenella* forms heteromorphic coils of hyphae of various width (hc+), arbusculate structures (a) as well as vesicles (v), often within one cortex cell. Degeneration begins with the arbusculate structures (a–); the thicker hyphae tend to persist longer. (**g**) Flowers of *B. hexaptera* emerging only a few centimeters above the soil surface. (**h**) Preserved flower (1 cm long) of *B. hexaptera*. (**i**) Root system of *B. hexaptera* comprised of vermiform roots (r), about 1.2 mm thick, also with the tendency to radiate at the base of the shoot (s), resulting in a coralloid appearance

Fig. 4.6 (continued) (**d**) Subterranean system of *G. divaricatus*, very similar to that of *A. aphylla* (see (**c**)) with short rhizome (rh) and exposed endodermis (en). (**e**) Top view of a flower of *G. divaricatus* (courtesy of H and PJM Maas). (**f**) Flower of *H. gentianoides* (courtesy of H and PJM Maas). (**g**) Subterranean system of *H. gentianoides*, also with a short rhizome (rh, 4 mm long) continuous with the shoot (s). The rhizome bears filiform roots (r); light coloration indicates fungal colonization. (**h**) Top view of a flower of *H. gentianoides* (courtesy of H and PJM Maas). (**i**) Preserved inflorescence of *C. congestus*. (**j**) Preserved single flower (9 mm long) of *C. congestus*. (**k**) Subterranean system of *C. congestus* with the shoot (s) continuous with a slightly tuberous rhizome (rh), 9 mm long and 2.5 mm thick, bearing filiform roots (r) leafy *Burmannias* and achlorophyllous, scale-leaved species exist (Jonker 1938; Maas et al. 1986; Leake 1994), suggesting an evolutionary trend towards mycoheterotrophy. All other genera are fully mycoheterotrophic. The monotypic genus *Desmogymnosiphon* (Guinea Lopez 1946) is most probably a *Gymnosiphon* species (compare to Maas et al. 1986).

4.5.3.1 *Apteria, Campylosiphon, Dictyostega, Gymnosiphon, Hexapterella, Marthella, Miersiella* (Figs. 4.5 and 4.6)

Except for *Campylosiphon congestus* and the pantropical *Gymnosiphon*, all these genera are exclusively neotropical (Jonker 1938; Maas et al. 1986). All species have the same basic architecture for their underground parts. The aerial shoots are continuous with rhizomes, densely covered by scale leaves. These scale leaves are conspicuously fringed in *Dictyostega* (Imhof 2001, Fig. 4.5g), which has led to the hypothesis they might ecologically replace the missing root hairs (Goebel and Süssenguth 1924; Maas et al. 1986). The rhizomes can be longer (up to 7.5 cm in e.g., *Miersiella umbellata*, Maas et al. 1986, up to 4 cm in *Dictyostega orobanchoides*, Imhof 2001, Fig. 4.5c) or rather short (e.g., *Apteria aphylla* (Fig. 4.6c), Uphof 1929, *Gymnosiphon longistylus*, Hepper 1968, *G. divaricatus* (Fig. 4.6d), Maas et al. 1986, *Hexapterella gentianoides* (Fig. 4.6g)), and can be slightly tuberous (e.g., *Campylosiphon purpurascens*, Maas et al. 1986, *Campylosiphon congestus*, Fig. 4.6i–k). Many filiform, less than 0.5 mm thick, sparsely branched roots arise from the axils of the scales. Species with short rhizomes, therefore, have a star-like root system (Fig. 4.6c, d+g), but roots also emerge as tufts on longer rhizomes (Imhof 2001, Fig. 4.5c). As a peculiar exception in this group of species, *Gymnosiphon afro-orientalis* develops little tubers of unknown origin beside scale leaves and filiform roots at the short rhizome (Cheek 2009), superficially reminiscent of those found in *Afrothismia* (e.g., Fig. 4.4b), but fundamentally differing in being distinct from the filiform roots.

Anatomically, these roots are characterized by a much reduced central cylinder with one central enlarged tracheary element surrounded by a ring of much smaller tracheary elements, and a pericycle (e.g., Fig. 4.5f). The tertiary endodermis is conspicuously reinforced (e.g., *Marthella trinitatis*, erroneously called *Burmannia capitata* by Johow 1885, *Gymnosiphon refractus* (formerly *Cymbocarpa refracta*, Merckx 2008), treated under two different synonyms by Johow 1889 and Goebel and Süssenguth 1924, *Apteria aphylla*, Uphof 1929). In transverse sections of a *Dictyostega orobanchoides* root, the fortification of a single endodermal cell may even be wider than the entire central cylinder (Imhof 2001, Fig. 4.5f). This reinforcement protects the essential connection to the shoot. In fact, the thin-walled cortex tissue is often found to be disrupted (Figs. 4.5d and 4.6c, d+g) whereas the central strand is even hard to disconnect using forceps (Imhof 2001, see section on *Petrosavia* for interpretation).

The two to three parenchyma layers and, in particular, the often large-celled persistent epidermis (Johow 1889; Imhof 2001 and unpublished observations) are colonized by coils of hyphae (Uphof 1929), vesicles, as well as arbuscular-like structures, often all together within a single cell (Imhof 2001). The fungal material often appears amorphous, suggesting a digestion process (Fig. 4.5e, f).

Dictyostega orobanchoides also has fungal colonization in the scale leaves (Fig. 4.5h) as well as in the interstitials of their imbricate arangement along the rhizome (Fig. 4.5g–i), but not in the rhizome axis. These hyphae and vesicles do not show signs of degeneration, and it has been hypothesized that they serve as a refugium for the fungus, which in turn enhances the rhizomosphere with the appropriate mycobiont (Imhof 2001, see Fig. 4.5j for a schematic view). It can be interpreted as a strategy for a sustained use from the fungus, analogous to the often complex colonization pattern in other MH plants (e.g., *Voyria, Afrothismia, Triuris, Sciaphila*). More investigations might clarify the possible general relevance of rhizomes and their scale leaves for the mycorrhiza in other Burmanniaceae.

Franke et al. (2006) found several *Glomus*-group A species and an Acaulosporaceae in *Campylosiphon congestus* (treated as *Burmannia congesta*). Also, *Dictyostega orobanchoides* is associated with *Glomus*-group A species (Merckx et al. 2010), as are *Apteria* (Courty et al. 2011) and *Gymnosiphon* spp. (Dirk Redecker, pers. comm. cited in Leake 2005; Courty et al. 2011).

4.5.3.2 *Burmannia* (Fig. 4.7)

Burmannia species are more diverse with respect to their subterranean structures than their sister genera. Although they are sometimes similar to the latter (e.g., *B. championii*, Ernst and Bernard 1911), they also can have thicker roots also arising from rhizomes (e.g., *B. larseniana*, Zhang and Saunders 1999) or even vermiform, up to 2.6 mm thick roots and no (visible) rhizomes (e.g., *Burmannia candida*, Smith 1911, *B. liukiuensis*, Terashita and Kawakami 1991, *B. tenella*, Imhof 1999b, Fig. 4.7b, *B. hexaptera*, Imhof unpublished, Fig. 4.7i). Others have tuberous organs of uncertain nature (*Burmannia hunanensis*, Liu et al. 2001), with filiform roots. However, more taxonomic investigation may result in new classifications resolving some of this subterranean diversity, as in fact, *Burmannia congesta*, having a tuberous rhizome, only recently was attributed to *Campylosiphon* (Fig. 4.6i–k) by molecular and morphological data (Merckx 2008; Maas-van de Kamer and Maas 2010).

Root anatomy is also diverse. Epidermal cells may be conspicuously enlarged (Johow 1889; Ernst and Bernard 1911, 1912; Bernard and Ernst 1914) or not (Colozza 1910; Ernst and Bernard 1911; Imhof 1999b, Fig. 4.7d). Depending on the variability of root thickness, the cortex parenchyma layers can be from three to many (Janse 1896; Ernst and Bernard 1911; Larsen 1963; Terashita and Kawakami 1991; Imhof 1999b, Fig. 4.7d), and can be uniform (Imhof 1999b) or heteromorphic (Ernst and Bernard 1911) or with lacunae (Johow 1889; Malme 1896a; Colozza 1910). Similar to the other genera of the family, the endodermis has obvious tertiary reinforcements and the central cylinder is much reduced (e.g., Malme 1896a; Ernst and Bernard 1911; Terashita and Kawakami 1991; Imhof 1999b, Fig. 4.7d).

In species with filiform roots, the mycorrhizal fungus colonizes epidermal cells (Johow 1889; Ernst and Bernard 1911), whereas in the species with thick roots, the cortex parenchyma cells are colonized (Meyer 1909; Ernst and Bernard 1911; Terashita and Kawakami 1991; Imhof 1999b, Fig. 4.7d). In the thick roots of *Burmannia tenella*, hyphal coils, vesicles and arbuscular-like structures may occur together in a single cell (Fig. 4.7e, f). A colonization pattern with compartmentation of root tissue similar to other MH plants is not obvious (Fig. 4.7d). However, a selective digestion of the thinner, arbusculate hyphae but not the thicker hyphae within cells (Imhof 1999b, Fig. 4.7e, f), seems to allow a sufficient spread of the colonization within the cortex parenchyma by the latter, while carbon and nutrients are obtained through digestion of the former.

The only *Burmannia* species which has been investigated for the identity of its mycorrhizal fungus is *B. hexaptera* (Fig. 4.7g–i). It is mycorrhizal with *Glomus*-group A species (Franke et al. 2006; Merckx and Bidartondo 2008).

4.5.4 Triuridaceae (Figs. 4.8–4.10)

Fossil specimens of this exclusively achlorophyllous family from the Upper Cretaceous (ca. 90 mya) are the oldest unequivocal monocotyledonous remnants known (Gandolfo et al. 2002). Eleven genera are grouped in three tribes, the Sciaphilae are pantropical, Triurideae neotropical (Maas-van de Kamer and Weustenfeld 1998), and Kupeaeae only occur in tropical Africa (Cheek 2003b). All genera except for *Sciaphila* and *Andruris* (included in *Sciaphila* by van de Meerendonk 1984) contain only one to three species. The affiliation of the family was long uncertain (Rübsamen-Weustenfeld 1991; Maas-van de Kamer 1995; Maas-van de Kamer and Weustenfeld 1998). Today, molecular methods have assigned them to the Pandanales, which is supported by structural features (Furness et al. 2002; Rudall and Bateman 2006).

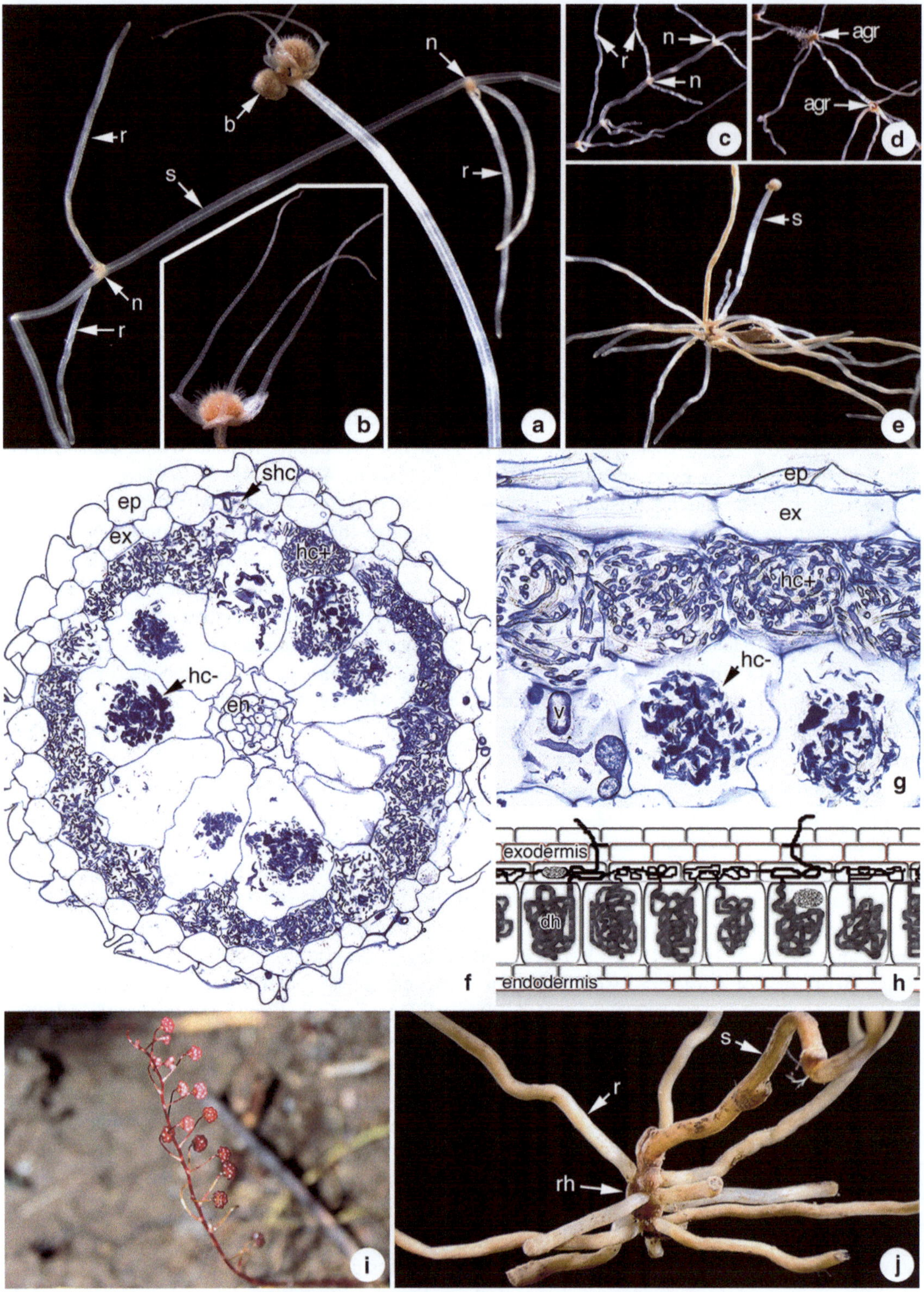

Fig. 4.8 (**a–i**) *Triuris hyalina*, (**j–k**) *Sciaphila ledermannii* (Triuridaceae). (**a**) Subterranean shoot (s) of *T. hyalina* with two nodes (n) bearing paired roots (r) in the axils of the scale leaves. The female inflorescence has one mature flower and one bud (b). (**b**) Female flower of *T. hyalina* with the characteristic tail-like tepal appendages.

With only few exceptions, the subterranean organs of Triuridaceae are rather uniform. The epiterrestrial shoots are continuous plagiotropically to orthotropically with subterranean shoot segments with various internode lengths without increasing in diameter (Fig. 4.8a+c, d). In addition to occasional side shoots, the axils of the nodal scale leaves bear pairs (sometimes solitary) of long filiform roots about as thick as the shoots (e.g., van de Meerendonk 1984; Maas and Rübsamen 1986; Maas and Maas-van de Kamer 1989), which can be glabrous (e.g., *Triuris hyalina*, Imhof 1998, *Triuridopsis intermedia*, Franke et al. 2000, Fig. 4.8a+c–e), sparsely hairy (e.g., several *Sciaphila* spp., Schlechter 1913, *Lacandonia schismatica*, Martinez and Ramos 1989) to conspicuously pilose (e.g., *Soridium spruceanum*, Miers 1852. several *Sciaphila* spp., Johow 1889; Hemsley 1907; Larsen 1972, *Andruris* spp., Schlechter 1913, *Seychellaria madagascariensis*, Fig. 4.10c). Usually the scale leaves and, consequently, the root pairs are spaced along the subterranean shoot (Fig. 4.8a+c), but there can also be dense clumps of filiform roots seemingly radiating from a single origin (e.g., several *Sciaphila*, *Triuris*, and *Peltophyllum* spp., Larsen 1972; Maas and Rübsamen 1986, Fig. 4.8d, e+j, *Seychellaria madagascariensis*, Fig. 4.10c), sometimes occurring in two or three tiers along the subterranean shoot (Fig. 4.8d). There are also a few species with more stout roots but also showing a star-like arrangement at the base of the shoot, namely the three species of the Kupeaeae (Cheek et al. 2003; Cheek 2003b, Fig. 4.10d–f), but also *Sciaphila polygyna* (Imhof 2004, Fig. 4.9a–d). *Sciaphila ledermannii* (Fig. 4.8i) has an intermediate root thickness (Fig. 4.8j). The star-like root aggregations by filiform or stout roots, even if they appear superficially very different, all follow the same developmental pattern, that is maximally one pair of shoot-borne roots per node, but are formed by the initiation of a side shoot from the scale leaf axil directly bearing a next node with scale leaf, giving rise to another pair of roots and a side shoot and so forth. The side shoots often do not elongate, which explains the abundance of roots (see details in Imhof 1998, 2004).

The tendency towards aggregations of thick and short roots seems to be characteristic for mycoheterotrophic plant families (Leake 1994, Imhof 2010, this chapter). Hence, the quite recent discovery of this feature in the Triuridaceae (Cheek et al. 2003; Imhof 2003, 2004, see Figs. 4.9d and 4.10f) was not too surprising.

The root anatomy of Triuridaceae is quite uniform also. Internal to the epidermis, there is a

Fig. 4.8 (continued) The apocarpous gynoecium is about 1.5 mm wide. (**c**) Subterranean shoots of *T. hyalina* with spaced nodes (n) where paired roots (r) arise from each scale leaf axil giving it a ladder-like appearance. The roots are uniformly 0.4 mm thick. (**d**) Each node seen in (**c**) may develop aggregations of paired roots (agr) as explained in text. (**e**) An aggregation of roots seen in (**d**) results in a star-like root system. At this stage, it may have already borne several flowering shoots (detached). A new shoot (s), 2 cm long, bearing a flower bud has developed. (**f**) Transverse section through a *T. hyalina* root measuring 0.4 mm in diameter. The epidermis (ep) is mostly free of hyphae and the exodermis (ex) is a barrier to the fungus except for the short cells (shc) with thickend outer tangential walls serving as passage cells. The outer cortex parenchyma layer bears dense hyphal coils (hc+) which do not become digested but may collapse when older. The middle parenchyma layer consists of enlarged cells containing mostly amorphous clumps of hyphal masses (hc–). The inner cortex layer of much smaller cells is free of hyphae. The endodermis (en) is only slightly suberized. (**g**) Longitudinal section through a *T. hyalina* root showing epidermis (ep), exodermis (ex), the dense hyphal coils (hc+) in the outer and the degenerated ones (hc–) in the middle cortex parenchyma layer. Occasionally vesicles (v) may occur in both layers. (**h**) Schematic view of the colonization pattern in *T. hyalina*: after penetration of epidermis and a short cell of the exodermis the hyphae start to coil and decrease their diameter while spread longitudinally and tangentially within the outer cortex parenchyma. The dense coils of narrow hyphae (see (**g**)) send branches into the middle parenchyma layer where they degenerate to amorphous clumps. Vesicles may occur in both layers. The red marked cells are impenetrable to the fungus. (**i**) Inflorescence of *S. ledermannii* showing female flowers. (**j**) Subterranean system of *S. ledermannii* consisting of a short rhizome (rh) continuous with the epiterrestrial shoot (s). The rhizome bears filiform roots (r) in this specimen up to 9 cm long and 0.8 mm thick

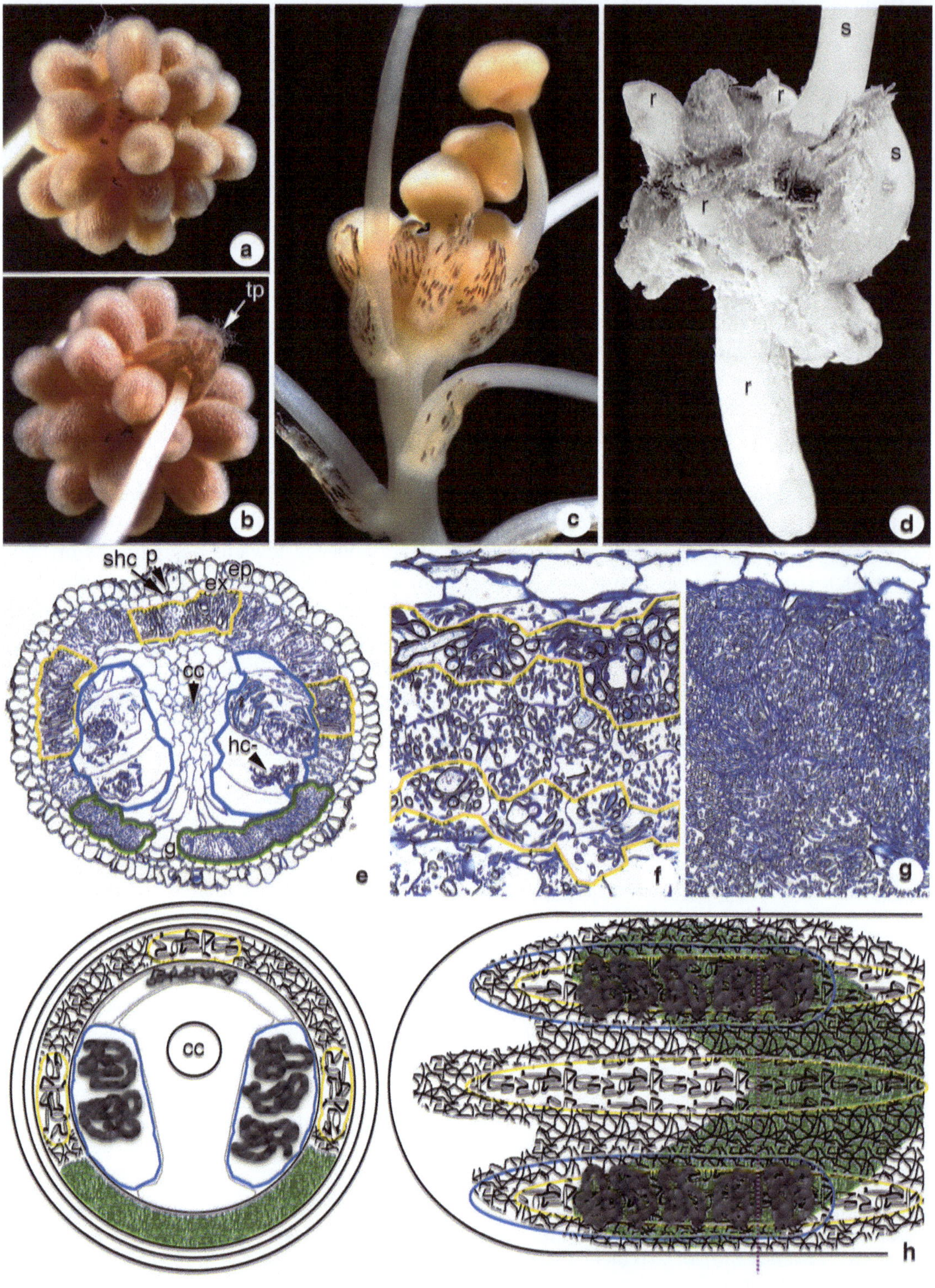

Fig. 4.9 *Sciaphila polygyna* (Triuridaceae). (**a**) Top view of a female flower of *S. polygyna* with its numerous carpels in fruiting stage (about 3 mm wide). (**b**) Same flower from the lower side showing the tepals (tp) with hair tufts. (**c**) Apex of an inflorescence of *S. polygyna* with numerous flower buds. (**d**) Subterranean system of *S. polygyna* with

suberized exodermis (Fiebrig 1921; Imhof 1998, 2003) and two (Johow 1889; Tomlinson 1982), three (Fiebrig 1921; Imhof 1998), to several (Imhof 2003) cortical parenchyma layers. The endodermis and/or pericycle may be reinforced (Poulsen 1886, 1890b; Johow 1889, Milanez and Meira 1943; Larsen 1963; Tomlinson 1982) or not (Malme 1896b; Imhof 1998, 2003). The central cylinder is much reduced. Very characteristic is the second cortex parenchyma layer, which mostly consists of conspicuously enlarged cells (Poulsen 1886, 1890b; Johow 1889; Fiebrig 1921; Tomlinson 1982; Imhof 1998, 2003), with the exception of *Sciaphila thaidanica* according to Larsen (1963, Figs. 4.8f, g and 4.9e).

Mycorrhizal colonization was recognized very early (e.g., Poulsen 1886, 1890b; Johow 1889; Janse 1896), with additional information added later (Fiebrig 1921; Ohga and Sinoto 1932; Milanez and Meira 1943; Palacios-Mayorga and Pérez-Silva 1993), but details of the colonization pattern were described rather recently (Imhof 1998, 2003; Franke 1999). *Triuris hyalina* attains a sustained benefit from the endophytic fungus by maintaining the hyphae in a functional state in the first cortical parenchyma layer and digesting them only in the enlarged cells of the second parenchyma layer (Fig. 4.8f–h, see details in Imhof 1998). An unpublished diploma thesis on *Sciaphila purpurea* (Franke 1999) not only yielded detailed information on morphology, anatomy, and ecology of the reproductive parts, but also confirmed the distinction of undigested hyphae in the outer vs. the digestion of hyphae in the inner enlarged cells of the root cortex. Beyond that, the structural diversity of the mycorrhizal colonization pattern in *Sciaphila polygyna* (Fig. 4.9a–d) is much more complicated and certainly belongs to the most complex mycorrhizas known (Fig. 4.9e–h). It includes four different morphologies of hyphae occurring in four distinct root tissue compartments. Moreover, it shows a disparate colonization at the tip compared to the base as well as the dorsal vs. the ventral side of the root, creating a monosymmetrical (only one plane of symmetry) root in transverse and longitudinal sections (Fig. 4.9h, see details in Imhof 2003). The purpose of these complex structures, except for the strictly localized digestion in the "giant cells" for a sustained carbon influx (Imhof 2003), is not yet understood.

The fungus in *Sciaphila secundiflora* (Yamato 2001, treated as *S. tosaensis*, the two treated as being synonymous by Ohashi 2000) was determined by DNA sequencing to be a *Glomus* species (Glomeromycota). More recently, *S. secundiflora* (still called *S. tosaensis*) and *Andruris japonica* (treated as *Sciaphila japonica*) were described to associate with *Glomus*-group A fungi (Yamato et al. 2011b), the phylotypes extracted from each species being closely related to another but quite distant when compared between the two species. *Sciaphila ledermannii* was also found to be colonized by a species from

Fig. 4.9 (continued) thick roots (r, several have detached) radiating from the base of two shoots (s, one is detached). For the architecture of this root system, see Imhof (2004). (**e**) Transverse section through a central part of a 1.2 mm thick root of *S. polygyna* surrounded by epidermis (ep) and exodermis (ex) with short cells (shc) serving as the only passage cells for fungal penetrations (p). Root anatomy and mycorrhizal pattern are highly heteromorphic with cells of the fourth root layer being much larger ("giant cells," *blue border*) than others dislocating the central cylinder (cc) out of its central position to create a dorsiventral architecture of the root. Fungal material degenerates (hc–) only in the fourth layer. The third layer has loose hyphal coils with swellings (*yellow border*), coils without swellings (not marked) and very dense coils of thin hyphae (*green border*, *ventral side*). Colonization by dense coils follows a v-shaped pattern leaving a gap of colonization (g) in some parts of the roots (see *right hand side* of (**h**)). (**f**) Tangential section through the dorsal side of a *S. polygyna* root showing areas colonized by coils with many (*yellow border*) and with less swellings. (**g**) Tangential section through the ventral side of the third root layer at the same magnification as (**f**) indicating the differences of the three types of coils in the third layer. (**h**) Schematic view of the mycorrhizal colonization pattern in *S. polygyna* in transverse (*left*) and longitudinal view (*right*). The coloration corresponds to those in (**e**, **f**). Note that the different colonization morphotypes, the appearance of giant cells, the digestion of hyphae therein is also heteromorphic in the longitudinal view as it is in transverse view (see details in Imhof 2003)

Fig. 4.10 (**a–c**) *Seychellaria madagascariensis*, (**b–f**) *Kupea martinetugei*, (Triuridaceae) (**g, h**) *Geosiris aphylla*, (Iridaceae) (**i, j**) *Thismia panamensis* (Thismiaceae). (**a**) Female flowers of *S. madagascariensis* about 2 mm wide with the basal filiform styles projecting above the carpels. (**b**) Young male flower of *S. madagascariensis*.

Glomus-group A as well as by an *Acaulospora* sp. (Franke et al. 2006), whereas Merckx and Bidartondo (2008) detected only a *Glomus*-group A fungi in a specimen of *S. ledermannii* from Mount Cameroon. *Kupea martinetugei* was associated with two closely related fungi of *Glomus*-group A (Franke et al. 2006), confirmed by Merckx and Bidartondo (2008).

4.5.5 Corsiaceae (Fig. 4.11)

4.5.5.1 *Arachnitis*

Roots of the inconspicuous plant *Arachnitis uniflora* (Corsiaceae, Fig. 4.11a), one of two unusual mycoheterotrophs in the genus confined to a few locations in the southern hemisphere (Dimitri 1972; Cribb et al. 1994; Ibisch et al. 1996; Domínguez and Sérsic 2004), are short and fleshy, radiating from the shoot base, and lack root hairs. Reiche (1907) described colonization of peripheral parenchyma cells in roots by endotrophic mycorrhizal fungi whereas Colozza (1910) called the plant "parassita," adding "fors'anche saprofita?" (= perhaps saprophytic?) in brackets. More recently, Minoletti (1986) referred to the colonization pattern as an ectendomycorrhiza because both intercellular and intracellular hyphae were present in the outer cortex of roots. Molecular methods have proven that roots are colonized by an AM fungus belonging to *Glomus*-Group A (Bidartondo et al. 2002). However, details of the structural characteristics of the plant-fungus interaction are unlike other plant associations with *Glomus* spp. (Domínguez et al. 2006, 2009). Unusual branched structures with inflated ends (Fig. 4.11c–g) form in addition to hyphal coils (Fig. 4.11d) in the cortical cells of the plant's fleshy roots. Arbuscules do not form and vesicles (Fig. 4.11g) rarely occur. The function of the branched structures is unknown but they, along with the hyphal coils, may be involved in the transfer of sugars from fungus to root cells (Domínguez et al. 2009). *Arachnitis uniflora* also develops unusual asexual propagules (Fig. 4.11b) on its fleshy roots (Domínguez et al. 2006) that are colonized by fungi from the parent root before they detach. The propagules develop a shoot apical meristem and adventitious roots and ultimately new plants that presumably link to neighboring photosynthetic plants for their source of carbon (Domínguez et al. 2006, 2009)

4.5.5.2 *Corsia*

Only a general description of roots in *Corsia* species is given by van Royen (1972); no further specific information can be found in the taxonomic section of his monograph. The roots are filiform, unbranched, white to cream-colored, growing horizontally through the humus layer. Compared to the entire plant, they are "quite sizable" and "extend over considerable distance in many directions" (van Royen 1972). They arise from short, creeping rhizomes, with sheathing scale leaves (Williams 1946; van Royen 1972). Beccari's (1877) and Schlechter's (1905) drawings, however, show some root branches in *Corsia ornata* and *C. unguiculata*, respectively. Similarly, Cribb (1985), without discussing them, depicts branching roots in *C. pyramidata*, also arising from branching rhizomes a few centimeters below the soil surface. Jones and Gray (2008), describing the only Australian species *C. dispar*, explained this discrepancy as an oversight by van Royen, since many herbarium sheets of *Corsia* spp. in the herbarium of Canberra have branched roots as well. According to Jones and Gray (2008), the rhizome in *C. dispar* is about 4 mm

Fig. 4.10 (continued) (**c**) Root aggregation of *S. madagascariensis* (compare Fig. 4.8e) showing four shoots (s) and numerous pilose roots (r) up to 10 cm long and 0.6 mm thick. (**d**) Female inflorescence of *K. martinetugei.* (**e**) Male inflorescence of *K. martinetugei.* (**f**) Root system of *K. martinetugei* with several radiating, 1.5 mm thick and up to 7 mm long, roots (r) at the base of the shoot (s). (**g**) Subterranean parts of *G. aphylla* with a rhizomatous tuber (t) and filiform roots (r) at its base. (**h**) Preserved capitulum of *G. aphylla.* (**i**) Subterranean tuber (t) of *T. panamensis* with filiform roots (r) radiating from it. (**j**) Flower of *T. panamensis* (courtesy of H and PJM Maas)

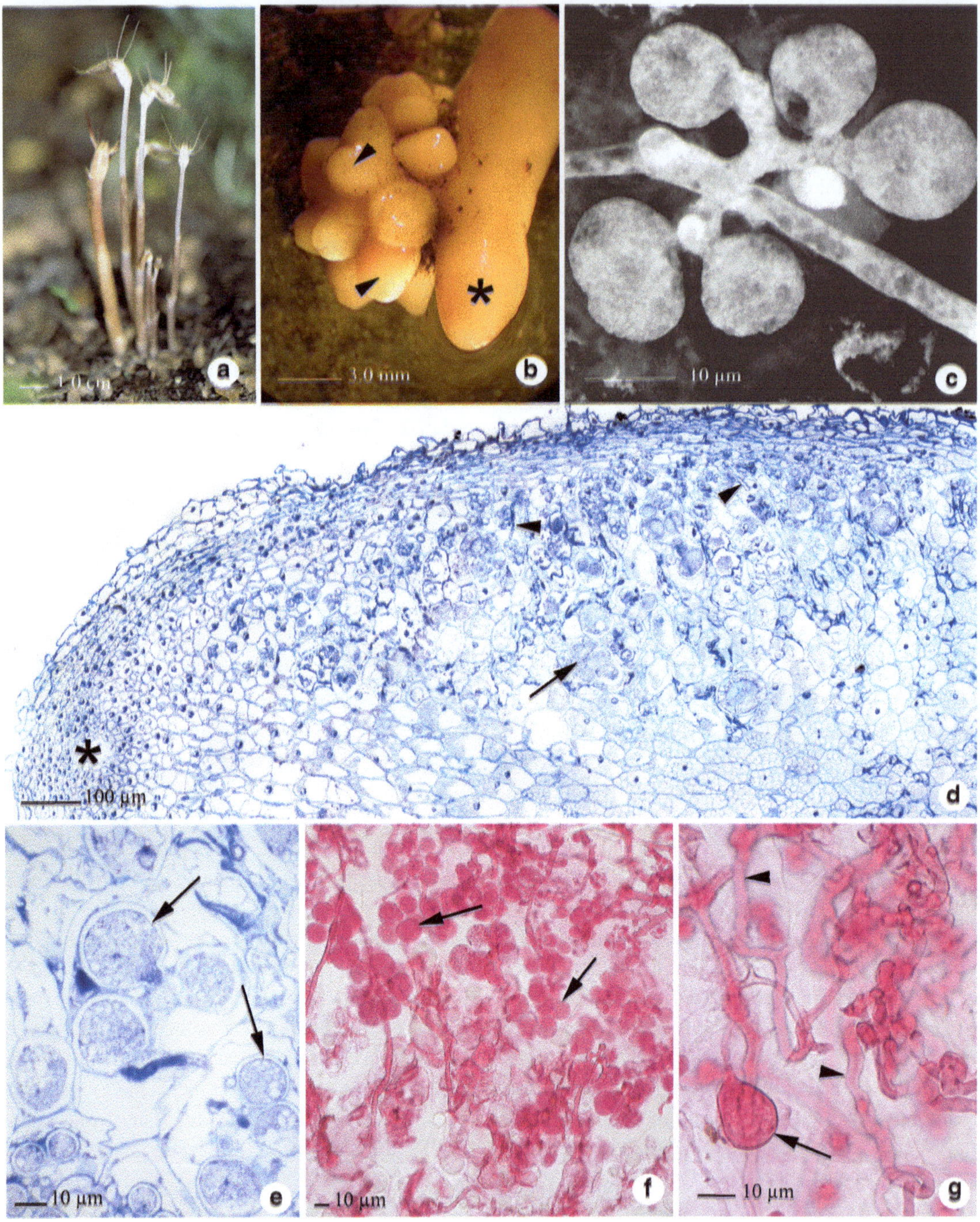

Fig. 4.11 *Arachnitis uniflora* (Corsiaceae). (**a**) Flowering stems of *Arachnitis uniflora*, each with a single flower. Image courtesy of Laura Domínguez. (**b**) Propagules (*arrowheads*) on a fleshy root (*asterisk*) of *A. uniflora*. (**c**) Confocal microscopy of intracellular branched hyphal structures of *Glomus*-Group A in a root of *A. uniflora*. (**d**) Longitudinal section of resin-embedded root of *A. uniflora* stained with toluidine blue O showing the apical meristem (*asterisk*), intracellular hyphae of *Glomus*-Group A (*arrowheads*), and intracellular branched hyphal structures (*arrow*). (**e**) Enlarged portion of a similar section of a *A. uniflora* root showing intracellular branched hyphal structures of *Glomus*-Group A (*arrows*). (**f**) Clearings of root cells of *A. uniflora* stained with acid fuchsin showing numerous intracellular branched hyphal structures of *Glomus*-Group A (*arrows*). (**g**) Clearings of root cells of *A. uniflora* stained with acid fuchsin showing intracellular hyphae (*arrowheads*) and a vesicle (*arrow*) of *Glomus*-Group A

thick and grows in annual increments. In contrast to the information given by van Royen (1972), Cribb (1985) and Jones and Gray (2008), a new variety *C. purpurata var. wiakabui* (Takeuchi and Pipoly 1998), later considered to be a separate species (Jones and Gray 2008), has a conspicuously tuberous rhizome bearing the roots (interpreted from the drawing in Takeuchi and Pipoly 1998). This is partly reminiscent of the third genus of the family, *Corsiopsis*, discussed below. No anatomical studies exist on this genus which could elucidate its mycorrhiza.

4.5.5.3 *Corsiopsis*

The monotypic *Corsiopsis chinensis* is only known from a single herbarium specimen (Zhang et al. 1999). The original description is of an ellipsoid rhizome 12–15 mm long and 5 mm in width, the drawing showing it in an orthotropous orientation. Roots were not seen.

4.5.6 Orchidaceae (Fig. 4.12)

The family Orchidaceae has the largest number of mycoheterotrophic genera of any plant family, with approximately 35 % of more than 500 fully mycoheterotrophic angiosperm species recognized (Leake 1994; Merckx et al. 2009; Imhof 2010). It is impossible to characterize the subterranean structures of all mycoheterotrophic orchid species (see Rasmussen 1995 for a thorough discussion) but a few examples will demonstrate the variability. Some species (e.g., *Cyrtosia javanica*) have rhizomes bearing fleshy adventitious roots, others (e.g., *Epipogium aphyllum*; *Corallorhiza* spp., *Rhizanthella garderi*) with rhizomes only, and others (e.g., *Wullschlaegelia calcarata*) with roots, some of which are modified as tubers.

Regardless of the nature of the underground structures, the majority of achlorophyllous orchid species are associated with fungi that form intracellular hyphal coils (pelotons) similar to those in photosynthetic orchids. These can develop within the majority of root or rhizome cortical cells and sometimes even in scale leaf tissue (Groom 1895c) and have an ephemeral existence since they undergo digestion by host cells (Smith and Read 2008). This process, termed tolypophagy (Burgeff 1932), can be repeated with recolonization by pelotons of cells containing hyphal remnants and subsequent digestion of these. Often, the cortex parenchyma is divided into an outer "Pilzwirtsschicht" (fungus host layer), where the coils do not degenerate, and an inner "Pilzverdauungsschicht" (fungus digestion layer), where digestion takes place (Magnus 1900; Burgeff 1932). Moreover, in some mycoheterotrophic species (e.g., *Gastrodia* spp.), a process called ptyophagy occurs (Burgeff 1932; Wang et al. 1997; Rasmussen 2002). While keeping the fungus host cell layers, this is characterized by only short hyphae penetrating the single-layered and particularly voluminous digestion cells and releasing their contents into it without coiling (Janse 1896; Burgeff 1932; Campbell 1962, 1963, 1964). As such, it very much resembles the "hyphal pegs" in monotropoid mycorrhizas (Lutz and Sjolund 1973; Duddridge and Read 1982). It is open to speculation if this can be interpreted as an evolutionary progression within orchid mycorrhiza, from non-differentiated colonization pattern (see e.g., Peterson et al. 2004), over the tissue compartmentation in host and digestion layers, to the ptyophagy as a special type of the latter in few MH orchids. More structural work is needed to elucidate this, but since arbuscular mycorrhizas and ectomycorrhizas seem to have undergone evolutionary progression (Imhof 2009), it would be surprising if this is not the case in orchid mycorrhizas.

Because of their "dust seeds," consisting of a rudimentary embryo and limited storage reserves, all orchid species (Fig. 4.12a) growing in native habitats require a suitable fungal partner to germinate and for the subsequent development of the protocorm (Peterson et al. 1998, 2004). The intracellular fungal hyphal coils (pelotons) are essential features for metabolite transfer into developing protocorms (Fig. 4.12b) and roots (Fig. 4.12c). All orchid species can, therefore, be considered to be mycoheterotrophic during this early stage of their life cycle (Leake 2004). The fungi involved are basidiomycete anamorphs such as *Ceratorhiza*, *Epulorhiza* and *Moniliopsis* which are capable of enzymatically reducing

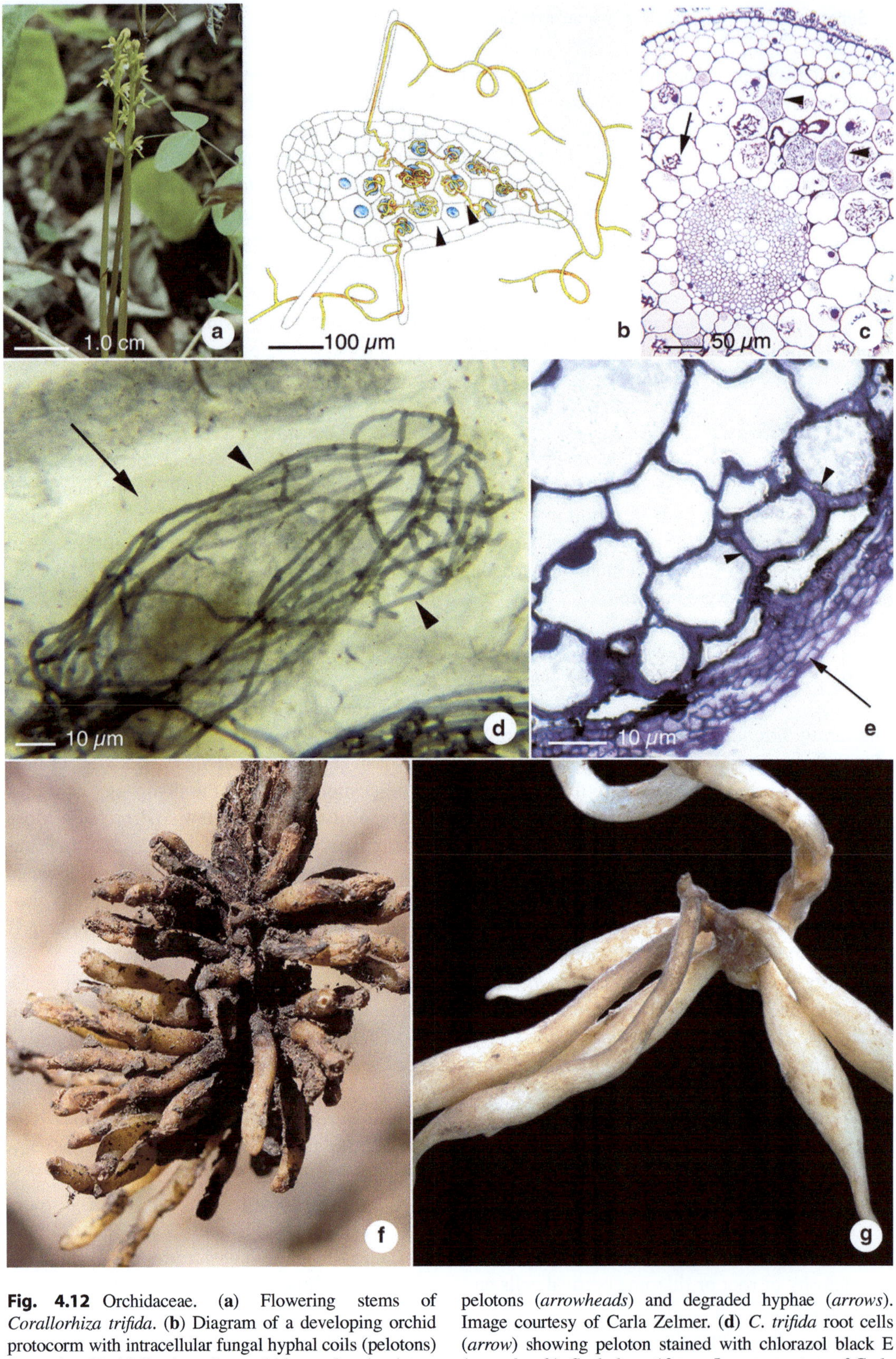

Fig. 4.12 Orchidaceae. (**a**) Flowering stems of *Corallorhiza trifida*. (**b**) Diagram of a developing orchid protocorm with intracellular fungal hyphal coils (pelotons) (*arrowheads*). (**c**) Section of an orchid root showing intact pelotons (*arrowheads*) and degraded hyphae (*arrows*). Image courtesy of Carla Zelmer. (**d**) *C. trifida* root cells (*arrow*) showing peloton stained with chlorazol black E (*arrowheads*). Scale bar = 10 μm. Image courtesy of Carla

complex carbohydrates to simple sugars that are not only used for fungal growth but are also transferred to developing protocorms to enable seedling establishment to occur.

Although the majority of orchid species develop photosynthetic adult plants, a considerable number of genera remain dependent on mycorrhizal fungi for carbon compounds throughout their life cycle and therefore continue to be mycoheterotrophs. In these situations, developing seedlings link to photosynthetic plant species via fungal mycelium, mostly belonging to members of the Basidiomycota (Taylor and Bruns 1997). Zelmer and Currah (1995) demonstrated that the fungus isolated from roots of *Corallorhiza trifida*, although not identified, formed pelotons in *Corallorhiza trifida* root cells (Fig. 4.12d) and typical ectomycorrhizas with lodgepole pine (*Pinus contorta*, Fig. 4.12e). It was recently demonstrated by Zimmer et al. (2008) that the fungal symbiont associated with *C. trifida* is a *Tomentella* sp. (Thelephoraceae). Another example in which seeds of *Neottia nidus-avis* (also a non-photosynthetic orchid) were enclosed in seed packets and placed either near adult plants or at some distance from them in a beech (*Fagus sylvatica*) woodland, McKendrick et al. (2002) were able to show that the genus *Sebacina* (anamorph, *Epulorhiza*) was the symbiont involved in the stimulation of seed germination. Adult plants of the mycoheterotroph *Neottia nidus-avis* remain associated primarily with the basidiomycete family Sebacinaceae (Selosse et al. 2002) whereas *Cephalanthera austinae* (another mycoheterotroph) associates with members of the Thelephoraceae (Taylor and Bruns 1997). Recently, Ogura-Tsujita and Yukawa (2008) reported the extreme specificity of the mycoheterotrophic orchid *Eulophia zollingeri* with the fungal symbiont, *Psathyrella candolleana* in the Agaricales (Basidiomycetes). In contrast, mycoheterotrophic species within the genus *Epipactis* have been reported to associate not only with members of the Basidiomycota but also with members of the Ascomycota, including *Tuber* (truffle) species (Selosse et al. 2004).

4.5.7 Iridaceae (*Geosiris*, Fig. 4.10g, h)

Unlike the other larger families comprising both autotrophic and mycoheterotrophic species (Orchidaceae, Burmanniaceae, Gentianaceae, Polygalaceae, Ericaceae), Iridaceae do not comprise morphologically intermediate species with reduced photosynthetic surface or amount of chlorophyll. *Geosiris aphylla* (Fig. 4.10g, h) and the recently described *G. albiflora* (Goldblatt and Mannings 2010) are the only mycoheterotrophic exceptions in the entire family. Systematically, *Geosiris* has been treated as a member of Iridaceae, Burmanniaceae or a family of its own (see Rübsamen-Weustenfeld et al. 1994). Within the Iridaceae, it has been considered as Nivenioideae (Goldblatt 1990; Goldblatt et al. 1987, 1998), but recently a position in its own subfamily Geosiridoideae, as suggested earlier (e.g., Thorne 1983), has been confirmed (Goldblatt et al. 2008).

Geosiris aphylla has an orthotropous, cormlike, oval to elongate rhizome with numerous scale leaves. The flowering shoots arise from the apical tip of this tuber-like organ, whereas at its base numerous filiform roots develop (Fig. 4.10g), similar to the base of onion bulbs. Anatomically, the rhizome consists of a wide cortex parenchyma surrounding a vascular cylinder with occasional gaps due to leaf and bud traces. A thin-walled epidermis and a fortified endodermis

Fig. 4.12 (continued) Zelmer and Randy Currah. (**e**) The same fungus isolated from *C. trifida* root cells and inoculated on *Pinus contorta* roots formed typical ectomycorrhizas with a mantle (*arrow*) and Hartig net (*arrowheads*). Scale bar=25 μm. Image courtesy of Carla Zelmer and Randy Currah. (**f**) Subterranean system of *Neottia nidus-avis*, consisting of numerous roots, 1–2.5 cm long and 2 mm thick, emerging from a short orthotropous rhizome. (**g**) Subterranenean system of *Wullschlaegelia calcarata*, with spindle-shaped root tubers (max. 2 cm long and 2 mm thick) at a short rhizome around the vascular cylinder are present and some isolated amphivasal (xylem around phloem) bundles are found in the ground tissue of the central pith. The parenchyma cells, particularly in the pith, contain starch grains (Goldblatt et al. 1987). The roots also have a thin-walled epidermis, the cortex has four layers of parenchyma cells and a strongly fortified tertiary endodermis (Goldblatt et al. 1987). As pointed out previously (see *Petrosavia*), this again corroborates the view of Imhof (2010), who considers a strong tertiary endodermis as one of the common adaptations of monocotyledonous MH plants in order to secure the essential linkage of roots and shoots.

The only anatomical work on *G. aphylla* (Goldblatt et al. 1987), aside from Rübsamen-Weustenfeld et al. (1994) studying embryology, does not mention any fungal colonization of roots or rhizomes. This is rather curious, since a carbon source for this non-photosynthetic plant is mandatory. A parasitic mode of life (*sensu* Weber 1993) is highly unlikely because, in contrast to the roughly 4,500 eudicotyledonous parasitic plants, monocots have never been found to be parasitic (Raynal-Roques and Paré 1998; Heide-Jørgensen 2008). Possibly, Goldblatt et al. (1987) may have overlooked the mycorrhizal structures. Therefore, further anatomical investigations focusing on the putative mycorrhiza of this species are necessary.

4.6 Eudicots

4.6.1 Polygalaceae (*Epirixanthes*, Fig. 4.13)

Epirixanthes from Southeast Asia is the only genus in the Polygalaceae entirely devoid of chlorophyll, although there are other species in *Polygala* and *Salomonia* (e.g., *Polygala setacea* (southeast USA) and *Salomonia ciliata* (Southeast Asia and northern Australia) that also show reductions in photosynthetic surface). The taxonomic accounts (e.g., Smith 1912; Ridley 1922; Backer and van den Brink 1963; van der Meijden 1988; Hsieh et al. 1995; Pendry 2010) of the six species of *Epirixanthes* do not yield information on the subterranean organs. However, there are two older and one contemporary study on the mycorrhizal roots of *E. papuana*, *E. elongata*, and *E. cylindrica* (Penzig 1901; van der Pijl 1934; Imhof 2007).

The rhizome of *E. papuana* and *E. elongata* is only a few millimeters long and continuous with the aerial shoot. The scale leaf axils give rise to sparsely branched filiform roots that are up to 12 cm long and have a maximum diameter of 0.65 mm (Imhof 2007, Fig. 4.13a+d). A primary root was never found. Since the rhizome is short, the roots seem to be radiating from the shoot

Fig. 4.13 (continued) (**c**) Inflorescence of *E. papuana*. (**d**) Subterranean system of *E. elongata* similar to that of *E. papuana* seen in (**a**). This specimen has basal shoot ramifications. Labels as in (**a**). (**e**) Longitudinal section through the cortex of an *E. papuana* root showing a part of the straight, cascading hyphae (ch) in the outer parenchyma, coiled hyphae (hc+) in layer 2 (l2) and degenerated coils (hc−) in layer 1 (layers counted from the endodermis). (**f**) Longitudinal section through parenchyma layers 1 (l1) and 2 (l2) of a *E. elongata* root. Layer 2 contains initially straight hyphae sending hyphal branches (hb1) into layer 1 where they immediately degenerate to amorphous clumps (hc−). (**g**) Tangential section external to the central cylinder through an *E. papuana* root showing 2 cell rows of each layer 1 (l1), 2 (l2) and 3 (l3). Hyphae in layer 2 remain functional (hc+) and send branches centripetally into layer 1 (hb1) as well as centrifugally into layer 3 (hb3), both of which digest the fungal material (hc−). A part of the cascading hyphae coming from the outer cortex layers is also visible (ch). (**h**) Transverse section through an *E. papuana* root. The epidermis (ep) as well as the outer three cortex parenchyma layers are not colonized by hyphal coils, layer 1 (l1) and 3 (l3) contain degenerated coils whereas layer 2 (l2 and *dotted line*) contains functional hyphae. The central cylinder inside the endodermis (en) is largely composed of lignified fibers. (**i**) Schematic view of the mycorrhizal colonization pattern in *Epirixanthes* spp.. After penetration, the hyphae grow straight in a cascading manner through the outer cortex (1), retain the straight growth when reaching layer 2 but send branches into layer 1 for digestion (2), start to coil hyphae in layer 2 when the mycorrhization proceeds (3) and then also send branches in layer 3 for digestion (4)

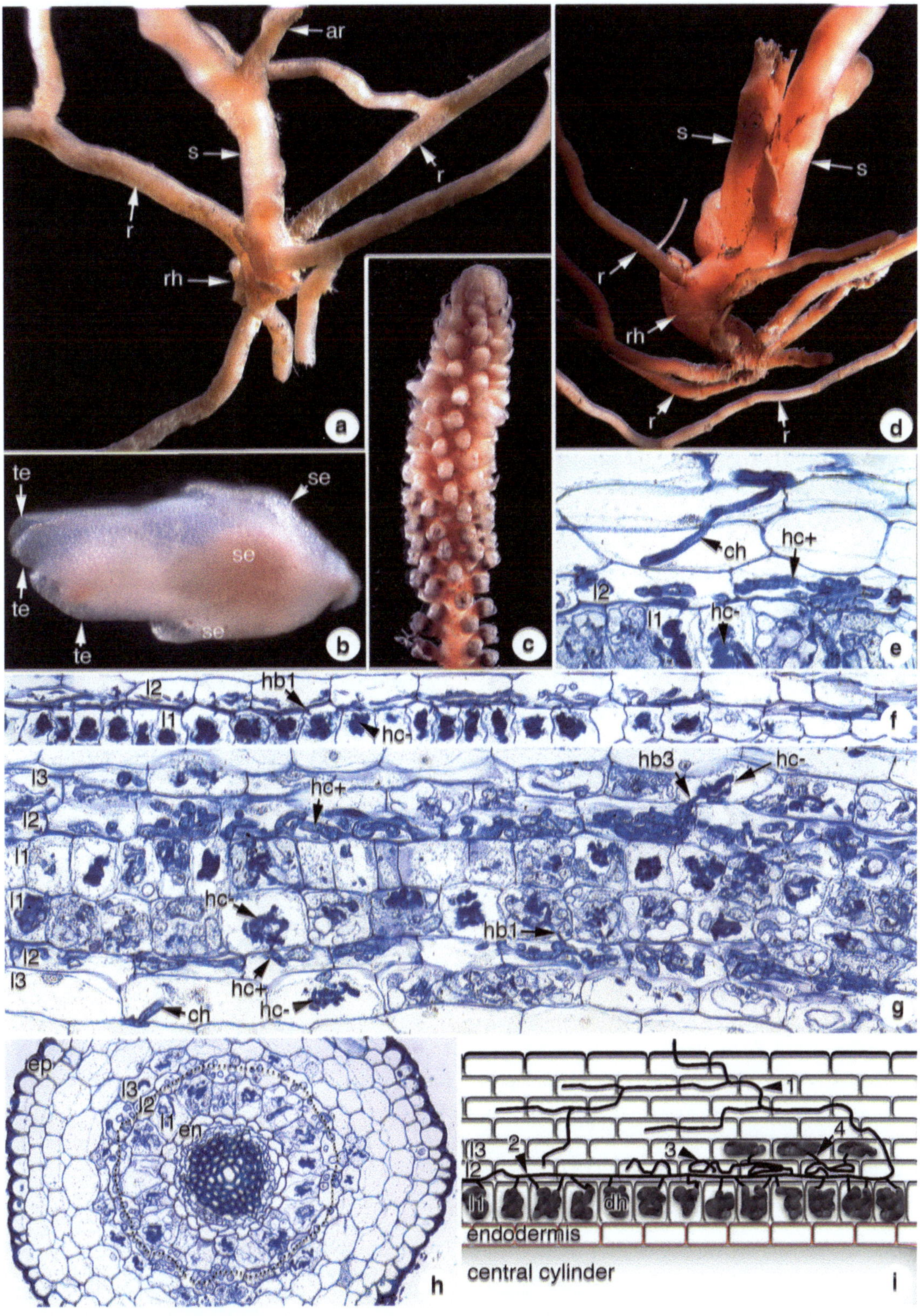

Fig. 4.13 *Epirixanthes* spp. (Polygalaceae). (**a**) Subterranean system of *E. papuana* with approximately 0.6 mm thick roots (r) arising from a short rhizome (rh) continuous with the epiterrestrial shoot (s). Additional roots (ar) may develop along the shoot where it is connected to the soil. (**b**) Isolated flower of *E. papuana* (spirit material), little more than 2 mm long, with three tepals (te). Not all of the five sepals (se) are visible.

base, but additional roots may develop along the shoot when it is still covered by soil or litter substrate (Imhof 2007, Fig. 4.13a). *Epirixanthes cylindrica* has thicker roots (up to 0.75 mm diameter) and the rhizome bearing the roots is longer (Penzig 1901). Roots of *Epirixanthes* have a triarch central cylinder with many lignified fibers, a pericycle, a suberized endodermis, up to seven cell layers of cortex parenchyma, and an epidermis (Fig. 4.13h). The cells of the innermost cortex parenchyma layer are larger in radial but shorter in longitudinal direction than the other cells (Penzig 1901; van der Pijl 1934; Imhof 2007, Fig. 4.13e, f+h).

Penzig (1901) recognized the coiled intracellular hyphae, their degeneration stages, especially in parenchyma layers 1 and 3 when counted outwards from the endodermis, and the nearly fungus-free outer cortex layers. Van der Pijl (1934) added the observation that the hyphae in layer two grow in longitudinal direction and send hyphal branches into the inner layer for digestion. The whole colonization pattern, however, is more complicated and only perceivable when considering sequential sections. After penetration, the hyphae grow straight through the cells of the outer cortex, branch repeatedly and spread coarsely in this root segment until they reach layer 2 (Fig. 4.13e, layer 1 being the innermost cortex parenchyma layer). There the hyphae keep growing straight (Fig. 4.13f), but later develop hyphal branches that coil within the cells (Fig. 4.13f, g). Directly after having reached layer 2, lateral hyphae enter the anatomically distinct layer 1 where they immediately swell and degenerate (Fig. 4.13e, f). In a later stage, layer 3 is also colonized from coiled hyphae in layer 2, again with immediate degeneration in layer 3 (Fig. 4.13g, h). In contrast, the hyphae in layer 2 as well as the straight hyphae in the outer cortex, remain alive for nearly the lifetime of the root (Fig. 4.13g, h). This rather complicated colonization pattern is interpreted as a reasonable strategy in order to have maximum as well as sustained benefit from the few fungal penetration events. It includes a coarse (outer cortex colonization) as well as a fine scale distribution mode (colonization in layer 2), specialized cells for digestion, and tissue to keep the fungus alive (schematic view on Fig. 4.13i, details see Imhof 2007).

4.6.2 Ericaceae (Monotropoideae, Figs. 4.14–4.17)

Eleven genera of mycoheterotrophic species are now recognized in the Monotropoideae: *Allotropa*, *Cheilotheca*, *Hemitomes*, *Hypopitys*, *Monotropa*, *Monotropastrum*, *Monotropsis*, *Pityopus*, *Pleuricospora*, *Pterospora* and *Sarcodes*, with several species endemic to a particular continent (Wallace 1975). Molecular phylogeny work has revealed that *Monotropa uniflora* is more closely related to *Monotropastrum humile*, whereas *Monotropa hypopitys* seem to be sister of *Pityopus californicus* (Bidartondo and Bruns 2001; Tsukaya et al. 2008). Therefore, some standard taxonomies (Stevens et al. 2004; Seybold 2011) have erected *Hypopitys* as a separate genus from *Monotropa*, and now use *Hypopitys monotropa* coined by Crantz (1766).

The minute seeds of members of the Monotropoideae have underdeveloped embryos and minimal nutritive tissue and therefore depend

Fig. 4.14 (continued) mycorrhiza with mantle (*asterisk*), fungal peg (*arrow*) and flask-shaped cystidia (*arrowheads*). (**g**) Scanning electron micrograph of large calcium oxalate crystals (*arrows*) among flask-shaped cystidia. (**h**) Freehand transverse section of root showing mantle (*asterisk*) and labyrinthine branching of Hartig net (*arrowheads*). (**i**) Longitudinal section of resin-embedded root stained with Toluidine blue O showing the apical meristem (*asterisk*), and the mantle covering the root apex. (**j**) Paradermal section of resin-embedded root stained with Toluidine blue O showing labyrinthine branching of Hartig net hyphae and fungal pegs in transverse section (*arrows*). (**k**) Higher magnification of a longitudinal section of resin-embedded root stained with Toluidine blue O showing mantle (*asterisk*), Hartig net (*arrowheads*) and fungal peg (*arrow*). (**l**) Transmission electron micrograph showing detail of the fungal peg with finger-like wall depositions (*arrows*)

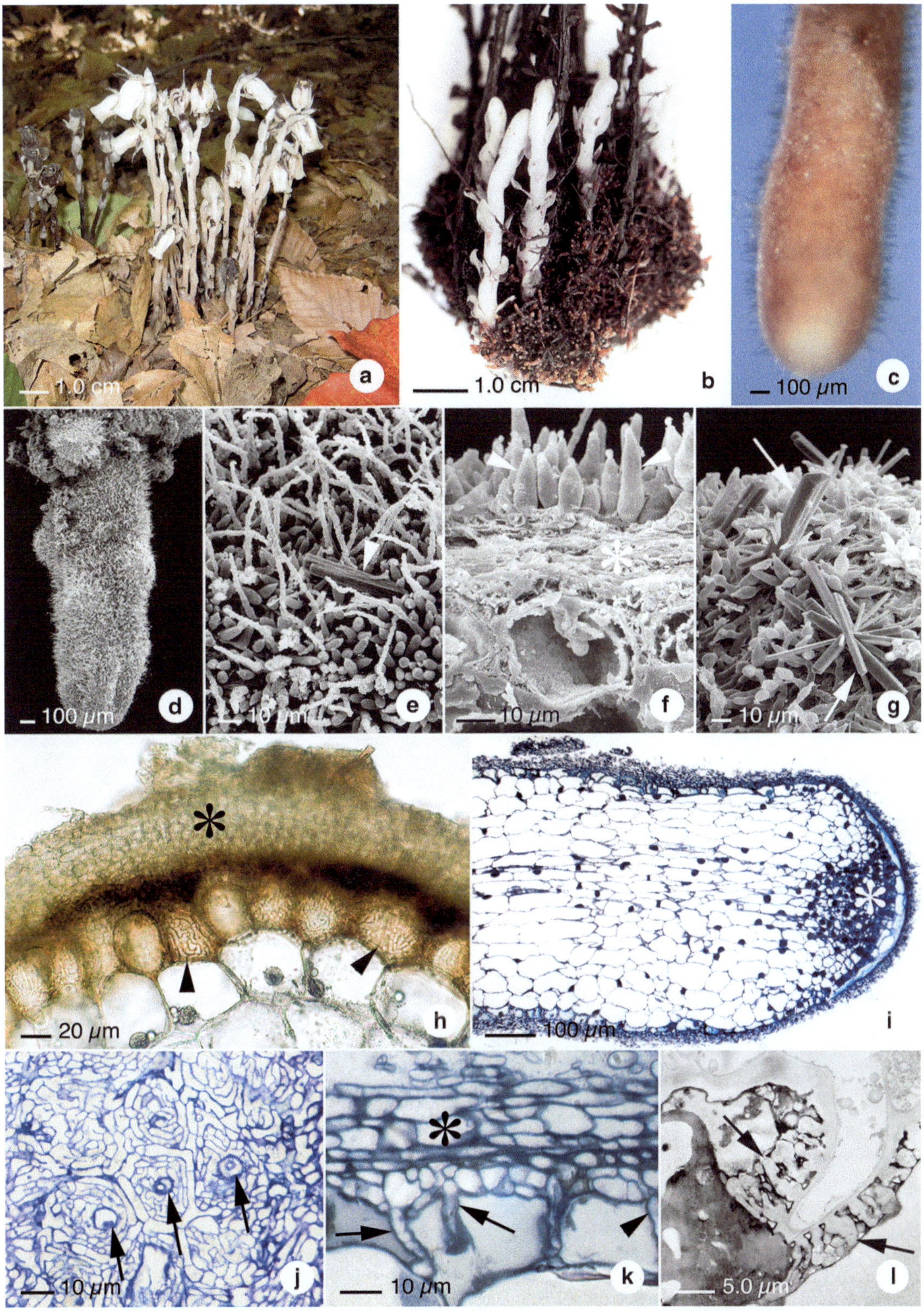

Fig. 4.14 *Monotropa uniflora* (Ericaceae/Monotropoideae). (**a**) Cluster of flowering stems in a hardwood forest in southern Ontario, Canada. (**b**) Young shoots and associated root ball. (**c**) Mycorrhizal root tip showing compact mantle with cystidia (*arrowheads*). Photo courtesy of Brent Young. (**d**) Scanning electron micrograph of a root tip showing cystidia. (**e**) Higher magnification scanning electron micrograph of portion of a mycorrhiza with a calcium-oxalate crystal (*arrowhead*) among cystidia. (**f**) Scanning electron micrograph of a fracture of a

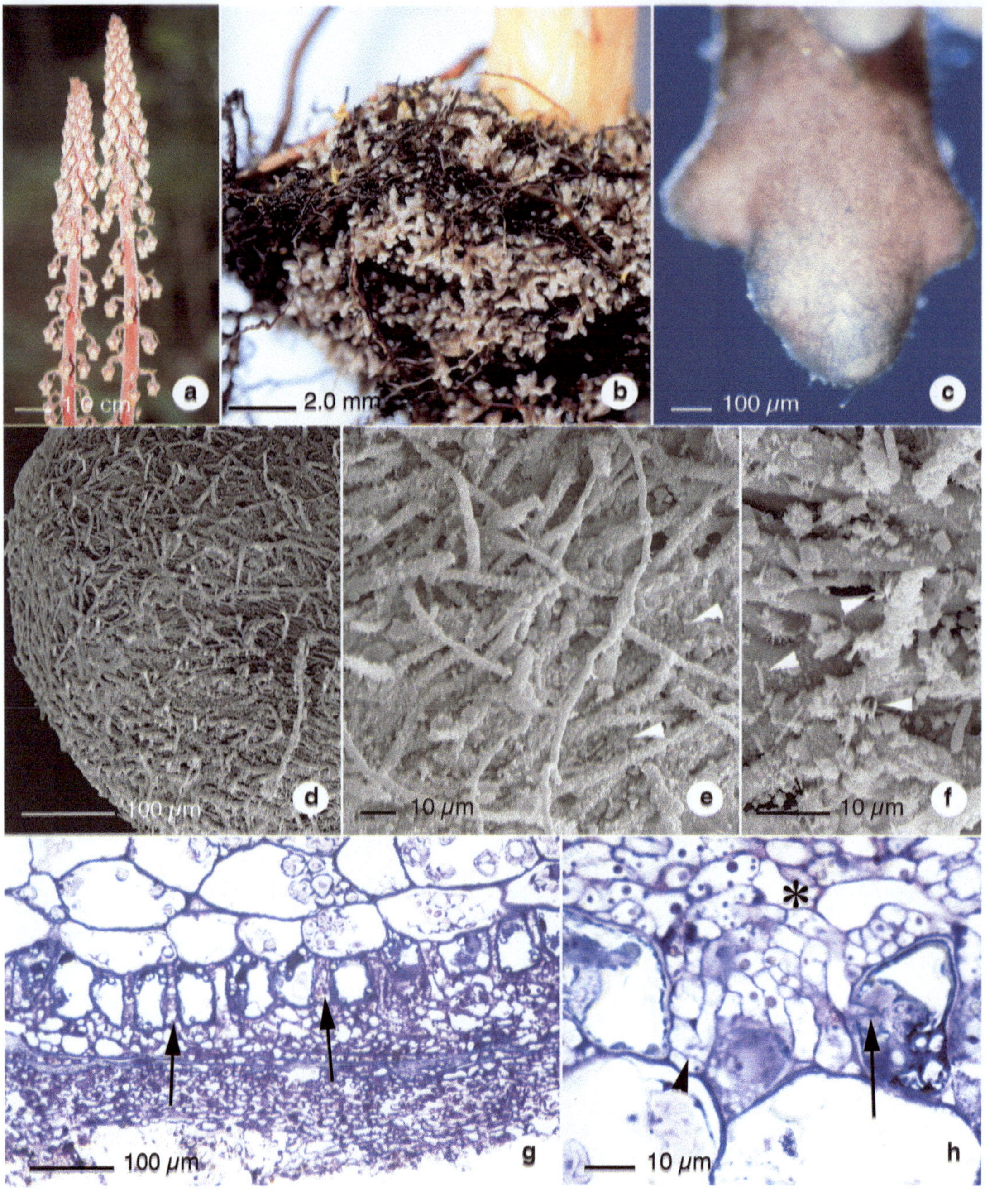

Fig. 4.15 *Pterospora andromedea* (Ericaceae/Monotropoideae). (**a**) Two flowering stems in the boreal forest in British Columbia, Canada. (**b**) Root ball showing mycorrhizal root tips. (**c**) Branched root tip showing colored mantle characteristic of a *Rhizopogon* sp. (**d**) Scanning electron micrograph of a portion of mantle showing compact hyphae. (**e**) Higher magnification scanning electron micrograph showing irregular hyphae and abundant small crystals (*arrowheads*). (**f**) Higher magnification scanning electron micrograph showing details of various crystals (*arrowheads*). (**g**) Longitudinal section of a root showing the thick mantle (*asterisk*) and Hartig net (*arrows*). (**h**) Longitudinal section of a root showing the inner mantle (*asterisk*), Hartig net (*arrowhead*) and fungal peg (*arrow*) penetrating the radial epidermal cell wall

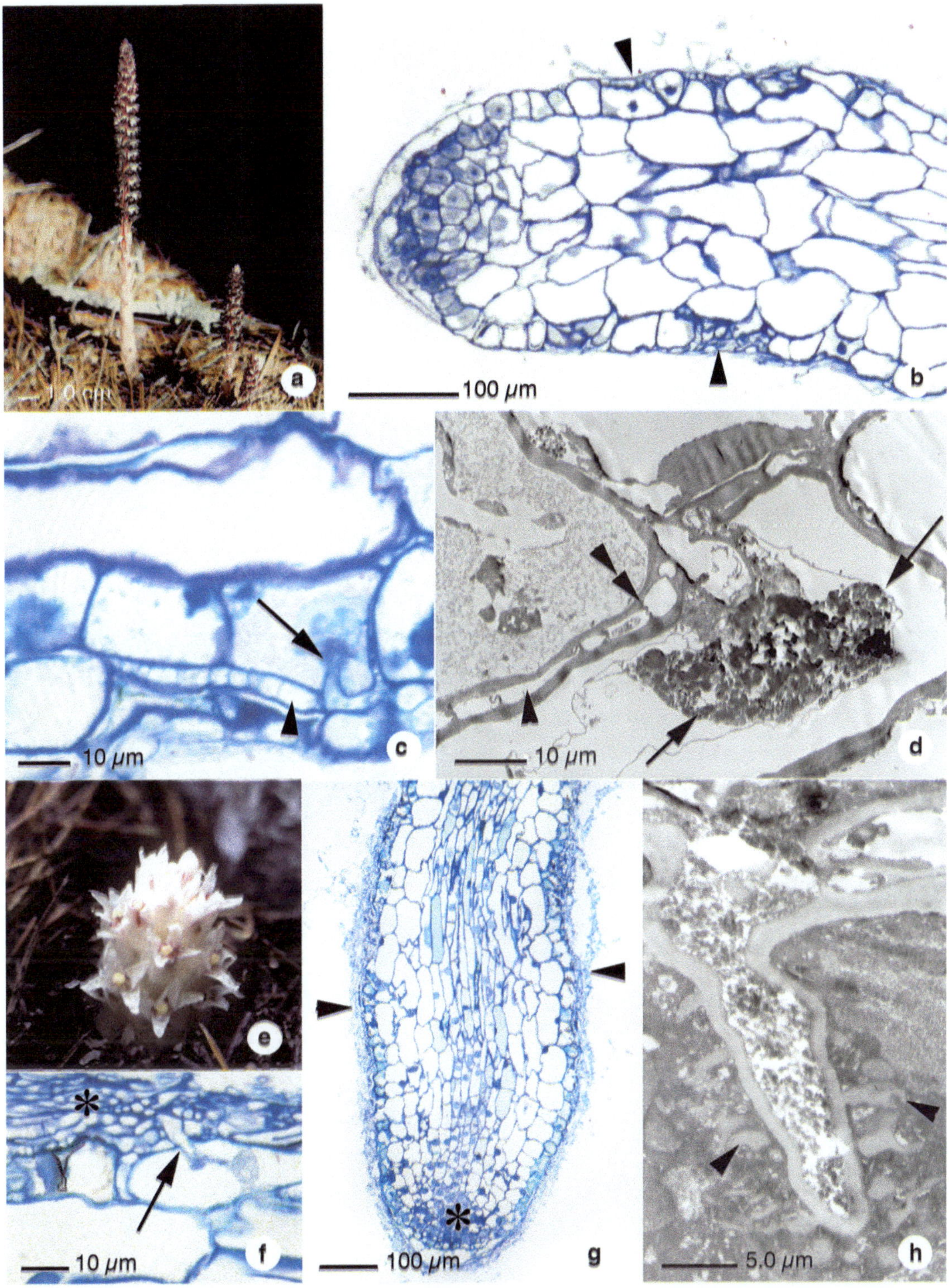

Fig. 4.16 (**a**, **b**) *Allotropa virgata*, (**e–h**) *Pleuricospora fimbriolata*, (Ericaceae/Monotropoideae). (**a**) Flowering shoots of *Allotropa virgata*. (**b**) Longitudinal section of a root showing a small apical meristem (*asterisk*) and some fungal colonization (*arrowheads*). (**c**) Higher magnification showing Hartig net (*arrowhead*) and fungal peg (*arrow*). (**d**) Transmission electron micrograph showing Hartig net (*arrowhead*) and detail of the fungal peg with finger-like wall depositions (*arrows*). (**e**) Emerged shoot with flowers of *Pleuricospora fimbriolata*. Photo courtesy of Dan Luoma. (**f**) Longitudinal section of a root showing the apical meristem (*asterisk*) and well-developed mantle (*arrowheads*). (**g**) Mantle (*asterisk*) and fungal peg (*arrow*) penetrating the outer tangential wall of an epidermal cell. (**h**) Transmission electron micrograph showing a fungal peg with finger-like wall depositions (*arrowheads*)

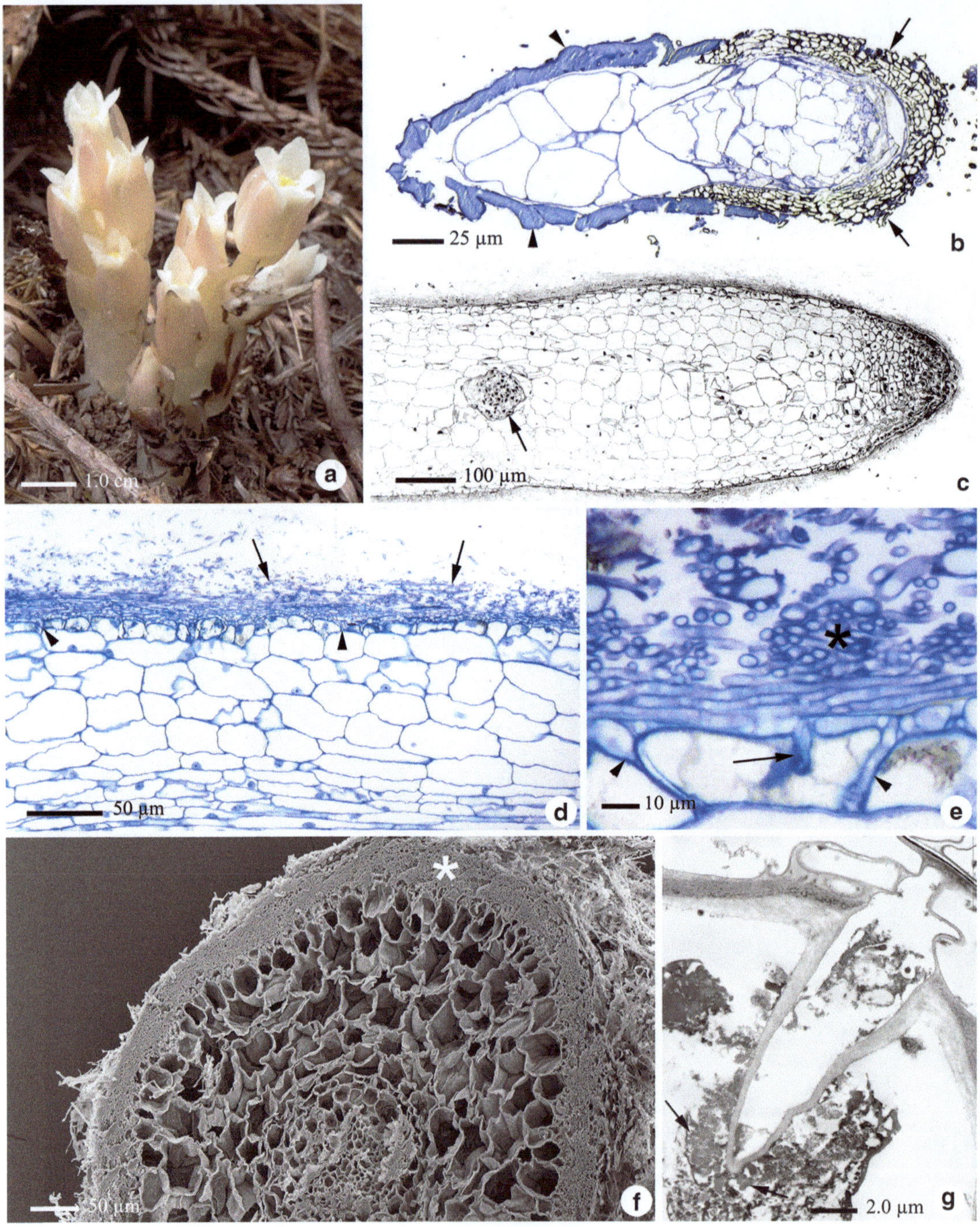

Fig. 4.17 *Pityopus californicus* (Ericaceae/Monotropoideae). (**a**) Flowering stems of *Pityopus californicus*. Photo courtesy of Barry Rice. (**b**) Developing embryo with multilayered mantle (*arrows*) on developing root. Remnants of seed coat are obvious (*arrowheads*). (**c**) Longitudinal section of an older root with mantle covering the apex. An emerging lateral root (*arrow*) is evident. (**d**) Higher magnification showing mantle (*arrows*) and Hartig net (*arrowheads*). (**e**) Detail of mantle (*asterisk*), Hartig net (*arrowheads*), and fungal peg (*arrow*) penetrating the tangential wall of an epidermal cell. (**f**) Scanning electron micrograph of fractured root showing a thick mantle (*asterisk*). (**g**) Transmission electron micrograph showing a fungal peg with finger-like wall depositions (*arrows*)

on the presence of a suitable fungus to provide sugars and perhaps other nutrients needed for germination and seedling establishment (Bruns and Read 2000; Leake et al. 2004; Smith and Read 2008). Massicotte et al. (2007) have shown that, in *Pityopus californicus*, fungal hyphae become associated with germinating seeds and form a mantle as the embryo begins to elongate. Later, a mantle, Hartig net, and fungal pegs form in the developing root. Mycoheterotrophy is therefore established very early in the life cycle of these plant species. Their root systems vary among species, ranging from large root balls comprised of numerous mycorrhizal roots (e.g., *Monotropa*, *Pterospora*), to more diffuse root systems with mycorrhizal roots distributed more randomly (e.g., *Pleuricospora*, *Monotropsis*, *Allotropa*). Hirce and Finocchio (1972) described in detail the remarkably compact root system and anatomy of *M. uniflora* and concluded that it represents a variation of the normal dicotyledonous condition. They documented a decrease in anatomical complexity of first order (hexarch stelar configuration of vascular tissue, several centimeters long and up to 1.4 mm thick, linking adjacent plants) over second order (up to 8 mm long and 0.85 mm thick) to the third order roots (protostelic arrangement, max. 4 mm long and 0.5 mm thick), even if all roots are densely covered with a mantle (and are presumably active). At the other extreme, *Allotropa* exhibits a more loose system of elongated rhizomes with first and second order adventitious roots (Massicotte et al. 2010), and likewise in *Monotropsis odorata* (treated as *Cryptophila pudica*) "the root system resembles a slender, mostly repeated many-branched, creeping rhizome" (Wolf 1922). Compared to *Allotropa*, *Pleuricospora* seems to have a slightly more condensed subterranean system (Massicotte et al. 2010).

In the Monotropoideae (formerly Pyrolaceae), a progressive compaction of the root system, from fibrous roots (e.g., *Allotropa*) to coralloid roots (e.g., *Pleuricospora*) to tight rootballs (e.g., *Monotropa*) has been hypothesized as reflecting a progressive dependence on epiparasitic mycotrophy (Furman and Trappe 1971), although this remains to be tested physiologically.

Structurally, monotropoid mycorrhizas resemble ectomycorrhizas in that a mantle and a Hartig net in this case confined to the epidermis, form (Peterson et al. 2004). However, they possess a unique feature, the invasion of epidermal cells by short hyphae originating from the Hartig net or inner mantle. These structures, referred to as fungal pegs (Lutz and Sjolund 1973; Duddridge and Read 1982; Robertson and Robertson 1982; Peterson and Massicotte 2004; Peterson et al. 2004) form either along the outer tangential wall of epidermal cells, or at the base of the radial wall of epidermal cells. Host cells respond by depositing additional cell wall material, in finger-like projections, around each peg. It has been hypothesized (Lutz and Sjolund 1973; Duddridge and Read 1982; Massicotte et al. 2005) that these structures, resembling "transfer cells" in other plant species, may be involved in nutrient transfer between the fungus and root cells although there is no experimental evidence to support this. In these systems, the Hartig net likely also plays a role in nutrient transfer but this needs to be confirmed. Kuga-Uetake et al. (2004) have shown the close association of microtubules with the fungal pegs in *M. uniflora*.

All of these species form monotropoid mycorrhizas with various fungal genera. The vast majority of fungi colonizing monotropoid roots are basidiomycetes and most of them have been identified using molecular approaches (Cullings et al. 1996; Lefevre 2002; Kretzer et al. 2000; Bidartondo and Bruns 2001, 2002). In the following paragraphs, we explore these critical features for five genera of Monotropoideae.

4.6.2.1 *Monotropa uniflora* (Fig. 4.14)

Monotropa uniflora, a northern hemisphere species (Fig. 4.14a), along with the Asian *Monotropastrum humile* (Yokoyama et al. 2005; Yamada et al. 2008; Matsuda et al. 2011) have a strong affinity for fungi in the family Russulaceae, including many species of *Russula* such as *R. brevipes*, *R. decolorans*, *R. nitida*, as well as *Lactarius* spp. (Young et al. 2002, Bidartondo 2005; Bidartondo and Bruns 2005; Yang and Pfister 2006). *Hypopitys monotropa* (=*Monotropa hypopitys*), in contrast, forms mycorrhizas mostly with fungal species in the Tricholomataceae. Excavated plants of *M. uniflora* reveal well-developed root balls (Fig. 4.14b), packed with mycorrhizal tips of Russulaceae (Fig. 4.14c), most forming numerous cystidia in the outer mantle (Fig. 4.14d). Scanning electron microscopy of the mantle surface has shown that cystidia can take various forms including, awl-shaped (Fig. 4.14e) and flask-shaped (Fig. 4.14f). As well, frequent calcium-oxalate crystals may be present (Fig. 4.14g). The mantle (Fig. 4.14h, i) and Hartig net (Fig. 4.14h) are easily observed but sectioning of mycorrhizal roots is required to show the presence and structure of fungal pegs (Fig. 4.14j–l).

4.6.2.2 *Pterospora andromedea* (Fig. 4.15)

Pterospora andromedea (Fig. 4.15a) and *Sarcodes sanguinea* (not shown), are confined to western North America and appear to associate almost exclusively with the section of *Rhizopogon* (Rhizopogonaceae) encompassing *R. ellenae*, *R. salebrosus* and *R. arctostaphyli* (Kretzer et al. 2000; Bidartondo and Bruns 2001, 2002; Taylor et al. 2002; Dowie et al., 2011). Large root balls (Fig. 4.15b) dominated with *Rhizopogon* mycorrhizas (Fig. 4.15c) are evident on excavated plants. A compact mantle with crystal inclusions of variable dimensions and shapes (Fig. 4.15d–f) is characteristic of the mycorrhizas of this species when viewed by scanning electron microscopy. Light microscopy shows a thick mantle, a well-developed Hartig net (Fig. 4.15g) and a fungal peg apparatus penetrating the radial epidermal cell wall (Fig. 4.15h).

4.6.2.3 *Allotropa virgata* and *Pleuricospora fimbriolata* (Fig. 4.16)

Allotropa virgata (Fig. 4.16a) forms mycorrhizas exclusively with *Tricholoma magnivelare* (Lefevre 2002; Taylor et al. 2002), and is one of the most specific host-fungal monotropoid symbiosis documented so far. Light microscopy reveals sporadic colonization at the root surface with a thin mantle (Fig. 4.16b) and a fungal peg penetrating radial walls of epidermal cells (Fig. 4.16c, d). *Pleuricospora fimbriolata* (Fig. 4.16e) parasitizes the fungal species *Gautieria monticola* (Bidartondo and Bruns 2001, 2002), a truffle forming species belonging to the Gomphaceae (Humpert et al. 2001). Typically, a well-developed mantle envelops the root (Fig. 4.16f) and fungal pegs, penetrating the outer tangential walls of the epidermis (Fig. 4.16g), are obvious. Characteristic finger-like wall depositions are found on the fungal peg (Fig. 4.16h).

4.6.2.4 *Pityopus californicus* (Fig. 4.17)

Pityopus californicus (Fig. 4.17a) also forms mycorrhizas mostly with fungal species in the Tricholomataceae, in this case *Tricholoma myomyces* (Bidartondo and Bruns 2005). However, a developmental study on young mycorrhizal embryos of *P. californicus* suggests other fungi are present in earlier stages (Fig. 4.17b) that are presumably replaced at later stages by *T. myomyces* (Fig. 4.17c, Massicotte et al. 2007). Mature mycorrhizal roots typically show a thick mantle (Fig. 4.17d + f), a well-developed Hartig net and a fungal peg, penetrating the outer tangential wall of epidermal cells (Fig. 4.17e). Small finger-like projections can be seen on the fungal peg (Fig. 4.17g).

4.6.3 Gentianaceae (Figs. 4.18–4.20)

In the Gentianaceae, 25 species in four genera are mycoheterotrophic. Additional to the genera covered here, two others are also considered to be at least partially "saprophytic" (Johow 1889; Knoblauch 1894; Gilg 1895; Holm 1897, 1906; Perrot 1898; Wood and Weaver 1982). The monotypic *Obolaria virginica* has scale-like leaves along the lower stem with larger spatulate-obdeltoid leaves towards the inflorescence. The fleshy stem and leaves are purplish-green. The roots are coralloid and mycorrhizal (Holm 1897; Gillett 1959; Wood and Weaver 1982), like many of the *Voyria* species described below. *Bartonia* comprises four species, *B. virginica*, *B. verna*, *B. paniculata*, the latter of which has two subspecies (Gillett 1959), and *B. texana* (Correll 1966). All species have only scale leaves but an overall greenish appearance. Compared to *Obolaria*

Fig. 4.18 (**a–c**) *Voyria tenuiflora*, (**d, e**) *Voyria obconica*, (**f–h**) *Voyria spruceana*, (**i–k**) *Exochaenium oliganthum* (Gentianaceae). (**a**) *V. tenuiflora* in its natural habitat. (**b**) Coralloid shaped root system of *V. tenuiflora* with branched roots (r) clumped at the base of a shoot (s). (**c**) Subterranean organs of *V. tenuiflora* showing the tendency to radiating roots (r) at the base of a shoot (s). The roots can be up to 1 mm thick and several centimeters long. (**d**) *V. obconica* in its natural habitat (courtesy of H and PJM Maas). (**e**) Subterranean system of *V. obconica* with stout, up to 1.5 mm thick and 1 cm long roots (r) at the base of a shoot (s). (**f**) Characteristic fringed, tail-like thecae appendages (ap) of *V. spruceana*. (**g**) Coralloid shaped root system of *V. spruceana*, also having the tendency for a star-like structure (this specimen measuring 14 mm in maximal extension). (**h**) Preserved flower (1.2 cm long) of *V. spruceana*. (**i**) Vermiform to filiform root of *E. oliganthum* with thickenings up to 0.8 mm where a light brown coloration indicates fungal colonization (fc). (**j**) *E. oliganthum* tends to develop radiating roots (r) at the base of a shoot (s). (**k**) Two preserved flowers (7 mm long) of *E. oliganthum*

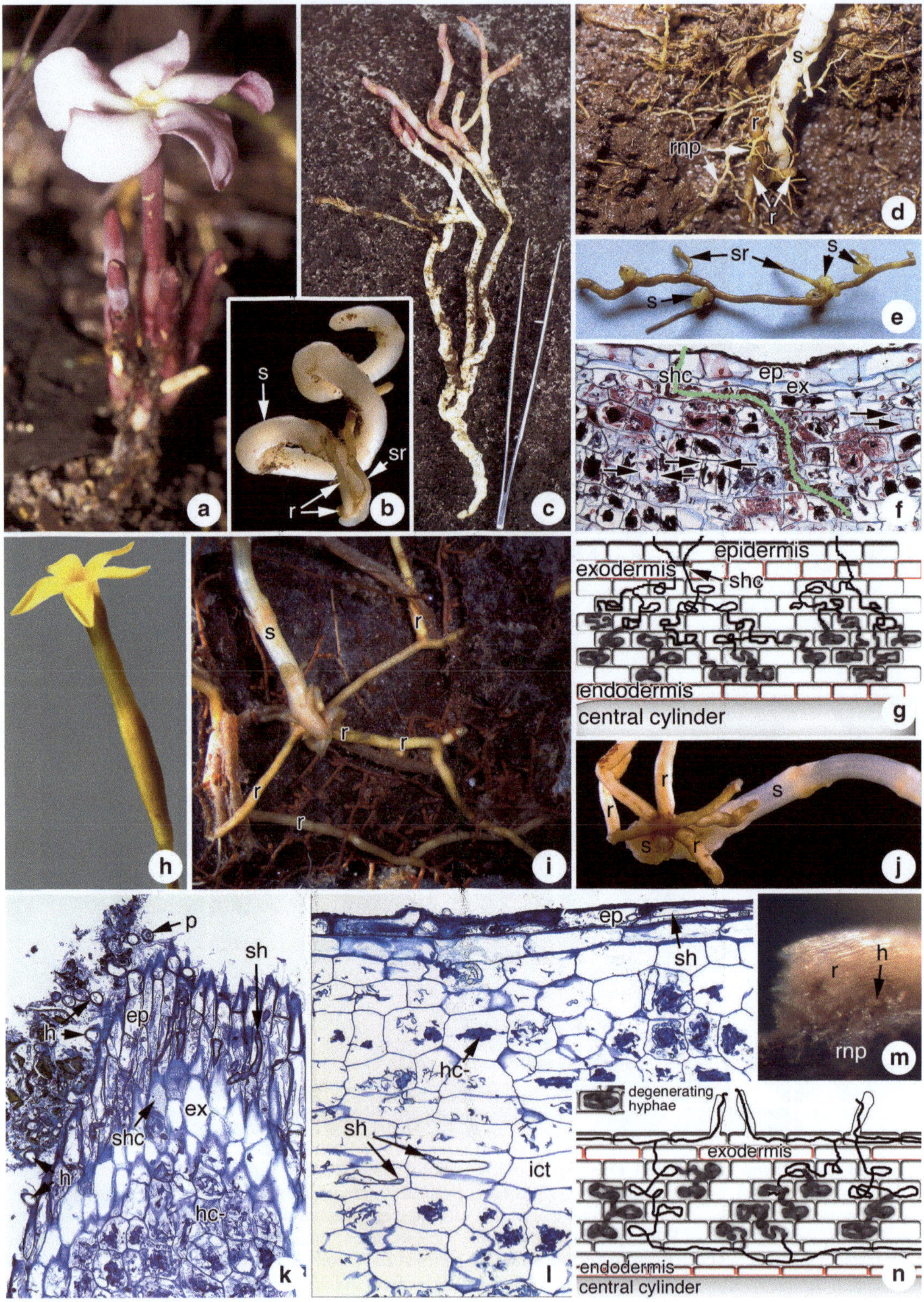

Fig. 4.19 (**a–g**) *Voyria truncata*, (**h–n**) *Voyria aphylla* (Gentianaceae). (**a**) Epiterrestrial part of *V. truncata*. (**b**) Subterranean shoot (s) of *V. truncata*, spirally bent due to soil obstructions. The shoot arises from the axil between a main root (r) and a side root (sr). (**c**) Complete specimen of *V. truncata* extracted from the soil, basally arising from

virginica, the *Bartonia* species are more delicate, but also have fleshy, sparsely branched mycorrhizal root systems (Holm 1906). Recent physiological investigations using a stable isotope distribution approach found strong indications for a partial mycoheterotrophy in *Obolaria virginica* and *Bartonia virginica* (Cameron and Bolin 2010).

4.6.3.1 *Exacum*

Cotylanthera was originally a genus comprising four achlorophyllous species, which was suspected to be closely related to the large genus *Exacum* (e.g., Raynal 1967a, Klackenberg 1985, 2002, Yuan et al. 2003); the genus has now been formally transferred into *Exacum* by Klackenberg (2006). The taxonomic accounts (e.g., Miquel 1856; Gray 1871; Clarke 1885; Gilg 1895; Lace 1914; Hara 1975) mostly do not refer to the subterranean parts, but there are two rather old but quite detailed morphological descriptions of the roots for *Exacum tenue* (Janse 1896; Figdor 1897, treated as *Cotylanthera tenuis*). Janse (1896) described the roots as tufted around the stem base. Such star-like root systems are a common feature of most MH plants and interpreted as a strategy to decentralize carbohydrate and nutrient transport: a root system with many but hierarchically equivalent roots can better compensate for the failure of some of its elements than few but high-capacity roots, as is the case in most allorhizic root systems (Imhof 2010). Combining the information given by Figdor (1897) and Imhof et al. (1994), the former showing a seedling of *E. tenue*, the latter describing the ontogeny of another gentian (*Voyria tenella*), the star-like root system of *E. tenue* is probably generated by a primary root, developing ray-like lateral roots, and only then does this structure give rise to a root-borne shoot. Later, further root-borne shoots are a means of vegetative propagation. Root hairs, if present, are much reduced (Figdor 1897). A considerable portion of the stem, up to half its length, may also be subterranean, often bent due to

Fig. 4.19 (continued) a plagiotropous root (detached). Only the upper, reddish branches were epiterrestrial and, hence, appeared superficially as clustered but distinct individuals. (**d**) Subterranean shoot (s) of *V. truncata* arising from a runner-like, plagiotropic root (r), intermingled with roots of neighboring plants (rnp). (**e**) Runner-like root (7 cm long and up to 2 mm thick) of *V. truncata* extracted from the soil with several side roots (sr) in the axils of which one or at most two root-borne shoots (s) develop. (**f**) Longitudinal section through a *V. truncata* root showing the epidermis (ep) and multilayered cortex with intracellular hyphal coils in various stages of degradation. The *green dotted line* indicates a course of colonization from penetration to the inner cortex. The passage through a short cell (shc) of the exodermis (ex) is not visible on this section but present on the subsequent one (not shown). The pattern of newly inserted cell walls (*arrows*) indicate an ongoing primary thickening. (**g**) Schematic view of the mycorrhizal colonization pattern in *V. truncata*. After penetration of the epidermis and a short cell (shc) of the exodermis as the only passage cells, the hyphae grow in a coiling manner from cell to cell deeper into the cortex. Extent of hyphal degradation increases with cortex depth. (**h**) Flower of *V. aphylla*. (**i**) Shoot (s) of *V. aphylla* arising from a net of runner-like, up to 0.5 mm thick roots (r), much smaller than in *V. truncata*. (**j**) Radiating roots (r) at the base of a shoot (s) of *V. aphylla*. (**k**) Tangential section through the cortex of a *V. aphylla* root showing the epidermis (ep) with straight, nondegenerated hyphae (sh), the exodermis (ex) with only the short cells (shc, anatomically not particular distinct) being used as passage cells for the hyphae, and the cortex parenchyma with often degenerated hyphal coils (hc–). Root hairs (h) only occur where organic material is attached to the root. Fungal penetrations (p) mostly happen via the root hairs. (**l**) Tangential section just external to the central cylinder through a *V. aphylla* root, showing the epidermis (ep) with straight hyphae (sh) and mostly degenerated hyphal coils (hc–) in the cortex parenchyma. However, straight hyphae (sh) also occur in the innermost parenchyma layers (ic), linked to the hyphae in the epidermis by nondegenerated coiled hyphae (not shown). (**m**) Root of *V. aphylla* (r) attached to a root of a neighboring plant (rnp). Root hairs (h) develop only at such root to root connections. (**n**) Schematic view of the mycorrhizal colonization pattern in *V. aphylla*. After penetration of root hairs, the hyphae grow straight in the epidermis, cross the exodermis via passage cells (miss the red coloration), build coils in the outer cortex parenchyma which partly degenerate but also reach the innermost cortex parenchyma layers, where they again grow in a straight manner along the root axis. From these inner straight hyphae which do not become digested, branches grow back into the outer cortex to build coils for digestion

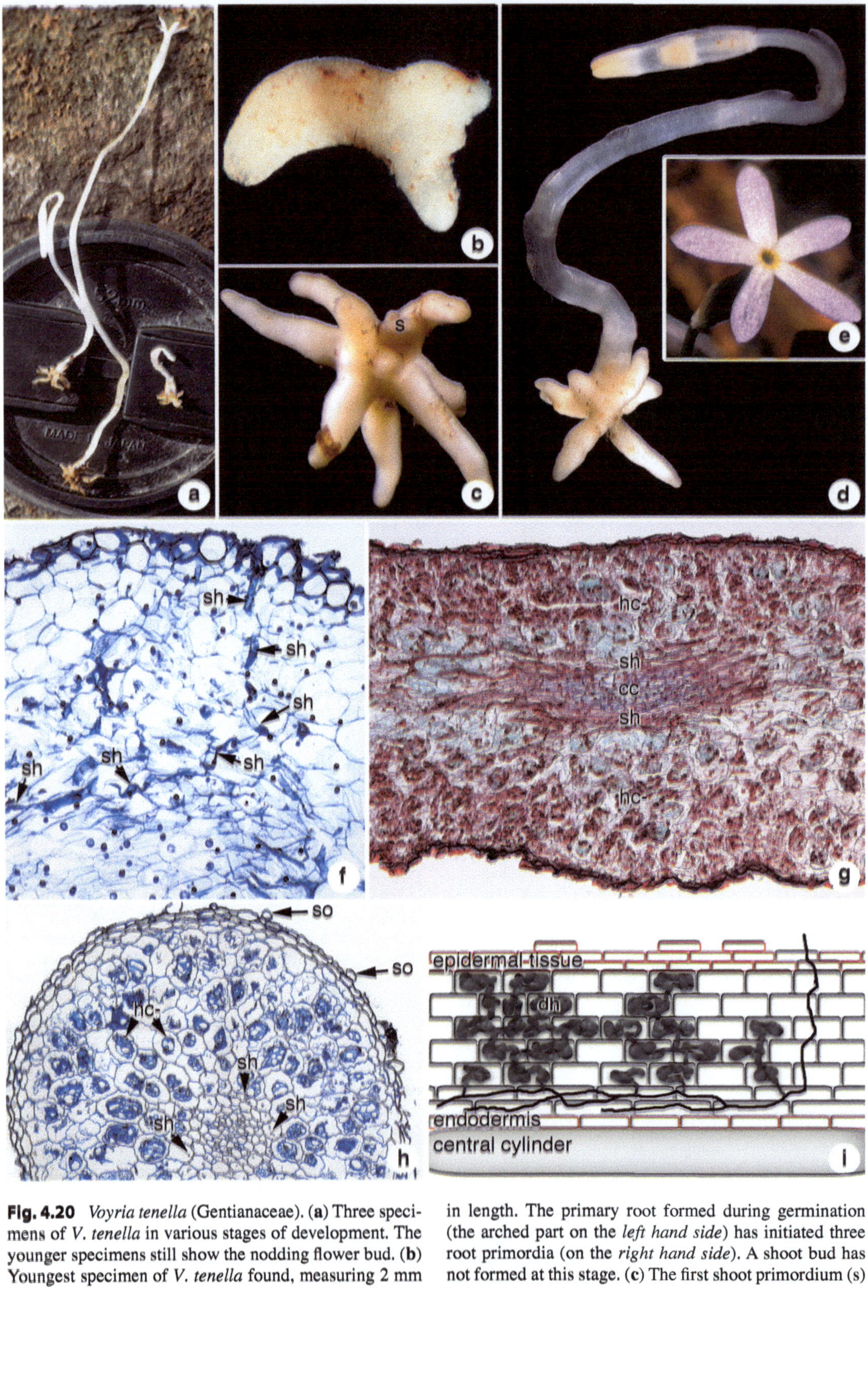

Fig. 4.20 *Voyria tenella* (Gentianaceae). (**a**) Three specimens of *V. tenella* in various stages of development. The younger specimens still show the nodding flower bud. (**b**) Youngest specimen of *V. tenella* found, measuring 2 mm in length. The primary root formed during germination (the arched part on the *left hand side*) has initiated three root primordia (on the *right hand side*). A shoot bud has not formed at this stage. (**c**) The first shoot primordium (s)

obstacles in the soil (Figdor 1897). Similar observations were made in *Voyria truncata* (Imhof and Weber 1997) and *Triuris hyalina* (Imhof 1998). The roots are several centimeters long and irregularly thickened, with the thicker parts three (Figdor 1897) to four times (Janse 1896) wider than the thinner ones. Tangential cell divisions in the outer cortex increase the number of parenchyma cell layers in those parts where fungal hyphae are found within the inner cortex cells. Uncolonized root segments, thus, have only four parenchyma layers whereas the mycorrhizal parts can have up to eight (Janse 1896). Cells in the course of division do not become colonized, neither does the epidermis, exodermis nor the first cortical parenchyma layer. The hyphae within the cells are coiled (Figdor 1897) and show local swellings. They form "sporangioles," a term used by Janse (1896) for degenerating hyphae. Figdor (1897) also speaks of clumped masses of dead hyphae. A prolonged primary thickening of the root as described by Janse (1896) has also been seen in *Voyria truncata*, although there the cell divisions happen throughout the cortex, irrespective of being colonized or not (Imhof and Weber 1997).

No information is given as to whether the subterranean part of the stem is colonized by the fungus. For comparison, more recent studies in *Voyria* species did not find hyphae in shoot tissues (Imhof and Weber 1997, 2000; Imhof 1997, 1999c), although Svedelius (1902) reports hyphae in aerial stems of *V. tenella* (treated as *Leiphaimos azurea*).

4.6.3.2 *Exochaenium* (Fig. 4.18i–k)

Exochaenium oliganthum (previously *Sebaea oligantha* Kissling et al. 2009; Kissling 2012) from Central Africa is the only species in this genus that is achlorophyllous (Raynal 1967a, Kissling et al. 2009). However, representatives with little photosynthetic surface such as *E. debilis*, *E. rara* and *E. pulsilla* (Marais and Verdoorn 1963) suggest that partial mycoheterotrophy may also be present in other species of the genus. As is often the case, the subterranean organs are neglected in taxonomic accounts (Gilg 1899; Robyns 1962). Raynal (1967a), recognizing this lack, described the roots as radiating from the stem base, being few (her drawing shows five at the shoot base), sparsely branched, terete, carnose, and up to 0.5 mm thick (Fig. 4.18i, j). This description is very similar to that for achlorophyllous *Exacum* species discussed above. More exciting is Raynal's (1967a) finding of additional, almost subterranean cleistogamous flowers covered by the leaf litter of the soil surface. In contrast to the straight epiterrestrial shoots, the stems and pedicels of these "subterranean" flowers are mostly positively geotropic, coiled, and intermingled with each other, but basically develop the same flowers and fruits as the aerial counterparts, except for some reductions in floral structures. The fruits, due to positive geotropism, are geocarpic (Raynal 1967a).

Information on the roots additional to that mentioned above is very scarce. Professor Mangenot did investigate the mycorrhiza of

Fig. 4.20 (continued) appears only after the development of a characteristically radiating root system (this specimen is 4 mm wide in its maximal extension). (**d**) Young specimen of *V. tenella*, the shoot is 1 mm thick. (**e**) Flower of *V. tenella*. (**f**) Longitudinal section through a young root of *V. tenella* showing the penetration and the subsequent direct growth of the straight hypha (sh) towards the inner root cortex where it proceeds along the central cylinder. (**g**) Longitudinal section through a mature root of *V. tenella* showing the straight hyphae (sh) in the inner cortex parenchyma around the central cylinder (cc) and the degenerated coils of hyphae (hc–) in the outer cortex. (**h**) Transverse section through a root of *V. tenella* (0.8 mm in diameter). The straight hyphae (sh) in the inner cortex parenchyma are visible as small circles. The epidermal tissue (et) consist of 2–3 layers of smaller cells never colonized by the fungus except for penetration points. Their outermost cells slough off (so) and are replaced by derivates of the layers underneath. The obvious digestion of hyphal coils (hc–) takes place in the majority of the cortex parenchyma. (**i**) Schematic view of the mycorrhizal colonization pattern in *V. tenella*. After penetration, the hyphae grows straighly towards the inner cortex layers which are longitudinally elongated and proceed therein along the central cylinder. Branches of these inner hyphae grow back into the outer cortex and degenerate (dh)

Exochaenium oliganthum, but never published it. Raynal (1967a) reports his findings communicated to her, which revealed fungal coils within the cells very similar to the conditions in *Neottia nidus-avis* (Orchidaceae). However, we know today that orchid mycorrhizas (e.g., Smith and Read 2008) and the AMs in gentians (e.g., Imhof 1999c) are only superficially alike. Recently, molecular methods have detected a *Glomus*-group A endophyte in *Exochaenium oliganthum*, which seems to be highly specific for this species (Franke et al. 2006).

4.6.3.3 *Voyriella*

There are roughly 30 publications mentioning the monotypic *Voyriella*, many of them only as part of an enumeration of gentianaceous genera. The approximately dozen of taxonomic or geographic accounts mostly lack information on subterranean organs. In fact, there are only five statements on the roots of *Voyriella parviflora*: "radice fibrosa" (Miquel 1851), "roots on the transition from filiform to coralloid shape" (Johow 1889, translated from German), "filiform" (Jonker 1936), and "30 mm long and 0.3 mm thick" (Maas and Ruyters 1986). The latter statement is simply restated by Pires O'Brien (1997) in a plant checklist of the Jari river in Brazil. There is no figure showing the roots of *V. parviflora*, but its mycorrhizal fungus has been identified by 18S rDNA sequencing. *V. parviflora* seem to be highly specific to a basal clade of *Glomus*-group A of the Glomeromycota (Bidartondo et al. 2002).

4.6.3.4 *Voyria* (Figs. 4.18–4.20)

There is considerable information available for the roots of several of the 19 *Voyria* species. The earliest observations by Aublet (1775) on *Voyria rosea* indicated that it has irregularly tuberous roots of the size of a fist and become eaten by the indigenous people (Garipons of the Guianas) after being cooked in a coal fire, tasting similar to potatoes. However, although *V. rosea* has roots up to 40 mm long and 15 mm (!) thick, it appears more as a loosely coralloid root system rather than a tuberculous one (Maas and Ruyters 1986). Either the specimen seen by Aublet had densely intermingled roots only appearing like a tuber of that size, or Aublet confused it with another plant. Aublet (1775) called the achlorophyllous genus after the Garipon name of that edible plant, *Voyria*. There is still some confusion on the nature of the subterranean parts of *Voyria* spp. as they are sometimes erroneously called rhizomes (e.g., Süssenguth 1937; Jonker 1936; Fukarek et al. 1994; Pringle 1995). Possibly, since the root-borne shoots (Imhof et al. 1994; Imhof and Weber 1997; Imhof 1997) often have to grow through a considerable layer of soil before they reach the surface (Fig. 4.19b–d, Imhof and Weber 1997, see before under *Exacum*), they might have been misinterpreted as orthotropous rhizomes.

Considerable differences, most probably representing evolutionary steps, occur within this genus. Paralleled by the reduction of floral (Oehler 1927; Maas and Ruyters 1986) and shoot anatomical features (Johow 1885; Solereder 1908; Oehler 1927; ter Welle 1986), the root systems can also be arranged according to reduction particularly in root length. Roots of *Voyria truncata* (a primitive member of the genus), can presumably be several meters long, growing horizontally and runner-like (Fig. 4.19e) as deep as 20 cm beneath the soil surface (Fig. 4.19c, d), and up to 2 mm thick, frequently branched, and give rise to two root-borne shoots in the axils of side roots (Imhof et al. 1994; Imhof and Weber 1997, Fig. 4.19e). Hence, seemingly distinct specimens above soil can well belong to the same individual, either by shoot ramifications already in the soil (Fig. 4.19c), or because of different root sprouts originating from the same root (Fig. 4.19e). Field observations recognized *V. truncata* shoots (Fig. 4.19a–d) emerging like a chain of beads for several meters, denoting the root course in the soil (Imhof et al. 1994). Aublet (1775) also reports on roots of *V. rosea* being a foot deep in the soil.

In contrast, the most advanced representative of the genus, *Voyria tenella*, has a small, star-like root system, shallowly rooted or only loosely connected to the litter substrate (Imhof et al.

1994, Fig. 4.20a–e). Ontogenetic studies revealed a root to appear first after germination (Fig. 4.19b) and only after a small star-like root system form does a first shoot arise (Imhof et al. 1994, Fig. 4.20c). In spite of the ray-like appearance, the roots do not grow evenly in all directions. In fact, there is a single pole of growth representing the primary root which develops side roots at very short distances creating the globose structure (Imhof 1997).

In conclusion, although the two root systems of *V. truncata* and *V. tenella* seem to be very different, the root system of *V. tenella* is easily interpreted as an extremely abbreviated root system of *V. truncata*. This notion is supported by intermediate root systems linking *V. truncata* and *V. tenella*, e.g., *V. aphylla* (Imhof 1999c, Fig. 4.19h–j), *V. rosea* (Maas and Ruyters 1986), *V. chionea* (Progel 1865), and *V. obconica* (Imhof and Weber 2000, Fig. 4.18d, e). The only African representative, *V. primuloides*, which is considered to be sister to *V. chionea* (Albert and Struwe 1997), is the only species in the genus with prominent root hairs (Raynal 1967b).

The root anatomy and mycorrhizal colonization patterns in *Voyria* also support the progression proposed above. *Voyria truncata* has an almost identical root anatomy to a young *Gentiana lutea* (Perrot 1898), including a distinct dimorphic exodermis with short and long cells (Fig. 4.19f, g) and a substantial central cylinder with lignified tracheary elements, but lacking secondary thickening. Instead, a prolonged primary thickening is established (Fig. 4.19f), which constantly enhances the essential tissue for fungal colonization and is interpreted as an important adaptation to its mycoheterotrophy (Imhof and Weber 1997). The type of arbuscular mycorrhiza is also very similar to other gentians (Neumann 1934; Jacquelinet-Jeanmougin and Gianinazzi-Pearson 1983), except for the lack of lateral arbuscules formed from the coiled intracellular hyphae, typical for a *Paris*-type AM. The degradation of hyphal coils in the cells, best explained as a digesting process of the plant to absorb carbon and nutrients from the hyphae, happens after about 15 cell passages (Imhof and Weber 1997, Fig. 4.19f).

The roots of *V. aphylla* (Fig. 4.19h, i) have all anatomical elements of *V. truncata* but are reduced in their size and there is a tendency for a radiating formation of side roots at the shoot bases (Fig. 4.19i, j). The mycorrhizal associations are also similar. However, some new features have been acquired: (1) a longitudinal spread of straight hyphae within the epidermis as well as the innermost cortex layers (Fig. 4.19k, l) and (2) the development of root hairs only where roots of neighboring plants are attached (Imhof 1999c, Fig. 4.19m). Whereas the hyphae in roots of *V. truncata* are largely restricted regarding the cortical spread, the nondegenerating hyphae in the epidermis of *V. aphylla* are able to reach more distant segments of the root as well. Direct hyphal bridges from attached roots of neighboring plants are frequent sources of fungal root penetrations in several *Voyria* species. The locally developed root hairs of *V. aphylla* increase this contact zone and, in fact, receive most of the external fungal penetrations (Imhof 1999c; Fig. 4.19k+m). Hyphal passage through the exodermis still is exclusively via the short cells, although those are not as anatomically distinct as they are in *V. truncata* (Imhof and Weber 1997). Digestion of fungal material takes place in the cortex parenchyma except for its innermost layers, where a still imperfect internal spread along the central cylinder can be seen. The latter feature foreshadows the highly efficient colonization pattern of the further derived *Voyria tenella*, *V. obconica* and *V. flavescens* (Fig. 4.20i).

The root anatomy of the most advanced *V. tenella* had been investigated by Johow (1885) and Vigodsky-de Philippis (1938, under the synonym *Leiphaimos brachyloba*). Both stress the reduced character of the vascular system, the lack of suberization of the endodermis in *V. tenella*, and the voluminous root cortices (more details in Imhof 1997, Fig. 4.20a–e). Johow (1885) also recognized the coiled fungal mycelium within the cortex cells, but called it "parasitic" and, in accepting Drude's hypothesis (1873), assigned its presence to the attraction through a "particularly rich flow of organic nutrients" (translated from German) he assumed to be a compulsory attribute of the roots of this "saprophytic" plant. He was aware of Kamienski's (1882) new notion of a symbiotic association between fungus and plant, yet with cautious criticism. Vigodsky-de Philippis (1938) called the hyphae a "micelio micorrhizico." However, these classical papers did not recognize the specialized colonization pattern. Root penetrating hyphae initially grow straight towards the innermost cortex layers and proceed in longitudinal direction along the reduced central cylinder. From there, hyphal branches grow back into the outer cortex parenchyma where they begin to coil, quickly inflate, and finally collapse into amorphous clumps (see details in Imhof 1997; Imhof and Weber 2000; Franke 2002, Fig. 4.20f–h). By this means, a sustained use of only few external penetrations of the fungus is attained, maintaining the hyphae alive in the inner cortex and only digesting branches of them in the outer cortex: an intraradical fungus garden (Imhof 1997, Fig. 4.19i). In summary, within *Voyria*, we can retrace not only morphological and anatomical reductions but also the evolutionary progression of a mycorrhiza (compare Figs. 4.19g, n and 4.20i), resulting in a highly efficient system to benefit from a fungus.

The mycorrhizal fungi of several *Voyria* species have been determined by molecular identification methods. Almost all of the endophytes belong to *Glomus*-group A of the Glomeromycota. However, *Voyria* spp. seem to not be as specific regarding their mycorrhizal associates as other mycoheterotrophic species (Bidartondo et al. 2002; Merckx et al. 2010).

Since mycoheterotrophic plants are very difficult to cultivate, the unexpected emergence of a *Voyria* species in the Botanical Garden in Hamburg (Germany) shall be briefly reported here. As an epiphytic stowaway on the trunk of a tree fern (*Alsophila salvinii*) imported in the mid seventies, a yellow *Voyria* (perhaps *V. aphylla*, which is known to grow also epiphytically, Groenendijk et al. 1997) was discovered around 1980. Unfortunately, it died in 1983, when its host fern tree was placed outside during summer (Poppendieck 1997).

4.7 Selected Species of Questionable Trophic Status

4.7.1 *Buxbaumia* spp. (Bryophyta, Fig. 4.1k)

The genus *Buxbaumia* is comprised of 12 species in the northern hemisphere (Crosby et al. 2000; Goffinet et al. 2008). In contrast to the majority of mosses, the up to 3 cm high sporophyte is the prominent phase of this genus and consists of a bulky, oblique-oval capsule on top of the seta (Eastwood 1939). This appearance has led to its enchanting vernacular names, e.g., Elfcap Moss, Humpbacked Elves, Bug-on-a-stick (Fig. 4.1k). While most species have greenish sporophytes (Udar et al. 1971; Ligrone et al. 1982; Stone 1983; Düll and Düll-Wunder 2008), the sporophytes of *Buxbaumia aphylla* and *B. minakatae* (Okamura 1911; Iwatsuki and Sharp 1967) seem to be largely devoid of chlorophyll. The gametophyte of *Buxbaumia aphylla* is minute and achlorophyllous (Goebel 1892; Dening 1928; Mueller 1972; Hancock and Brassard 1974). The possibly perennial protonema (Steven and Long 1989), consisting of single-lined threads of thick walled cells (Mueller 1972), which may form velvety mats (McClymont 1950), is green but of questionable trophic relevance (Haberlandt 1886; Eastwood 1939). Early stages of the sporophyte may show some green color (Goebel 1892; Dening 1928; Eastwood 1936; Hancock and Brassard 1974; Schoepe and Philippi 2000; van Rompu and Stieperaere 2002), but many recent as well as older publications consider it to have a heterotrophic mode of life (e.g., Haberlandt 1886; Eastwood 1936; Mueller 1972; Watson and Dallwitz 2005 onwards; Düll and Düll-Wunder 2008). It is clear therefore that the relation of auto- vs. heterotrophy in the whole genus is completely unknown and should be studied.

In contrast to Eastwood (1939), who believed in direct absorption by *Buxbaumia* of organic substances from humus or neighboring green mosses, we know today that plants are unable to do that directly, but are either parasitic (Kuijt 1969; Weber 1993) or mycoheterotrophic when lacking chlorophyll (Leake 1994, 2005). However, no information is yet available on mycorrhizal, endophytic, or parasitic associations, which could possibly explain the strange habit of these mosses. The ultrastructural investigation on the green *B. piperi* did not find any association with fungi (Ligrone et al. 1982). Neither did the detailed description (Stone 1983) on the foot and vaginula of *B. novae-zelandiae* mention fungal hyphae, but explained the dense indumentum with anastomosing elements as "rhizoidal outgrowth from the epidermal cells of the fertile axis." Udar et al. (1971) also show drawings of longitudinal and transverse sections of the tomentose and slightly tuberous basis of the sporophyte of *B. himalayensis*, which superficially resembles an ectomycorrhiza. Unfortunately, the authors did not comment on this feature at all. However, Haberlandt (1886) has shown drawings of what he interpreted to be rhizoids. These "rhizoids" could well be fungal hyphae, and Haberlandt (1886) even explicitly remarked on their striking similarity to hyphae, particularly because of their frequent anastomoses, which, as far as we know, is not characteristic for moss rhizoids. Similarly, Goebel (1892) reports of anastomosing as well as achlorophyllous protonemata, which even show clamp-like connections (line drawing in Goebel 1892). In any case, contemporary studies providing photographic micrographs instead of drawings on the micromorphology and anatomy of *Buxbaumia aphylla* as well as stable isotope investigations (see Chap. 8) in order to elucidate the trophic mode, are urgently needed.

4.7.2 *Pyrola picta* (*Pyrola aphylla*, Ericaceae)

Pyrola aphylla was first described by James Edward Smith 1814 in Abraham Rees' Cyclopaedia (Vol. 29, No. 7), calling it "leafless." Other authors as well, such as de Candolle (1838) and Hooker (1840), described this species as having no leaves. However, Nuttall (1843, cited by Holm 1898) detected leaves of this species, and Holm (1898) ascertained subterranean shoots connecting apparently leafless specimens with rosettes of green leaves, proving that they belong to the same individual. Also Andres (1914) noted that *P. picta* can have few to no leaves. The same observation was made by Camp (1940), who revisited herbarium sheets and argued for a close relationship, if not identity, of *Pyrola aphylla*, *P. picta* and *P. dentata*. More recently, Haber (1987) eventually merged the three species within the highly variable *Pyrola picta* Sm.

In addition to the subterranean shoots (stolons) already mentioned, *P. picta* also has horizontally growing, runner-like, branching roots. Root-borne shoots as well as adventitious roots from the stolons can develop (Holm 1898). The only specific investigations on the mycorrhiza of *P. picta* is by Largent et al. (1980), calling it arbutoid and ericoid (both seen in specimens of *P. picta* var. *picta*) and ericoid (in *P. picta* var. *aphylla*). Other authors have called the mycorrhizas of *Pyrola* spp. arbutoid (Robertson and Robertson 1985; Massicotte et al. 2008; Vincenot et al. 2008), ectendomycorrhizal (Wang and Qiu 2006), or pyroloid (Cullings 1996). It also has been considered as a linking mycorrhizal type between arbutoid and ericoid mycorrhiza in a new classification of mycorrhizas (Imhof 2009).

Because *P. picta* has green leaves, and the extreme, leafless variant of *P. picta*, *P. aphylla*, still has chlorophyll in the shoot bark (Holm 1898), it actually should not be fully mycoheterotrophic. However, Hynson et al. (2009) found characteristic stable isotope signatures typical for mycoheterotrophic plants in *Pyrola aphylla* specimen. Interestingly, *Pyrola picta* with leaves, although being the same species taxonomically, did not show signs of mycoheterotrophy according to carbon stable isotope signatures (Hynson et al. 2009). Hence, the trophic status of this species is ambiguous. Possibly, dependency on the fungal carbon is not determined by the species in the taxonomical sense but on the actual ability for assimilation in a particular specimen.

4.8 Trends, Conclusions, and Future Directions

MH plants have distinctive structural necessities in contrast to autotrophic species due to their mycorrhizal dependence for carbon supply. Secondary growth of roots, for example, is deleterious since it sheds the primary tissue, which alone can host the indispensable mycobiont. Moreover, the primary tissue, less important in autotrophic plants, must be present in sufficient quantity. Most importantly, intracellular (in contrast to intercellular) mycorrhizal colonization is a major prerequisite. In fact, there is no mycoheterotroph having an *Arum*-type AM or an ectomycorrhiza, both of which are characterized by predominantly intercellular hyphal growth. Obviously, the transfer of nutrients and carbohydrates provided by those mycorrhizal types is not sufficient to support achlorophylly. There also must be a high probability to become colonized by an appropriate fungus, keeping in mind that MH plants are often quite specific with respect to their endophyte (e.g., Kretzer et al. 2000; Taylor et al. 2002; Bidartondo et al. 2002; Franke et al. 2004; Ogura-Tsujita and Yukawa 2008). A widely branched allorhizic root system seems to be suitable for this, but in turn, is susceptible to functional failure of large parts by only a single blocking, collapsing or disconnection event in a proximal segment. This is particularly critical when secondary growth for securing the connection is impossible. In any case, the transfer of carbon to the reproductive parts must be either short or reliably assured. These challenges are reasons for the following convergent evolutionary trends concerning subterranean organs of MH plants in unrelated plant families:

1. Star-like root systems consisting of many roots radiating from the base of the shoot, either created by root-borne shoots or shoot-borne roots, reduce the risk of becoming disconnected to a major part of the root system (e.g., Figs. 4.2g, 4.4b, 4.6c+g+k, 4.7b, 4.8e+j, 4.10c+f+g+i, 4.12f+g, 4.13a+d, 4.14b, 4.15b, 4.18b+e+g+j, 4.19j, 4.20a).
2. Short and thick roots shorten the transport distance of carbon to the shoot while retaining the tissue volume of long and thin roots (e.g., Figs. 4.3a, 4.4b, 4.7b, 4.9d, 4.10f, 4.12f+g, 4.14b, 4.15b, 4.17e+g, 4.19c).
3. Specialized colonization pattern that enables a sustained use of a few fungal penetrations counterbalance the reduced probability to become colonized in short and thick roots compared to filiform roots (e.g., Figs. 4.3i, 4.4i, 4.5j, 4.8h, 4.9h, 4.13h, 4.14k, 4.16d, 4.19n, 4.20i).
4. Strong reinforcement of thin roots, either by tertiary endodermae (in monocots) or the development of multicellular fibrous tissue, protect the carbon supply of the shoot (e.g., Figs. 4.5d+f, 4.6c + d, 4.13h).

The contradicting needs for a large root surface for high infection probability and short distances for carbon transport, has been discussed as the "mycoheterotroph's dilemma" (Imhof 2010) and supposedly has shaped much of the subterranean organs in MH plants during evolution. As an effect, advanced MH plants within a family have stout, clumped roots and (in orchids) rhizomes mostly with a specialized fungal colonization pattern. This trend is best exemplified in *Voyria* (Gentianaceae, Imhof 1999c) and in Ericaceae (Furman and Trappe 1971). Gentianales and Ericaceae especially, having two fundamentally distinct groups of mycorrhiza (AM group vs. ECM group, Imhof 2009) but both show evolutionary reductions from trees to achlorophyllous herbs (Henderson 1919; Imhof 1999c) including changes in mycorrhizal pattern, turn out to be a textbook example for convergent evolution. In Triuridaceae (Imhof 2003), Burmanniaceae (Imhof 2001), *Thismia* (this chapter) and *Afrothismia* (both Thismiaceae, Imhof 2006), and Orchidaceae (Furman and Trappe 1971), this trend is partly detectable, but further investigations are necessary for more support. Research on taxa-like *Geosiris* (Iridaceae), *Corsia* (Corsiaceae), *Kupea* (Triuridaceae), *Haplothismia* (Thismiaceae) and others for which nothing is known concerning the fungal structures, will also help to understand the evolution of mycoheterotrophy. Moreover, given that 15 investigated vascular MH plants associated with AM fungi (i.e., Monotropoideae and orchids excluded) revealed 13 different mycorrhizal colonization patterns, there is a considerable chance for more fascinating novelties. In orchid mycorrhizas, although belonging to the oldest fields of mycorrhizal research (e.g., Schleiden 1845), comparatively little is known on the two existing types: tolypophagy and ptyophagy (Burgeff 1932; Wang et al. 1997; Rasmussen 2002; Imhof 2009). Since the latter type was found exclusively in achlorophyllous orchids so far (e.g., Janse 1896; Campbell 1963, 1964; Wang et al. 1997), an examination of the mycorrhizal structures of other MH orchids is highly desired.

In conclusion, mycoheterotrophy is based on a number of specializations with respect to morphology and anatomy of the underground parts, and, most importantly, on the evolution of sophisticated mycorrhizal pattern.

Acknowledgements Thanks to Mori Thomann for his help with the Japanese paper of K. Watanabe, as well as Jesper Hansen for translating parts of the Danish articles by V. A. Poulsen.

References

Abe C, Akasawa Y (1989) A new species of *Oxygyne* (Burmanniaceae) found in Shikoku, Japan. J Jap Bot 64:161–164

Agardh JG (1858) Theoria systematis plantarum, vol 99. Gleerup, Lundae

Airy Shaw HK (1952) A new genus and species of Burmanniaceae from South India. Kew Bull 2:277–279

Albert VA, Struwe L (1997) Phylogeny and classification of *Voyria* (saprophytic Gentianaceae). Brittonia 49:466–479

Andres H (1914) Studien zur speziellen Systematik der Pirolaceen II. Allg Bot Zeitschr 20:109–117

APG (Angiosperm Phylogeny Group) (2009) An update of the angiosperm phylogeny group classification for the orders and families of flowering plants: APG III. Bot J Linn Soc 161:105–121

Aublet JBCF (1775) Voyria. In: Histoire des Plantes de la Guiane Françoise, vol 1. P.-F. Didot jeune, London, Paris, p 208–211

Backer CA, Bakhuizen van den Brink RC (1963) Polygalaceae. In: Flora of Java. Noordhoff, Groningen, p 196–201

Beccari O (1877) Descrizione di una nuova e singolare pianta parassita: Gen. *Corsia* Becc. Malesia 1:238–240

Benson-Evans K (1952) The distribution of sex in *Cryptothallus*. Nature 169:39

Bernard C, Ernst A (1910) Beiträge zur Kenntnis der Saprophyten Javas. II. Äussere und innere Morphologie von Thismia javanica J. J. S. Ann Jard Bot Buitenzorg 23:36–46

Bernard C, Ernst A (1911) Beiträge zur Kenntnis der Saprophyten Javas. V. Anatomie von Thismia clandestina Miq. und Thismia versteegii Sm. Ann Jard Bot Buitenzorg 24:61–69

Bernard C, Ernst A (1914) Beiträge zur Kenntnis der Saprophyten Javas. XIV. Äussere und innere Morphologie von Burmannia tuberosa Becc. Ann Jard Bot Buitenzorg 28:102–118

Bidartondo MI, Bruns TD (2001) Extreme specificity in epiparasitic Monotropoideae (Ericaceae): widespread phylogenetic and geographical structure. Mol Ecol 10:2285–2295

Bidartondo MI, Bruns TD (2002) Fine-level mycorrhizal specificity in the Monotropoideae (Ericaceae): specificity for fungal species groups. Mol Ecol 11: 557–569

Bidartondo MI, Redecker D, Hijri I, Wiemken A, Bruns TD, Dominguez L, Sérsic A, Leake J, Read DJ (2002) Epiparasitic plants specialized on arbuscular mycorrhizal fungi. Nature 419:389–392

Bidartondo MI, Bruns TD, Weiß M, Sérgio C, Read DJ (2003) Specialized cheating of the ectomycorrhizal symbiosis by an epiparasitic liverwort. Proc Roy Soc Lond B 270:835–842

Bidartondo MI (2005) The evolutionary ecology of mycoheterotrophy. New Phytol 167:335–352

Bidartondo MI, Bruns TD (2005) On the origins of extreme mycorrhizal specificity in the Monotropoideae (Ericaceae): performance trade-offs during seed germination and seedling development. Mol Ecol 14: 1549–1560

Bidartondo MI, Duckett JG (2009) Conservative ecological and evolutionary patterns in liverwort-fungal symbioses. Proc Roy Soc Lond B 277:485–492

Bierhorst DW (1953) Structure and development of the gametophyte of *Psilotum nudum*. Am J Bot 40: 649–658

Bierhorst DW (1968) Observations on *Schizaea* and *Actinostachys* spp., including *A. oligostachys*, sp. nov. Am J Bot 55:87–108

Bierhorst DW (1971) Morphology of vascular plants. Macmillan, New York

Bonfante P, Perotto S (1995) Strategies of arbuscular mycorrhizal fungi when infecting host plants. New Phytol 130:3–21

Boudier P, Bardat J, Perrera S (1999) *Cryptothallus mirabilis* v. Malmborg (Aneuraceae, Hepaticopsida) dans le Perche d'Eve-et-Loire (France). Cryptogamie Bryol 20:189–196

Boullard B (1957) La mycotrophie chez les Ptéridophytes. Sa fréquence, ses caractères, sa signification. Le Botaniste 41:5–185

Boullard B (1979) Considérations sur la symbiose fongique chez les Ptéridophytes. Syllogeus 19:1–58. National Museum of Natural Sciences, Ottawa

Bruchmann H (1885) Das Prothallium von *Lycopodium*. Bot Centralbl 21:23–28 + 309–313

Bruchmann H (1906) Über das Prothallium und die Sporenpflanze von *Botrychium lunaria* Sw. Flora 96:203–230

Bruchmann H (1910) Die Keimung der Sporen und die Entwicklung der Prothallien vor *Lycopodium clavatum* L., *L. annotinum* L. und *L. selago* L. Flora 101:220–267

Brundrett MC (2002) Coevolution of roots and mycorrhizas of land plants. New Phytol 154:275–304

Bruns TD, Read DJ (2000) In vitro germination of nonphotosynthetic, myco-heterotrophic plants stimulated

by fungi isolated from the adult plants. New Phytol 148:335–342
Burgeff H (1932) Saprophytismus und Symbiose. Studien an tropischen Orchideen. G. Fischer, Jena
Burgeff H (1938) Mycorrhiza. In: Verdoom F (ed) Manual of pteridology. M. Nijhoff, The Hague, pp 159–196
Burgeff H (1943) Problematik der Mycorhiza. Naturwissenschaften 31:558–567
Caddick LR, Rudall PJ, Wilkin P (2000) Floral morphology and development in Dioscoreales. Fedd Repert 111:189–230
Cameron DD, Bolin JF (2010) Isotopic evidence of partial mycoheterotrophy in the Gentianaceae: *Bartonia virginica* and *Obolaria virginica* as case studies. Am J Bot 97:1272–1277
Cameron KM, Chase MW, Rudall PJ (2003) Recircumscription of the monocotyledonous family Petrosaviaceae to include *Japonolirion*. Brittonia 55:214–225
Camp WH (1940) Aphyllous forms in *Pyrola*. Bull Torrey Bot Club 67:453–465
Campbell DH (1908) Symbiosis in fern prothallia. Am Nat 42:154–165
Campbell EO (1962) The mycorrhiza of Gastrodia cunninghamii Hook. f. Trans Roy Soc New Zealand Bot 1:289–296
Campbell EO (1963) *Gastrodia minor* Petrie, an epiparasite of Manuka. Trans Roy Soc New Zealand 2:73–81
Campbell EO (1964) The fungal association in a colony of *Gastrodia sesamoides* R. Br. Trans Roy Soc New Zealand 2:237–246
Campbell EO (1968) An investigation of *Thismia rodwayi* F. Muell. and its associated fungus. Trans Roy Soc New Zealand 3:209–219
Chantanaorrapint S (2008) *Thismia angustimitra* (Thismiaceae), a new species from Thailand. Blumea 53:524–526
Cheek M (2003a) A new species of *Afrothismia* (Burmanniaceae) from Kenya. Kew Bull 58:951–955
Cheek M (2003b) Kupeaeae, a new tribe of Triuridaceae from Africa. Kew Bull 58:939–949
Cheek M (2009) Burmanniaceae. In: Timberlake JR, Marins ES (eds) Flora Zambesiaca, vol 12(2). Flora Zambesiaca Managing Committee, London, pp 141–148
Cheek M, Williams S, Etuge M (2003) *Kupea martinetugei*, a new genus and species of Triuridaceae from western Cameroon. Kew Bull 58:225–228
Chen X, Tamura MN (2000) Petrosavia. In: Flora of China, vol 24. Science Press, Beijing, p 77
Cherrier J-F, Gondran M, Woltz P, Vogt G (1992) Parasitisme interspécifique chez les Gymnospermes. Données inédites chez deux Podocarpaceae endémiques Néo-Calédoniennes. Rev Cytol Biol Végét Bot 15:65–87
Chiang PY, Hsieh T-H (2011) *Thismia huangii* (Thismiaceae), a new species from Taiwan. Taiwania 56:138–142
Clarke CB (1885) Gentianaceae. In: Hooker JD (ed) Flora of British India, vol 4. Kent, L. Reeve, pp 93–132
Coleman DG (1936) *Sarcosiphon rodwayi* in Australia. Victorian Naturalist 52:163–166
Colozza A (1910) Contributo allo studio anatomico delle Burmanniaceae. Boll Soc Ital 1910:106–115
Correll DS (1966) Two new plants in Texas. Wrightia 3:188–191
Courty P-E, Walder F, Boller T, Ineichen K, Wiemken A, Rousteau A, Selosse M-A (2011) Carbon and nitrogen metabolism in mycorrhizal networks and mycoheterotrophic plants of tropical forests: a stable isotope analysis. Plant Physiol 156:952–961
Crantz HJN (1766) 1096. *Hypopitys*. In: Institutiones Rei Herbariae juxta nutum Naturae Digestae ex Habitu, Tomus II: 467. Ioannis Pauli Kraus, Viennensis (Vianden, Luxemburg)
Cribb AB (1995) The saprophytic flowering plant, *Thismia yorkensis* sp. nov., from Australia. Queensland Nat 33:51–56
Cribb P (1985) The saprophytic genus *Corsia* in the Solomon Islands. Kew Mag 2(3):320–323
Cribb PJ, Wilkin P, Clements M (1994) Corsiaceae: a new family for the Falkland Islands. Kew Bull 50:171–172
Cronquist A (1988) The evolution and classification of flowering plants, 2nd edn. New York Botanical Garden, New York
Crosby MR, Magill RE, Allen B, He S (2000) A checklist of the Mosses. Missouri Botanical Garden, St. Louis
Cullings KW, Szaro TM, Bruns TD (1996) Evolution of extreme specialization within a lineage of ectomycorrhizal epiparasites. Nature 379:63–66
Cullings KW (1996) Single phylogenetic origin of ericoid mycorrhizae within the Ericaceae. Can J Bot 74: 1896–1909
Darnell-Smith G (1917) The gametophyte of *Psilotum*. Trans Roy Soc Edinb 52:79–91
Dauby G, Parmentier I, Stévart T (2008) *Afrothismia gabonensis* sp. nov. (Burmanniaceae) from Gabon. Nord J Bot 25:268–271
Davis R (1976) A light and electron microscope investigation of the fungus-gametophyte association of *Psilotum nudum* (L.) Beauv. M.A. Thesis, Arcata, California/USA, Biological Sciences: Humboldt State University
de Candolle AP (1838) Ordo CXVI: Pyrolaceae. In: Prodromus systematis naturalis regni vegetabilis, vol 7. Treuttel et Würtz, Paris, p 772–776
de Laubenfels DJ (1959) Parasitic conifer found in New Caledonia. Science 130:97
de Laubenfels DJ (1972) Gymnospermes. In: Aubreville A, Leroy JF (eds) Flore de la Nouvelle-Calédonie et Dépendances. Museum d'Histoire Naturelle, Paris, p 4
Dening K (1928) Entwicklungsgeschichtliche Untersuchungen am Gametophyten von *Buxbaumia aphylla* (L.). Verh Naturhist Vereines Preuss Rheinl Westphalens 85:306–444
Denis MM (1919) Sur quelques thalles d'*Aneura* dépourvus de chlorophylle. Comptes rendus hebdomadaires des séances de l'Académie des. Sciences 168:64–66

Dimitri MJ (1972) Una nueva del genero *Arachnitis* Phil. (Corsiaceae). Rev Fac Agron Univ Nac La Plata 48:37–45

Domínguez L, Sérsic A, Melville L, Peterson RL (2006) 'Prepackaged symbioses': propagules on roots of the myco-heterotrophic plant *Arachnitis uniflora*. New Phytol 169:191–198

Domínguez LS, Sérsic A (2004) The southernmost myco-heterotrophic plant, *Arachnitis uniflora*: root morphology and anatomy. Mycologia 96:1143–1151

Domínguez LS, Melville L, Sérsic A, Faccio A, Peterson RL (2009) The mycoheterotroph *Arachnitis uniflora* has a unique association with arbuscular mycorrhizal fungi. Botany 87:1198–1208

Dowie NJ, Hemenway JJ, Trowbridge SM, Miller SL (2011) Mycobiont overlap between two myco-heterotrophic genera of Monotropoideae (*Pterospora andromedea* and *Sarcodes sanguinea*) found in the Greater Yellowstone Ecosystem. Symbiosis 54:29–36

Drude O (1873) Die Biologie von *Monotropa hypopitys* L. und *Neottia nidus avis* L. unter vergleichender Hinzuziehung anderer Orchideen. Dieterich'sche Univ. Buchdruckerei, Göttingen

Duckett JG, Renzaglia KS, Pell K (1990) Desiccation causes the proliferation of multicellular hairs, but not mucilage papillae, in *Cryptothallus mirabilis* (Hepatophyta): a correlated light and electron microscopy study. Can J Bot 68:697–706

Duddridge JA, Read DJ (1982) An ultrastructural analysis of the development of mycorrhizas in *Monotropa hypopitys* L. New Phytol 92:203–214

Düll R, Düll-Wunder B (2008) Moose einfach und sicher bestimmen: ein illustrierter Exkursionführer zu den Arten Deutschlands und angrenzender Länder. Quelle & Meyer, Wiebelsheim

Eastwood SK (1936) Notes on *Buxbaumia aphylla* (Linnaeus) Hedwig. Bryologist 39:127–129

Eastwood SK (1939) On the moss genus *Buxbaumia*. Trillia 10:28–35

Ernst A, Bernard C (1911) Beiträge zur Kenntnis der Saprophyten Javas. VIII. Äussere und innere Morphologie von Burmannia candida Engl. und Burmannia championii Thw. Ann Jard Bot Buitenzorg 24:84–95

Ernst A, Bernard C (1912) Beiträge zur Kenntnis der Saprophyten Javas. XI. Äussere und innere Morphologie von Burmannia coelestis Don. Ann Jard Bot Buitenzorg 26:224–232

Esau K (1965) Planzenanatomie. Wiley, New York

Feild TS, Brodribb TJ (2005) A unique mode of parasitism in the conifer coral tree *Parasitaxus ustus* (Podocarpaceae). Pl Cell Environ 28:1316–1325

Fiebrig C (1921) Fanerógamas saprófitas: Triuris mycoides sp. nov. Revista Jard Bot Mus Hist Nat Paraguay 1:164–165

Figdor W (1897) Ueber *Cotylanthera* Bl. Ein Beitrag zur Kenntnis tropischer Saprophyten. Ann Jard Bot Buitenzorg 14:213–239

Franke T (1999) Untersuchungen zur Biologie von *Sciaphila purpurea* Benth. (1855) (Triuridaceae, Monocotyledoneae). Diploma Thesis, München: Ludwig Maximilian Universität

Franke T, Beenken L, Hahn C (2000) *Triuridopsis intermedia* spec. nov. (Triuridaceae), a new myco-heterotrophic plant from Bolivia. Pl Syst Evol 225:141–144

Franke T (2002) The myco-heterotrophic *Voyria flavescens* (Gentianaceae) and its associated fungus. Mycol Prog 1:366–376

Franke T, Sainge MN, Agerer R (2004) A new species of *Afrothismia* (Burmanniaceae; Tribe: Thismieae) from the western foothills of Mount Cameroon. Blumea 49:451–456

Franke T (2004) *Afrothismia saingei* (Burmanniaceae, Thismieae), a new myco-heterotrophic plant from Cameroon. Syst Geogr Pl 74:27–33

Franke T, Beenken L, Döring M, Kocyan A, Agerer R (2006) Arbuscular mycorrhizal fungi of the *Glomus-group* A lineage (Glomerales; Glomeromycota) detected in myco-heterotrophic plants from tropical Africa. Mycol Prog 5:24–31

Fukarek F, Hammer K, Hanelt P, Keller J, Kruse J, Gladis T (1994) Urania Pflanzenreich—Blütenpflanzen 2. Urania, Leipzig

Furman TE, Trappe JM (1971) Phylogeny and ecology of mycotrophic achlorophyllous angiosperms. Quart Rev Biol 46:219–225

Furness CA, Rudall PJ, Eastman A (2002) Contribution of pollen and tapetal characters to the systematics of Triuridaceae. Pl Syst Evol 235:209–218

Gandolfo MA, Nixon KC, Crepet WL (2002) Triuridaceae fossil flowers from the upper cretaceous of New Jersey. Am J Bot 89:1940–1957

Gifford EM, Foster AS (1996) Morphology and evolution of vascular plants, 3rd edn. W. H. Freeman, New York

Gilg E (1895) Gentianaceae. In: Engler A, Prantl K (eds) Die Natürlichen Pflanzenfamilien, IV. Teil, 2. Abteilung. W. Engelmann, Leipzig, pp 50–108

Gilg E (1899) Gentianaceae africanae—Beiträge zur Kenntnis der Gentianaceae II. Bot Jahrb 26:86–110

Gillett JM (1959) A revision of *Bartonia* and *Obolaria* (Gentianaceae). Rhodora 61:43–62

Goebel K (1892) Archegoniatenstudien. 1. Die einfachste Form der Moose. Flora 76:92–104

Goebel K, Süssenguth K (1924) Beiträge zur Kenntnis der südamerikanischen Burmanniaceen. Flora 117:55–90

Goffinet B, Buck WR, Shaw AJ (2008) Morphology, anatomy, and classification of the Bryophyta. In: Goffinet B, Shaw AJ (eds) Bryophyte biology, 2nd edn. Cambridge Univ. Press, Cambridge, pp 55–138

Goldblatt P, Rudall P, Cheadle VI, Dorr LJ, Wiliams CA (1987) Affinities of the madagascan endemic *Geosiris*, Iridaceae or Geosiridaceae. Bull Mus Nat Hist Paris 4 Serie Adansonia Section B 9; 3:239–248

Goldblatt P (1990) Phylogeny and classification of Iridaceae. Ann Missouri Bot Gard 77:607–627

Goldblatt P, Manning JC, Rudall P (1998) Iridaceae. In: Kubitzki K (ed) The families and genera of vascular plants, III flowering plants—monocotyledons, Lilianae (except Orchidaceae). Springer, Berlin, pp 295–333

Goldblatt P, Manning JC (2010) Iridaceae—*Geosiris albiflora* (Geosiridoideae), a new species from the Comoro Archipelago. Bothalia 40:169–171

Goldblatt P, Rodriguez A, Powell MP, Davies TJ, Manning JC, van der Bank M, Savolainen V (2008) Iridaceae 'Out of Australasia'? Phylogeny, biogeography, and divergence time based on plastid DNA sequences. Syst Bot 33:495–508

Gray A (1871) Characters of a new genus consisting of two species of parasitic Gentianeae. J Linn Soc Bot 11:22–23

Gregory P (2006) Plant roots. Blackwell, Oxford

Groenendijk JP, van Dulmen ATJ, Bouman F (1997) The "forest floor" saprophytes *Voyria spruceana* and *V. aphylla* (Gentianaceae) growing as epiphytes in Colombian Amazonia. Ecotropica 3:129–131

Groom P (1895a) On a new saprophytic monocotyledon. Ann Bot 9:45–58

Groom P (1895b) On *Thismia aseroe* (Beccari) and its mycorhiza. Ann Bot 9:327–361

Groom P (1895c) Contribution to the knowledge of monocotyledonous saprophytes. J Linn Soc Bot 31: 149–215

Guinea Lopez E (1946) Novedades sistemáticas de mi primer viaje a la Guinea continental espanola (verano de 1945). Anales Jard Bot Madrid 6:463–485

Haber E (1987) Variability, distribution and systematics of *Pyrola picta* s.l. (Ericaceae) in Western North America. Syst Bot 12:324–335

Haberlandt G (1886) Beiträge zur Anatomie und Physiologie der Laubmoose. Jahrb Wiss Bot (Pringsheim) 17:359–498

Hancock JA, Brassard GR (1974) Phenology, sporophyte production, and life history of *Buxbaumia aphylla* in Newfoundland, Canada. Bryologist 77:501–513

Hara H (1975) A new species of *Cotylanthera* (Gentianaceae) from Philippines, with a conspectus of the genus. J Jap Bot 50:321–328

Heide-Jørgensen H (2008) Parasitic flowering plants. Brill, Leiden

Hemsley W (1907) Two new Triuridaceae, with some remarks on the genus *Sciaphila*, Blume. Ann Bot 21:71–77

Henderson MW (1919) A comparative study of the structure and saprophytism of the Pyrolaceae and Monotropaceae with reference to their derivation from the Ericaceae. Contr Bot Lab Morris Abor Univ Pennsylvania 5:42–109

Hepper FN (1968) Burmanniaceae. In: Flora of West Tropical Africa, vol 3, 2nd edn. The Crown Agents for Oversea Governments and Administration, London, p 176–180

Hirce EG, Finocchio AF (1972) Stem and root anatomy of *Monotropa uniflora*. Bull Torrey Bot Club 99:89–94

Holloay JE (1921) Further studies on the prothallus, embryo, and young sporophyte of *Tmesipteris*. Trans Proc N Z Inst 53:386–422

Holm T (1897) *Obolaria virginica* L.: a morphological and anatomical study. Ann Bot 11:369–383

Holm T (1898) *Pyrola aphylla*: a morphological study. Bot Gaz 25:246–254

Holm T (1906) *Bartonia*, Muehl. An anatomical study. Ann Bot 20:441–448

Hooker WJ (1840) Flora Boreali-Americana, vol 2. Henry G. Bohn, London

Hsieh C-F, Hsieh T-H, Lai I-L (1995) *Epirixanthes elongata* Bl.—a new record to the Flora of Taiwan. Taiwania 40:381–384

Humpert AJ, Muench EL, Giachini AJ, Castellano MA, Spatafora JW (2001) Molecular phylogenetics of *Ramaria* and related genera: evidence from nuclear large subunit and mitochondrial small subunit rDNA sequences. Mycologia 93:465–477

Hynson NA, Preiss K, Gebauer G, Bruns TD (2009) Isotopic evidence of full and partial myco-heterotrophy in the plant tribe Pyroleae (Ericaceae). New Phytol 182:719–726

Ibisch P, Neinhuis C, Rojas NP (1996) On the biology, biogeography, and taxonomy of *Arachnitis* Phil. nom. cons. (Corsiaceae) in respect to a new record from Bolivia. Willdenowia 26:321–332

Imhof S, Weber HC, Gómez LD (1994) Ein Beitrag zur Biologie von *Voyria tenella* Hook. und *Voyria truncata* (Standley) Standley & Steyermark (Gentianaceae). Beitr Biol Pflanzen 68:113–123

Imhof S, Weber HC (1997) Root anatomy and mycotrophy (AM) of the achlorophyllous *Voyria truncata* (Standley) Standley & Steyermark (Gentianaceae). Bot Acta 110:127–134

Imhof S (1997) Root anatomy and mycotrophy of the achlorophyllous *Voyria tenella* Hooker (Gentianaceae). Bot Acta 110:298–305

Imhof S (1998) Subterranean structures and mycotrophy of the achlorophyllous *Triuris hyalina* Miers (Triuridaceae). Can J Bot 76:2011–2019

Imhof S (1999a) Anatomy and mycotrophy of the achlorophyllous *Afrothismia winkleri* (Engl.) Schltr. (Burmanniaceae). New Phytol 144:533–540

Imhof S (1999b) Subterranean structures and mycorrhiza of the achlorophyllous *Burmannia tenella* Benth. (Burmanniaceae). Can J Bot 77:637–643

Imhof S (1999c) Root morphology, anatomy and mycotrophy of the achlorophyllous *Voyria aphylla* (Jacq.) Pers. (Gentianaceae). Mycorrhiza 9:33–39

Imhof S, Weber HC (2000) Root structures and mycorrhiza of the achlorophyllous *Voyria obconica* Progel (Gentianaceae). Symbiosis 29:201–211

Imhof S (2001) Subterranean structures and mycotrophy of the achlorophyllous *Dictyostega orobanchoides* (Hook.) Miers (Burmanniaceae). Rev Biol Trop 49:239–247

Imhof S (2003) A dorsiventral mycorrhizal root in the achlorophyllous *Sciaphila polygyna* (Triuridaceae). Mycorrhiza 13:327–332

Imhof S (2004) Morphology and development of the subterranean organs of the achlorophyllous *Sciaphila polygyna* (Triuridaceae). Bot J Linn Soc 146:295–301

Imhof S (2006) Two distinct fungi colonize roots and rhizomes of the myco-heterotrophic *Afrothismia gesnerioides* (Burmanniaceae). Can J Bot 84:852–861

Imhof S (2007) Specialized mycorrhizal colonization pattern in achlorophyllous *Epirixanthes* spp. (Polygalaceae). Plant Biol 9:786–792

Imhof S, Sainge MN (2008) Ontogeny of the mycoheterotrophic species *Afrothismia hydra* (Burmanniaceae). Bot J Linn Soc 157:31–36

Imhof S (2009) Arbuscular, ecto-related, orchid mycorrhizas—three independent structural lineages towards mycoheterotrophy: implications for classification? Mycorrhiza 19:357–363

Imhof S (2010) Are monocots particularly suited to develop mycoheterotrophy? In: Seberg O, Petersen G, Barfod A, Davis JI (eds) Diversity, phylogeny and evolution in the monocotyledons. Aarhus University Press, Kopenhagen, pp 11–23

Iwatsuki Z, Sharp AJ (1967) The bryogeographic relationships between eastern Asia and North America. J Hattori Bot Lab 30:152–170

Jacquelinet-Jeanmougin S, Gianinazzi-Pearson V (1983) Endomycorrhizas in the Gentianaceae. I. The fungi associated with Gentiana lutea L. New Phytol 95:663–666

Janse JM (1896) Les endophytes radicaux de quelques plantes javanaises. Ann Jard Bot Buitenzorg 14:53–212

Jessop JP (1979) Petrosavia. In: van Steenis CGGJ (ed) Flora Malesiana, Series 1, vol 9. Martinus Nijhoff, The Hague, pp 198–200

Johow F (1885) Die chlorophyllfreien Humusbewohner West-Indiens, biologisch-morphologisch dargestellt. Jahrb Wiss Bot 16:415–449

Johow F (1889) Die chlorophyllfreien Humuspflanzen nach ihren biologischen und anatomisch-entwicklungsgeschichtlichen Verhältnissen. Jahrb Wiss Bot 20:475–524

Jones DL, Gray B (2008) *Corsia dispar* D. L. Jones & B. Gray (Corsiaceae), a new species from Australia, and a new combination of *Corsia* for a New Guinea taxon. Austrobaileya 7:717–722

Jonker FP (1936) Gentianaceae. In: Pulle AA (ed) Flora of Suriname. Brill, Leiden, pp 400–430

Jonker FP (1938) A monograph of the Burmanniaceae. Meded Bot Mus Herb Rijks Univ Utrecht 51:1–279

Kamienski F (1882) Les organes végétatifs du *Monotropa hypopithys* L. Mém Soc Nat Sc Natur et Math Cherbourg 24:5–40

Kidston R, Lang WH (1921) On old red sandstone plants showing structure, from the Rhynie chert bed, Aberdeenshire. Trans Roy Soc Edinb 52:855–901

Kiew R (1999) *Thismia goodii* (Burmanniaceae), the blue-capped *Thismia*, a new species from Borneo. Gard Bull Singap 51:179–182

Kissling J (2012) Taxonomy of *Exochaenium* and *Lagenias*: two resurrected genera of tribe Exaceae (Gentianaceae). Syst Bot 37:238–253

Kissling J, Yuan Y-M, Küpfer PM, Mansion G (2009) The polyphyletic genus *Sebaea* (Gentianaceae): a step forward in understanding the morphological and karyological evolution of the Exaceae. Mol Phylogenet Evol 53:734–748

Klackenberg J (1985) The genus *Exacum* (Gentianaceae). Opera Bot 84:1–144

Klackenberg J (2002) Tribe Exaceae. In: Struwe L, Albert V (eds) Gentianaceae: systematics and natural history. Cambridge University Press, Cambridge, pp 66–108

Klackenberg J (2006) *Cotylanthera* transferred to *Exacum* (Gentianaceae). Bot Jahrb Syst 126:477–481

Knoblauch E (1894) Beiträge zur Kenntniss der Gentianaceae. Bot Centralbl 60:321–401

Köpke E, Musselman LJ, de Laubenfels DJ (1981) Studies on the anatomy of *Parasitaxus ustus* and its root connections. Phytomorph 31:85–92

Kottke I, Nebel M (2005) The evolution of mycorrhiza-like associations in liverworts: an update. New Phytol 167:330–334

Kretzer AM, Bidartondo MI, Grubisha LC, Spatafora JW, Szaro TM, Bruns TD (2000) Regional specialization of *Sarcodes sanguinea* (Ericaceae) on a single fungal symbiont from the *Rhizopogon ellenae* (Rhizopogonaceae) species complex. Am J Bot 87:1778–1782

Kuga-Uetake Y, Purich M, Massicotte HB, Peterson RL (2004) Host microtubules in the Hartig net region of ectomycorrhizas, ectendomycorrhizas, and monotropoid mycorrhizas. Can J Bot 82:938–946

Kuijt J (1969) The biology of parasitic flowering plants. Univ. Calif. Press, Berkeley

Kutschera L, Lichtenegger E (1982) Wurzelatlas mitteleuropäischer Grünlandpflanzen, Teil 1: Monocotyledonae. Fischer, Stuttgart

Kutschera L, Lichtenegger E (1992) Wurzelatlas mitteleuropäischer Grünlandpflanzen, Teil 2: Dicotyledonae. Fischer, Stuttgart

Kutschera L, Lichtenegger E, Sobotik M (2009) Wurzelatlas der Kulturpflanzen gemäßigter Gebiete mit Arten des Feldgemüsebaues. DLG Verlags-GmbH, Frankfurt

Lace JH (1914) Decades Kewenses—767. *Cotylanthera caerulea*. Bull Misc Inform 1914:150–156

Lang WH (1899) The prothallus of *Lycopodium clavatum* L. Ann Bot 13:279–317

Lang WH (1902) On the prothalli of *Ophioglossum pendulum* and *Helminthostachys zeylanica*. Ann Bot 16:23–56

Largent DL, Sugihara N, Wishner C (1980) Occurrence of mycorrhizae on ericaceous and pyrolaceous plants in northern California. Can J Bot 58:2274–2279

Larsen K (1963) Studies in the flora of Thailand. 14. Cytological studies in vascular plants of Thailand. Dansk Bot Ark 20:211–275

Larsen K (1965) Thismiaceae. Dansk Bot Ark 23:171–174

Larsen K (1972) Triuridaceae. In: Smitinand T, Larsen K (eds) Flora of Thailand, vol 2, Part 2. Asrct Press, Bangkok, pp 175–176

Larsen K, Averyanov LV (2007) *Thismia annamensis* and *Thismia tentaculata*, two new species of Thismiaceae from central Vietnam. Rheedea 17:13–19

Lawson AA (1917) The prothallus of *Tmesipteris tannensis*. Trans Roy Soc Edinb 51:785–794

Leake JR (1994) Tansley Review No. 69. The biology of myco-heterotrophic ('saprophytic') plants. New Phytol 127:171–216

Leake JR (2004) Myco-heterotroph/epiparasitic plant interactions with ectomycorrhizal and arbuscular mycorrhizal fungi. Curr Opin Plant Biol 7:422–428

Leake JR, McKendrick SL, Bidartondo M, Read DJ (2004) Symbiotic germination and development of the myco-heterotroph *Monotropa hypopitys* in nature and its requirement for locally distributed *Tricholoma* spp. New Phytol 163:405–423

Leake JR (2005) Plants parasitic on fungi: unearthing the fungi in myco-heterotrophs and debunking the 'saprophytic' plant myth. Mycologist 19:113–122

Lefevre CK (2002) Host associations of *Tricholoma magnivelare*, the American matsutake. Dissertation, Corvallis, Department of Forest Science: Oregon State University

LePage BA, Currah RS, Stockey RA, Rothwell GW (1997) Fossil ectomycorrhizae from the middle Eocene. Am J Bot 84:410–412

Ligrone R, Gambardella R, Castaldo R, Giordano S, de Lucia Sposito ML (1982) Gametophyte and sporophyte ultrastructure in *Buxbaumia piperi* Best (Buxbaumiales, Musci). J Hattori Bot Lab 52:465–499

Ligrone R, Pocock K, Duckett JG (1993) A comparative ultrastructural study of endophytic basidiomycetes in the parasitic achlorophyllous hepatic *Cryptothallus mirabilis* and the closely allied photosynthetic species *Aneura pinguis* (Metzgeriales). Can J Bot 71:666–679

Liu K-M, Long C-L, Liu Y-T (2001) *Burmannia hunanensis* (Burmanniaceae), a new species from Hunan, China. Ann Bot Fennici 38:211–214

Lutz RW, Sjolund RD (1973) *Monotropa uniflora*: ultrastructural details of its mycorrhizal habit. Am J Bot 60:339–345

Maas H, Maas PJM (1987) A new *Thismia* (Burmanniaceae) from French Guiana. Brittonia 39:376–378

Maas PJM, Rübsamen T (1986) Triuridaceae. Flora Neotropica Monogr 40:1–55

Maas PJM, Ruyters P (1986) *Voyria* and *Voyriella* (Saprophytic Gentianaceae). Flora Neotropica Monogr 41:1–93

Maas PJM, Maas-van de Kamer H, Benthem JV, Snelders HCM, Rübsamen T (1986) Burmanniaceae. Flora Neotropica Monogr 42:1–189

Maas PJM, Maas-van de Kamer H (1989) Triuridaceae. In: Görts-van Rijn ARA (ed) Flora of the Guianas, series A: Phanerogams, vol 5. Koeltz Scientific Books, Koenigstein, pp 3–16

Maas-van de Kamer H (1995) Triuridiflorae—Gardner's delight? In: Rudall PJ, Cribb PJ, Cutler DF, Humphries CJ (eds) Monocotyledons: systematics and evolution. Royal Botanic Gardens, Kew, pp 287–301

Maas-van de Kamer H, Maas PJM (2010) Thismiaceae. In: Sosef MSM (ed) Flore du Gabon, vol 41. Margraf, Weikersheim, pp 60–62

Maas-van de Kamer H, Weustenfeld T (1998) Triuridaceae. In: Kubitzki K (ed) The families and genera of vascular plants, III flowering plants—monocotyledons, Lilianae (except Orchidaceae). Springer, Berlin, pp 452–458

Maas-van de Kamer H (2003) *Afrothismia gesnerioides*, another new species of *Afrothismia* (Burmanniaceae) from tropical Africa. Blumea 48:475–478

Magnus W (1900) Studien an der endotrophen Mycorrhiza von *Neottia nidus avis* L. Jahrb Wiss Bot 35:205–272

Makino T (1903) Observations of the flora of Japan. Bot Mag (Tokyo) 17:144–152

Malme GO (1896a) Die Burmannien der ersten Regnell'schen Expedition. Ein Beitrag zur Kenntnis der amerikanischen Arten dieser Gattung. Bih. Kongl. Svenska Vetensk.-Akad. Handl. 22 (Afd. III, No. 8):1–32

Malme GO (1896b) Über *Triuris lutea* (Gardn.) Benth. et Hook. Bih. Kongl. Svenska Vetensk.-Akad. Handl. 21 (Afd. III, No. 14):1–16

Marais W, Verdoorn IC (1963) Gentianaceae. In: Dyer RA, Codd LE, Rycroft HB (eds) Flora of Southern Africa, vol 26. Government Printer, Pretoria, pp 171–243

Martinez E, Ramos CH (1989) Lacandoniaceae (Triuriadales): una nueva familia de Mexico. Ann Missouri Bot Gard 76:128–135

Massicotte HB, Melville LH, Peterson RL (2005) Structural features of mycorrhizal associations in two members of the Monotropoideae, *Monotropa uniflora* and *Pterospora andromedea*. Mycorrhiza 15:101–110

Massicotte HB, Melville LH, Tackaberry LE, Peterson RL (2007) *Pityopus californicus*: structural characteristics of seed and seedling development in a myco-heterotrophic species. Mycorrhiza 17:647–653

Massicotte HB, Melville LH, Tackaberry LE, Peterson RL (2008) A comparative study of mycorrrhizas in several genera of Pyroleae (Ericaceae) from western Canada. Botany 86:610–622

Massicotte HB, Melville LH, Peterson RL, Tackaberry LE, Luoma DL (2010) Structural characteristics of root-fungus associations in two mycoheterotrophic species, *Allotropa virgata* and *Pleuricospora fimbriolata* (Monotropoideae), from southwest Oregon, USA. Mycorrhiza 20:391–397

Matsuda Y, Okochi S, Katayama T, Yamada A, Ito S-I (2011) Mycorrhizal fungi associated with *Monotropastrum humile* (Ericaceae) in central Japan. Mycorrhiza 21:569–576

McClymont JW (1950) Notes on *Buxbaumia*. Bryologist 53:176–277

McKendrick SL, Leake JR, Taylor DL, Read DJ (2002) Symbiotic germination and development of the myco-heterotrophic orchid *Neottia nidus-avis* in nature and its requirement for locally distributed *Sebacina* spp. New Phytol 154:233–247

McLennan EI (1958) *Thismia rodwayi* F. Muell. and its endophyte. Aust J Bot 6:25–37

Merckx V (2008) Myco-heterotrophy in Dioscoreales: systematics and evolution. Dissertation, Leuven, Faculteit Wetenschappen: K.U. Leuven

Merckx V, Schols P, Maas-van de Kamer H, Maas P, Huysmans S, Smets E (2006) Phylogeny and evolution of Burmanniaceae (Dioscoreales) based on nuclear and mitochondrial data. Am J Bot 93:1684–1698

Merckx V, Bidartondo MI (2008) Breakdown and delayed cospeciation in the arbuscular mycorrhizal mutualism. Proc Roy Soc Lond B 275:1029–1035

Merckx V, Bidartondo MI, Hynson NA (2009) Mycoheterotrophy: when fungi host plants. Ann Bot 104:1255–1261

Merckx VSFT, Janssens SB, Hynson NA, Specht CD, Bruns TD, Smets E (2012) Mycoheterotrophic interactions are not limited to a narrow phylogenetic range of arbuscular mycorrhizal fungi. Mol Ecol 21:1524–1532

Merckx V, Stöckel M, Fleischmann A, Bruns TD, Gebauer G (2010) 15N and 13C natural abundance of two mycoheterotrophic and a putative partially mycoheterotrophic species associated with arbuscular mycorrhizal fungi. New Phytol 188:590–596

Mesler MR (1975) The gametophytes of *Ophioglossum palmatum* L. Am J Bot 62:982–992

Mesler MR (1976) Gametophytes and young sporophytes of *Ophioglossum crotalophoroides* Walt. Am J Bot 63:443–448

Meyer K (1909) Untersuchungen über *Thismia clandestina*. Bull Soc Imp Naturalistes Moscou 1909:1–18

Miers J (1852) On the family of Triuridaceae. Trans Linn Soc Lond 21:43–59

Miers J (1866) On *Myostoma*, a new genus of the Burmanniaceae. Trans Linn Soc Lond 25:461–476

Milanez FR, Meira E (1943) Observacoes sobre '*Triuris alata*' Brade. Arq Serv Florest 2:51–61

Minoletti OML (1986) *Arachnitis uniflora* Phil. una curiosa monocotiledonea de la Flora Chilena. Bol Soc Biol Concepcíon 57:7–20

Miquel FAW (1851) Gentianeae parasiticae. In: Stirpes surinamenses selectae. Arnz & Comp, Leiden, p 146–151

Miquel FAW (1856) Gentianae. In: Flora van Nederlandsch Indie, vol 2. Fried. Fleischer, Leipzig, p 554–564

Mueller DMTJ (1972) Observations on the ultrastructure of *Buxbaumia* Protonema. Plasmodesmata in the cross walls. Bryologist 75:63–68

Neumann G (1934) Über die Mykorrhiza in der Gattung *Gentiana*. Zentralbl Bakteriol 2 Abt 89:433–458

Nuttall T (1843) Notices of new or rare plants in the natural orders Lobeliaceae, Campanulaceae, Vaccinieae, Ericaceae collected in a journey over the continent of North America. Trans Am Phil Soc 8:271

Oehler E (1927) Entwicklungsgeschichtlich-zytologische Untersuchungen an einigen saprophytischen Gentianaceen. Planta 3:641–729

Ogura-Tsujita Y, Yukawa T (2008) High mycorrhizal specificity in a widespread mycoheterotrophic plant, *Eulophia zollingeri* (Orchidaceae). Am J Bot 95:93–97

Ohashi H (2000) A new treatment in *Sciaphila* (Triuridaceae) in Taiwan and Japan. Taiwania 45:351–354

Ohga T, Sinoto Y (1932) Cytological studies on *Sciaphila japonica* Mak. II On pollen and embryo sac-development. III. On micorhiza. Bot Mag (Tokyo) 46:311–315

Okamura S (1911) Neue Beiträge zur Moosflora Japans. Bot Mag (Tokyo) 25:30–34

Palacios-Mayorga P, Pérez-Silva E (1993) Endotrophic mycorrhiza in *Lacandonia schismatica*, Lacandoniaceae, a new saprophytic Angiospermae family. Revista Latinoamer Microbiol 35:65–69

Pendry CA (2010) *Epirixanthes compressa* Pendry, a new mycoheterotrophic species of Polygalaceae from Thailand. Thai For Bull 38:184–186

Penzig O (1901) Beiträge zur Kenntniss der Gattung Epirrhizanthes Bl. Ann Jard Bot Buitenzorg 17:142–170

Perrot E (1898) Anatomie comparée des Gentianacées. Ann Sci Nat Bot Sér 7 16:105–292

Petersen PM (1972) *Cryptothallus mirabilis* Malmb. found on Disko Island, West Greenland (69°15′ N., 53°34′ W.). Lindbergia 1:189–190

Peterson RL, Howarth MJ, Whittier DP (1981) Interactions between a fungal endophyte and gametophyte cells in *Psilotum nudum*. Can J Bot 59:711–720

Peterson RL (1992) Adaptations of root structure in relation to biotic and abiotic factors. Can J Bot 70:661–675

Peterson RL, Uetake Y, Zelmer C (1998) Fungal symbioses with orchid protocorms. Symbiosis 25:29–55

Peterson RL, Massicotte HB (2004) Exploring structural definitions of mycorrhizas, with emphasis on nutrient-exchange interfaces. Can J Bot 82:1074–1088

Peterson RL, Massicotte HB, Melville LH (2004) Mycorrhizas: anatomy and cell biology. NRC Research Press, Ottawa

Pfeiffer NE (1914) Morphology of *Thismia americana*. Bot Gaz 57:122–135

Pires O'Brien MJ (1997) Annotated check-list of the flora in the Jari river region: I. Saprophytic parasites. Boletim-do-Museu-Paraense-Emilio-Goeldi-Serie-Botanica 13:49–67

Pocock K, Duckett JG (1984) A comparative ultrastructural analysis of the fungal endophytes in *Cryptothallus mirabilis* Malm. and other British thalloid hepatics. J Bryol 13:227–233

Polomski J, Kuhn N (1998) Wurzelsysteme. Paul Haupt, Bern

Poppendieck H-H (1997) Eine *Voyria*-Art als 'blinder Passagier' im Botanischen Garten Hamburg. Palmengarten 61:30

Poulsen VA (1886) Bidrag til Triuridaceernes Naturhistorie. Vidensk Meddel Dansk Naturhist Foren Kjøbenhavn 36–38:161–179

Poulsen VA (1890a) *Thismia Glaziovii* nov. sp. Overs Kongel Danske Vidensk Selesk Forh Medlemmers Arbeider 1890:18–38

Poulsen VA (1890b) *Triuris major* sp. nov. Et Bidrag til Triuridaceernes Naturhistorie. Bot Tidsskr 17:293–306

Pringle JS (1995) Gentianaceae. In: Harling G, Andersson L (eds) Flora of Ecuador, vol 53. Council for Nordic Publications in Botany, Copenhagen, pp 3–131

Progel A (1865) Gentianaceae: *Voyria*. In: Martius (ed) Flora Brasiliensis, vol 6, C.F.P.v. Frid. Fleischer, Leipzig, p 219–226

Rasmussen HN (1995) Terrestrial orchids. Froseed to mycotrophic plant. Cambridge Univ. Press, Cambridge

Rasmussen HN (2002) Recent developments in the study of orchid mycorrhiza. Pl Soil 244:149–163

Raynal A (1967a) Sur un *Sebaea* africain saprophyte (Gentianaceae). Adansonia 7:207–219

Raynal A (1967b) Étude critique des genres *Voyria* et *Leiphaimos* (Gentianaceae et révision des *Voyria* d'Afrique). Adansonia 7:53–71

Raynal-Roques A, Paré J (1998) Biodiversité des phanérogames parasites: leur place dans la classification systématique. Adansonia 20:313–322

Read DJ, Duckett JG, Francis R, Ligrone R, Russell A (2000) Symbiotic fungal associations in 'lower' land plants. Phil Trans R Soc Lond B 355:815–831

Reiche K (1907) Grundzüge der Pflanzenverbreitung in Chile. In: Engler A, Drude O (eds) Die Vegetation der Erde. Sammlung Pflanzengeographischer Monographien VIII. W. Engelmann, Leipzig

Remy W, Taylor TN, Hass H, Kerp H (1994) Four hundred-million-year-old vesicular arbuscular mycorrhizae. Proc Natl Acad Sci U S A 91:11841–11843

Ridley HN (1922) Polygalaceae. In: Flora of the Malay Peninsula, vol 1. Reeve & Co, London, p 137–149

Robertson DC, Robertson JA (1982) Ultrastructure of *Pterospora andromedea* Nuttal and *Sarcodes sanguinea* Torrey mycorrhizas. New Phytol 92:539–551

Robertson DC, Robertson JA (1985) Ultrastructural aspects of *Pyrola* mycorrhizae. Can J Bot 63(108):9–1098

Robyns A (1962) Deux *Exochaenium* nouveaux (Gentianaceae) d'Afrique tropicale. Bull Jard Bot Etat Bruxelles 32:471–475

Rudall PJ, Bateman RM (2006) Morphological phylogenetic analysis of Pandanales: testing contrasting hypotheses of floral evolution. Syst Bot 31:223–238

Rübsamen-Weustenfeld T (1991) Morphologische, embryologische und systematische Untersuchungen an Triuridaceae. Biblioth Bot 140:1–113

Rübsamen-Weustenfeld T, Mukielka V, Hamann U (1994) Zur Embryologie, Morphologie und systematischen Stellung von *Geosiris aphylla* Baillon (Monocotyledoneae—Geosiridaceae/Iridaceae). Mit einigen embryologischen Daten zur Samenanlage von *Isophysis tasmanica* (Hook.) T. Moore (Iridaceae). Bot Jahrb Syst 115:475–545

Sainge MN, Franke T (2005) A new species of *Afrothismia* (Burmanniaceae) from Cameroon. Nord J Bot 23:299–303

Sasidharan N, Sujanapal P (2000) Rediscovery of *Haplothismia exannulata* Airy Shaw (Burmanniaceae) from its type locality. Rheedea 10:131–134

Saunders RMK (1996) Typification of the name *Thismia fumida* Ridl. (Burmanniaceae). Taxon 45:107–110

Schiffner V (1934) Über *Cryptothallus mirabilis* Malmb. Ann Bryol 7:165–166

Schlechter R (1905) Heft 2: Microspermae. In: Schumann K, Lauterbach K (eds) Nachträge zu Flora der Deutschen Schutzgebiete in der Südsee. Gebr. Borntraeger, Leipzig, pp 71–234

Schlechter R (1913) Neue Triuridaceae Papuasiens. Bot Jahrb Syst 49:70–89

Schlechter R (1919) Eine neue papuasische Burmanniaceae. Bot Jahrb Syst 55:202–203

Schleiden MJ (1845) Grundzüge der Wissenschaftlichen Botanik, 2. Aufl. Leipzig, Wilhelm Engelmann

Schmid E, Oberwinkler F (1993a) Mycorrhiza-like interaction between the achlorophyllous gametophyte of *Lycopodium clavatum* L. and its fungal endophyte studied by light and electron microscopy. New Phytol 124:69–81

Schmid E, Oberwinkler F (1993b) Light and electron microscopy of the host-fungus interaction in the achlorophyllous gametophyte of *Botrychium lunaria*. Can J Bot 72:182–188

Schneckenburger S (1999) Einige bemerkenswerte Coniferen Neukaledoniens. Mitt Deutsch Dendrol Ges 84:5–21

Schoepe G, Philippi G (2000) Buxbaumiidae. In: Nebe M, Philippi G (eds) Die Moose Baden-Württembergs. Ulmer, Stuttgart, pp 90–97

Schüßler A, Schwarzott D, Walker C (2001) A new fungal phylum, the Glomeromycota: phylogeny and evolution. Mycol Res 105:1413–1421

Selosse M-A, Weiß M, Jany J-L, Tillier A (2002) Communities and populations of sebacinoid basidiomycetes associated with the achlorophyllous orchid *Neottia nidus-avis* (L.) L.C.M. Rich. and neighbouring tree ectomycorrhizae. Mol Ecol 11: 1831–1844

Selosse M-A, Faccio A, Scappaticci G, Bonfante P (2004) Chlorophyllous and achlorophyllous specimens of *Epipactis microphylla* (Neottieae, Orchidaceae) are associated with ectomycorrhizal Septomycetes, including Truffles. Microbial Ecol 47:416–426

Sergio C, Seneca A (1997) The first report of *Cryptothallus mirabilis* Malmborg (Hepaticae, Aneuraceae) in southern Europe (Portugal). Cryptogamie Bryologie Lichénologie 18:213–215

Sergio C, Garcia C (1999) *Cryptothallus mirabilis* Malmb. (Aneuraceae, Hepaticae) in Portugal. New ecological and phytogeographical data. Haussknechtia Beiheft 9:343–346

Seybold S (2011) Schmeil-Fitschen—Flora von Deutschland und der angrenzenden Länder, 95. Aufl. Quelle & Meyer, Wiebelsheim

Sinclair WT, Mill RR, Gardner MF, Woltz P, Jaffré T, Preston J, Hollingsworth ML, Ponge A, Möller M (2002) Evolutionary relationships of the New Caledonian heterotrophic conifer, *Parasitaxus usta* (Podocarpaceae), inferred from chloroplast *trnL-F* intron/spacer and nuclear rDNA ITS2 sequences. Pl Syst Evol 233:79–104

Sjörs H (1949) Nagra växter funna huvudsakligen ar 1948. Botaniska Notiser 1949:95–103

Smith AR, Pryer KM, Schuettpelz E, Korall P, Schneider H, Wolf PG (2006) A classification for extant ferns. Taxon 55:705–731

Smith JJ (1911) Beiträge zur Kenntnis der Saprophyten Javas. VII. Zur Systematik von Burmannia candida Engl. und Burmannia championii Thw. Ann Jard Bot Buitenzorg 24:79–82

Smith JJ (1912) Neue papuanische Pflanzen I. Fedd Repert 10:486–488

Smith SE, Read DJ (2008) Mycorrhizal symbiosis, 3rd edn. Acad. Press, San Diego

Solereder H (1908) Gentianae. In: Systematische Anatomie der Dicotyledonen. Ergänzungsband. F. Enke, Stuttgart, p 220–229

Stant MY (1970) Anatomy of *Petrosavia stellaris* Becc., a saprophytic monocotyledon. Bot J Linn Soc 63:147–161

Steven G, Long DG (1989) An update on the status of *Buxbaumia aphylla* on bings in central Scotland. Bot J Scotland 45:389–395

Stevens PF (2001) onwards. Angiosperm Phylogeny Website. http://www.mobot.org/MOBOT/research/APweb/

Stevens PF, Luteyn J, Oliver EGH, Bell TL, Brown EA, Crowden RK, George AS, Jordan GJ, Ladd P, Lemson K, McLean CB, Menadue Y, Pate JS, Stace HM, Weiller CM (2004) Ericaceae. In: Kubitzki K (ed) The families and genera of vascular plants, vol 6. Springer, Heidelberg, pp 145–194

Stockey RA, Rothwell GW, Accy HD, Currah RS (2001) Mycorrhizal association of the extinct conifer *Metasequoia milleri*. Mycol Res 105:202–205

Stone BC (1980) Rediscovery of *Thismia clavigera* (Becc.) F. v. M. (Burmanniaceae). Blumea 26:419–425

Stone G (1983) *Buxbaumia* in Australia, including one new species, *Buxbaumia thorsborneae*. J Bryol 12:541

Sun B-Y, Kim MH, Kim CH, Park C-W (2001) *Mankyua* (Ophioglossaceae): a new fern genus from Cheju Island, Korea. Taxon 50:1019–1024

Süssenguth K (1937) Einige bemerkenswerte Pflanzen des Münchner Staatsherbars. Repert Spec Nov Regni Veg 42:43–49

Svedelius N (1902) Zur Kenntnis der saprophytischen Gentianaceen. Kongl Svenska Vetenskapsakad Handl Bih 28(3) No. 4:3–16

Swatzell LJ, Powell MJ, Kiss JZ (1996) The relationship of endophytic fungi to the gametophyte of the fern *Schizaea pusilla*. Int J Plant Sci 157:53–62

Takeuchi W, Pipoly JJ III (1998) New flowering plants from southern New Ireland, Papua New Guinea. Sida 18:161–168

Takhtajan A (1997) Diversity and classification of flowering plants. Columbia Univ. Press, New York

Taylor DL, Bruns TD (1997) Independent, specialized invasions of ectomycorrhizal mutualism by two non-photosynthetic orchids. Proc Natl Acad Sci U S A 94:4510–4515

Taylor DL, Bruns TD, Leake JR, Read DJ (2002) Mycorrhizal specificity and function in myco-heterotrophic plants. In: van der Heijden MGA, Sanders I (eds) Mycorrhizal ecology. Springer, Berlin, pp 375–413

Taylor TN, Remy W, Hass H, Kerp H (1995) Fossil arbuscular mycorrhizae from the early Devonian. Mycologia 87:560–573

ter Welle BJH (1986) Anatomy. In: Maas PJM, Ruyters P (eds) *Voyria* and *Voyriella* (Saprophytic Gentianaceae). Flora Neotropica Monograph 41:25–29

Terashita T, Kawakami Y (1991) An endomycorrhizal fungus of *Burmannia liukiuensis*. Trans Mycol Soc Japan 32:207–215

Thorne RF (1983) Proposed new realignments in the angiosperms. Nord J Bot 3:85–117

Thorne RF (1992) Classification and geography of the flowering plants. Bot Rev 58:225–348

Tomlinson PB (1982) Helobiae (Alismatidae). In: Metcalfe CR (ed) Anatomy of the monocotyledons, vol 7. Clarendon, Oxford, pp 1–559

Treub M (1885) Études sur les Lycopodiacées. Ann Jard Bot Buitenzorg 5:87–139

Treub M (1890) Études sur les Lycopodiacées. Ann Jard Bot Buitenzorg 8:1–37

Tsukaya H, Yokoyama J, Imaichi R, Ohba H (2008) Taxonomic status of *Monotropastum humile*, with special reference to *M. humile* var. *glaberrimum* (Ericaceae, Monotropoideae). J Plant Res 121:271–278

Udar R, Srivastava SC, Kumar D (1971) A new species of *Buxbaumia* Hedwig, *B. himalayensis* Udar, Srivastava et Kumar, from Deoban, Western Himalayas. Trans Brit Bryol Soc 6:266–269

Uphof JCT (1929) Beiträge zur Kenntnis der Burmanniacee *Apteria aphylla* (Nutt.). Barnh Österr Bot Z 78:71–80

van de Meerendonk JPM (1984) Triuridaceae. In: van Steenis CGGJ (ed) Flora Malesiana, Ser. I, vol 10. Kluwer Academic Publishers, Dordrecht, pp 109–121

van der Meijden R (1988) Polygalaceae. In: van Steenis CGGJ, de Wilde WJJO (eds) Flora Malesiana, vol 10. Kluwer Academic Publishers, Dordrecht, pp 455–539

van der Pijl L (1934) Die Mycorrhiza von *Burmannia* und *Epirrhizanthes* und die Fortpflanzung ihres Endophyten. Recueil Trav Bot Néerl 31:761–779

van Rompu W, Stieperaere H (2002) Fate of sporophytes in a Belgian population of *Buxbaumia aphylla*. J Bryol 24:83–85

van Royen P (1972) Sertulum papuanum 17. Corsiaceae of New Guinea and surrounding areas. Webbia 27:223–255

Vigodsky-de Philippis A (1938) Studio morfologico ed anatomica di Leipaimos brachyloba' (Griseb.) Urb. var. 'cumbrensis' Urb. et Ekm. Nuovo Giorn Bot Ital 45:190–194

Vincenot L, Tedersoo L, Richard R, Horcine H, Kõljalg U, Selosse M-A (2008) Fungal associates of *Pyrola rotundifolia*, a mixotrophic Ericaceae, from two Estonian boreal forests. Mycorrhiza 19:15–25

von Malmborg S (1933) *Cryptothallus* nov. gen., ein saprophytisches Lebermoos. Ann Bryol 6:122–123

von Malmborg S (1934) Weiteres über die Gattung Cryptothallus. Ann Bryol 7:108–110

Wallace GD (1975) Studies of the Monotropoideae (Ericaceae). Taxonomy and distribution. Wasmann J Biol 33:1–88

Wang B, Qiu Y-L (2006) Phylogenetic distribution and evolution of mycorrhizas in land plants. Mycorrhiza 16:299–363

Wang H, Wang Z, Zhang F, Liu J, He X (1997) A cytological study on the nutrient-uptake mechanism of a saprophytic orchid *Gastrodia elata*. Acta Bot Sin 39:500–504

Wapstra M, French B, Davies N, O'Reilly-Wapstra J, Peters D (2005) A bright light on the dark forest floor: observations on fairy lanterns *Thismia rodwayi* F. Muel. (Burmanniaceae) in Tasmanian forests. Tasmanian Naturalist 127:2–18

Warming E (1901) Sur quelques Burmanniacées recueillies au Brésil par le Dr. A. Glaziou. Oversigt over Kgl. Danske Vidensk. Selsk. Forhand 1901:173–188

Watanabe K (1944) Morphologisch-biologische Studien über *Miyoshia Sakuraii* Makino. J Jap Bot 20:85–93

Watson L, Dallwitz MJ (2005) onwards. The moss families of the British Isles. Version: 21st June 2009. http://delta-intkey.com

Weber HC (1993) Parasitismus von Blütenpflanzen. Wissenschaftl. Buchges, Darmstadt

Wickett NJ, Goffinet B (2008) Origin and relationships of the myco-heterotrophic liverwort *Cryptothallus mirabilis* Malmb. (Metzgeriales, Marchantiophyta). Bot J Linn Soc 156:1–12

Wiehle W, Berg C, Grolle R (1989) *Cryptothallus mirabilis* Malmborg neu in Mitteleuropa. Herzogia 8:107–124

Williams LO (1946) Notes on the family Corsiaceae. Botanical Museum Leaflets Harvard University 12(5):179–182, Pl. 24

Williams S (1950) The occurrence of *Cryptothallus mirabilis* v. Malmb. in Scotland. Trans Brit Mycol Soc 1:357–366

Winther JL, Friedman WE (2007) Arbuscular mycorrhizal symbionts in *Botrychium* (Ophioglossaceae). Am J Bot 94:1248–1255

Winther JL, Friedman WE (2008) Arbuscular mycorrhizal associations in Lycopodiaceae. New Phytol 177:790–801

Wolf W (1922) Notes on Alabama plants. Am Midl Nat 8:104–127

Woltz P, Stockey RA, Gondran M, Cherrier J-F (1994) Interspecific parasitism in the Gymnosperms: unpublished data on two endemic New Caledonian Podocarpaceae using scanning electron microscopy. Acta Bot Gallica 141:731–746

Wood CE, Weaver RE (1982) The genera of the Gentianaceae in the Southeastern United States. J Arnold Arbor 63:441–487

Woodward CL, Berry PE, Maas-van de Kamer H, Swing K (2007) *Tiputinia foetida*, a new mycoheterotrophic genus of Thismiaceae from Amazonian Ecuador, and a likely case of deceit pollination. Taxon 56:157–162

Yahara T, Tsukaya H (2008) *Oxygyne yamashitae*, a new species of Thismiaceaefrom Yaku Island, Japan. Acta Phytotax Geobot 59:97–104

Yamada A, Kitamura D, Setoguchi M, Matsuda Y, Hashimoto Y, Matsushita N, Fukuda M (2008) *Monotropastrum humile* var. *humile* is associated with diverse ectomycorrhizal Russulaceae fungi in Japanese forests. Ecol Res 23:983–993

Yamato M (2001) Identification of a mycorrhizal fungus in the roots of achlorophyllous *Sciaphila tosaensis* Makino (Triuridaceae). Mycorrhiza 11:83–88

Yamato A, Yagame T, Shimomura N, Iwase K, Takahashi H, Ogura-Tsujita Y, Yukawa T (2011a) Specific arbuscular mycorrhizal fungi associated with non-photosynthetic *Petrosavia sakuraii* (Petrosaviaceae). Mycorrhiza 21:631–639

Yamato A, Yagame T, Iwase K (2011b) Arbuscular mycorrhizal fungi in roots of non-photsynthetic plants, *Sciaphila japonica* and *Sciaphila tosaensis* (Triuridaceae). Mycoscience 52:217–233

Yang S, Pfister DH (2006) *Monotropa uniflora* plants of eastern Massachusetts form mycorrhizae with a diversity of russulacean fungi. Mycologia 98:535–540

Yang S-Z, Saunders RMK, Hsu C-J (2002) *Thismia taiwanensis* sp. nov. (Burmanniaceae tribe Thismieae): first record of the tribe in China. Syst Bot 27:485–488

Yokoyama J, Fukuda T, Tsukaya H (2005) Molecular identification of the mycorrhizal fungi of the epiparasitic plant *Monotropastrum humile* var. *glaberrimum* (Ericaceae). J Plant Res 118:53–56

Yokoyama J, Koizumi Y, Yokota M, Tsukaya H (2008) Phylogenetic position of *Oxygyne shinzatoi* (Burmanniaceae) inferred from 18S rDNA sequences. J Plant Res 121:27–32

Young BW, Massicotte HB, Tackaberry LE, Baldwin QF, Egger KN (2002) *Monotropa uniflora*: morphological and molecular assessment of mycorrhizae retrieved from sites in the Sub-Boreal Spruce biogeoclimatic zone in central British Columbia. Mycorrhiza 12:75–82

Yuan Y-M, Wohlhauser S, Möller M, Chassot P, Mansion G, Grant J, Küpfer P, Klackenberg J (2003) Monophyly and relationships of the tribe Exaceae (Gentianaceae) inferred from nuclear ribosomal and chloroplast DNA sequences. Mol Phylogenet Evol 28:500–517

Zelmer CD, Currah RS (1995) Evidence for a fungal liaison between *Corallorhiza trifida* (Orchidaceae) and *Pinus contorta* (Pinaceae). Can J Bot 73:862–866

Zhang DX, Saunders RM (1999) *Burmannia larseniana* (Burmanniaceae): a new species from Thailand. Nord J Bot 19:241–244

Zhang DX, Saunders RMK, Hu CM (1999) *Corsiopsis chinensis* gen. et sp. nov. (Corsiaceae): first record of the family in Asia. Syst Bot 24(3):311–314

Zimmer K, Meyer C, Gebauer G (2008) The ectomycorrhizal specialist orchid *Corallorhiza trifida* is a partial myco-heterotroph. New Phytol 178:395–400

Evolution and Diversification 5

Vincent S.F.T. Merckx, Constantijn B. Mennes, Kabir G. Peay, and József Geml

5.1 Introduction

The ability to obtain carbon from soil fungi is an exceptional trophic strategy in plants. The bizarre appearance of fully mycoheterotrophic plants immediately sparks questions about their evolutionary history. How and when did they originate? Are all mycoheterotrophic plant lineages closely related? What did their ancestors look like? A quick glance through the list of plant families that contain fully mycoheterotrophic species reveals that mycoheterotrophy must have evolved multiple times independently during land plant evolution and that similarities in habit and ecological interactions of many groups of mycoheterotrophs thus are the result of remarkable convergent evolution (Chap. 2). The molecular revolution in plant systematics has helped to advance our understanding of land plant evolution considerably, including the evolutionary framework behind shifts from autotrophy to full mycoheterotrophy. In combination with the substantial progress that has been made in our knowledge of the ecology of mycoheterotrophic plants and their fungal partners, we are now able to derive hypotheses of the evolutionary framework leading to mycoheterotrophy.

V.S.F.T. Merckx (✉) • C.B. Mennes • J. Geml
Naturalis Biodiversity Center, Leiden University, P.O. Box 9514, 2300 RA Leiden, The Netherlands
e-mail: vincent.merckx@naturalis.nl

K.G. Peay
Department of Biology, Stanford University, 371 Serra Mall, Stanford, CA 94305, USA

5.2 Reconstructing the Evolutionary History of Mycoheterotrophy

The evolutionary trajectory toward a fully mycoheterotrophic mode of life can only be studied when solid phylogenetic frameworks are available. Despite our expanding knowledge of the evolutionary relationships in most plant families, our understanding of the affinities of many mycoheterotrophic plant lineages is somewhat lagging behind. Indeed, in many cases, several obstacles cause the identification of relatives of full mycoheterotrophs to be a phylogenetic challenge. First, many mycoheterotrophic plants are rare, difficult to find, or only grow at remote localities (Chap. 2). In some extreme cases, particular species are known from a single collection only. Obtaining study material is therefore often the first difficulty to overcome when trying to unravel the evolutionary history of mycoheterotrophs. Second, the evolution of full mycoheterotrophy is often associated with the loss of morphological key characters, leaving few reliable characters with which to infer evolutionary relationships. In such cases, the use of DNA sequence data for unraveling phylogenetic history provides a promising endeavor. Unfortunately, the evolution of mycoheterotrophy is not only accompanied by

V.S.F.T. Merckx (ed.), *Mycoheterotrophy: The Biology of Plants Living on Fungi*,
DOI 10.1007/978-1-4614-5209-6_5,

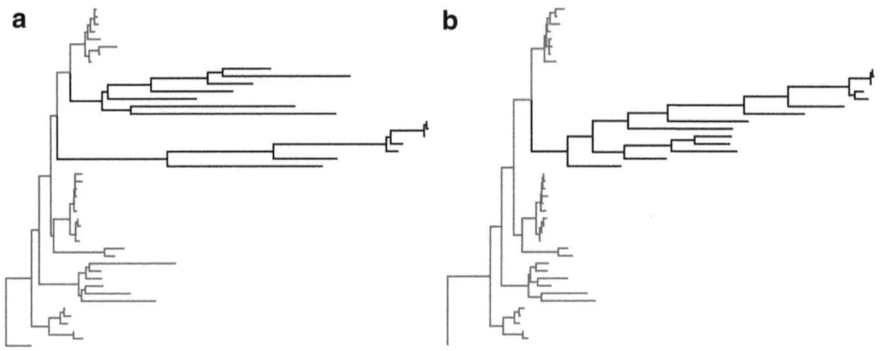

Fig. 5.1 An example of how increased substitution rates in mycoheterotrophic lineages can lead to bias in phylogenetic reconstruction. (**a**) Most optimal maximum likelihood tree of a selection of Dioscoreales species based on nuclear 18S rDNA and mitochondrial *atp1* data. Branches leading to taxa of Thismiaceae are shown in *black* and clearly display elevated substitution rates in comparison with other Dioscoreales lineages. This analysis suggests that Thismiaceae are a paraphyletic group. (**b**) One of the 1,824 most parsimonious trees obtained with maximum parsimony analysis of the same Dioscoreales dataset. In this analysis, Thismiaceae (black lineages) are monophyletic, but this was shown to be the result of a long-branch attraction artifact (Figure adapted from Merckx et al. (2009))

physiological, anatomical, and morphological adaptations but also by high rates of molecular evolution. Chloroplast DNA data is the preferred choice for phylogenetic reconstruction of plants. However, including nonphotosynthetic full mycoheterotrophs in phylogenies based on plastid DNA data poses many difficulties. Some achlorophyllous species still contain amplifiable plastid DNA, but the resulting sequences are often problematic in alignment and analyses due to elevated substitution rates (Chase et al. 1993; Caddick et al. 2002). Furthermore, as we discuss further in this chapter, these high rates of molecular evolution are not restricted to the chloroplast genome but are often present in nuclear and mitochondrial genes as well. The occurrence of substantial heterogeneity in rates of molecular evolution hampers sequence alignment and may also cause bias in phylogenetic reconstruction. Particularly, long phylogenetic branches can positively mislead parsimony analyses, causing the wrong tree to be estimated with increasing confidence as more characters are added. This phenomenon is called long-branch attraction (Felsenstein 1978). Model-based phylogenetic reconstruction methods are less sensitive—but not immune—to long-branch attraction (e.g., Gaut and Lewis 1995; Huelsenbeck 1995). Long-branch attraction has been shown to prevent accurate phylogenetic reconstruction of Dioscoreales using parsimony analyses of nuclear and mitochondrial data due to the excessively increased substitution rates in the mycoheterotrophic lineages (Merckx et al. 2009; Fig. 5.1). To minimize potential effects of long-branch attraction in phylogenetic inference of mycoheterotrophic plant lineages, it is suggested to (1) sequence DNA regions that are less prone to rate heterogeneity (nuclear and mitochondrial DNA regions may be better candidates than chloroplast regions), (2) maximize taxon sampling so that long phylogenetic branches are "broken up," and (3) use different phylogenetic reconstruction methods, for example, results obtained with parsimony methods should be compared to results of model-based methods (maximum likelihood and Bayesian methods). When long-branch attraction is suspected to have influenced the results, its influence can be assessed by repeating the analysis without particular long branches (see Bergsten 2005) or by performing topology tests, such as the Shimodaira–Hasegawa test (Shimodaira and Hasegawa 1999) or the Swofford–Olsen–Waddell–Hillis test (Swofford et al. 1996).

Despite these drawbacks, the phylogenetic position of many mycoheterotrophic groups has been successfully inferred using DNA data, sometimes with surprising results, which in turn urged a reevaluation of morphological characters. For example, the mycoheterotrophic family Triuridaceae was previously considered closely related to the alismatid families (Maas-van de Kamer 1995; Maas-van de Kamer and Weustenfeld 1998), but phylogenetic hypotheses based on molecular data surprisingly placed them in Pandanales (Rudall and Bateman 2006).

5.3 Evolution Toward Mycoheterotrophy

5.3.1 Origins of Mycoheterotrophy

Precise information about the closest relatives of some mycoheterotrophic plant lineages is still lacking. Nevertheless, all available evidence suggests that plants capable of mycoheterotrophy evolved from autotrophic, mycorrhizal ancestors. Indeed, taxonomic and phylogenetic evidence suggests that full mycoheterotrophs living on arbuscular mycorrhizal fungi are most closely related to chlorophyllous arbuscular mycorrhizal plants. For example, mycoheterotrophic species in the genera *Voyria*, *Voyriella*, *Exacum*, and *Exochaenium*, as well as the putative partially mycoheterotrophic species in *Obolaria* and *Bartonia* are all part of Gentianaceae, of which the majority of the species are autotrophic and form arbuscular mycorrhiza (Struwe and Albert 2002; Yuan et al. 2003; Wang and Qiu 2006; Matthews et al. 2009; Cameron and Bolin 2010). Other arbuscular mycorrhizal mycoheterotrophs are imbedded in, or most closely related to, clades of arbuscular mycorrhizal autotrophs as well, for example, Burmanniaceae (Merckx et al. 2006), Thismiaceae (Merckx et al. 2008), *Geosiris* (Goldblatt et al. 2008), and *Petrosavia* (Cameron et al. 2003).

Similar observations have been made for plants that perform mycoheterotrophic interactions through ectomycorrhizal fungi. Phylogenetic evidence shows that an ectomycorrhizal fungal association capable of simultaneously infecting the roots of neighboring plants was probably in place in the common ancestor of the mycoheterotroph liverwort *Aneura mirabilis* and its most closely related species (Kottke and Nebel 2005; Wickett and Goffinet 2008). In Ericaceae, the phylogenetic relationships remain unclear, but it is safe to infer that all monotropoid mycoheterotrophs evolved from photosynthetic plants that formed ectomycorrhizas (Bidartondo 2005). Similarly, in Pyroleae, phylogenetic hypotheses suggest the full mycoheterotroph *Pyrola aphylla* evolved from autotrophic ancestors (Liu et al. 2010).

Also in Orchidaceae, there is ample evidence that fully mycoheterotrophic species are embedded in clades of green orchids (e.g., Molvray et al. 2000). It is well known that orchids form mycorrhizas. However, while an association with saprotrophic fungi is probably the ancestral state for Orchidaceae (Yukawa et al. 2009), many fully mycoheterotrophic species of orchids associate with ectomycorrhizal fungi (Chap. 7). Some evidence suggests that the establishment of a symbiosis with ectomycorrhizal fungi in these lineages served as preadaptation for the evolution of the full mycoheterotrophy (Motomura et al. 2010; see further).

Thus ultimately the evolutionary start point of mycoheterotrophy is mycorrhizal autotrophy, that is, an interaction in which a photosynthetic plant has a mutualistic association with mycorrhizal fungi. An evolutionary breakdown then leads to a shift from this mutualistic plant–fungus association to a situation in which the plant exploits its mycorrhizal fungi.

5.3.2 Evolutionary Stability of the Mycorrhizal Mutualism

In general, the breakdown of mutualisms is expected because they are theoretically unstable (Axelrod and Hamilton 1981; Bull and Rice 1991; Sachs et al. 2004). Given the selfish interest of individuals, why expend resources to benefit another species when resources could be redirected for one's own fitness (Kiers and van der Heijden 2006)? However, mutualisms are ubiquitous in nature, and many mutualisms have an ancient origin, which suggest that they are evolutionary stable. A prime example is the arbuscular mycorrhizal mutualism, which encompasses more than 80% of all land plants and is believed to have played a major role in the invasion of the land by plants (Smith and Read 2008). About 3% of seed plant species have mutualistic interactions with ectomycorrhizal fungi, but while this number is relatively small, the global importance of ectomycorrhizal plants is greatly increased by their disproportionate occupancy of terrestrial habitats. Furthermore, ectomycorrhizal plant–fungus associations have a long history as well, which demonstrates its evolutionary stability (Smith and Read 2008; Hibbett and Matheny 2009).

In one-to-one plant–fungal symbiont interactions, fungal symbionts will increase their own fitness by helping plants grow, and vice versa. However, mycorrhizas are diffuse symbioses because a mycorrhizal plant typically associates simultaneously with multiple fungi and a mycorrhizal fungus often associates simultaneously with multiple plants (Giovannetti et al. 2004; Lian et al. 2006). This can select for parasitism by exploitation of the benefits provided by others while avoiding the costs of supplying resources. Plants may be able to "enforce" cooperation of mycorrhizal fungi, as has been observed in plant–pollinator (Pellmyr and Huth 1994; Goto et al. 2010; Jandér and Herre 2010) and legume–rhizobium interactions (Kiers et al. 2003, 2006; Simms et al. 2006). However, in these systems, sanction mechanisms rely on a single host interacting with multiple partners. In contrast, the mycorrhizal symbiosis involves interactions with multiple fungi and multiple host plants, and it is not clear whether sanctions could operate in the same way (Kiers et al. 2011).

In the arbuscular mycorrhizal mutualism, in vitro experiments have shown that both plants and AM fungi are able to detect variation in the resources supplied by their partners, allowing them to adjust their own resource allocation accordingly (Bever et al. 2009; Kiers et al. 2011). The system therefore seems to represents a "biological market," in which reciprocal rewards stabilize cooperation (Schwartz and Hoeksema 1998; Kiers and van der Heijden 2006; Kiers et al. 2011). In addition, spatial structure of AM fungi within the plant root system may enhance the stability of the AM mutualism, as less beneficial AM fungi proliferate in spatially well-mixed environments (Bever et al. 2009).

5.3.3 Breakdown of the Mycorrhizal Symbiosis

Given the high costs for both partners in the mycorrhizal mutualism, the diffuse character of the partnership and the inability to assess the costs of particular associations at the time of the initial infection, a breakdown of the mycorrhizal symbiosis seems inevitable (Bruns et al. 2002; Kiers and van der Heijden 2006). Theoretically breakdown of mutualisms leads to extinction, abandonment, or exploitation (Sachs and Simms 2006). There are no straightforward examples of extinction in the mycorrhizal symbiosis. Evidence of mycorrhizal mutualists reverting to autonomy appears in the phylogenetic records of plants, while whether or not such reversals occurred in fungi is still debated. Multiple lineages of angiosperm plants, presumed to have arbuscular mycorrhizal ancestors, have switched to a non-mycorrhizal condition (e.g., species of the families Brassicaceae, Caryophyllaceae, Proteaceae, and Cyperaceae; Smith and Read 2008). In fungi, early phylogenetic hypotheses suggested that diverse lineages of free-living (saprotrophic) fungi were nested within ancestrally ectomycorrhizal clades (Hibbett et al. 2000). Hibbett et al. (2000) used parsimony reconstructions of trophic traits and tested internal node reconstructions with a likelihood model for evolution of the mycorrhizal trait. The two models the authors compared were a nonreversible model (only gain of ectomycorrhizal habit, no loss) and a model in which gains and losses were equally probable. The nonreversible model was significantly worse, and the authors estimated that nine losses of the ectomycorrhizal habit have occurred in the lineages sampled. However, the same original dataset was reanalyzed by Bruns and Shefferson (2004), and they found that all except one reversals were converted to gains when assuming that acquiring ectomycorrhizal habit was twice as probable as losing it. Indeed, while still arbitrary, assuming similar probabilities of the different switches may be biologically more justifiable, given that reverting to free-living lifestyle requires the activation of numerous metabolic pathways involved in obtaining carbon and may be, therefore, more difficult to achieve. The remaining one reversal might have been caused by questionable state assignment for the ambiguous taxon involved (*Lentaria byssoides*) and/or may be an artifact of limited taxon sampling and too little genetic variation in the gomphoid lineage the taxon belongs to. Free-living Glomeromycota fungi are not known.

Breakdown of a mutualism into exploitation occurs when individuals obtain commodities offered by mutualists but that provide fewer commodities in return or even, in the case of pure cheaters, no commodities at all (Ferrière et al. 2007). While it remains to be demonstrated that mycoheterotrophic plants provide no commodities to their associated fungi, they seem to qualify as the result of an evolutionary breakdown of the mycorrhizal mutualism. An important aspect of this evolutionary trajectory is the ability of a mycorrhizal fungus to associate simultaneously with multiple host plants. In general, multi-partner interactions appear to help maintain newly evolved exploiters (Sachs and Simms 2006). For example, in the fig–fig wasp symbiosis, all non-pollinating wasps coexist with pollinating species, and coexisting pairs on a host tree species are not sister taxa (Pellmyr and Huth 1994; Pellmyr et al. 1996; Machado et al. 2001). In the mycorrhizal symbiosis, multi-partner interactions are of vital importance for the evolution of mycoheterotrophs. Since a mycorrhizal fungus needs an association with an autotrophic plant to obtain carbohydrates, an exclusive one-to-one interaction between a mycorrhizal fungus and a fully mycoheterotrophic plant cannot exist. An important exception to this multipartite obligation for the maintenance of mycoheterotrophic interactions is mycoheterotrophic orchids that associate with saprotrophic fungi. Although in this case, the persistence of the system relies on ability of the fungus to access dead or decaying organic matter.

5.3.4 Intermediate Evolutionary Steps Between Autotrophy and Full Mycoheterotrophy

Some groups of fully mycoheterotrophic plants consist of evolutionary isolated lineages that diverged from their most closely related extant autotrophic relatives tens of million years ago. In these cases, the large evolutionary gap between autotrophic and fully mycoheterotrophic species prevents us to study the putative transitional steps that occurred in the evolution toward full mycoheterotrophy. Examples are the families Thismiaceae and Triuridaceae that exclusively contain fully mycoheterotrophic species and of which no closely related autotrophic or partially mycoheterotrophic relatives are known. Other fully mycoheterotrophic species, particularly in Orchidaceae, do have closely related green relatives. The fact that some of these green relatives are able to obtain carbon from root-associated fungi as well (partial mycoheterotrophy) provides an interesting evolutionary perspective on the shift to full mycoheterotrophy. In particular, Motomura et al. (2010) reported that full mycoheterotrophic species of *Cymbidium* probably evolved from partially mycoheterotrophic ancestors, which suggests that in this case, full mycoheterotrophy evolved gradually rather than through a direct shift from autotrophy to full mycoheterotrophy. Partial mycoheterotrophs with close relationships to full mycoheterotrophs also exist in other orchid genera, such as *Cephalanthera* (Bidartondo et al. 2004; Julou et al. 2005; Abadie et al. 2006) and *Corallorhiza* (Zimmer et al. 2008; Cameron et al. 2009). In addition, initial mycoheterotrophy is ubiquitous among orchid species (Bernard 1899; Rasmussen 1995), and thus, it is likely that partially mycoheterotrophic orchids originated from initially mycoheterotrophic ancestors. A scenario for the evolution toward full mycoheterotrophy in Orchidaceae therefore seems to include shifts from initial mycoheterotrophy to partial mycoheterotrophy and from partial mycoheterotrophy to full mycoheterotrophy (Fig. 5.2). Whether this is a universal model for the evolution of full mycoheterotrophy in plants remains unclear. Initial and partial mycoheterotrophic ancestors of extant full mycoheterotrophic species in other families may have disappeared over the course of evolution. A similar pattern in the evolution of mycoheterotrophy seems to be present in Pyroleae (Ericaceae). Species of Pyroleae produce dust seeds, and some evidence suggests that germination and early development depends on fungal nutrition (Leake

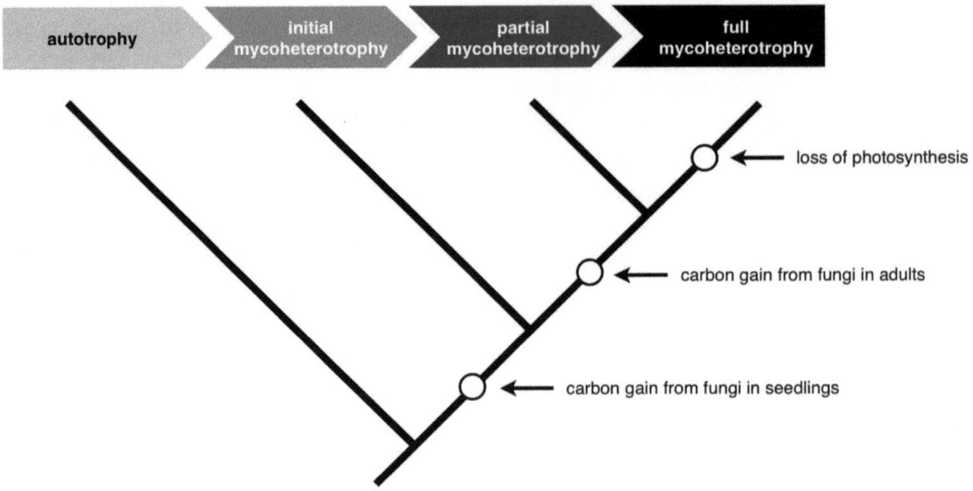

Fig. 5.2 Model for the evolution of mycoheterotrophy in orchids. Key innovations are indicated on the tree

1994; Smith and Read 2008; Eriksson and Kainulainen 2011). Moreover, in *Pyrola,* partial mycoheterotrophy has been detected in the relatives of the full mycoheterotroph *Pyrola aphylla* (Tedersoo et al. 2007; Zimmer et al. 2007; Hynson et al. 2009), which indicates that partial mycoheterotrophy preceded full mycoheterotrophy in the evolution of *Pyrola*. Some evidence suggests that partial mycoheterotrophy occurs in arbuscular mycorrhizal Gentianaceae as well, although not in lineages closely related to full mycoheterotrophs (Cameron and Bolin 2010). In general, autotrophic Gentianaceae are known to be difficult to germinate under laboratory conditions. This has been attributed to specific abiotic requirements (Bouman et al. 2002), but it is also possible that a symbiotic interaction with a mycorrhizal fungus is necessary for seedling development (Ramsbottom 1922). These observations show that Gentianaceae deserve further interest as they may include undiscovered initial and partial mycoheterotrophs.

For some partially mycoheterotrophic orchid species, non-chlorophyllous phenotypes are known. From an evolutionary point of view, these "albino" variants offer a fascinating model for the shift from partial to full mycoheterotrophy. Achlorophyllous specimens have been reported for *Epipactis microphylla* (Selosse et al. 2004), *E. helleborine* (Salmia 1989), *Cephalanthera longifolia* (Abadie et al. 2006), and *C. damasonium* (Julou et al. 2005). Stable isotope analyses indicate that they are fully mycoheterotrophic and thus obtain all of their carbon from associated fungi. They seem to have lost the ability to perform photosynthesis and respire less than chlorophyllous individuals (Julou et al. 2005). In some cases, forms with variegated leaves are known, demonstrating that there is a continuum of leaf chlorophyll concentrations between green and albino individuals. For *Cephalanthera damasonium,* there is a linear relationship between leaf chlorophyll concentrations and the proportional reliance on fungi as a carbon source (Stöckel et al. 2011). While these non-chlorophyllous individuals are able to persist for multiple years, they are extremely rare and produce less seeds than their chlorophyllous phenotypes. Thus, they seem to have a reduced fitness, and consequently, they are sometimes regarded to represent an evolutionary stage of a failed transition from partial mycoheterotrophy to full mycoheterotrophy (Selosse and Roy 2009; see also Chap. 8).

5.3.5 Evolutionary Drivers of Mycoheterotrophy

What drives plants toward mycoheterotrophy? This is one of the most compelling questions about mycoheterotrophy, and while yet unanswered, some data allow us to at least speculate about the evolutionary drivers behind the shift to a mycoheterotrophic mode of life.

In orchids and probably also in Pyroleae, the occurrence of initial mycoheterotrophy seems to be the first step in the evolutionary trajectory toward full mycoheterotrophy. Dependence on fungal carbon for germination and early development reduces the requirement of maternal resources. This means that smaller seeds can be produced at lower cost and thus in larger numbers. Indeed, initially mycoheterotrophic orchids and Pyroleae, as well as most fully mycoheterotrophic plant species in general, produce large amounts of very small, dust-like seeds (Eriksson and Kainulainen 2011). Having many seeds promotes dispersal and increases the likelihood of reaching microsites with suitable host fungi. From an evolutionary point of view, this involves a paradox: seeds are small because they have to be so numerous to successfully reach suitable hosts, yet they can only be so small because they completely depend on their hosts for successful establishment. It is unlikely that extremely small seeds evolved before a dependency on mycorrhizal fungi because that would make an impossible way of living. But on the other hand, evolution of a high dependency on (a narrow range of) host fungi before reduction of seed size and increase of seed number is equally unlikely. Thus, both trends of increased host dependence and reduction of seed size must have evolved gradually, reciprocally reinforcing each other (Eriksson and Kainulainen 2011). Moreover, at least in some cases, increased dependence on host fungi seems to be coupled with an increased specialization toward narrow ranges of fungi which further increases the need for maximal seed dispersal capabilities.

Thus, reduction of seed size and fungal dependency are tightly coupled, and neither trend seems to be the evolutionary driver of the other. Eriksson and Kainulainen (2011) argue that evolution of dust seeds in mycoheterotrophic flowering plant lineages coincided with the development of closed-canopy forests. Initial mycoheterotrophy would thus allow plants to establish in forest understory habitats where there is strong completion for light. Indeed, several authors have argued that mycoheterotrophy evolved as an adaption to living in deep shade (e.g., Leake 1994; Bidartondo 2005). Although initial mycoheterotrophy may also help plants to establish in nitrogen-limited habitats (Read and Perrez-Moreno 2003) or serve as an escape from water stress. Initial mycoheterotrophic Lycopodiaceae gametophytes, for example, sexually reproduce undercover in moist soil and litter environments that provide protection from desiccation (Leake et al. 2008). Thus, several factors may contribute to the evolution of initial mycoheterotrophy in plants.

As mentioned above, partial and full mycoheterotrophy is often seen as an adaptation to low-light environments. This seems evident since nearly all fully mycoheterotrophic plants grow in the dark understory of primary forests, while partially mycoheterotrophic plants are often found in more open forests, where irradiance is higher (Selosse and Roy 2009). Indeed, studies using stable isotope natural abundances suggest that light availability plays a major role in the degree of mycoheterotrophy at least in a few partially mycoheterotrophic orchids and Pyroleae (Zimmer et al. 2007; Liebel et al. 2010; Preiss et al. 2010; Hynson et al. 2012; see also Chap. 8). However, differences in ^{13}C abundances of partially mycoheterotrophic populations growing under different environmental conditions cannot always be explained by differences in irradiance only (e.g., Zimmer et al. 2007; Hynson et al. 2012). Thus, in these cases, light availability is not the only driver of the heterotrophy level. It has been suggested that C gain in partially mycoheterotrophic plants may in some cases arise as a "side-product" of N and P gain from mycorrhizal fungi (Selosse et al. 2009; Chap. 8). Thus, the evolution toward increasing levels of mycoheterotrophy may be driven by multiple factors. However, since fully mycoheterotrophic plants successfully abandoned photosynthesis *and* mostly occur in habitats where light availability is extremely low, it can be hypothesized that light availability is one of the most important drivers behind the evolution of mycoheterotrophy.

5.4 Phylogenetic Aspects

5.4.1 Number of Origins

Our increasing knowledge about the phylogenetic relationships of fully mycoheterotrophic plants allows us to estimate the number of origins of a fully mycoheterotrophic mode of life in land plants. Hereby it is essential to assume that reversals to partial mycoheterotrophy or autotrophy are impossible. There is no direct evidence that supports this assumption. However, the relaxed evolutionary constraints on photosynthesis genes of fully mycoheterotrophic plants leads to a profound increase of substitution rates, frameshift mutations, and even the loss of complete genes (Barrett and Freudenstein 2008; Wickett et al. 2008a, 2008b; Delannoy et al. 2011). Therefore, it seems highly unlikely that the reduced plastid genomes of fully mycoheterotrophic plants are able to regain their essential role in functional photosynthesis.

By adding up estimates of the number of independent origins of full mycoheterotrophy in distinct clades of plants, we find evidence of at least 47 origins of full mycoheterotrophy in land plants (or 46 if *Parasitaxus* is not classified as a mycoheterotroph) (Table 5.1). This number is probably an underestimation due to the uncertain phylogenetic position of several mycoheterotrophic species and genera in Burmanniaceae, Orchidoideae, and Epidendroideae. Thus, full mycoheterotrophy has likely more than four times the number of evolutionary origins compared with haustoria-forming holoparasitic plants. The number of evolutionary origins of mycoheterotrophy *sensu lato*—that is, the ability to obtain carbon from mycorrhizal fungi—is probably lower. There may be only a single origin of initial mycoheterotrophy in Orchidaceae and Ericaceae, and this is perhaps also the case for Burmanniaceae and Gentianaceae. On the other hand, initial and partial mycoheterotrophy may remain to be detected in plant groups that do not include fully mycoheterotrophic species. Undiscovered initial mycoheterotrophy may be present in diverse plant families with species that produce dust seeds, such as Rubiaceae, Buddlejaceae, and Gesneriaceae (Eriksson and Kainulainen 2011), and partial mycoheterotrophy may occur in plant families that are adapted to grow in the low-light conditions of forest understories (Selosse and Roy 2009). Also, mycoheterotrophy evolved independently in lycophytes and ferns, of which several lineages have putative fully mycoheterotrophic gametophytes.

5.4.2 Diversity

Mycoheterotrophy occurs in almost every major lineage of land plants (Fig. 5.3). Mycoheterotrophic interactions are unknown in the hornworts and in the mosses. The latter may be explained by the putative nonmycorrhizal status of mosses (Smith and Read 2008). Here we provide a short overview of mycoheterotrophic interactions in major lineages of land plants. See Chap. 2 for a detailed overview.

5.4.2.1 Liverworts

In liverworts, which comprises ca. 5,000 species, at least one species (*Aneura mirabilis*) is fully mycoheterotrophic (Bidartondo et al. 2003). Another closely related species awaits investigation (Crum and Bruce 1996). *A. mirabilis* is specialized on *Tulasnella* species that form ectomycorrhizas with surrounding trees (Bidartondo et al. 2003). Other Aneuraceae species associate nearly exclusively with *Tulasnella* species as well (Bidartondo and Duckett 2010). This interaction is unique because Aneuraceae are the only thalloid group with basidiomycetous mycobionts and the only liverwort group known to have *Tulasnella* symbionts (Preussing et al. 2010).

5.4.2.2 Lycophytes

Lycophytes comprise ca. 1,200 species in three families: Lycopodiaceae, Isoëtaceae, and Selaginellaceae (Christenhusz et al. 2011). Several of the ca. 300 species of Lycopodiaceae have achlorophyllous gametophytes that are presumably mycoheterotrophic on arbuscular mycorrhizal fungi (Winther and Friedman 2008; Chap. 2).

5.4.2.3 Ferns

The gametophytes of several fern species in the families Ophioglossaceae, Psilotaceae,

Table 5.1 Overview of independent occurrence of full mycoheterotrophy in land plants

Clade	Number of origins of full mycoheterotrophy	Associated fungi (see Chap. 6)	References
Aneuraceae	1[a]	Basidiomycota (*Tulasnella*)	Wickett and Goffinet (2008)
Podocarpaceae	1[b]	Unknown	Sinclair et al. (2002); Feild and Bodribb (2005)
Petrosaviaceae	1	Glomeromycota	Cameron et al. (2003)
Triuridaceae	1	Glomeromycota	Rudall and Bateman (2006)
Burmanniaceae	≥8[c]	Glomeromycota	Merckx et al. (2008)
Thismiaceae	1–2[d]	Glomeromycota	Merckx and Bidartondo (2008)
Corsiaceae	1[e]	Glomeromycota	Neyland and Hennigan (2003); Davis et al. (2004)
Iridaceae	1[f]	Glomeromycota	Goldblatt et al. (2008)
Orchidaceae—Vanilloideae	2–3[g]	Ascomycota and Basidiomycota	Cameron (2009)
Orchidaceae—Orchidoideae	≥9[h]	Basidiomycota	See Chap. 2
Orchidaceae—Epidendroideae	≥14[i]	Ascomycota and Basidiomycota	Freudenstein (1994); Chase et al. (2003); Freudenstein et al. (2004); Rothacker (2007), Roy et al. (2009); Pedersen et al. (2009)
Polygalaceae	1[j]	Glomeromycota	Eriksen (1993)
Ericaceae	2–3[k]	Ascomycota and Basidiomycota	Kron (1996); Bidartondo and Bruns (2002)
Gentianaceae	4[l]	Glomeromycota	Albert and Struwe (1997); Struwe and Albert (2002); Yuan et al. (2003); Kissling et al. (2009)

[a]*Aneura mirabilis* is the only fully mycoheterotrophic species in this family

[b]*Parasitaxus usta* is the only fully heterotrophic species in this family, but it is debatable whether it is a true mycoheterotroph

[c]The high number of independent origins of full mycoheterotrophy in this family is due to the polyphyletic status of chlorophyllous *Burmannia* species

[d]It is unclear whether Thismiaceae are monophyletic. A paraphyletic status for the family would mean two separate origins of full mycoheterotrophy

[e]Corsiaceae may not be monophyletic as suggested by Neyland and Hennigan (2003). In that case, more than a single origin of full mycoheterotrophy might be needed to explain the evolution of full mycoheterotrophy

[f]*Geosiris*, which includes two closely related species, is the only fully mycoheterotrophic genus in Iridaceae

[g]Phylogenetic evidence indicates that full mycoheterotrophy evolved independently in *Lecanorchis* and the common ancestor of the *Erythrorchis–Pseudovanilla–Galeola–Cyrtosia* clade. If *Pseudovanilla* species are not considered full mycoheterotrophs, three origins are needed to explain the evolution of full mycoheterotrophy in Vanilloideae

[h]Full mycoheterotrophy probably evolved separately in the genera *Arthrochilus*, *Brachycorythis*, *Corybas*, *Cryptostylis*, *Cystorchis*, *Odontochilus*, *Platanthera*, and *Platythelys* because these genera also contain chlorophyllous species. Due to the uncertain phylogenetic position of *Burnettia*, *Chamaegastrodia*, *Danhatchia*, *Degranvillea*, and *Rhizanthella*, it remains unknown how many independent shifts they represent

[i]Due to taxonomy and phylogenetic placement probably independent origins of full mycoheterotrophy in *Neottia*, *Cephalanthera*, *Cremastra*, *Cymbidium*, *Corallorhiza*, *Stereosandra*, *Tropidia*, *Epipogium*, *Malaxis*, *Hexalectris*, *Eulophia*, *Yoania*, the common ancestor of *Aphyllorchis* and *Limodorum*, and Gastrodieae (including *Didymoplexiella*, *Didymoplexis*, *Gastrodia*, *Uleiorchis*, *Auxopus*, *Neoclementia*). Due to the uncertain phylogenetic position of *Kalimantanorchis*, *Pogoniopsis*, *Risleya*, *Silvorchis*, and *Stereosandra*, it remains unclear how many independent shifts they represent

[j]Probably a single origin of full mycoheterotrophy in the genus *Epirixanthes*

[k]Independent origins in the tribes Pyroleae, Monotropeae, and Pterosporeae, although a single origin in the ancestor of the two latter tribes is also possible

[l]Independent origins for *Voyria*, *Voyriella*, and mycoheterotrophic *Exochaenium* and *Exacum* species

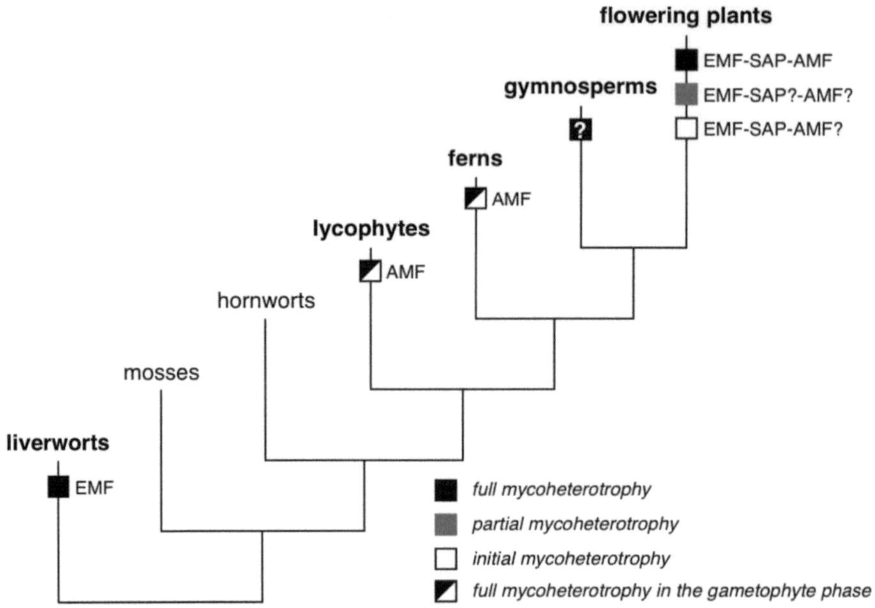

Fig. 5.3 Mycoheterotrophic interactions in land plants. Phylogenetic relationships of land plants based on Palmer et al. (2004) and Forrest et al. (2006). Occurrence of different types of mycoheterotrophy is plotted on the tree with symbols shown in the legend. The identity of fungi involved is indicated. *AMF* arbuscular mycorrhizal fungi; *EMF* ectomycorrhizal fungi; *SAP* saprotrophic fungi. *Question marks* indicate putative interactions (Based on data obtained from Chap. 2)

Gleicheniaceae, and Schizeaceae are achlorophyllous and presumably mycoheterotrophic on arbuscular mycorrhizal fungi (Chap. 2). Further study on mycoheterotrophy of achlorophyllous gametophytes of ferns is needed.

5.4.2.4 Gymnosperms

The species *Parasitaxus usta* (Podocarpaceae) sprouts from roots and trunks of another podocarp, *Falcatifolium taxoides*. Fungi are thought to play an important role in this unique interaction, and therefore, the plant may be a mycoheterotroph rather than a holoparasite. However, the identity and the exact role of the fungi involved in this interaction remain undetermined.

5.4.2.5 Angiosperms

There are 10 angiosperm families with fully mycoheterotrophic species, and full mycoheterotrophy probably evolved at least 45 times independently within flowering plants (Table 5.1). In terms of number of species, most fully mycoheterotrophic angiosperms are monocots (ca. 468 species), while full mycoheterotrophy occurs in only 47 species of eudicots (Fig. 5.4). In contrast, haustorial parasitism has 11 independent origins in flowering plants and is limited to ca. 390 species of eudicots (Barkman et al. 2007; Heide-Jørgensen 2008).

In ca. 236 species, full mycoheterotrophy is supported by exploitation of arbuscular mycorrhizal fungi (Glomeromycota). All full mycoheterotrophic species in Ericaceae, and most in Orchidaceae, exploit ectomycorrhizal fungi. Some tropical fully mycoheterotrophic orchids rely on an association with saprotrophic fungi. However, for many species of full mycoheterotrophs, in particular, tropical species of Orchidaceae, the identity of their fungal associates remains to be investigated. Partial mycoheterotrophy has been detected in Orchidaceae and Ericaceae. In most cases, it seems to occur through ectomycorrhizal fungi, although some orchids may receive limited amounts of carbon through saprotrophic fungi (see Chap. 8). Putative partial mycoheterotrophy through arbuscular mycorrhizal fungi has been reported in Gentianaceae (Cameron and Bolin 2010). All three families also contain fully mycoheterotrophic species. Many putatively partially mycoheterotrophic taxa, also in

Fig. 5.4 Species diversity of full mycoheterotrophy in angiosperms. Phylogenetic relationships are based on APG (2009). Orders with fully mycoheterotrophic species are highlighted in *bold*. The families to which these species belong are written after the dash. The *gray bars* represent the number of fully mycoheterotrophic species in each order. Species numbers were derived from Chap. 2

other families, remain to be investigated. Chlorophyllous species that are closely related to fully mycoheterotrophic species are prime candidates for undiscovered partial mycoheterotrophy. Similarly, initial mycoheterotrophy has only been recorded in Orchidaceae, mostly with saprotrophic fungi and more rarely with ectomycorrhizal fungi. However, initial mycoheterotrophy may be present in other lineages that produce dust seeds (Eriksson and Kainulainen 2011).

5.4.3 Timing

Fossil evidence to trace the evolutionary history of mycoheterotrophy is lacking. There is only one series of fossils that might be assigned to an extant mycoheterotrophic lineage. These fossils are from the Upper Cretaceous (about 90 Ma) and show affinities with extant Triuridaceae (Gandolfo et al. 2002). However, it remains questionable whether these fossilized plants were in fact mycoheterotrophic (Gandolfo et al. 2002) and members of the Triuridaceae (Furness et al. 2002). Nevertheless, based on the taxonomic and phylogenetic data that is available, it is obvious that the fully mycoheterotrophic mode of life has evolved during various epochs of plant diversification. A relatively recent origin of full mycoheterotrophy can be inferred for species that are imbedded in genera that also contain autotrophic species. Examples include *Exacum* and *Exochaenium* in Gentianaceae (Yuan et al. 2003; Kissling et al. 2009) and *Cymbidium* in Orchidaceae (Motomura et al. 2010). For widespread, species-rich clades that contain only fully mycoheterotrophic species, it is likely that full mycoheterotrophy evolved in the common ancestor of these species and thus represents a much older origin of full mycoheterotrophy (e.g., Triuridaceae, Thismiaceae).

Due to the difficulties associated with molecular dating methods, in particular, the use of fossil calibration points, the absence of fossil data for mycoheterotrophic groups, the paucity of data, and the increased substitution rates of mycoheterotrophic clades, only very few mycoheterotrophic lineages have been dated using these methods. In Dioscoreales, fully mycoheterotrophic lineages are estimated to have diverged from as early as the Cretaceous to the Miocene (Merckx et al. 2010). The fully mycoheterotrophic genus *Wullschlaegelia* (Orchidaceae) has been estimated to have an Eocene (Ramirez et al. 2007) or Oligocene (Gustafsson et al. 2010) origin. Based on the results of Yuan et al. (2003), mycoheterotrophic *Exacum* species (Gentianaceae) diverged not before the Miocene. However, a divergence time estimate of the origin of a fully mycoheterotrophic clade ("stem node" age) is not an accurate estimate for the origin of mycoheterotrophy, since it has to be assumed that mycoheterotrophy could evolve between the origin of the mycoheterotrophic clade and the beginning of the diversification of the clade ("crown node") (Fig. 5.5a). This assumption considerably broadens age estimates for the evolution of mycoheterotrophy. But even when crown node ages are considered as upper boundaries for the evolution of mycoheterotrophy, some shifts from autotrophy to mycoheterotrophy must have occurred several tens of million years ago. The diversification of Thismiaceae, for example, probably started during the Eocene; therefore, a shift to mycoheterotrophy in the ancestor of this clade has occurred at least around that time.

In some clades, full mycoheterotrophs may be preceded by partial mycoheterotrophs or initial mycoheterotrophs. In these cases, the first occurrence of mycoheterotrophy predates the origin of fully mycoheterotrophic species (Fig. 5.5b). For example, under the assumption that most orchids are initial mycoheterotrophs, the origin of mycoheterotrophy in Orchidaceae may be traced back to the common ancestor of all Orchidaceae species. As a result, the evolution of mycoheterotrophy in Orchidaceae may have started during the Cretaceous (Ramirez et al. 2007). Despite the difficulties to accurately estimate the timing of the evolution of mycoheterotrophy, the examples above indicate that for at least some groups of mycoheterotrophs, the origin of mycoheterotrophy is relatively ancient. This demonstrates that mycoheterotrophic lineages were able to emerge, persist, and diverge over a considerable amount of time and thus are remarkably evolutionarily stable.

5.4.4 Diversification

Clades of fully mycoheterotrophic plants generally contain only few species. In some cases, this may reflect their relatively recent origin. For example, the mycoheterotrophic species in the

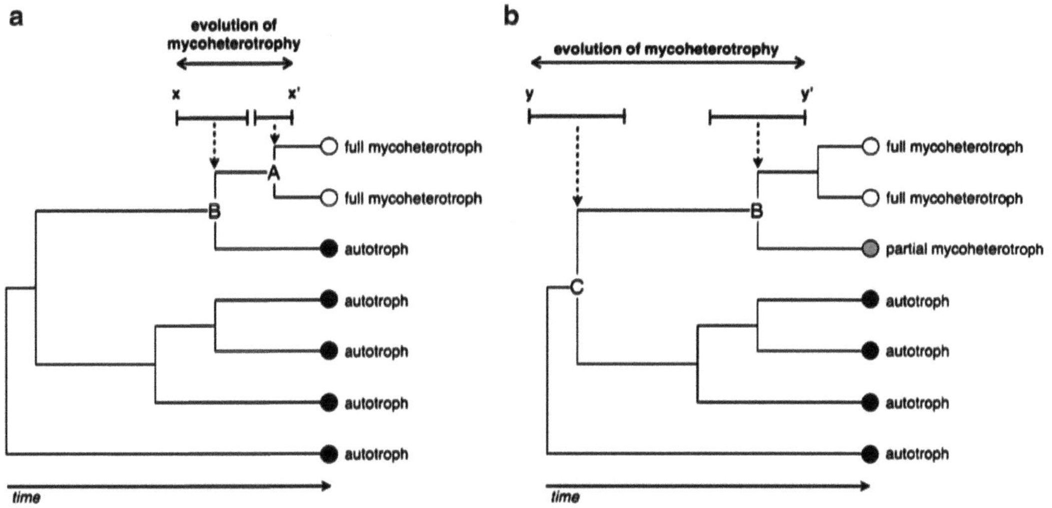

Fig. 5.5 Estimating the age of the origin of mycoheterotrophy using molecular clock methods. (**a**) When a monophyletic clade of fully mycoheterotrophic species is assessed, it can be assumed that full mycoheterotrophy has evolved somewhere between the node where this mycoheterotrophic clade branched from its autotrophic ancestors ("B") and the node where diversification started ("A"). Age estimates of these nodes will encompass 95% confidence intervals. When these are taken into account, evolution of mycoheterotrophy occurred between time x and x'. (**b**) Similarly in this example, with a clade of partial and full mycoheterotrophs, it can be derived that mycoheterotrophy evolved between time y and y'. Accurate taxon sampling is essential in this context. Failure to include extant lineages (particularly earliest diverging mycoheterotrophic species and most closely related autotrophs) will influence the upper and lower boundaries of the age estimates

genera *Exacum* and *Exochaenium* (Gentianaceae) and several Orchidaceae genera may have diverged in relatively recent times (see above). However, some early diverging lineages of full mycoheterotrophic plants (e.g., *Geosiris*, *Petrosavia*) are species poor as well. Relatively speciose lineages include Thismiaceae (ca. 65 species) and Gastrodieae (ca. 47 species), but even in these clades, species numbers are not particularly high. This suggests that diversification rates in lineages of fully mycoheterotrophic plants are low, extinction rates are high, or a combination of both. Diversification through time has only been studied in detail for Burmanniaceae, for which it has been suggested that the diversification and radiation of the pantropical genera *Burmannia* and *Gymnosiphon* significantly increased during the Eocene (Merckx et al. 2008; Fig. 5.6). The beginning of the Eocene was characterized by high global temperatures, which allowed tropical rain forests to expand and reach high levels of diversity (Jaramillo 2002; Jaramillo et al. 2006). Dense closed-canopy rainforest is the prime habitat for mycoheterotrophic species of Burmanniaceae. The increase in rainforest area may have resulted in an increase in mycoheterotrophic plant diversity in tropical regions during the Eocene. Larger regions can support more species, which enhance both regional and local diversity by reducing the risk of extinction and increasing niche opportunities (Leigh et al. 2004; Linder 2008). Subsequent cooling during the end of the Eocene caused a global retraction of rainforest and a decrease in plant diversity (Jaramillo et al. 2006), perhaps leading to a decrease in mycoheterotrophic plant diversity as well. It remains to be investigated whether this also influenced the diversity of other tropical mycoheterotrophic groups such as Triuridaceae and Thismiaceae, which were likely to be present in rainforest during that period as well (Gandolfo et al. 2002; Merckx et al. 2010). In Orchidaceae, major diversification is also estimated to have occurred during the early Eocene (Ramirez et al. 2007), although dates suggested by Gustafsson et al. (2010) place orchid diversification during the cooler period at the end of the Eocene and into the Oligocene.

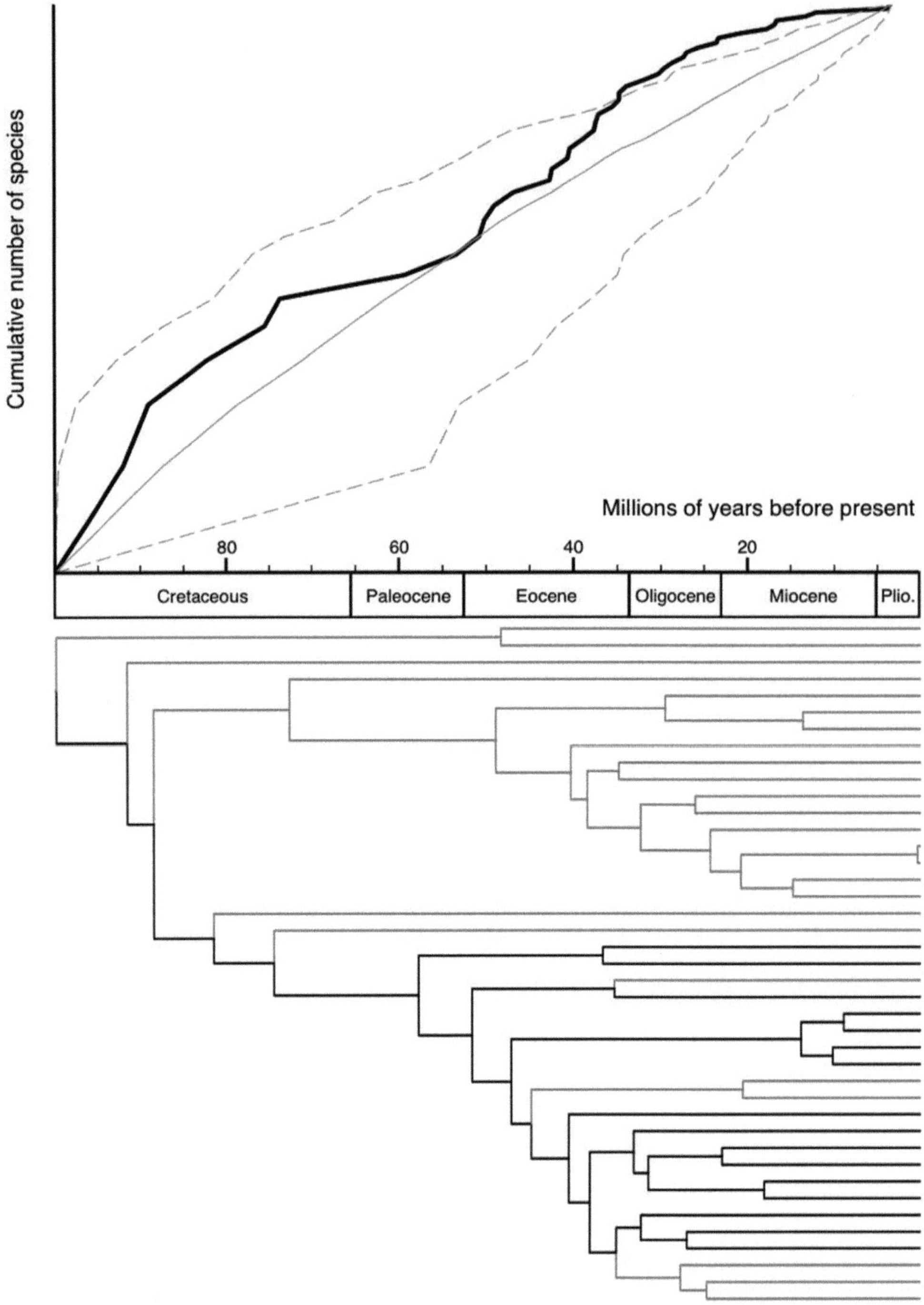

Fig. 5.6 An example of species diversification through time estimated from a molecular phylogeny. Semilogarithmic lineage-through-time (LTT) plot of Burmanniaceae (*black*) calculated from the chronogram shown below. A simulated LTT plot with 95% confidence intervals under a constant death–birth rate of 0.5 is shown in gray for comparison. *Gray* branches in the chronogram represent fully mycoheterotrophic species (Adapted from Merckx et al. (2008))

The species diversity of mycoheterotrophic lineages is not solely to be explained by ancient fluctuations in diversification rates. In some lineages, the occurrence of closely related species over a more or less continuous and restricted geographic range suggests that the diversity in these lineages is the result of a recent rapid diversification. The species in the *Hexalectris spicata* species complex, for example, are likely to have evolved only recently (Kennedy and Watson 2010; Kennedy et al. 2011). Other candidates for recent rapid diversification include the *Corsia* species in New Guinea, *Didymoplexiella* species on Borneo, and *Lecanorchis* species in Japan.

The fragmented but often widespread distribution patterns of many mycoheterotrophic lineages (Chap. 3) suggest that extinction may have played a significant role in the current distribution and diversity of mycoheterotrophic plants. The comparative low diversity of mycoheterotrophic plants in African rain forests, for example, is likely the result of the degrading effect of significant climatic changes during the Cenozoic, and more recently to Pleistocene climatic fluctuations, resulting in a drier climate throughout most of equatorial Africa (Chap. 3). However, the fossil record of mycoheterotrophic plant is virtually nonexistent and thus prevents the detection of extinction events in these lineages. Also, episodes of rapid extinction do not leave a clear signal in phylogenies based on extant species which makes it hard to estimate extinction rates based on a phylogenetic approach (Harvey et al. 1994; Purvis 2008; Tarver and Donoghue 2011).

5.5 Common Evolutionary Trends

5.5.1 Morphology

The shift from autotrophy to full mycoheterotrophy is accompanied by profound morphological changes, and despite the very diverse range of families and genera of mycoheterotrophic plants, these show some quite remarkable parallel evolutionary forms.

5.5.1.1 Subterranean Morphology

The subterranean parts of mycoheterotrophic plants show considerable trends of convergent evolution, consistent with a change in function from organs of absorption to organs of storage: (1) root hairs are mostly absent, (2) there is trend toward stout, clumped roots and rhizomes mostly with a specialized fungal colonization pattern, and (3) an increased width of the root cortex often accommodates mycorrhizal infection and stores of carbohydrates and other materials obtained from the fungal host (Leake 1994). See Chap. 4 for a detailed overview of root morphology of mycoheterotrophic plants.

5.5.1.2 Shoots

Many fully mycoheterotrophic species have exceptionally slender and threadlike stems, resulting in a hyaline appearance. Vascular tissues are often reduced to a single narrow cylinder of bicollateral bundles or, minimally, to four or six narrow bundles in the cortex. Most species lack secondary thickening, and their stems are succulent and brittle (e.g., *Rhizanthella*) or hyaline and slender (e.g., *Gymnosiphon*). Lignification is generally confined to a narrow ring of xylem vessels, or rarely only a few scalariform xylem vessels are present (e.g., *Voyria tenella*). Phloem is present in very small amounts and then mainly as parenchyma with sieve tubes frequently recorded as narrow and possibly with adherent sieve plates (Leake 1994).

5.5.1.3 Leaves

In fully mycoheterotrophic plants, leaves no longer serve a useful function. As a result, leaves are typically reduced to widely spaced achlorophyllous scales on the inflorescence axis. Occasionally, they are present only on underground rhizomes or tubers or even totally absent. The vascular supply to the leaf scales is often reduced to a single trace or may be absent. Stomata are generally absent as well, although some species retain vestigial stomata on their leaves and shoots (Leake 1994).

5.5.1.4 Seeds

An extreme reduction of seed size and seed complexity is one of the most common and significant modifications seen in fully mycoheterotrophic plants. Most species of mycoheterotrophic plants have extremely small seeds, which are commonly described with the term "dust seeds." For example, some of the smallest seeds known are found in the genus *Voyria* in the Gentianaceae (Maas and Ruyters 1986) although the seeds of some orchids are equally reduced in size (Arditti and Ghani 2000). This reduction in seed size is coupled with a reduction of endosperm and a lack of differentiation of the embryo at maturity. Some parasitic plant families are also characterized by the occurrence of very small seeds (e.g., Orobanchaceae, Rafflesiaceae, Balanophoraceae, Hydnoraceae). And dust seeds are also present in a few angiosperm lineages that have not been associated with heterotrophic interactions, such as Rubiaceae, Buddlejaceae, and Gesneriaceae (Eriksson and Kainulainen 2011). The small dust seeds of mycoheterotrophic plants are often produced in very large numbers. A single capsule of the mycoheterotrophic orchid *Galeola altissima* contains about 18,000 seeds (Arditti and Ghani 2000). And *Pterospora andromedea* (Ericaceae) produces 2,000–4,800 seeds per capsule (Bakshi 1959). However, not all full mycoheterotrophic plants produce large numbers of seeds; individuals of *Rhizanthella gardneri* (Orchidaceae), for example, only produce 20–50 seeds (George and Cooke 1981).

5.5.2 Biomass Reduction

In general, full mycoheterotrophic species are considerably smaller than their most closely related autotrophic/partially mycoheterotrophic species. In many cases, this reduction in biomass cannot solely be explained by the absence of leaves. Perhaps biomass production of fully mycoheterotrophic plants is limited by the amount of carbon they can obtain from their associated fungal networks. Or it may be the result of positive selection toward smaller plants because these can minimize the amount of carbon needed from the fungus, which makes it easier to subvert the mycorrhizal mutualism while remaining undetected.

5.5.3 Habitats

Fully mycoheterotrophic plants generally occur in shaded habitats, like the ground layer of closed-canopy forest (Chap. 3). These habitats are often characterized by a lack of understory plants, which is attributed to the absence of sufficient light for photosynthetic plants to survive. Partially mycoheterotrophic orchids, Ericaceae, and Gentianaceae often grow in forest habitats as well, although partially mycoheterotrophic orchids and gentians (*Bartonia*) also can be found in open vegetations, such as bogs and meadows (Matthews et al. 2009; Girlanda et al. 2011). Interestingly, Preiss et al. (2010) demonstrated that light availability is a major determinant of the degree of mycoheterotrophy in two partially mycoheterotrophic species of *Cephalanthera* (Orchidaceae). These observations provide support for a strong correlation between irradiance levels and dependence on fungal carbon. Therefore, an evolutionary switch from autotrophy to full mycoheterotrophy seems to be accompanied by a shift toward more shaded habitats.

5.5.4 Reduction of the Chloroplast Genome

In holoparasitic plants, the loss of photosynthesis has led to a loss of various chloroplast genes (Bungard 2004; Barbrook et al. 2006). Similar trends are observed in fully mycoheterotrophic plants, although only few taxa have been studied in detail. The RuBisCO large-subunit gene (*rbcL*) was amplified in a few mycoheterotrophic Dioscoreales and showed significantly increased substitution rates (Caddick et al. 2002). Davis et al. (2004) noticed that *rbcL* could not be amplified from species of *Arachnitis* (Corsiaceae), *Thismia* (Thismiaceae), *Lacandonia*, *Sciaphila*, and *Triuris* (Triuridaceae). Cameron (2004) reported putative pseudogenes of *rbcL* in *Cyrtosia* but not for *Erythrorchis* and *Pseudovanilla* (Orchidaceae). The evolution of *rbcL* was studied in detail in *Corallorhiza* orchids by Barrett and Freudenstein (2008). Pseudogenes were detected in some lineages of *Corallorhiza* but not in others.

This is probably indicative of the early stages of pseudogene formation and the result of a recent switch from autotrophy to mycoheterotrophy. The maturaseK gene (*matK*) is probably a pseudogene in species of *Corallorhiza* as well. But because large 5′ deletions and frameshift indels occur in closely related leaf-bearing species of *Aplectrum* and *Oreorchis*, major changes in the functionality of *matK* apparently precede the transition to full mycoheterotrophy (Freudenstein and Senyo 2008).

To date, complete chloroplast genomes have been sequenced of only three species of fully mycoheterotrophic plants: *Aneura mirabilis* (Aneuraceae), *Rhizanthella gardneri*, and *Neottia nidus-avis* (Orchidaceae) (Wickett et al. 2008a; Delannoy et al. 2011; Logacheva et al. 2011). In *A. mirabilis*, all *ndh* genes are either absent or pseudogenes. Five of 15 *psb* genes are pseudogenes, as are 2 of 6 *psa* genes and 2 of 6 *pet* genes. In addition, pseudogenes of *cysA*, *cysT*, *ccsA*, and *ycf 3* were also detected. The remaining genes retained intact open reading frames. Chlororespiratory genes are the most affected of any functional category in *A. mirabilis*, with the partial or complete loss of all genes (Wickett et al. 2008a). All gene losses and pseudogenes are limited to *Aneura mirabilis* and were not found in closely related photosynthetic species of *Aneura* suggesting that they are likely correlated with the loss of photosynthesis in this liverwort (Wickett et al. 2008b). With 108,007 base pairs and a structure that is remarkably collinear with its distant relative, *Marchantia polymorpha*, the plastid genome of *A. mirabilis*, probably represents a genome in the early stages of decay following the relaxation of selection pressures. In contrast, the plastid genome of *Rhizanthella gardneri* consists of only 59,190 base pairs and contains only 37 genes. It is the least gene-rich plastid genome known apart from the fragmented plastid genome of some dinoflagellates. In comparison with the plastid genome of *Phalaenopsis aphrodite* (Orchidaceae), an estimated 70% of the original genes are lost or transferred to the nucleus. These missing genes include those coding for the plastid-encoded RNA polymerase (PEP), the maturase-like protein MatK, all of the genes required for photosynthesis (encoding subunits of photosystem I, photosystem II, cytochrome b_6f complex, and ATP synthase), as well as 6 genes encoding ribosomal proteins and 27 genes encoding tRNAs. Despite rampant gene loss, the plastid genome of *R. gardneri* retains a minimal set of protein-encoding and tRNA genes and appears to be the basis of a functioning gene expression system, with transcription, splicing, and RNA editing all detected and translation likely (Delannoy et al. 2011). The plastid genome of *Neottia nidus-avis* was also found to be reduced in both genome size and gene content. However, these reductions are not as drastic as in *Rhizanthella*: the plastome of *Neottia nidus-avis* lacks all genes encoding photosynthetic proteins and RNA polymerase subunits, but retains most genes of the translational apparatus (Logacheva et al. 2011).

5.5.5 High Substitution Rates in Nuclear and Mitochondrial Genomes

High rates of molecular evolution have also been observed in nuclear and mitochondrial genomes of mycoheterotrophic plants. Merckx et al. (2006, 2009) reported that branch lengths in 18S rDNA gene trees of mycoheterotrophic Dioscoreales (particularly in Thismiaceae) are up to 6.5-fold longer than those of related autotrophic species (see also Yokoyama et al. 2008). Lemaire et al. (2011) measured 18S rDNA substitution rates of heterotrophic species across the phylogeny of the angiosperms and observed that branch lengths of many included mycoheterotrophic species are significantly longer than those of related autotrophic lineages. Substitution rates of 18S rDNA are particularly high in mycoheterotrophic species of Thismiaceae, Corsiaceae, Orchidaceae (*Rhizanthella*), and Triuridaceae (*Kupea*). However, in many other lineages of mycoheterotrophic plants, no significantly faster substitution rates of 18S rDNA were detected. In lineages where substitution rates of 18S rDNA are high, only few mutations occur in major functional and structural regions of the small ribosomal molecule, suggesting that the efficiency of the translational apparatus in nonphotosynthetic plants has not been affected (Lemaire et al. 2011). Although data is more limited, high substitution rates have been demonstrated to occur in some mitochondrial genes of fully mycoheterotrophic species as well (Merckx et al. 2006, 2009).

Interestingly high substitution rates in nuclear and mitochondrial genomes have also been observed in holoparasitic plants and have, for example, frustrated attempts to infer the phylogenetic position of the world's largest flower, *Rafflesia* (Barkman et al. 2004). The underlying causes of accelerated substitution rates in heterotrophic plants are unknown, but numerous hypotheses have been proposed. The long-term effects of a small effective population size resulting in a genetic bottleneck effect (Wu and Li 1985), the influence of a short generation time and the correlated higher number of mutation-generating reproductive events (Wu and Li 1985), an increased tolerance of mutations due to a relaxation of selective constraints, variations in mutation rate (Sniegowski et al. 2000), DNA repair efficiency (Modrich and Lahue 1996), and speciation rates (Barraclough and Savolainen 2001) are possible factors that trigger high substitution rates in mycoheterotrophic and parasitic plants. However, none of these hypotheses can unequivocally explain higher substitution rates in parasitic plants (Nickrent and Starr 1994; dePamphilis et al. 1997; Young and dePamphilis 2005) and mycoheterotrophic plants (Lemaire et al. 2011).

5.5.6 Mycorrhizal Specificity

Both the arbuscular and the ectomycorrhizal symbiosis are generally characterized by low specificity between plants and fungi. In contrast to autotrophic mycorrhizal plants, however, mycoheterotrophic plants often show high specificity toward fungi even though the fungi remain generalists (e.g., Taylor and Bruns 1997; Bidartondo et al. 2002, 2003; Chap. 6). In the Monotropoideae (Ericaceae), five related mycoheterotrophic plant lineages are each specialized on one of five distantly related basidiomycete fungal lineages (Bidartondo 2005). Species of mycoheterotrophic orchids often show specificity toward various lineages of either ectomycorrhizal or saprotrophic fungi (e.g., Taylor and Bruns 1997; Ogura-Tsujita and Yukawa 2008). Similarly, specificity toward narrow lineages of fungi is also observed in mycoheterotrophs that are living on arbuscular mycorrhizal fungi, such as *Arachnitis uniflora* (Corsiaceae) (Bidartondo et al. 2002), *Kupea martinetugei* (Triuridaceae) (Franke et al. 2006), *Afrothismia* (Thismiaceae) (Merckx and Bidartondo 2008), and *Petrosavia sakuraii* (Petrosaviaceae) (Yamato et al. 2011). High specificity toward host fungi is also observed in non-angiosperm mycoheterotrophs (Bidartondo et al. 2003; Winther and Friedman 2007, 2008, 2009). See Chap. 6 for a detailed overview on the interactions between mycoheterotrophs and their host fungi.

Association with compatible fungi is essential for the establishment of a fully mycoheterotrophic plant. In absence of their specific fungus, many mycoheterotrophic plants will not germinate or develop (Bruns and Read 2000). Even if germination is triggered by a close relative of the host fungus, the seedling may not survive past the early stages of development (Bidartondo and Read 2008). In an evolutionary context, highly specialized full mycoheterotrophs have evolved from ancestors with more generalist fungal associations. There are two mechanisms that may drive this evolutionary trend toward increased fungal specificity: (1) a mycoheterotrophic plant selects from the potential fungal community the best target to meet its nutrient demands, and (2) a mycoheterotrophic plant, because of its increasingly parasitic interaction with fungi, is "denied" access to most members of the fungal community except for a few fungal lineages that fail to detect or exclude the plant (Bruns et al. 2002; Egger and Hibbett 2004; Bidartondo 2005; Merckx et al. 2009).

Mycorrhizal specificity of mycoheterotrophic plants is not always targeted toward a single fungal lineage. Some mycoheterotrophic species associate with distinct sets of fungal lineages. For example, *Corallorhiza maculata* (Orchidaceae) associates with exclusive sets of fungi, spanning ca. 22 described species across the Russulaceae (Taylor et al. 2004). It was demonstrated that *C. maculata* consists of several distinct fungal host-associated races, and even when different orchid genotypes were found growing in close proximity, they maintained their distinct fungal associations (Taylor et al. 2004). Similarly, the closely related species *Corallorhiza striata* targets a narrow breadth of fungi in the genus *Tomentella* (Thelephoraceae)—specifically *T. fuscocinerea*. A population-level analysis of *C. striata* and its mycorrhizal associates across the entire distribution range of the orchid shows that distinct orchid clades associate with divergent sets of fungi. The associations between *C. striata* and its mycorrhizal fungi are strongly determined by geography and phylogenetic affinity of the orchid populations (Barrett et al. 2010). Levels of mycorrhizal specificity are also high in species of *Hexalectris* (Orchidaceae). The phylogenetic breadth of mycorrhizal associations between the species ranges from specificity toward a single narrow clade of ectomycorrhizal fungi (e.g., *H. brevicaulis*, *H. grandiflora*) to specificity toward multiple clades of different fungal groups in *H. spicata*. However, fungal associations of *H. spicata* showed geographic variation (Kennedy et al. 2011). In Ericaceae, the fully mycoheterotrophic species *Monotropa uniflora* has a widespread distribution in North America. The species specializes on various Russulaceae species within *Russula*, *Lactarius*, and *Martellia*, but fungal associations on population level appear to be geographically structured as well (Bidartondo and Bruns 2001; Bidartondo 2005). Specifically, *M. uniflora* populations in Oregon associated exclusively with *Russula brevipes*, while a single population in Vermont, USA, associated with three *Russula* species groups and *Lactarius theiogalus*. These examples demonstrate that the interactions between mycoheterotrophs and their mycorrhizal fungi may resemble a geographic mosaic (Thompson 2005). The geographic mosaic theory of coevolution hypothesizes that three processes are the primary drivers of coevolutionary dynamics: geographic selection mosaics, coevolutionary hot spots, and trait remixing. Selection mosaics occur when natural selection on interactions varies among different communities. While this has yet to be tested for mycoheterotrophic plants, fitness was found to vary in vitro across different fungus–orchid combinations in the orchid *Tolumnia variegata*, suggesting potential for natural selection to act on these relationships (Otero et al. 2005). Hot spots are communities in which interacting species have reciprocal effects on fitness and are often embedded within surrounding communities in which interspecific selection affects only one or neither species (cold spots). Finally, a combination of gene flow, random genetic drift, and extinction/colonization dynamics continually reshapes the genetic landscape over which future selection takes place (trait remixing) (Gomulkiewicz et al. 2000; Thompson 2005). This tripartite coevolutionary process should produce three general ecological patterns: different combinations of coevolved traits in different regions, local maladaptation within some interactions, and few geographically uniform coevolved traits (Thompson 2005).

While high mycorrhizal specificity is often regarded as a hallmark of mycoheterotrophic plants, not all full mycoheterotrophs show high specificity toward fungi. Examples are *Pyrola aphylla* (Ericaceae) (Hynson and Bruns 2009), *Wullschlaegelia aphylla* (Martos et al. 2009), and species of *Aphyllorchis* (Orchidaceae) (Roy et al. 2009). This indicates that identifying a specific fungus that meets the plant's demands need not be the initiating process in the subversion of the mycorrhizal mutualism. Also, high mycorrhizal specificity is not restricted to fully mycoheterotrophic plants. Many green species of orchids (although are all initial mycoheterotrophs and some perhaps partially mycoheterotrophic) show specificity toward mycorrhizal fungi. Specialization trends within some of these lineages appear to be phylogenetically conserved as well (Shefferson et al. 2007, 2010; Jacquemyn et al. 2011).

5.5.7 Host Shifts

Broad shifts of fungal associations in the evolutionary trajectory toward full mycoheterotrophy have probably occurred multiple times in Orchidaceae: while an association with saprotrophic fungi is likely the ancestral state for all orchids (Yukawa et al. 2009), many fully and partially mycoheterotrophic orchids associate with ectomycorrhizal fungi (see Chaps. 6 and 8).

This "host" switch has been described in detail for the genus *Cymbidium* (Orchidaceae) by Motomura et al. (2010). Early diverging *Cymbidium* species are presumably autotrophic as adults and generally have associations with saprotrophic fungi only (mainly Tulasnellaceae). The partially mycoheterotrophic species diverging from this group harbor both Tulasnellaceae and ectomycorrhizal fungi (Russulaceae and others), while the latest diverging branch in this clade comprises the fully mycoheterotrophic species *C. macrorhizon* and *C. aberrans* that associate exclusively with the ectomycorrhizal fungi. These results indicate that the nutritional shift from autotrophy to mycoheterotrophy through partial mycoheterotrophy in *Cymbidium* may correlate with shifts in associated fungi from saprotrophic to ectomycorrhizal fungi. Hashimoto et al. (2012) recently reported a remarkable congruent fungal shift in a Japanese *Pyrola* species (Ericaceae), of which adult plants are supposedly capable of partial mycoheterotrophy. Adult *Pyrola asarifolia* plants, like other Pyroleae, associate with diverse ectomycorrhizal fungal species that are a subset of those that are forming mycorrhiza with surrounding trees. Conversely, seedlings of *Pyrola asarifolia* specifically associate with a lineage of Sebacinales B, which probably derive carbon through saprotrophy (see Chap. 8).

Similarly, in the liverwort family Aneuraceae, an association with tulasnelloid symbionts is a secondary acquisition of recent origin following the loss of the ancestral arbuscular mycorrhizal symbiosis (Bidartondo and Duckett 2010). The evolution of mycoheterotrophy in *Aneura mirabilis* then led to a subsequent specialization toward narrow *Tulasnella* lineages (Bidartondo et al. 2003; Bidartondo and Duckett 2010). In these examples, evolution of mycoheterotrophy is thus preceded by a secondary gain of associations with fungal lineages, which are in a later evolutionary step exploited to obtain carbon.

Shifts of fungal associations within lineages of full mycoheterotrophic plants have not yet been documented, and the mycoheterotrophs' common preference toward narrow fungal lineages suggests that such shifts may be rare. However, the tendency of many species of monotropes and orchids to specialize toward multiple distant lineages of mycorrhizal fungi may be interpreted as the result of radical host jumps. Another possible explanation for this pattern is high lineage attrition creating vast phylogenetic gaps that then appear as host jumps (Bidartondo 2005). Only by studying mycorrhizal specialization in a phylogenetic context it is possible to distinguish host jumps from lineage attrition (Box 5.1). Although the lack of extant species that represent in-between evolutionary steps of this mycorrhizal specialization process may prevent the uncovering of the underlying evolutionary pattern.

Box 5.1 Analysis of Mycorrhizal Interactions in an Evolutionary Framework: A Primer

Fully mycoheterotrophic plants often show extreme specificity in their associations with fungi. Since all fully mycoheterotrophic plants evolved from autotrophic mycorrhizal ancestors, and autotrophic mycorrhizal plants generally show low specificity in their interaction with mycorrhizal fungi, this extreme specificity is the result of an evolutionary specialization process. Estimating effects of evolutionary relatedness on mycorrhizal associations requires information on phylogenetic relationships and plant–fungal interactions. Depending on the questions asked and the available data, phylogenetic patterns can be used to infer different evolutionary processes shaping mycorrhizal interactions (Fig. 5.7).

Mycorrhizal Associations and Plant Phylogenetic Relatedness

Do plant species that are phylogenetically related show similar patterns of mycorrhizal specificity? This question can be answered when a well-supported plant phylogeny is available as well as information on the mycorrhizal fungi that are associated with each plant species in the phylogeny. Plant species can be

(continued)

Box 5.1 (continued)

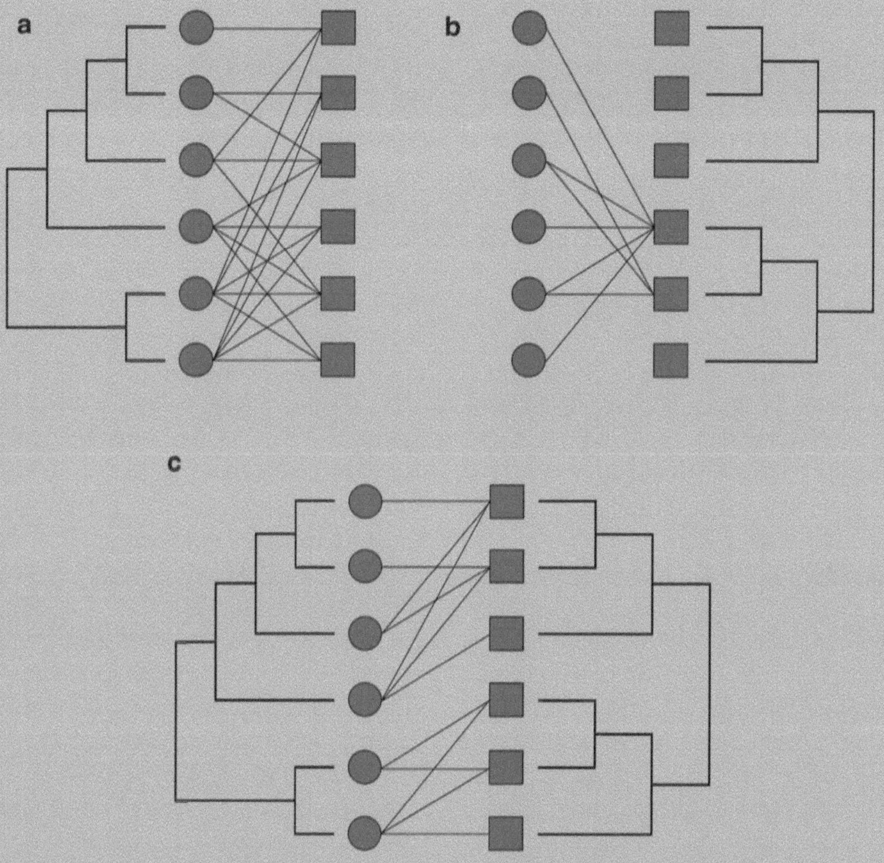

Fig. 5.7 Hypothetical models showing examples of mycorrhizal interactions between plants (*circles*) and fungi (*squares*) influenced by phylogenetic relationships of plants, fungi, or both. (**a**) Example in which the degree of mycorrhizal specificity of host plants is clearly influenced by the phylogenetic relationships between the plants, shown as a progressive loss of associated fungi moving from the *bottom* to the *top* of the phylogeny. (**b**) Example in which plants in a clade associate with a phylogenetically restricted set of fungi. (**c**) Example in which plant–fungal interactions are influenced by the phylogenetic relationships of both plants and fungi

scored according to the number of associated fungi (genotypes or fungal OTU's) or according to the mean pairwise genetic or phylogenetic distance among the associated fungal taxa (e.g., Nei and Tajima 1981). The former method assesses mycorrhizal specificity purely as a function of number of associated fungi per species, without taking the phylogenetic distance between the associated fungi into account. This method results in discrete character states. Calculating genetic distances defines mycorrhizal specificity as a function of phylogenetic breadth of associated fungi, but may be misleading when species tend to specialize on multiple, distant fungal lineages. This method results in continuous character states. Both the number of associated fungal OTU's and distances among the associated fungi can be mapped on the plant phylogeny and used reconstruct evolutionary patterns of mycorrhizal specificity. For this purpose, a wide variety of methods for ancestral state reconstruction are available, for example, in the software packages Mesquite (Maddison and Maddison 2011) and APE (Paradis et al. 2004). In addition, there are different methods

(continued)

Box 5.1 (continued)

to test whether the variation in mycorrhizal specificity between the plant species is influenced by their phylogenetic relationships, such as phylogenetic autocorrelation tests (Gittleman and Kot 1990), and measurements of phylogenetic signal (e.g., *K* statistic of Blomberg et al. 2003). Several of these tests are available in the R packages APE and Picante (Kembel et al. 2010). These methods have been used to study the evolution of mycorrhizal specificity, for example, in the orchid genera *Cypripedium* (Shefferson et al. 2007), *Goodyera* (Shefferson et al. 2010), and *Orchis* (Jacquemyn et al. 2011).

A drawback of the methodology described above is that it ignores the actual identity of the fungal partners. Related plant species may be associated with the same number of fungal taxa or with a similar phylogenetic range of fungi, yet they may actually grow with completely different sets of mycorrhizal fungi. To test whether phylogenetically related plant species interact with similar fungi, plant phylogenetic distances can be regressed against metrics of fungal community dissimilarity using a Mantel test. The dissimilarity (or distance) in fungal community structure between plant species pairs can be calculated as $1-S$, in which S is the fungal community similarity between species pairs. There are a number of different community similarity statistics available in the ecological literature (reviewed in Koleff et al. 2003; Anderson et al. 2010), but they are generally based on a ratio of shared vs. total fungal taxa hosted between plant species. Differences in the number of associated fungal taxa per plant species may affect estimates of community similarity, and therefore, a partial Mantel test controlling for a number of associated fungi or choice of a metric that controls for differences in richness (e.g., β_{sim}, Koleff et al. 2003) may be more appropriate in most cases. The ZT software package (Bonnet and Van de Peer 2002) and the R package APE can be used to run these Mantel tests, and the R package Vegan (Oksanen et al. 2011) can be used to calculate most metrics of community similarity.

Mycorrhizal Associations and Fungal Phylogenetic Relatedness

Are mycorrhizal interactions evolutionarily conserved across the fungal phylogeny? In theory, the same methods as described above for mycorrhizal interactions in the context of plant relationships can be applied to fungal phylogenies. However, for most fungal taxa, we lack data about the range of associated plant species. Nevertheless, we can test the degree to which fungi associating with particular plant species tend to be close phylogenetic relatives. These tests can be performed at different ecological and evolutionary scales. For example, a fungal phylogeny may represent all lineages of a particular clade of fungi, regardless of their occurrence or their ecology, and may be used to infer whether a particular group of plants shows a preference for fungi based on the phylogenetic relatedness of the latter. Or a fungus phylogeny may be restricted to all mycorrhizal fungi detected at a particular research site and can be used to test whether mycorrhizal associations of particular plants at this site are influenced by fungal phylogenetic relatedness. Choosing the appropriate pool of species is critical to matching the scale of analysis to the scale of the question being asked. A popular method to examine the phylogenetic structure of a "community" (in this case, a set of mycorrhizal fungi detected in a particular sample of plant species) is the calculation of the net relatedness index (NRI) and nearest taxon index (NTI) (Webb et al. 2002). NRI is a standardized measure of the mean pairwise phylogenetic distance of taxa in a sample and quantifies overall clustering of taxa on a tree, while NTI is a measure of the phylogenetic distance to the nearest co-occurring taxon for each taxon in the sample and quantifies the extent of terminal clustering, independent

(continued)

Box 5.1 (continued)

of deep-level clustering (Webb et al. 2002). The NTI and NRI indices are measured as standardized effect sizes based on a null model generated by random permutation of a phylogeny that contains an appropriate pool of species for the given question. Positive NTI and NRI values indicate that species within a sample are more closely related than expected by chance (phylogenetic clustering), and negative values indicate that species within a sample are less related than might be expected by chance (phylogenetic evenness or overdispersion). Phylogenetic clustering and overdispersion are generally interpreted as the result of evolutionary trait conservation or trait convergence as the result of ecological processes such as niche partitioning and competition. However, care must be used when interpreting results as multiple processes can result in patterns of clustering or overdispersion (Kraft et al. 2007). These methods are implemented in the software program Phylocom (Webb et al. 2011) and the R package Picante and have been used by Merckx et al. (2012) to investigate the phylogenetic clustering of arbuscular mycorrhizal fungi detected in mycoheterotrophic interactions. Alternatively, multivariate statistics can be used to analyze phylogenetic distance between fungi and their ecological role (e.g., presence/absence in mycoheterotrophic interactions).

Plant–Fungal Networks

The methods described above evaluate mycorrhizal interactions either from a plant or fungus phylogenetic perspective. A more integrative approach estimates the influence of both phylogenies on a plant–fungal interaction network. This can be achieved by calculating the phylogenetic relationships of both plants and their interacting mycorrhizal fungi and assessing the mycorrhizal network from a classic host–parasite coevolutionary perspective. A simple approach is to calculate congruency between plant and fungal phylogenies (e.g., Penny and Hendy 1985; Farris et al. 1994) and has been used by Bidartondo and Bruns (2002) for Monotropoideae and their associated fungi and Barrett et al. (2010) for *Corallorhiza–Tomentella* interactions. The latter study also applied a more advanced congruence/incongruence statistical test implemented in ParaFit (Legendre et al. 2002). Several other methodologies to analyze host–parasite associations are available (reviewed by Stevens 2004). However, many of these approaches are based on a host–parasite co-speciation model (e.g., Charleston and Page 2002; Ronquist 2002), which may be not an appropriate model to explain the observed branching patterns of mycorrhizal specialization processes. When congruence between both phylogenies is high, co-speciation of congruent plant–fungus nodes can be tested by comparing their absolute ages obtained with molecular clock techniques. When age estimates for congruent nodes overlap, they may be the result of synchronous co-speciation events. When age estimates do not overlap, the pattern may be explained by delayed co-speciation of phylogenetic tracking events, as was observed in fully mycoheterotrophic *Afrothismia* species and their associated *Glomus* fungi (Merckx and Bidartondo 2008).

Food-web approaches are probably more appropriate to explore complex plant–fungus mycorrhizal interactions than host–parasite co-speciation models. A relevant method for evaluating mycorrhizal interactions is that of Ives and Godfray (2006), which calculates the strength of the phylogenetic signal of the two phylogenies on the interaction matrix. It is implemented in the R package Picante and has been used to study phylogenetic effect on *Orchis*–Tulasnellaceae interactions by Jacquemyn et al. (2011).

5.5.8 Phylogenetic Tracking

It has been argued that once an appropriate fungal partner has been found, a mycoheterotroph fine-tunes its physiology to adapt to that particular fungus, and it is therefore largely incapable of host jumps to distantly related fungi (Bidartondo and Bruns 2002). Subsequent speciation, mycorrhizal specialization, and a lack of host switching may force the plants to track the fungus phylogeny. Such phylogenetic tracking has been observed between species of *Afrothismia* (Thismiaceae) and *Glomus* group A fungi (Merckx and Bidartondo 2008). Phylogenetic tracking and host loyalty has been extensively studied in Monotropoideae, where it resulted in extreme mycorrhizal specialization, most extensively toward *Tricholoma* and Russulaceae (Bidartondo 2005).

5.6 Conclusions

Many millions of years of evolution have produced an astonishing array of land plants. Fully mycoheterotrophic plants are perhaps among the most remarkable of these. They clearly demonstrate that the mycorrhizal mutualism, which in case of the AM mutualism has been stable for over 400 million years, can be successfully cheated. A large majority of land plants associate with mycorrhizal fungi, and therefore, it is perhaps not surprising that the ability to cheat this interaction arose many times independently during plant evolution. However, it remains unknown why particular families of mycorrhizal plants contain many fully mycoheterotrophic species, while mycoheterotrophy is absent in most other mycorrhizal plant families. Lineages of fully mycoheterotrophic plants always comprise relatively few species, and those species are often rare and have scattered geographic distributions (see also Chap. 3). This may reflect limited ecological success, at least under recent climatic and biotic conditions. However, detailed evolutionary histories of ancient mycoheterotrophic lineages may reveal pronounced diversification and geographic dispersal—and thus increased ecological success—during past epochs of plant evolution. The drivers behind the evolution toward mycoheterotrophy remain unknown as well, although several lines of evidence suggests that mycoheterotrophy may provide an evolutionary escape route from intense competition for light in forest understory habitats.

References

Abadie J, Püttsepp Ü, Gebauer G, Faccio A, Bonfante P, Selosse MA (2006) *Cephalanthera longifolia* (Neottieae, Orchidaceae) is mixotrophic: a comparative study between green and nonphotosynthetic individuals. Can J Bot 84:1462–1477

Albert VA, Struwe L (1997) Phylogeny and classification of *Voyria* (saprophytic Gentianaceae). Brittonia 49:466–479

Anderson MJ, Crist TO, Chase JM, Vellend M, Inouye BD, Freestone AL, Sanders NJ, Cornell VH, Comita LS, Davies KF, Harrison SP, Kraft HJB, Stegen JC, Swenson NG (2010) Navigating the multiple meanings of β diversity: a roadmap for the practicing ecologist. Ecol Lett 14:19–28

APG (2009) An update of the Angiosperm Phylogeny Group classification for the orders and families of flowering plants: APG III. Bot J Linn Soc 161:105–121

Arditti J, Ghani AKA (2000) Numerical and physical properties of orchid seeds and their biological implications. New Phytol 145:367–421

Axelrod R, Hamilton WD (1981) The evolution of cooperation. Science 211:1390–1396

Bakshi TS (1959) Ecology and morphology of *Pterospora andromedea*. Bot Gaz 120:203–217

Barraclough TG, Savolainen V (2001) Evolutionary rates and species diversity in flowering plants. Evolution 55:677–683

Barbrook AC, Howe CJ, Purton S (2006) Why are plastid genomes retained in non-photosynthetic organisms? Trends Plant Sci 11:101–108

Barkman TJ, Lim S-H, Salleh KM, Nais J (2004) Mitochondrial DNA sequences reveal the photosynthetic relatives of *Rafflesia*, the world's largest flower. Proc Natl Acad Sci USA 101:787–792

Barkman TJ, McNeal JR, Lim S-H, Coat G, Croom HB, Young ND, dePamphilis CW (2007) Mitochondrial DNA suggests at least 11 origins of parasitism in angiosperms and reveals genomic chimerism in parasitic plants. BMC Evol Biol 7:248

Barrett CF, Freudenstein JV (2008) Molecular evolution of rbcL in the mycoheterotrophic coralroot orchids (*Corallorhiza* Gagnebin, Orchidaceae). Mol Phylogenet Evol 47:665–679

Barrett CF, Freudenstein JV, Taylor DL, Kõljalg U (2010) Rangewide analysis of fungal associations in the fully mycoheterotrophic *Corallorhiza striata* complex (Orchidaceae) reveals extreme specificity on ectomycorrhizal *Tomentella* (Thelephoraceae) across North America. Am J Bot 97:628–643

Bergsten J (2005) A review of long-branch attraction. Cladistics 21:163–193

Bernard N (1899) Sur la germination du *Neottia nidus-avis*. Compte Rendu Hebdomadaire des Séances de l'Académie des Sciences, Paris 128:1253–1255

Bever JD, Richardson SC, Lawrence BM, Holmes J, Watson M (2009) Preferential allocation to beneficial symbiont with spatial structure maintains mycorrhizal mutualism. Ecol Lett 12:13–21

Bidartondo MI (2005) The evolutionary ecology of myco-heterotrophy. New Phytol 167:335–352

Bidartondo MI, Bruns TD (2001) Extreme specificity in epiparasitic Monotropoideae (Ericaceae): widespread phylogenetic and geographic structure. Mol Ecol 10:2285–2295

Bidartondo MI, Bruns TD (2002) Fine-level mycorrhizal specificity in the Monotropoideae (Ericaceae): specificity for fungal species groups. Mol Ecol 11:557–569

Bidartondo MI, Bruns TD, Weiß M, Sérgio C, Read DJ (2003) Specialized cheating of the ectomycorrhizal symbiosis by an epiparasitic liverwort. Proc R Soc Lond B 270:835–842

Bidartondo MI, Burghardt B, Gebauer G, Bruns TD, Read D (2004) Changing partners in the dark: isotopic and molecular evidence of ectomycorrhizal liaisons between forest orchids and trees. Proc R Soc Lond B 271:1799–1806

Bidartondo MI, Duckett JG (2010) Conservative ecological and evolutionary patterns in liverwort-fungal symbioses. Proc R Soc Lond B 277:485–492

Bidartondo MI, Read DJ (2008) Fungal specificity bottlenecks during orchid germination and development. Mol Ecol 17:3707–3716

Bidartondo MI, Redecker D, Hijiri I, Wiemken A, Bruns TD, Dominguez LS, Sérsic A, Leake JR, Read DJ (2002) Epiparasitic plants specialized on arbuscular mycorrhizal fungi. Nature 419:389–392

Blomberg SP, Garland T, Ives AR (2003) Testing for phylogenetic signal in comparative data: behavioral traits are more labile. Evolution 57:2147–2156

Bonnet E, Van de Peer Y (2002) ZT: a software tool for simple and partial Mantel tests. J Stat Softw 7:1–12

Bouman F, Cobb L, Devente N, Goethals V, Maas PJM, Smets E (2002) The seeds of Gentianaceae. In: Struwe L, Albert VA (eds) Gentianaceae: systematics and natural History. Cambridge Univ. Press, Cambridge, pp 498–572

Bruns TD, Bidartondo MI, Taylor DL (2002) Host specificity in ectomycorrhizal communities: what do the exceptions tell us? Integ Comp Biol 42:352–359

Bruns TD, Read DJ (2000) In vitro germination of non-photosynthetic, myco-heterotrophic plants stimulated by fungi isolated from the adult plants. New Phythol 148:335–342

Bruns TD, Shefferson RP (2004) Evolutionary studies of ectomycorrhizal fungi: recent advances and future directions. Can J Bot 82:1122–1132

Bull JJ, Rice WR (1991) Distinguishing mechanisms for the evolution of cooperation. J Theor Biol 149:63–74

Bungard RA (2004) Photosynthetic evolution in parasitic plants: insight from the chloroplast genome. Bioessays 26:235–247

Caddick LR, Rudall PJ, Wilkin P, Hedderson TA, Chase MW (2002) Phylogenetics of Dioscoreales based on combined analyses of morphological and molecular data. Bot J Linn Soc 138:123–144

Cameron DD, Bolin JF (2010) Isotopic evidence of partial mycoheterotrophy in the Gentianaceae: *Bartonia virginica* and *Obolaria virginica* as case studies. Am J Bot 97:1272–1277

Cameron DD, Preiss K, Gebauer G, Read DJ (2009) The chlorophyll-containing orchid *Corallorhiza trifida* derives little carbon through photosynthesis. New Phytol 183:358–364

Cameron KM (2004) Utility of plastid *psaB* gene sequences for investigating intrafamilial relationships within Orchidaceae. Mol Phylogenet Evol 31:1157–1180

Cameron KM (2009) On the value of nuclear and mitochondrial gene sequences for reconstructing the phylogeny of vanilloid orchids (Vanilloideae, Orchidaceae). Ann Bot 10:377–385

Cameron KM, Chase MW, Rudall P (2003) Recircumscription of the monocotyledonous family Petrosaviaceae to include *Japonolirion*. Brittonia 55:214–225

Charleston MA, Page RDM (2002) TREEMAP 2.0ß: a Macintosh program for the analysis of how dependent phylogenies are related, by cophylogeny mapping. Available from http://evolve.zoo.ox.ac.uk/software/TreeMap/main.html

Chase MW, Barrett RL, Freudenstein JV, Cameron KM (2003) DNA data and Orchidaceae systematics: a new phylogenetic classification. In: Dixon KM, Kell SP, Barrett RL, Cribb PJ (eds) Orchid conservation. Natural History publications, Kota Kinabalu, Sabah, Malaysia, pp 69–89

Chase MW, Soltis DE, Olmstead RG, Morgan D, Les DH, Mishler BD, Duvall MR, Price RA, Hills HG, Qiu Y-L, Kron KA, Rettig JH, Conti E, Palmer JD, Manhart JR, Sytsma KJ, Michaels HJ, Kress WJ, Karol KG, Clark WD, Hedrén M, Gaut BS, Jansen RK, Kim K-J, Wimpee CF, Smith JF, Furnier GR, Strauss SH, Xiang Q-Y, Plunkett GM, Soltis PS, Swensen SM, Williams SE, Gadek PA, Quinn CJ, Equiarte LE, Golenberg E, Learn GH, Graham SW, Barrett SCH, Dayanandan S, Albert VA (1993) Phylogenetics of seed plants: an analysis of nucleotide sequences from the plastid gene *rbcL*. Ann Missouri Bot Gard 80:528–580

Christenhusz MJM, Zhang X-C, Schneider H (2011) A linear sequence of extant families and genera of lycophytes and ferns. Phytotaxa 19:7–54

Crum HA, Bruce J (1996) A new species of *Cryptothallus* from Costa Rica. Bryologist 99:433–438

Davis JI, Stevenson DW, Petersen G, Seberg O, Campbell LM, Freudenstein JV, Goldman DH, Hardy CR, Michelangeli FA, Simmons MP, Specht CD, Vergara-Silva F, Gandolfo MA (2004) A phylogeny of the monocots, as inferred from *rbcL* and atpA sequence variation, and a comparison of methods for calculating jackknife and bootstrap values. Syst Bot 29:467–510

Delannoy E, Fujii S, Colas des Francs C, Brundrett M, Small I (2011) Rampant gene loss in the underground orchid *Rhizanthella gardneri* highlights evolutionary constraints on plastid genomes. Mol Biol Evol 28:2077–2086

dePamphilis CW, Young ND, Wolfe AD (1997) Evolution of plastid gene *rps2* in a lineage of hemiparasitic and holoparasitic plants: many losses of photosynthesis and complex patterns of rates variation. Proc Natl Acad Sci USA 94:7367–7372

Egger KN, Hibbett DS (2004) The evolutionary implications of exploitation in mycorrhizas. Can J Bot 82:1110–1121

Eriksen B (1993) Phylogeny of the Polygalaceae and its taxonomic implications. Pl Syst Evol 186:33–55

Eriksson O, Kainulainen K (2011) The evolutionary ecology of dust seeds. Perspect Plant Ecol Evol Systemat 13:73–87

Farris JS, Källersjö M, Kluge AG, Bult C (1994) Testing significance of incongruence. Cladistics 10: 315–319

Feild TS, Bodribb TJ (2005) A unique mode of parasitism in the conifer coral tree *Parasitaxus ustus* (Podocarpaceae). Plant Cell Environ 28:1316–1325

Felsenstein J (1978) Cases in which parsimony or compatibility methods will be positively misleading. Syst Zool 27:401–410

Ferrière R, Gauduchon M, Bronstein JL (2007) Evolution and persistence of obligate mutualists and exploiters: competition for partners and evolutionary immunization. Ecol Lett 10:115–126

Forrest LL, Davis EC, Long DG, Crandall-Stotler BJ, Clark A, Hollingsworth ML (2006) Unraveling the evolutionary history of the liverworts (Marchantiophyta): multiple taxa, genomes and analyses. Bryologist 109:303–334

Franke T, Beenken L, Döring M, Kocyan A, Agerer R (2006) Arbuscular mycorrhizal fungi of the *Glomus*-group A lineage (Glomerales; Glomeromycota) detected in myco-heterotrophic plants from tropical Africa. Mycol Prog 5:24–31

Freudenstein JV (1994) Gynostemium structure and relationships of the Corallorhizinae (Orchidaceae: Epidendroideae). Pl Syst Evol 193:1–19

Freudenstein JV, Senyo DM (2008) Relationships and evolution of *matk* in a group of leafless orchids (*Corallorhiza* and Corallorhizinae; Orchidaceae: Epidendroideae). Am J Bot 95:498–505

Freudenstein JV, van den Berg C, Goldman DH, Kores PJ, Molvray M, Chase MW (2004) An expanded plastid DNA phylogeny of Orchidaceae and analysis of jackknife branch support strategy. Am J Bot 91:149–157

Furness CA, Rudall P, Eastman A (2002) Contribution of pollen and tepatal characters to the systematics of Triuridaceae. Pl Syst Evol 235:209–218

Gandolfo MA, Nixon KC, Crepet WL (2002) Triuridaceae fossil flowers from the Upper Cretaceous of New Jersey. Am J Bot 89:1940–1957

Gaut BS, Lewis PO (1995) The success of maximum likelihood in the four-taxon case. Mol Biol Evol 12:152–162

George A, Cooke J (1981) *Rhizanthella*: the underground orchid of Western Australia. In: Lawler L, Kerr RD (eds) Proceedings of the Orchid Symposium, Thirteenth International Botanical Congress. Harbour Press, Sydney, Australia, pp 77–78

Giovannetti M, Sbrana C, Avio L, Strani P (2004) Patterns of below-ground plant interconnections established by means of arbuscular mycorrhizal networks. New Phytol 164:175–181

Girlanda M, Segreto R, Cafasso D, Liebel HT, Rodda M, Ercole E, Cozzolino S, Gebauer G, Perotto S (2011) Photosynthetic Mediterranean meadow orchids feature partial mycoheterotrophy and specific mycorrhizal associations. Am J Bot 98:1148–1163

Gittleman JL, Kot M (1990) Adaptation: statistics and a null model for estimating phylogenetic effects. Syst Zool 39:227–241

Goldblatt P, Rodriguez A, Powell MP, Davies TJ, Manning JC, van der Bank M, Savolainen V (2008) Iridaceae 'Out of Australasia'? Phylogeny, biogeography, and divergence time based on plastid DNA sequences. Syst Bot 33:495–508

Gomulkiewicz R, Thompson JN, Holt RD, Nuismer SL, Hochberg ME (2000) Hot spots, cold spots, and the geographic mosaic theory of coevolution. Am Nat 156:156–174

Goto R, Okamoto T, Kiers ET, Kawakita A, Kato M (2010) Selective flower abortion maintains moth cooperation in a newly discovered pollination mutualism. Ecol Lett 13:321–329

Gustafsson ALS, Verola CF, Antonelli A (2010) Reassessing the temporal evolution of orchids with new fossils and a Bayesian relaxed clock, with implications for the diversification of the rare South American genus *Hoffmannseggella* (Orchidaceae: Epidendroideae). BMC Evl Biol 10:177

Harvey PH, May RM, Nee S (1994) Phylogenies without fossils. Evolution 48:523–529

Hashimoto Y, Fukukawa S, Kunishi A, Suga H, Richard F, Sauve M, Selosse M-A (2012) Mycoheterotrophic germination of *Pyrola asarifolia* dust seed reveals convergences with germination in orchids. New Phytol. Published online before print. doi:10.1111/j.1469-8137.2012.04174.x

Heide-Jørgensen HS (2008) Parasitic flowering plants. Koninklijke Brill NV, Leiden, The Netherlands

Hibbett D, Matheny PB (2009) The relative ages of ectomycorrhizal mushrooms and their plant hosts estimated using Bayesian relaxed molecular clock analyses. BMC Biol 7:13

Hibbett DS, Gilbert L-B, Donoghue MJ (2000) Evolutionary instability of ectomycorrhizal symbioses in basidiomycetes. Nature 407:506–508

Huelsenbeck JP (1995) The performance of phylogenetic methods in simulation. Syst Biol 44:17–48

Hynson NA, Bruns TD (2009) Evidence of a myco-heterotroph in the plant family Ericaceae that lacks mycorrhizal specificity. Proc R Soc B 276:4053–4059

Hynson NA, Mambelli S, Amend AS, Dawson TE (2012) Measuring carbon gains from fungal networks in understory plants from the tribe Pyroleae (Ericaceae):

a field manipulation and stable isotope approach. Oecologia 169:307–317
Hynson NA, Preiss K, Gebauer G, Bruns TD (2009) Isotopic evidence of full and partial myco-heterotrophy in the plant tribe Pyroleae (Ericaceae). New Phytol 182:719–726
Ives AR, Godfray HC (2006) Phylogenetic analysis of trophic associations. Am Nat 168:E1–E14
Jacquemyn H, Merckx V, Brys R, Tyteca D, Cammue B, Honnay O, Lievens B (2011) Analysis of network architecture reveals phylogenetic constraints on mycorrhizal specificity in the genus *Orchis* (Orchidaceae). New Phytol 192:518–528
Jandér KC, Herre EA (2010) Host sanctions and pollinator cheating in the fig tree–fig wasp mutualism. Proc R Soc Lond B 277:1481–1488
Jaramillo CA (2002) Response of tropical vegetation to Paleogene warming. Paleobiology 28:222–243
Jaramillo CA, Rueda MJ, Mora G (2006) Cenozoic plant diversity in the neotropics. Science 311: 1893–1896
Julou T, Burghardt B, Gebauer C, Berveiller D, Damesin C, Selosse MA (2005) Mixotrophy in orchids: insights from a comparative study of green individuals and nonphotosynthetic individuals of *Cephalanthera damasonium*. New Phytol 166:639–653
Kembel SW, Cowan PD, Helmus MR, Cornwell WK, Morlon H, Ackerly DD, Blomberg SP, Webb CO (2010) Picante: R tools for integrating phylogenies and ecology. Bioinformatics 26:1463–1464
Kennedy AH, Taylor DL, Watson LE (2011) Mycorrhizal specificity in the fully mycoheterotrophic *Hexalectris* Raf. (Orchidaceae: Epidendroideae). Mol Ecol 20:1303–1316
Kennedy AH, Watson LE (2010) Species delimitations and phylogenetic relationships within the fully mycoheterotrophic *Hexalectris* (Orchidaceae). Syst Bot 35:64–76
Kiers ET, Duhamel M, Beesetty Y, Mensah JA, Franken O, Verbruggen E, Fellbaum CR, Kowalchuk GA, Hart MH, Bago A, Palmer TM, West SA, Vandenkoornhuyse P, Jansa J, Bücking H (2011) Reciprocal rewards stabilize cooperation in the mycorrhizal symbiosis. Science 333:880–882
Kiers ET, Rousseau RA, Denison RF (2006) Measured sanctions: legume hosts detect quantitative variation in rhizobium cooperation and punish accordingly. Evol Ecol Res 8:1077–1086
Kiers ET, Rousseau RA, West SA, Denison RF (2003) Host sanctions and the legume-rhizobium mutualism. Nature 425:78–81
Kiers ET, van der Heijden MGA (2006) Mutualistic stability in the arbuscular mycorrhizal symbiosis: exploring hypotheses of evolutionary cooperation. Ecology 87:1627–1637
Kissling J, Yuan Y, Küpfer P, Mansion G (2009) The polyphyletic genus *Sebaea* (Gentianaceae): a step forward in understanding the morphological and karyological evolution of the Exaceae. Mol Phylogenet Evol 53:734–748
Koleff P, Gaston J, Lennon JJ (2003) Measuring beta diversity for presence-absence data. J Anim Ecol 72:367–382
Kottke I, Nebel M (2005) The evolution of mycorrhiza-like associations in liverworts: an update. New Phytol 167:330–334
Kraft NJB, Cornwell WK, Webb CO, Ackerly DD (2007) Trait evolution, community assembly, and the phylogenetic structure of ecological communities. Am Nat 170:271–283
Kron KA (1996) Phylogenetic relationships of Empetraceae, Epacridaceae, Ericaceae, Monotropaceae, and Pyrolaceae: evidence from nuclear ribosomal 18s sequence data. Ann Bot 77:293–303
Leake JR (1994) The biology of myco-heterotrophic ('saprophytic') plants. New Phytol 127:171–216
Leake JR, Cameron DD, Beerling BJ (2008) Fungal fidelity in the myco-heterotroph-to-autotroph life cycle of Lycopodiaceae: a case of parental nurture? New Phytol 177:572–576
Legendre P, Desdevises Y, Bazin E (2002) A statistical test for host–parasite coevolution. Syst Biol 51: 217–234
Leigh EG, Davidar P, Dick CW, Puyravaud J-P, Terborgh J, ter Steege H, Wright SJ (2004) Why do some tropical forests have so many species of trees? Biotropica 36:447–473
Lemaire B, Huysmans S, Smets E, Merckx V (2011) Rate accelerations in nuclear 18S rDNA of myco-heterotrophic and parasitic angiosperms. J Plant Res 124:561–576
Lian C, Narimatsu M, Nara K, Hogetsu T (2006) *Tricholoma matsutake* in a natural *Pinus densiflora* forest: correspondence between above- and below-ground genets, association with multiple host trees and alteration of existing ectomycorrhizal communities. New Phytol 171:825–836
Liebel HT, Bidartondo MI, Preiss K, Segreto R, Stöckel M, Rodda M, Gebauer G (2010) C and N isotope signatures reveal constraints to nutritional modes in orchids of the Mediterranean and Macaronesia. Am J Bot 97:903–912
Linder HP (2008) Plant species radiations: where, when, why? Philos Trans R Soc Lond B 363:3097–3105
Liu Z-W, Zhou J, Liu E-D, Peng H (2010) A molecular phylogeny and a new classification of *Pyrola* (Pyroleae, Ericaceae). Taxon 59:1690–1700
Logacheva MD, Schelkunov MI, Penin AA (2011) Sequencing and analysis of plastid genome in myco-heterotrophic Orchid *Neottia nidus-avis*. Genome Biol Evol 3:1296–1303
Maas PJM, Ruyters P (1986) *Voyria* and *Voyriella* (saprophytic Gentianaceae). Fl Neotrop 41:1–93
Maas-van de Kamer H (1995) Triuridiflorae—Gardner's delight? In: Rudall PJ, Cribb PJ, Cutler DF, Humphries CJ (eds) Monocotyledons: systematics and evolution. Royal Botanic Gardens, Kew, pp 287–301
Maas-van de Kamer H, Weustenfeld T (1998) Triuridaceae. In: Kubitzki K (ed) The families and genera of flowering plants, vol 3. Springer, Berlin, pp 452–458

Machado CA, Jousselin E, Kjellberg F, Compton SG, Herre EA (2001) Phylogenetic relationships historical biogeography and character evolution of fig pollinating wasps. Proc R Soc Lond B 268:685–694

Maddison WP, Maddison DR (2011) Mesquite: a modular system for evolutionary analysis. Version 2.75. http://mesquiteproject.org

Martos F, Dulormne M, Pailler T, Bonfante P, Faccio A, Fournel J, Dubois M-P, Selosse M-A (2009) Independent recruitment of saprotrophic fungi as mycorrhizal partners by tropical achlorophyllous orchids. New Phytol 184:668–681

Matthews KG, Dunne N, York E, Struwe L (2009) A phylogenetic analysis and taxonomic revision of *Bartonia* (Gentianaceae: Gentianeae), based on molecular and morphological evidence. Syst Bot 34:162–172

Merckx V, Bakker FT, Huysmans S, Smets E (2009) Biass and conflict in phylogenetic inference of myco-heterotrophic plants: a case study in Thismiaceae. Cladistics 25:64–77

Merckx V, Bidartondo MI (2008) Breakdown and delayed cospeciation in the arbuscular mycorrhizal mutualism. Proc R Soc Lond B 275:1029–1035

Merckx V, Chatrou LW, Lemaire B, Sainge MN, Huysmans S, Smets E (2008) Diversification of myco-heterotrophic angiosperms: evidence from Burmanniaceae. BMC Evol Biol 8:178

Merckx V, Huysmans H, Smets E (2010) Cretaceous origins of myco-heterotrophy in Dioscoreales. In: Seberg O, Petersen G, Barfod AS, Davis JI (eds) Diversity, phylogeny, and evolution in the monocotyledons. Aarhus University Press, Arhus, pp 39–53

Merckx VSFT, Janssens SB, Hynson NA, Specht CD, Bruns TD, Smets EF (2012) Mycoheterotrophic interactions are not limited to a narrow phylogenetic range of arbuscular mycorrhizal fungi. Mol Ecol 21:1524–1532

Merckx V, Schols P, Maas-van de Kamer H, Maas P, Huysmans S, Smets E (2006) Phylogeny and evolution of Burmanniaceae (Dioscoreales) based on nuclear and mitochondrial data. Am J Bot 93:1684–1698

Modrich P, Lahue R (1996) Mismatch repair in replication fidelity, genetic recombination, and cancer biology. Annu Rev Biochem 65:101–133

Molvray M, Kores PJ, Chase MW (2000) Polyphyly of mycoheterotrophic orchids and functional influences of floral and molecular characters. In: Wilson KL, Morrison DA (eds) Monocots: systematics and evolution. CSIRO Publishing, Collingwood (Australia), pp 441–448

Motomura H, Selosse M, Martos F, Kagawa A, Yukawa T (2010) Mycoheterotrophy evolved from mixotrophic ancestors: evidence in *Cymbidium* (Orchidaceae). Ann Bot 106:573–581

Nei M, Tajima F (1981) DNA polymorphism detectable by restriction endonucleases. Genetics 97:145–163

Neyland R, Hennigan M (2003) A phylogenetic analysis of large-subunit (26S) ribosome DNA sequences suggests that the Corsiaceae are polyphyletic. New Zealand J Bot 41:1–11

Nickrent DL, Starr EM (1994) High rates of nucleotide substitutions in nuclear small-subunit (18S) rDNA from holoparasitic flowering plants. J Mol Evol 39:62–70

Ogura-Tsujita Y, Yukawa T (2008) High mycorrhizal specificity in a widespread mycoheterotrophic plant, *Eulophia zollingeri* (Orchidaceae). Am J Bot 95:93–97

Oksanen J, Blanchet FG, Kindt R, Legendre P, O'Hara RB, Simpson GL, Solymos P, Stevens MHH, Wagner H (2011) Vegan: community ecology package. R package version 1.17-7

Otero JT, Bayman P, Ackerman JD (2005) Variation in mycorrhizal performance in the epiphytic *orchid Tolumnia variegata in vitro*: the potential for natural selection. Evol Ecol 19:29–43

Palmer JD, Soltis DE, Chase MW (2004) The plant tree of life: an overview and some points of view. Am J Bot 91:1437–1445

Paradis E, Claude J, Strimmer K (2004) APE: analyses of phylogenetics and evolution in R language. Bioinformatics 20:289–290

Pedersen HA, Watthana S, Roy M, Suddee S, Selosse M (2009) *Cephalanthera exigua* rediscovered: new insights in the taxonomy, habitat requirements and breeding system of a rare mycoheterotrophic orchid. Nord J Bot 27:460–468

Pellmyr O, Huth CJ (1994) Evolutionary stability of mutualism between yuccas and yucca moths. Nature 372:257–260

Pellmyr O, Leebens-Mack J, Huth CJ (1996) Non-mutualistic yucca moths and their evolutionary consequences. Nature 380:155–156

Penny D, Hendy MD (1985) The use of tree comparison metrics. Syst Zool 34:75–82

Preiss K, Adam IK, Gebauer G (2010) Irradiance governs exploitation of fungi: fine-tuning of carbon gain by two partially myco-heterotrophic orchids. Proc R Soc Lond B 277:1333–1336

Preussing M, Nebel M, Oberwinkler F, Weiss M (2010) Diverging diversity patterns in the *Tulasnella* (Basidiomycota, Tulasnellales) mycobionts of *Aneura pinguis* (Marchantiophyta, Metzgeriales) from Europe and Ecuador. Mycorrhiza 20:147–159

Purvis A (2008) Phylogenetic approaches to the study of extinction. Annu Rev Ecol Evol Syst 39:301–319

Ramirez SR, Gravendeel B, Singer RB, Marshall CR, Pierce NE (2007) Dating the origin of the Orchidaceae from a fossil orchid with its pollinator. Nature 448:1042–1045

Ramsbottom J (1922) Orchid mycorrhiza. Percy Lund, Humphries & Co. Ltd., London

Rasmussen HN (1995) Terrestrial orchids: from seed to mycotrophic plant. Cambridge University Press, Cambridge

Read DJ, Perrez-Moreno J (2003) Mycorrhizas and nutrient cycling in ecosystems—a journey towards relevance? New Phytol 157:475–492

Ronquist F (2002) TreeFitter. Version 1.2. Available from http://morphbank.ebc.uu.se/TreeFitter

Rothacker EP (2007) The primitive Epidendroideae (Orchidaceae): phylogeny, character evolution and the systematics of *Psilochilus* (Triphoreae). PhD Dissertation, The Ohio State University

Roy M, Watthana S, Stier A, Richard F, Vessabutr S, Selosse MA (2009) Two mycoheterotrophic orchids from Thailand tropical dipterocarpacean forests

associate with a broad diversity of ectomycorrhizal fungi. BMC Biol 7:51
Rudall PJ, Bateman RM (2006) Morphological phylogenetic analysis of Pandanales: testing contrasting hypotheses of floral evolution. Syst Bot 31:223–238
Sachs JL, Mueller UG, Wilcox TP, Bull JJ (2004) The evolution of cooperation. Q Rev Biol 79:135–160
Sachs JL, Simms EL (2006) Pathways to mutualism breakdown. Trends Ecol Evol 21:585–592
Salmia A (1989) General morphology and anatomy of chlorophyll-free and green forms of *Epipactis helleborine* (Orchidaceae). Ann Bot Fenn 26:95–105
Schwartz MW, Hoeksema JD (1998) Specialization and resource trade: biological markets as a model of mutualisms. Ecology 79:1029–1038
Selosse M-A, Dubois M-P, Alvarez N (2009) Do Sebacinales commonly associate with plant roots as endophytes? Mycol Res 113:1062–1069
Selosse M-A, Faccio A, Scappaticci G, Bonfante P (2004) Chlorophyllous and achlorophyllous specimens of *Epipactis microphylla* (Neottieae, Orchidaceae) are associated with ectomycorrhizal septomycetes, including truffles. Microb Ecol 47:416–426
Selosse MA, Roy M (2009) Green plants that feed on fungi: facts and questions about mixotrophy. Trends Plant Sci 14:64–70
Shefferson RP, Cowden CC, McCormick MK, Yukawa T, Ogura-Tsujita Y, Hashimoto T (2010) Evolution of host breadth in broad interactions: mycorrhizal specificity in East Asian and North American rattlesnake plantains (*Goodyera* spp.) and their fungal hosts. Mol Ecol 19:3008–3017
Shefferson RP, Taylor DL, Weiß M, Garnica S, McCormick MK, Adams S, Gray HM, McFarland JW, Kull T, Tali K, Yukawa T, Kawahara T, Miyoshi K, Lee Y (2007) The evolutionary history of mycorrhizal specificity among lady's slipper orchids. Evolution 61:1380–1390
Shimodaira H, Hasegawa M (1999) Multiple comparisons of log-likelihoods with applications to phylogenetic inference. Mol Biol Evol 16:114–116
Simms EL, Taylor DL, Povich J, Shefferson RP, Sachs JL, Urb ina M, Tausczik Y (2006) An empirical test of partner choice mechanisms in a wild legume–rhizobium interaction. Proc R Soc Lond B 273:77–81
Sinclair WT, Mill RR, Gardner MF, Woltz P, Jaffré T, Preston J, Hollingsworth ML, Ponge A, Möller M (2002) Evolutionary relationships of the New Caledonian heterotrophic conifer, *Parasitaxus usta* (Podocarpaceae), inferred from chloroplast *trnL-F* intron/spacer and nuclear rDNA ITS2 sequences. Pl Syst Evol 233:79–104
Smith SE, Read DJ (2008) Mycorrhizal symbiosis, 3rd edn. Academic, London
Sniegowski PD, Gerrish PJ, Johnson T, Shaver A (2000) The evolution of mutation rates: separating causes from consequences. Bioessays 22:1057–1066
Stevens J (2004) Computational aspects of host–parasite phylogenies. Brief Bioinform 5:339–349
Stöckel M, Meyer C, Gebauer G (2011) The degree of mycoheterotrophic carbon gain in green, variegated and vegetative albino individuals of *Cephalanthera damasonium* is related to leaf chlorophyll concentrations. New Phytol 189:790–796
Struwe L, Albert VA (2002) Gentianaceae: systematics and evolution. Cambridge University Press, Cambridge
Swofford DL, Olsen GJ, Waddell PJ, Hillis DM (1996) Phylogenetic inference. In: Hillis DM, Mortiz C, Mable BK (eds) Molecular systematics, 2nd edn. Sinauer Ass, Sunderland, MA, pp 407–514
Tarver JE, Donoghue PCJ (2011) The trouble with topology: phylogenies without fossils provide a revisionist perspective of evolutionary history in topological analyses of diversity. Syst Biol 60:700–712
Taylor DL, Bruns TD (1997) Independent, specialized invasions of the ectomycorrhizal mutualism by two non-photosynthetic orchids. Proc Natl Acad Sci USA 94:4510–4515
Taylor DL, Bruns TD, Hodges SA (2004) Evidence for mycorrhizal races in a cheating orchid. Proc R Soc Lond B 271:35–43
Tedersoo L, Pellet P, Kõljalg U, Selosse MA (2007) Parallel evolutionary paths to mycoheterotrophy in understorey Ericaceae and Orchidaceae: ecological evidence for mixotrophy in Pyroleae. Oecologia 151:206–217
Thompson JN (2005) The geographic mosaic of coevolution. University of Chicago Press, Chicago, IL
Wang B, Qiu YL (2006) Phylogenetic distribution and evolution of mycorrhizas in land plants. Mycorrhiza 16:299–363
Webb CO, Ackerly DD, McPeek MA, Donoghue MJ (2002) Phylogenies and community ecology. Annu Rev Ecol Evol Syst 33:475–505
Webb C, Ackerly D, Kembel S (2011) Phylocom user's manual. Version 4.2. Available from http://www.phylodiversity.net/phylocom/phylocom_manual.pdf
Wickett NJ, Goffinet B (2008) Origin and relationships of the myco-heterotrophic liverwort *Cryptothallus mirabilis* Malmb. (Metzgeriales, Marchantiophyta). Bot J Linn Soc 156:1–12
Wickett NJ, Fan Y, Lewis PO, Goffinet B (2008a) Distribution and evolution of pseudogenes, gene losses, and a gene rearrangement in the plastid genome of the nonphotosynthetic liverwort, *Aneura mirabilis* (Metzgeriales, Jungermanniopsida). J Mol Evol 67:111–122
Wickett NJ, Zhang Y, Hansen SK, Roper JM, Kuehl JV, Plock SA, Wolf PG, dePamphilis CW, Boore JL, Goffinet B (2008b) Functional gene losses occur with minimal size reduction in the plastid genome of the parasitic liverwort *Aneura mirabilis*. Mol Biol Evol 25:393–401
Winther JL, Friedman WE (2007) Arbuscular mycorrhizal symbionts in *Botrychium* (Ophioglossaceae). Am J Bot 94:1248–1255
Winther JL, Friedman WE (2008) Arbuscular mycorrhizal symbionts in Lycopodiaceae. New Phytol 177: 790–801
Winther JL, Friedman WE (2009) Phylogenetic affinity of arbuscular mycorrhizal symbionts in *Psilotum nudum*. J Plant Res 122:485–496
Wu C-I, Li W-H (1985) Evidence for higher rates of nucleotide substitution in rodents than in man. Proc Natl Acad Sci USA 82:1741–1745

Yamato M, Yagame T, Shimomura N, Iwase K, Takahashi H, Ogura-Tsujita Y, Yukawa T (2011) Specific arbuscular mycorrhizal fungi associated with non-photosynthetic *Petrosavia sakuraii* (Petrosaviaceae). Mycorrhiza 21:631–639

Yokoyama J, Koizumi Y, Yokota M, Tsukaya H (2008) Phylogenetic position of *Oxygyne shinzatoi* (Burmanniaceae) inferred from 18S rDNA sequences. J Pl Res 121:27–32

Young ND, dePamphilis CW (2005) Rate variation in parasitic plants: correlated and uncorrelated patterns among plastid genes of different function. BMC Evol Biol 5:16

Yuan YM, Wohlhauser S, Moller M, Chassot P, Mansion G, Grant J, Küpfer P, Klackenberg J (2003) Monophyly and relationships of the tribe Exaceae (Gentianaceae) inferred from nuclear ribosomal and chloroplast DNA sequences. Mol Phylogenet Evol 28:500–517

Yukawa T, Ogura-Tsujita Y, Shefferson RP, Yokoyama J (2009) Mycorrhizal diversity in *Apostasia* (Orchidaceae) indicates the origin and evolution of orchid mycorrhiza. Am J Bot 96:1997–2009

Zimmer K, Hynson NA, Gebauer G, Allen EB, Allen MF, Read DJ (2007) Wide geographical and ecological distribution of nitrogen and carbon gains from fungi in pyroloids and monotropoids (Ericaceae) and in orchids. New Phytol 175:166–175

Zimmer K, Meyer C, Gebauer G (2008) The ectomycorrhizal specialist orchid *Corallorhiza trifida* is a partial myco-heterotroph. New Phytol 178:395–400

6 Progress and Prospects for the Ecological Genetics of Mycoheterotrophs

D. Lee Taylor, Craig F. Barrett, Gemma E. Beatty, Sarah E. Hopkins, Aaron H. Kennedy, and Matthew R. Klooster

6.1 Introduction

Mycoheterotrophic plants provide fascinating evolutionary narratives. Examples include convergent evolution of traits such as miniscule seeds, shortened and thickened roots, narrow mycorrhizal specificity, and self-fertilization (Leake 1994). Another example is provided by the patterns of decay of the photosynthetic machinery (DePamphilis and Palmer 1990; Barrett and Freudenstein 2008; Delannoy et al. 2011; Logacheva et al. 2011). Finally, the dynamics of specialization and host-jumps among various clades of fungal hosts (Taylor et al. 2004; Kennedy et al. 2011) offer compelling cases for comparison with more "main-stream" parasites, such as lice (Hafner and Page 1995), phytophagous insects (Hawthorne and Via 2001), or rust fungi (Jarosz and Burdon 1991). Another interesting feature of mycoheterotrophs, which is perhaps less often considered, is that multiple traits, including most of those mentioned above, appear to be evolving rapidly across numerous independent lineages. For example, from the limited sampling available at present, it appears that sister species of mycoheterotrophic plants always target different fungal clades (Taylor and Bruns 1999; Bidartondo and Bruns 2001, 2002; Kennedy et al. 2011). This phenomenon is too widespread to be merely coincidence. But we are currently without any clear understandings of the evolutionary dynamics and selective pressures that underlie this phenomenon. Investigation of selective pressures and rapidly evolving traits falls principally under the purview of population genetics, quantitative genetics, genomics, and molecular evolution. When applied to ecological questions, these fields form the basis for ecological genetics, the subject of this review. Ecological genetics of mycoheterotrophic plants is only in its infancy, hence in this review we suggest some ideas about ways in which ecological genetics might enlighten the field of mycoheterotrophic plant research going forward. We also present case studies of several fully mycoheterotrophic

D.L. Taylor (✉) • S.E. Hopkins
Institute of Arctic Biology, University of Alaska Fairbanks, 311 Irving I Building, Fairbanks, AK 99775, USA
e-mail: ltaylor@iab.alaska.edu

C.F. Barrett
Department of Plant Biology, Cornell University, 412 Mann Library, Ithaca, NY 14853, USA

Department of Biological Sciences, California State University, 5151 State University Drive, Los Angeles, CA 90032, USA

G.E. Beatty
School of Biological Sciences, Queen's University Belfast, 97 Lisburn Road, Belfast, Northern Ireland, BT9 7BL, UK

A.H. Kennedy
National Identification Services, USDA-APHIS-PPQ-PHP, 10300 Baltimore Avenue, B-010A, Beltsville, MD 20705, USA

M.R. Klooster
Centre College, Young Hall-121, 600 West Walnut Street, Danville, KY 40422, USA

V.S.F.T. Merckx (ed.), *Mycoheterotrophy: The Biology of Plants Living on Fungi*,
DOI 10.1007/978-1-4614-5209-6_6, © Springer Science+Business Media New York 2013

species, to illustrate some emerging trends, and further questions raised by these early findings.

Mycoheterotrophic plants present several challenges as potential model systems in ecological genetics. The most significant is the fact that they cannot routinely be grown in cultivation. Hence, standard methods such as controlled crosses (and resulting tools, such as recombinant inbred lines), common gardens, and most experimental manipulations are difficult or impossible to apply to mycoheterotrophic plants. On the other hand, methods for the study of the ecological genetics of wild populations have advanced significantly in recent years (Travers et al. 2007; Baird et al. 2008; Nadeau and Jiggins 2010; Helyar et al. 2011; Baxter et al. 2011). Today, ecological genetics and genomics have much to offer the study of the biology of mycoheterotrophic plants. Mycoheterotrophic plants are usually rare plants, with patchy local distributions, yet sometimes rather wide geographic distributions (see Chap. 3) as well as occasional cleistogamy, and other forms of inbreeding (Chap. 7). Hence, studies of mating systems are fundamental to an understanding of the microevolution of mycoheterotrophic plants. However, traditional methods of studying breeding systems, such as observations of pollinator behavior and experimental manipulation of pollination, are not adequate to characterize mating systems in many mycoheterotrophic plants. To understand mating and gene-flow, molecular methods based on multiple, independent, highly variable markers offer the best way forward (e.g., Klooster and Culley 2010). Population genetics summary statistics (Table 6.1) such as Hardy–Weinberg, heterozygosity, F_{st} can then be calculated to infer patterns of gene-flow and answer key questions about the degree of genetic differentiation within and among populations at various spatial scales. Marker data can also be subject to methods such as parentage analysis (Blouin 2003) and population assignment (Pritchard et al. 2000) to provide additional insights into within-population mating patterns, as well as population boundaries where they are not obvious from morphology or geography. As described in several of the case studies below, even species boundaries are often ambiguous in mycoheterotrophic plants, due to highly variable floral morphologies and the occurrence of nearly continuously variable intermediate forms. An absence of gene-flow between populations based on molecular markers can provide strong evidence for the existence of distinct biological species. This contemporary perspective is complementary to a historical perspective based upon multilocus, sequence-based phylogenetic analyses (e.g., Kennedy and Watson 2010) for distinguishing lineages that are on independent evolutionary trajectories.

Understanding species and population boundaries and elucidating patterns of gene-flow and genetic variation set the stage for investigation of adaptation and natural selection. Numerous approaches have been developed, only a few of which will be mentioned here. One fundamental definition of evolution is change in allele frequencies in a population over time. Hence, investigation of the genetic constituency of mycoheterotrophic plant populations over life stages, generations, space and time can alert us to situations in which evolution has taken place. If, for example, certain alleles or genotypes are consistently lost in the transition from zygotes to adult mycoheterotrophic plants, selection against those genotypes is likely. A major emphasis in ecological genetics is the identification of genes that influence traits of interest, especially those that are related to fitness differences underlying ecological adaptation (Hohenlohe et al. 2010). Furthermore, reconstruction of the geographic and demographic histories of populations and species is a rapidly advancing, energetic field (Huang et al. 2011). All of these subdisciplines are becoming exponentially more informative as the genome-wide distribution of variable markers increases in a greater diversity of taxa. In tandem with improvements in molecular tools, advances in related "–omics" technologies are providing radical new opportunities to investigate functional genomics, even in non-model organisms. For example, next-generation sequencing of cDNAs can provide a detailed snapshot of gene expression in any organism (Mortazavi et al. 2008).

Table 6.1 Summary genetic statistics for selected mycoheterotroph species and varieties

Family (Subfamily)	Species	Sampling distribution	Natural forms/variants	Breeding system	*N*	*S*	π	*#h*	*hd*	He	Ho	F_{is} (f)	F_{st} (theta)	Data type	Citation
Ericaceae (Monotropoideae)	*Monotropa hypopitys*	Eastern North America	Red color form	Facultatively xenogamous; partial autogamy	41.9					0.25	0.33	−0.364–0.286	0,12	microsatellites	Klooster and Culley (2010)
			Yellow color form	Mixed breeding; primarily autogamous	31.6					0.2	0.14	−0.358–0.503	0.13–0.61		
		Northern Ireland	Yellow color form		141					0.492–0.507		0.339–0.497	n/a	microsatellites	Beatty and Provan (2011a)
Orchidaceae (Epidendroideae: Calypsoeae)	*Corallorhiza striata* complex (nuclear DNA)	Southwestern USA, Mexico	var. *vreelandii*	Facultatively xenogamous; partial autogamy	46	6	0.0016	3[a]	0.4	0.39	0.11	n/a	n/a	Nuclear intron sequences	(Barrett et al. 2010); Barrett and Freudenstein (2011)
		Northern USA, southern Canada	var. *striata*	Primarily xenogamous; some autogamy	75	10	0.004	4.33	0.51	0.54	0.5	n/a	n/a		
		Sierra Nevada, California, USA	Californian populations	Primarily xenogamous; some autogamy	34	18	0.0066	5.67	0.69	0.69	0.61	n/a	n/a		
		Mexico	*Corallorhiza involuta*	Primarily autogamous, xenogamy unobserved	7	0	0	1	0	no var	no var	no var	no var		
		Virginia, West Virginia, USA	*Corallorhiza bentleyi*	Primarily autogamous, xenogamy unobserved	6	0	0	1	0	no var	no var	no var	no var		
												0.04 (phi-is, overall)	0.45[b] (phi-ct, overall)		
	C. striata complex (plastid DNA)	Southwestern USA, Mexico	var. *vreelandii*	Facultatively xenogamous; partial autogamy	54	11	0.0011	14	0.86	n/a	n/a	n/a	n/a	Sequences from 2 plastid regions	Barrett et al. (2010); Barrett and Freudenstein (2011)
		Northern USA, southern Canada	var. *striata*	Primarily xenogamous; some autogamy	94	18	0.0019	15	0.8	n/a	n/a	n/a	n/a		
		Sierra Nevada, California, USA	Californian populations	Primarily xenogamous; some autogamy	39	4	0.0005	5	0.71	n/a	n/a	n/a	n/a		
		Mexico	*C. involuta*	Primarily autogamous, xenogamy unobserved	8	1	n/a	2	0.24	n/a	n/a	n/a	n/a		
		Virginia, West Virginia, USA	*C. bentleyi*	Primarily autogamous, xenogamy unobserved	7	0	0	1	0	n/a	n/a	n/a	n/a		
												n/a	0.45[b] (phi-ct, overall)		

(continued)

Table 6.1 (continued)

Family (Subfamily)	Species	Sampling distirbution	Natural forms/variants	Breeding system	*N*	*S*	π	*#h*	*hd*	He	Ho	F_{is} (f)	F_{st} (theta)	Data type	Citation
Orchidaceae (Epidendroideae: Bletiinae)	*H. arizonica*	Texas, New Mexico, Arizona		Mostly autogamous, some falcultatively xenogamous in Arizona	15	1	0.0002	2	0.16	n/a	n/a	n/a	0.9923 (4)	Sequences from 6 plastid regions	Unpublished, sequence data from Kennedy and Watson (2010)
	H. colemanii	Arizona		facultatively xenogamous	4	1	0.0002	2	0.5	n/a	n/a	n/a	0.9926		
	H. nitida	Sierra Madre Oriental, Mexico; Texas, New Mexico		mostly autogamous, some falcultatively xenogamous	5	2	0.000399	3	0,7	n/a	n/a	n/a	0.9923		
	H. parviflora	Western Mexico and northern Guatemala		Xenogamy observed, autogamy not observed	3	4	0.001197	3	1	n/a	n/a	n/a	0.9914		
	H. revoluta	Sierra Madre Oriental, Mexico; New Mexico, Texas; USA		Facultatively xenogamous	5	0	0	1	0	n/a	n/a	n/a	0.9935		
	H. spicata	Northern Mexico; Arizona east through Texas and the eastern US north to Virginia		Only xenogamy observed	11	3	0.0004	4	0.69	n/a	n/a	n/a	0.9917		
													Overall *F*st = 0.9923		
Orchidaceae (Epidendroideae: Calypsoeae)	*Corallorhiza maculata*	USA except the Great Plains and Southeast, Northern Canada		Primarily autogamous, facultatively xenogamous	47					0.575	0.140[b]	0.7622	0.2821	Microsatellites	Hopkins and Taylor (2011)
	Corallorhiza mertensiana	Western USA, Western Canada and Alaska		Partially autogamous; little data	10					0.527	0.575				

N sample size, *S* # segregating sites, π nucleotide diversity, *#h* number of alleles/haplotypes (nuclear DNA/organellar DNA), *hd* haplotype diversity, *He* expected heterozygosity, *Ho* observed heterozygosity

[a]=for *C. striata* nrDNA, mean of F3H intron II, RPB2 5′ region, RPB2 3′ region

[b]Indicates significant deviation from H–W frequency, F_{is} (*f*) = inbreeding coefficient, F_{st} = proportion of variation between demes (phi-ct = proportion of variation between groupings of CLADES (4 hierarchical levels))

6.2 Population Genetic Analysis of *Hypopitys monotropa* (Ericaceae)

The Monotropoideae are eudicots belonging to the Ericales, and include the fully mycoheterotrophic genera *Allotropa, Pleuricospora, Pterospora, Sarcodes, Monotropa, Hypopitys, Monotropastrum, Monotropsis, Pityopus* and *Hemitomes* (Chap. 2). Though the green genus *Pyrola* has traditionally been placed within the Monotropoideae, molecular systematic studies have not yet resolved the closest photosynthetic relatives to the fully mycoheterotrophic Monotropoideae (Cullings 2000; Kron et al. 2002).

Studies of the Monotropoideae have historically played a fundamental role in enhancing our understanding of the biology of mycoheterotrophic plants and their fungal associates. Specifically, *Hypopitys monotropa* Crantz (syn. *Monotropa hypopitys* L.), has functioned as a model system for investigating many aspects of the biology of mycoheterotrophic plants, with pioneering investigations including the nature of mycorrhizal infections (Björkman 1960), mycorrhizal symbioses (e.g., Bidartondo and Bruns 2002, 2005; Leake et al. 2004), developmental biology (Olson 1990, 1993), reproductive ecology (Klooster and Culley 2009), and the life history chronology of mycoheterotrophic plants from seed to reproductively mature adult (Leake et al. 2004; Bidartondo and Bruns 2005). Despite the importance of this species to our understanding of mycoheterotrophic plants, *H. monotropa* has also created much confusion among scientists since it was first described by Carl Linnaeus over 250 years ago as *M. hypopitys*. Ironically, the taxonomic fate of this species was predestined for quandary from its very inception, as even Linnaeus misspelled the species name "*hypopithys.*"

Taxonomists have long struggled to resolve the relationships among various color forms and morphs of *H. monotropa* throughout its circumboreal distribution. In fact, the species has undergone over 85 taxonomic rearrangements dating back to Linneaus' initial, errant circumscription, with classifications ranging from multiple species within the genus *Hypopitys*, to a single species within the genus *Monotropa*, with ascribed subspecies, variety, and form epithets. Much of this confusion arises from the lack of thorough ecological investigations in this system. Additionally, many taxonomic arguments have been constructed from assessments of pressed herbaria specimens, with the dramatic changes in color and loss of diagnostic morphological features of dried tissues clouding our resolution to effectively assign taxonomic identities. The taxonomic dilemma of *H. monotropa* was further investigated over the past 10 years through the use of molecular systematics, with studies by Cullings (2000), Bidartondo and Bruns (2002), and Neyland (2004). Despite these thoughtful attempts to elucidate the true taxonomic identity of *H. monotropa*, our understanding of the evolutionary ecology of the various color forms and morphs has remained unresolved.

A recent study of the reproductive ecology of two genera within the Monotropoideae (Klooster and Culley 2009) identified distinct ecological differences between color forms within *H. monotropa*. Specifically, this study reconfirmed the presence of discrete blooming periods presented by Neyland (2004), with the yellow color form exhibiting a summer blooming phenology (June–August) and the red color blooming in the fall (September–October). Also, breeding system differences were identified between populations of each color form, with the yellow form exhibiting a mixed breeding system with high rates of autogamous, self-pollination, and the red form approaching herkogamy (spatial separation between anthers and stigma) and facultative xenogamy (movement of pollen among genetically distinct plants; see Chap. 7). The functional data presented in this study supported the hypothesis that forms within *H. monotropa* possess some ecological differences that extend beyond natural plasticity in colouration and that may be attributable to genetic variation.

Because phylogenetic analyses had previously been somewhat unsuccessful at assigning a conclusive genetic identity to the color forms of *H. monotropa*, Klooster and Culley (Klooster and

Culley 2010) utilized population genetic techniques to investigate *H. monotropa*. Using seven populations consisting of red and yellow color forms of *H. monotropa* growing in sympatry and allopatry in the Ohio River Valley, USA, Klooster and Culley assessed levels of genetic variation within populations, genetic differentiation among populations, and genetic structuring by color form. Results from this investigation demonstrated low to moderate levels of within-population genetic variation, with higher variation present in the facultatively xenogamous, red color form, and relatively low levels occurring within the primarily autogamous yellow form (see Table 6.1). Furthermore, pair-wise comparisons of some populations of the yellow color form exhibited high levels of genetic differentiation, although this did not significantly correlate with geographic distance, demonstrating the possibility that various extrinsic and/or intrinsic factors may be contributing to genetic substructuring within the yellow color form. All pair-wise comparisons between yellow and red color forms growing in both allopatry and sympatry indicated high levels of genetic differentiation (Table 6.1). Finally, population genetic spatial analyses revealed high levels of genetic structuring by color form, indicating minimal gene-flow and strong genetic divergence between color forms, suggesting reproductive isolation and corresponding speciation between "color forms."

The synergistic culmination of historical, ecological, reproductive, and genetic analyses of *H. monotropa* conclusively demonstrates that forms of *H. monotropa* possess discrete traits which merit taxonomic recognition as separate species. Also, these multifaceted analyses illuminate "hidden" divergence between populations that might have otherwise gone ignored given the minimal and sometimes ambiguous morphological differences present in this system. In both of the aforementioned studies, minimal to no morphologically discrete features revealed the existence of intriguing genetic differences and informative patterns. Clearly, merging fine-scale population genetic analysis with broad scale phylogeographic assessments will provide a vastly enhanced understanding of the evolutionary history of *H. monotropa*, as we have only begun to scratch the surface of ecological genetic discoveries.

In 2011a, Beatty and Provan published a complimentary study examining peripheral populations of the yellow form of *H. monotropa* in Northern Ireland (Beatty and Provan 2011a). These remnant populations are located at the western edge of the species European range and are small and highly fragmented. Microsatellite analysis of the 21 extant populations, which occur in two separate areas, revealed high levels of genetic diversity, genetic structuring and inbreeding. *H. monotropa* is a highly self-compatible species (Klooster and Culley 2009) and with numbers of individuals in patches generally low, the high incidence of inbreeding is probably due to increased self-pollination. Reproduction in these populations was found to be predominantly sexual, but several small clones were detected. This is in contrast to *Orthilia secunda*, another member of the Monotropoideae that exhibits a similar distribution to *H. monotropa* in Northern Ireland, where all populations studied each comprised a single clone (Beatty et al. 2008). Although typical range-edge population dynamics (small and fragmented populations) were evident for both species, it is thought that the observed switch to clonal growth in *O. secunda* might reflect the response of this boreal species to global climate change.

A larger scale phylogeographic study by Beatty and Provan (Beatty and Provan 2011b) investigated the glacial history of *H. monotropa* in North America. The species exhibits an East–West disjunct distribution, with additional pockets of the species found in central North America. Phylogeographic analysis using the chloroplast *rps*2 gene, the nuclear internal transcribed spacer (ITS) region and eight microsatellite loci revealed that the present-day distribution is due to persistence in separate eastern and western refugia during the last glaciation. Patterns of genetic variation, namely levels of diversity and the occurrence of unique haplotypes, indicated two western refugia in Oregon and further north in the Alexander Archipelago. In the eastern part of the species range, refugia were identified in the south, around the area of the Carolinas, as well as further north in the "driftless region," just south of the ice sheets. Due to the parasitic nature of *H. monotropa*, presence of a host tree and associated fungal species would have been fundamental to existence in any refugial location, and palynological records confirmed the existence of host tree species in all areas during the glaciations. Unlike *H. monotropa*, *O. secunda*, which has a similar contemporary East–West disjunct distribution, was confined to exclusively western refugia during the LGM based on chloroplast sequencing and nuclear microsatellites (Beatty and Provan 2010). Its present-day distribution has resulted from eastward postglacial recolonization following the retreat of the ice sheets, and loss of central populations as forests were replaced by the grasslands of the Great Plains.

Phylogeographic studies in Europe again revealed some differences between the two species (Beatty and Provan 2011c). In this case, patterns of genetic variation generally corresponded more to the classic scenario of "southern richness vs. northern purity" observed in numerous European phylogeographic studies (Taberlet et al. 1998; Hewitt 1999; Provan and Bennett 2008). Southern refugia were identified in the Balkans/southeast Europe for *H. monotropa*, and in the French and Austrian Alps and Slovakia for *O. secunda*. Populations of *H. monotropa* in recolonized areas, however, were much more genetically depauperate than those of *O. secunda*, possibly indicating more northerly persistence of the latter species, which is cold-tolerant, during the glaciations. Models of future species distributions under climate change scenarios suggested that loss of rear-edge populations will have a disproportionally greater effect on *H. monotropa* than in *O. secunda*, since these populations harbor the majority of the genetic variation. A similar scenario has recently been reported in other species (Alsos et al. 2012; Provan and Maggs 2012), and these studies highlight the importance of taking into account the distribution of genetic variation across species ranges when considering the potential effects of climate change and population extinction (Hampe and Petit 2005).

The extrinsic and intrinsic mechanisms underlying the observed population genetic patterns and possible cryptic species in *H. monotropa* still remain primarily inferential. Continuing such analyses, coupled with a corresponding investigation into the identity and diversity of *Tricholoma* fungal associates will more completely elucidate both historic and contemporary factors influencing the evolutionary ecology of this fascinating symbiosis. Given the intrinsic reliance of mycoheterotrophic taxa upon their fungal associates, it is possible that a substantial portion of the population genetic patterns discovered in the aforementioned studies will correspond with equally intriguing evolutionary dynamics of interactions with their fungal partners. Also, human-mediated habitat destruction and fragmentation has likely exacerbated the degree of genetic structuring among populations, contributing to further reproductive isolation and divergence of populations. Consequently, ecological investigations coupled with genetic analyses are required to better understand the evolutionary processes behind speciation within and among lineages of the Monotropoideae.

6.3 Hybrid Origin of Monotropoid Taxa

The process of hybridization as a mechanism of speciation within the Monotropoideae has been proposed numerous times throughout the literature. Specifically, Wallace (1975) observed intermediate pinkish color forms of *H. monotropa* growing among yellow and red forms in the western United States, suggesting the possibility that hybridization may be responsible for producing intermediate, hybrid color forms. Also, Cullings (2000) found polyphyletic placement of *H. monotropa* within his phylogenetic reassessment and suggested the possibility of having sampled individuals arising from hybridization between *H. monotropa* and *Pterospora* growing in sympatry in Yellowstone National Park, Wyoming, USA or, alternatively, he argued that the samples represented a cryptic new species that is morphologically similar to *H. monotropa*.

It has also been suggested that *Pityopus californicus* represents a possible hybrid between the parent lineages of *Hemitomes congestum*, *Pleuricospora fibriolata*, and *H. monotropa* (G. Wallace pers. comm., in Cullings 2000; Neyland 2005). Given the high degree of morphological similarity between *P. californicus* and *H. monotropa*, both species were once classified within the genus *Monotropa* and *P. californicus* was later given the name *Hypopitys californica* (Eastwood 1897, 1902). Cullings (2000) and Neyland (2005) further entertained the suggestion that *Pityopus* arose from hybrid origin, although subsequent molecular analyses failed to conclusively support this theory.

Although hybridization remains a viable mechanism for speciation in the Monotropoideae and may account for some of the morphological and molecular variation observed within particular lineages, it has not yet been empirically shown to occur. Additionally, it is unclear how hybridization might impact the ability of a mycoheterotroph to successfully recruit and associate with fungal hosts. The many factors that have limited empirical assessment of hybridization in this and similar systems include high level of molecular divergence among lineages, discrete reproductive phenologies that preclude cross-pollination manipulations, and our inability to cultivate these mycoheterotrophic species in controlled settings.

6.4 *Hexalectris spicata* Species Complex

Hexalectris is a New World genus of nine fully mycoheterotrophic terrestrial orchid species belonging to the derived "higher" Epidendroideae, and appears to be most closely related to the photosynthetic genera *Basiphyllaea* and *Bletia*, collectively the Bletiinae (Goldman et al. 2002; Sosa 2007). Within *Hexalectris* are six species that comprise the *H. spicata* species complex and share a floral architecture that is distinctive from those of the remaining species, *H. brevicaulis*, *H. grandiflora*, and *H. warnockii* (Goldman et al. 2001). Morphological variation among members of the *H. spicata* complex is primarily in terms of floral characteristics, although height, thickness, and color are also informational. Floral variation within these species is relatively low; however higher for the more wide-ranging members, such as *H. arizonica* and *H. nitida* which each exhibit chasmogamous (open-flowered) and cleistogamous (closed-flowered) floral forms with an accompanying loss of rostellar flaps and reduced flower size.

Members of the *H. spicata* species complex range throughout the distribution of *Hexalectris*, which generally follows the mountainous regions of Mexico (Sierra Madre Occidental, Sierra Madre Oriental, Sierra Madre del Sur, and the Trans-Mexican Volcanic Belt) and their extensions into the United States. Observed species richness is at its height in these mountainous regions near the borders between Texas, USA and Coahuila, MX; and Arizona, USA and Sonora, MX. Overlaying species-level distributions while considering phylogenetic relationships within this complex (Kennedy and Watson 2010) reveals a high degree of sympatry between well-supported sister species and suggests that the Sierra Madre Oriental and the Sierra Madre Occidental have facilitated independent northward migration routes for this radiation and potential contact zones for hypothesized hybrid progenitors (Kennedy and Watson 2010).

Members of the *H. spicata* species complex persist mostly undetected due to rarity and inconspicuous and inconsistent flowering patterns across this distribution. Collectively this group exploits a wide array of habitats from tropical dry forests of western Mexico, oak-lined desert canyons of the Santa Rita Mountains, juniper woodlands of the Edwards Plateau, mixed conifer-hardwood forests on the peaks of the Chisos Mountains in the Big Bend, to the mixed mesophytic hardwood forests of Appalachia (Liggio and Liggio 1999; Coleman 2002; Kennedy and Watson 2010).

Independent and combined phylogenetic analyses of the nuclear ITS region and six plastid

DNA regions (*trnL*$^{(uaa)}$ intron (*trnL*), *trnL*$^{(uaa)}$–*trnF*$^{(gaa)}$ intergenic spacer (*trnL-F* IGS), *matK*, *psbA*, *atpA*, and *accD*) revealed strong support for the monophyly of this complex and six phylogenetic species, *H. spicata*, *H. arizonica*, *H. nitida*, *H. revoluta*, *H. colemanii*, and *H. parviflora* (Kennedy and Watson 2010). The derived phylogenetic position of the *H. spicata* complex within *Hexalectris*, a small number of morphological synapomorphies for each species, a low level of differentiation in ITS sequences, and poor support for the position of some clades in all trees suggests a recent, rapid, and continuing radiation in this lineage (Kennedy and Watson 2010). Hybridization has been proposed as an explanation for morphological intermediacy in floral traits in some species (Catling and Engel 1993), and for incongruent phylogenetic positions of some clades between plastid and nuclear gene trees (Kennedy and Watson 2010). However, strong evidence for hybridization remains elusive. For example, Kennedy and Watson (2010) detected no evidence for hybridization after statistically testing for recombination among ITS clones from putative hybrid species. Also, population genetic analyses of several plastid DNA regions reveals that genetic diversity is almost entirely distributed among species (Table 6.1), providing evidence that although hybridization may have occurred historically, each modern lineage is likely reproductively isolated and, therefore, an independent biological species (Kennedy et al. unpublished data).

A particularly intriguing example of potential historical hybridization may be found within the morphological species *H. spicata* s.l., which contains the cryptic phylogenetic species *H. spicata* and *H. arizonica* (Fig. 6.1; Kennedy and Watson 2010). *H. spicata* s.l. ranges from Virginia, south to central Florida, and west to Missouri and Arizona. Its varieties, *spicata* and *arizonica*, may be distinguished based on the presence or absence of a rostellar flap, respectively (Catling and Engel 1993). In the eastern portion of this species' range (i.e., east of Texas) only chasmogamous plants with a well-developed rostellum may be found, whereas cleistogamous and chasmogamous plants with a variety of rostellum development may be found throughout Texas and Arizona (Kennedy and Watson 2010). In Texas, differentiating between these taxa is clear; closed-flowered plants lack rostellar flaps and open-flowered plants have well-developed flaps. However in Arizona, plants may be found with almost any combination of these characteristics, including open flowers and no rostellar flap, making identification difficult. Phylogenetic analyses provided strong support for two clades, one containing only plants with open flowers and a well-developed rostellum from the eastern portion of the distribution and Texas (i.e., *H. spicata* s.s.), and another comprised exclusively of plants with cleistogamous flowers lacking a rostellum from throughout Texas and all plants from Arizona (i.e., *H. arizonica*). Despite this clarity, incongruent positions of the *H. arizonica* clade between nuclear and plastid gene trees suggested that this lineage may be the product of hybridization between *H. spicata* s.s. or *H. nitida* and some other member of the *H. spicata* species complex (Kennedy and Watson 2010).

Hexalectris species maintain long-lived rhizomes with highly reduced roots that are the sites for mycorrhizal colonization (Taylor et al. 2003). Members of the *H. spicata* complex associate nearly exclusively with fungi from Sebacinaceae subgroup A (Taylor et al. 2003; Kennedy et al. 2011), a group of ectomycorrhizal fungi known to simultaneously form endomycorrhizas with orchids (Selosse et al. 2002a, b, 2009). Specificity, as measured by the average genetic pair-wise distances between fungal associates of a particular *Hexalectris* species (i.e., π; Nei and Tajima 1981), revealed narrow associations within each species, ranging between 0.012 in *H. arizonica* and 0.047 in *H. spicata* (Kennedy et al. 2011). Interestingly, the two primarily self-pollinating species in this group, *H. arizonica* and *H. nitida*, are the most highly specialized toward their mycorrhizal fungi (0.012 and 0.021, respectively), suggesting that specialization toward mycorrhizal partnerships may be narrowed with the loss of genetic diversity that accompanies self-pollination (Kennedy et al. 2011). From a phylogenetic perspective, each species within this complex was identified to associate with a unique clade or group of clades within

Fig. 6.1 Photos of the phylogenetic species that comprise the *Hexalectris spicata* species complex (*sensu* Kennedy and Watson 2010). *H. spicata* (a-b, Harrison Co., Indiana, USA) and the open-flowered form of *H. arizonica* (c, Pima Co., Arizona, USA) are difficult to distinguish morphologically, but fortunately for identification purposes only the closed-flowered form of *H. arizonica* (d, Santa Cruz Co., Arizona, USA; e, Dallas Co., Texas, USA) is sympatric with *H. spicata*. *H. nitida* also has open- and closed-flowered forms (f, Brewster Co., Texas, USA; g, Dallas Co., Texas, USA; respectively). *H. parviflora* (h, Cuquío, Jalisco, Mexico), *H. colemanii* (i, Pima Co., Arizona, USA), and *H. revoluta* (j, Brewster Co., Texas, USA) complete this species complex. Photographs by Aaron H. Kennedy

Sebacinaceae subgroup A, suggesting that specialization is not only narrow within this complex, and narrower within each of its species, but that specificity is a rapidly evolving characteristic.

Finally, these data also revealed that many of the mycorrhizal fungi that *H. spicata* species complex members specialize on are widely distributed across North America and have largely sympatric distributions. This finding supported the conclusion that the geographic distributions of *Hexalectris* species may not be influenced by their fungal associates' distributions, a finding also made in *Cypripedium* (Shefferson et al. 2007). For example, the sebacinaceous fungi identified from *H. spicata* in North Carolina alone span nearly the entire breadth of associations formed by this species across its total geographic distribution, revealing wide distribution for these sebacinaceous fungi. Also, even though the extremely rare *H. colemanii* is restricted to only a few populations in southern Arizona, the fungi identified from two populations in highly similar habitats and only a few dozen kilometers apart were widely distant members of Sebacinaceae subgroup A, and identified from other members of the *H. spicata* species complex ranging from western Mexico to the eastern United States (Kennedy et al. 2011). These findings therefore also suggest that mycorrhizal host-jumps and preferences have evolved largely where *Hexalectris* species and their potential mycorrhizal fungi exist in sympatry.

6.5 The Genus *Corallorhiza*

Corallorhiza is a new world genus of about ten mycoheterotrophic species that falls within the Epidendroideae, with the closest leafy relatives belonging to the genera *Cremastra*, *Oreorchis* (both Asian), and *Aplectrum* (North American) (Freudenstein and Senyo 2008). *Corallorhiza* species occupy temperate forests across North America; only *Corallorhiza trifida* has spread widely across boreal regions, achieving a pan-Arctic distribution.

6.5.1 *Corallorhiza striata* Complex

From a phylogeographic perspective, *Corallorhiza striata* (Orchidaceae) is one of the most comprehensively studied fully mycoheterotrophic species complexes. This wide-ranging, North American, temperate-montane group displays extensive geographic variation in floral morphology (Fig. 6.2; Freudenstein 1997; Barrett and Freudenstein 2009, 2011). Using plastid DNA

Fig. 6.2 The *Corallorhiza striata* species complex. From the left, *Corallorhiza involuta* (Morelos, Mexico), *Corallorhiza bentleyi* (cleistogamous population, Virginia, USA), *C. striata vreelandii* (New Mexico, USA), *C. striata* from California, USA, and *C. striata striata* (the only known population in New York State, USA). Photographs by Craig Barrett and John Freudenstein

(*rbcL*, *rpl32-trnL*), Barrett and Freudenstein (2009) demonstrated that the *C. striata* complex also displays substantial genetic diversity ($n=84$ individuals). Four largely allopatric plastid DNA clades were identified across North America, and no populations were found to harbor members of more than one clade, even in potential contact zones. These clades are distributed as follows: clade (a) *Corallorhiza involuta* (Mexico) + *Corallorhiza bentleyi* (Virginia and West Virginia, USA), clade (b) Sierra Nevadan clade (California, USA), clade (c) *C. striata* var. *vreelandii* (southwestern USA, Mexico, and Newfoundland, Canada), and clade (d) *C. striata* var. *striata* (northern USA, Canada). Overall flower size in this complex appears to be clinal, roughly increasing with latitude (Freudenstein 1997). The incorporation of plastid DNA information gives a different perspective on morphological variation, with each clade corresponding to a distinct morphometric grouping.

The four plastid clades associate with overlapping—yet significantly divergent—members of a single, highly variable, ectomycorrhizal fungal taxon, *Tomentella fuscocinerea* (Fig. 6.3; Barrett et al. 2010). This fungal species may, in fact, be composed of 9–12 cryptic species, based on 3 % and 2.5 % nuclear ITS divergence criteria, respectively. Thus, the *C. striata* complex has highly specific nutritional preferences across the entire geographic range. Overall, gene trees based on plastid DNA (*C. striata* complex) and ITS (*T. fuscocinerea*) were incongruent but also significantly nonindependent, suggesting some level of cophylogeographic structure (Fig. 6.3). In particular, the plastid clade endemic to California associates almost exclusively with a single ITS clade of *T. fuscocinerea*; this finding was consistent across multiple populations along a ca. 600 km transect through the Sierra Nevada. Overall, patterns of subspecificity within the *C. striata* complex reveal a "geographic mosaic" (Thompson 1994, 2005), with orchid plastid types associating with divergent sets of *T. fuscocinerea* fungi in different biogeographic regions of North America. This study represents one of the most extensive phylogeographic investigations of host specificity for any mycoheterotroph to include both plant and fungal DNA, and has conservation implications for *C. striata* and the habitats in which these and several other mycoheterotrophic species occur. Analysis of the *C. striata* complex based on nuclear intron sequences (flavanone-3 hydroxylase, RNA polymerase II beta subunit) showed less geographic structuring than did plastid DNA, with alleles shared between plastid groupings/geographic regions. *C. bentleyi* (eastern USA) and *C. involuta* were genetically identical for all loci, forming a clade that was highly divergent from the remaining *C. striata* (= *sensu stricto*). A closer investigation of population structure within *C. striata* s.s. identified three geographically parapatric clusters, corresponding to var. *vreelandii*, var. *striata*, and Californian populations Interestingly, a few individuals of both var. *striata* and Californian populations displayed admixed multilocus genotypes, suggesting either limited gene-flow between groupings or residual ancestral polymorphism. Multilocus distance estimates of relationships within *C. striata sensu stricto* based on nuclear alleles indicated strong evidence for divergence between var. *vreelandii* and var. *striata*, with Californian individuals occupying intermediate positions relative to both. *C. striata*, *C. striata vreelandii*, and the Californian accessions (i.e., var. *californica* ined) are best described at the level of variety, thus comprising a widespread, highly variable (and geographically structured) species, *C. striata sensu stricto*. Furthermore, the varieties therein represent evolutionarily significant units (ESUs) for conservation purposes (*sensu* Dizon et al. 1992; Moritz 1994).

Based on cumulative integration of genetic, morphological, geographic, phenological, and reproductive-mode variation, there is evidence to suggest that the *C. striata* complex is composed of three species: *C. bentleyi*, *C. involuta*, and *C. striata* s.s. There is ample evidence for largely autogamous modes of reproduction in both *C. bentleyi* and *C. involuta* based on: (1) small, drab-colored, partially closed flowers (some populations of *C. bentleyi* are fully cleistogamous), (2) the tendency for all flowers in a raceme to set seed (even before anthesis), and (3) low plastid/

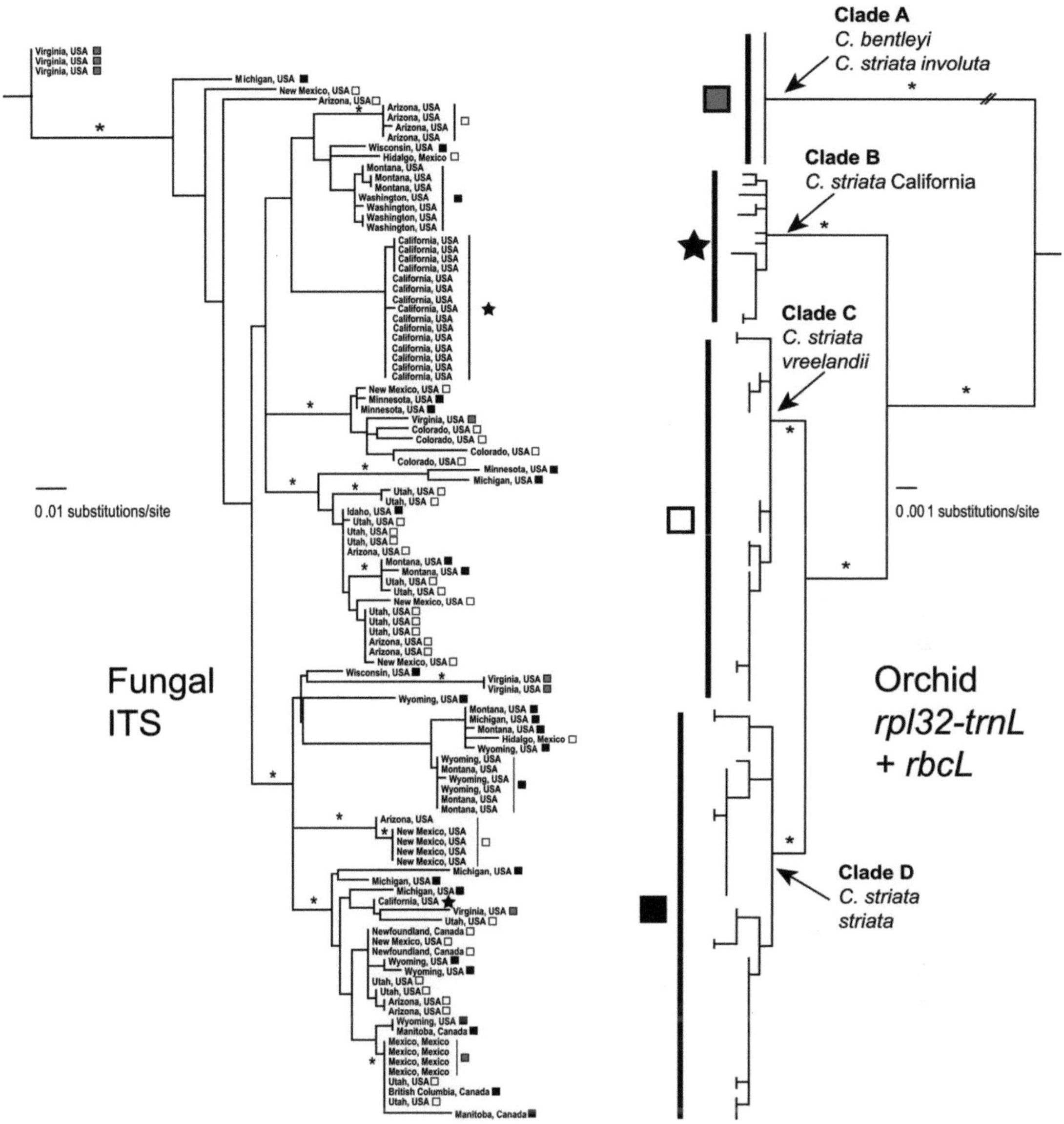

Fig. 6.3 Comparison of fungal internal transcribed spacer (ITS) and orchid plastid *rbcL*+*rpl32-trnL* (right) gene trees for a rangewide sampling of *Corallorhiza striata* complex populations across North America. Trees are represented by highest likelihood topologies under the GTR-Γ model in RAxML (Stamatakis 2006). Asterisks adjacent to branches indicate likelihood bootstrap values>90 % based on 2,000 pseudoreplicates. *Left scale bar* (fungi)=0.01 substitutions/site and right scale bar (orchids)=0.001 substitutions/site. Accessions in fungal ITS tree are coded by US/Mexican State, or by Canadian Province, and with symbols corresponding to orchid DNA clade (also found on orchid DNA tree). Clade A (*Corallorhiza bentleyi, Corallorhiza involuta*)=*gray squares*; Clade B (Californian *C. striata*)=*black stars*; Clade C (*C. striata vreelandii*)=*empty squares*; Clade D (*C. striata striata*)=black squares

nuclear genetic diversity. The latter observation may be a cause for conservation concern, in that adaptive potential (similar to neutral variation) could be reduced in autogamous, genetically depauperate species, rendering them unable to cope with future environmental fluctuations (e.g., climate change). Alternatively, they may represent highly adapted, fixed genotypes in populations that have been historically purged of deleterious alleles; this certainly deserves further study.

C. striata s.s. displays morphological and genetic attributes concordant with a xenogamous reproductive mode. Several observations have been made of visitation and pollinium transfer to/from flowers by ichneumonid wasps known to pollinate *C. striata* (Freudenstein 1997; C. Barrett, pers. observ.). *C. striata* s.s. is likely to be reproductively isolated from *C. involuta* and *C. bentleyi*, and the same is also true between the latter two entities based on their autogamous reproductive modes and the immense geographic distances separating them.

Members of the *C. striata* complex are of great conservation interest, because they are good candidates as both "indicator" and "umbrella" species. The former concept refers to the idea that *C. striata* represents a window into otherwise elusive soil processes; its fungal host (*T. fuscocinerea*) is rare and extremely ephemeral (in terms of sporocarp formation), having been collected very few times (U. Kõljalg, pers. comm.). Thus, *C. striata* is highly sensitive to habitat perturbations affecting soil microbial processes, and is typically only found in old-growth habitats. This *C. striata* complex is also composed of umbrella species in that any conservation decisions focused toward them will most certainly benefit co-occurring species, many of which are sensitive/rare as well (e.g., other orchids and mycoheterotrophs).

6.5.2 *Corallorhiza wisteriana–odontorhiza* Complex

Corallorhiza odontorhiza ranges from Nicaragua northward through Mexico into the eastern USA/Canada. *Corallorhiza wisteriana* is distributed from Chiapas, Mexico through the Sierras Madre Oriental and Occidental, into the USA (Freudenstein 1997). In the USA, the latter becomes disjunct between the southeastern USA and a portion of the Rocky Mountains to the west, separated by the Great Plains and arid regions of north-central Mexico (Freudenstein 1997). Both species have similar flowering times in Mexico (winter), but diverge phenologically in more northern populations. In the USA, *C. wisteriana* flowers in the spring, while *C. odontorhiza* flowers in the autumn. Both species show evidence for a largely autogamous reproductive mode (e.g., *C. odontorhiza* is mostly cleistogamous), but this remains to be further investigated in the group as a whole. In the eastern US portion of its range, *C. wisteriana* primarily occurs in deciduous forests, whereas in the West and in Mexico, it occurs in higher elevation conifer forests. Plastid DNA analyses based on *rpl32-trnL* and *trnL-F* indicate that *C. odontorhiza* forms a clade, with three Mexican accessions sister to those from eastern US populations (Freudenstein and Barrett, unpublished data). There seems to be little additional phylogeographic structure within this species, with multiple plastid types often occurring in the same population. Sister to all *C. odontorhiza* populations are Mexican accessions of *C. wisteriana*, with the remainder of *C. wisteriana* sister to this collective assemblage. Analyses based on plastid DNA within *C. wisteriana* from the USA ($n=40$ individuals; *rpl32-trnL* + *trnL-F*) indicate two divergent haplotypes, corresponding to eastern and western populations; however, these loci are invariant within populations. Patterns based on plastid DNA are consistent with an origin of the group in Mexico, followed by migration northward and divergence in flowering times. More variable markers will be needed to better address population genetics and patterns of gene-flow within and between both species.

C. wisteriana presents an intriguing example of allopatric genetic differentiation that is correlated with differences in habitat preferences and a major shift in fungal associations. First, both *C. odontorhiza* and western-US populations of *C. wisteriana* associate with members of the genus *Tomentella* (Thelephoraceae), while in the eastern USA, *C. wisteriana* associates with Russulaceae based on fungal ITS sequencing (Taylor 1997; J. Freudenstein and Barrett, unpublished data). The earliest-diverging lineages of *Corallorhiza* (*C. striata, C. trifida*) typically associate with Thelephoraceae (Taylor 1997; McKendrick et al. 2000b; Barrett et al. 2010), so association with Russulaceae in the eastern USA may represent a geographic/habitat-correlated host shift, but a broader genus-level investigation is warranted to determine the polarity of host shifts. It remains to be investigated whether these plastid DNA, habitat, and fungal host correlations persist toward the southern extent of the geographic range in Mexico.

6.5.3 *Corallorhiza maculata–mertensiana* Complex

Corallorhiza maculata, or the spotted coral root orchid, is one of the most abundant terrestrial orchids in North America (Coleman 1995) and, in particular, the most likely encountered species of *Corallorhiza* (Freudenstein and Doyle 1994). The species ranges from Mexico to Canada in western North America and can be found throughout the rest of the United States with the exception of the Great Plains region and the deep Southeast (Freudenstein 1997). In general, the plants are found in shaded areas of both deciduous and coniferous old-growth forests. This species can occur singly or in large clumps (Luer 1975). It has long been acknowledged that there is a large amount of floral variation within the species, with many populations encompassing morphologically dissimilar individuals (Freudenstein 1997). While this morphological variation can be quite useful as a proxy for genetic relatedness, it is not reliable enough to supplant the need for DNA sequencing and other molecular markers (Freudenstein 1997).

C. maculata is often referred to as the *C. maculata* complex because it is composed of several recognized varieties: *C. maculata* var. *maculata*, *C. maculata* var. *occidentalis* (both occurring throughout the entire range), and *C. maculata* var. *mexicana* (restricted to Mexico, and possibly bordering states) (Freudenstein 1992, 1997). This complex also includes other closely related species: *C. bulbosa* (Mexico), *C. macrantha* (Mexico), and *Corallorhiza mertensiana* (northwestern North America), the latter of which is likely to be the closest relative to the two aforementioned varieties of *C. maculata*. Analysis of the ITS and the *rbcL* gene, (Barrett and Freudenstein 2008) and other plastid DNA (Freudenstein and Doyle 1994) places *C. mertensiana* within the *C. maculata* clade. Chloroplast RFLP studies have indicated that *C. mertensiana* may be derived from *C. maculata* (Freudenstein and Doyle 1994). The range of *C. mertensiana* is almost entirely contained within the range of *C. maculata* and the two species are often found co-occurring (Taylor and Bruns 1999). In addition to different varieties of *C. maculata*, it has also been observed in the Midwestern and Eastern United States that "early" and "late" flowering forms exist, and correspond to vars. *maculata* and var. *occidentalis* (Freudenstein 1987). The phenology of these individuals is sufficiently distinct that cross-pollination between them is unlikely.

It has long been suspected that *C. maculata* is almost completely autogamous. While capable of outcrossing via pollinators, any flower that has not been pollinated via foreign pollen will self-fertilize, with plants consistently displaying nearly 100 % seed set. Personal observation (S. Hopkins) has provided evidence that plants whose emergence from the soil has been impeded due to a physical obstruction will still flower and experience full seed set underground. Despite its ability to self-pollinate, microsatellite analysis of parents and seeds is revealing more outcrossing than was previously thought to occur (Hopkins and Taylor 2011). Hypothesized pollinators include dance flies (Empididae), Bombyliid flies, and Acroceratid flies (Kipping 1971) and Luer (1975) published a photograph of *C. maculata* being visited by a bee in the genus *Andrena* with pollinia attached to its back.

The question of potential cross-pollination is an interesting one given that *C. maculata* is the first species complex in which diversification of fungal specificity had been shown to exist below the species level. *C. maculata* and *C. mertensiana* both associate with fungi in the Russulaceae, but have never been found to share a *Russula* species, even where the two orchids co-occur (Taylor and Bruns 1999). Furthermore, molecular methods were used to identify six distinct genotypes of *C. maculata* using three single nucleotide polymorphisms (SNPs) in the nuclear ITS region and the nuclear single-copy flavanone beta hydroxylase gene (Taylor et al. 2004). These genotypes, or races, have been shown to associate with separate clades of fungi within the family Russulaceae, with fairly little overlap in fungal associations across races (Taylor et al. 2004). Multiple races have been found growing in the same location but never utilizing the same fungal partners, demonstrating a genetic control of fungal associations that is not dependent on habitat, and occurs below the species level (Taylor et al. 2004). Current work utilizing 6 polymorphic microsatellite markers on adults and seeds from 15 putative populations with multiple morphotypes suggests that there are a minimum of 6 races present in this species and that considerable population structure exists. Furthermore, these microsatellites show that *C. mertensiana* is genetically distinct from *C. maculata*, despite their paraphyletic relationship in prior chloroplast studies. Preliminary data has indicated that only 1 of the ca. 100 *C. maculata* adults examined may be a hybrid between two different races and that less than 5 % of the seed capsules surveyed display any heterozygosity at the genotyped loci. Along with this distinct lack of gene-flow between populations, individuals of *C. maculata* display extreme fungal specificity. Recent work with microsatellite markers has yielded results in agreement with those above: A single clade of individuals of *C. maculata* will nearly always associate with only a single clade of fungi, or at most two clades of fungi, within the genus *Russula*. In some cases individuals of different genotypes will grow within centimeters of one another, however these different genotypes have never been found to associate with the same fungus or even the same clade of fungi. These results demonstrating fine-scale fungal preference suggest that mycorrhizal specificity is evolving rapidly, with changes occurring among very recently separated biological species or possibly even at the population level.

6.6 Emerging Patterns

The case studies presented above suggest several interesting trends in the ecological genetics of mycoheterotrophic plants. On the other hand, these examples are limited to only a few species, mostly from North American temperate forests (two *Hexalectris* species were sampled in subtropical forests). Therefore, the generality of the patterns discussed below remains to be determined. Nevertheless, the trends are tantalizing. First, we note that species and population boundaries are difficult to discern, even when careful morphological studies and informative molecular markers are combined (Barrett and Freudenstein 2009; Kennedy and Watson 2010). This observation is likely explained, in part, by the mixed mating systems seen in these mycoheterotrophic plants. Nearly all are self-fertile, many have explicit mechanisms for selfing such as cleistogamy (though all appear capable of outcrossing), all have small, widely scattered populations, and all display moderate to high levels of inbreeding. The specter of hybridization as a pathway to novel forms or species has been raised for several taxa, though proven in none. Thus, these plants have complex evolutionary histories, making species and population boundaries extremely dependent on what criteria are used and how they are weighed (Barrett and Freudenstein 2009; Kennedy and Watson 2010). When integrative approaches are applied, the best-studied taxa turn out to be species complexes encompassing one or several cryptic species (Kennedy and Watson 2010; Barrett and Freudenstein 2011; Hopkins and Taylor, unpublished data). From an evolutionary point of view, these plants offer exciting research opportunities, since their particular combination of population dynamics, symbioses and speciation patterns are unusual in the plant world.

Focusing in on the ESUs, host-races, and populations that have been uncovered, we find that genetically discrete demes are often maintained in sympatry or parapatry. Examples include the two color forms of *H. monotropa* (Klooster and Culley 2010), the three western ESUs within *C. striata sensu stricto* (Barrett and Freudenstein 2011), the host-races within *C. maculata* (Taylor et al. 2004), and the western sibling species within the *H. spicata* complex (Kennedy and Watson 2010). However, we do not interpret these patterns to suggest that geographic barriers are unimportant to the diversification of these lineages. There are several very clear examples of geographic patterning of genetic breaks, including East–West disjunctions and associated genetic divergence within *H. monotropa* (Beatty and Provan 2010) and *C. wisteriana*. Thus, we infer that genetic divergence in allopatry has played an important role in several of these lineages, although the spatial and temporal scales of allopatric diversification remain open questions.

Perhaps the most exciting trend that coincides with the discovery of these cryptic species and ESUs is that these entities have diverged in traits of considerable ecological interest. In some cases, different floral forms or mating systems (color, phenology, cleistogamy) distinguish the lineages. And, in some cases, mycorrhizal specificity has diverged between these closely related lineages. The best examples of the latter trend are the distinct *Sebacina* clades targeted by different members of the *H. spicata* complex (Kennedy et al. 2011) and the distinct *Russula* clades targeted by different *C. maculata* host-races (Taylor et al. 2004; Hopkins, unpublished). These rapid and ongoing evolutionary dynamics open the way for incisive studies of the genes and selective forces underlying ecological diversification. For example, do host-jumps to novel mycorrhizal taxa tend to occur in geographically isolated populations following rare long-distance dispersal? Or do increases in selfing, such as in transitions to cleistogamy, and associated genetic drift and low genetic diversity, precede the evolution of extremely narrow specialization, as suggested for *H. arizonica* and *H. nitida* (Kennedy et al. 2011)?

6.7 Future Questions and Approaches

In the past, developing molecular markers and other tools with which to carry out research on the ecological genetics of a non-model species has been a laborious, costly, and often untenable undertaking. Technological breakthroughs, particularly next-generation sequencing and computational tools, are allowing ecological genetics and genomics to be applied to any organism. We anticipate that this revolution will greatly expand our view of the evolution, ecology and function of mycoheterotrophic plants.

Currently, the major advantage provided by next-generation technologies to studies of non-model organisms is the much greater ease with which variable markers can be discovered and assayed. For example, highly polymorphic microsatellites have been the loci of choice for studies of gene-flow and population structure for the last 20 years. But it has been slow and costly to develop microsatellites for a single new species. Today, a fraction of a single pyrosequencing run, using either random genomic DNA (shotgun sequencing) or microsatellite-enriched genomic DNA, can rapidly yield thousands of loci containing microsatellite motifs (Guichoux et al. 2011). Assays of SNPs are beginning to rival microsatellites in popularity, as methods for genome-wide analysis become more routine. By tagging and pooling individuals prior to a shotgun sequencing run, numerous SNPs can be discovered. In a further refinement of this approach, genomic DNA is cut with a restriction enzyme before tagging, pooling and sequencing (RADseq; Davey and Blaxter 2010). This approach has been used to simultaneously discover and genotype tens of thousands of SNPs spread across the genome in non-model organisms. For studies at or above the species level, where a select set of loci are known to be informative, the target regions can be PCR amplified, ligated to adaptors, pooled and sequenced (De Leeneer et al. 2011), or enriched then sequenced. These target capture methods include molecular inversion probes (MIP; Porreca et al. 2007), solution hybrid selection (SMS; Elshire et al. 2011), and microarray-based genomic selection (MGS; Okou et al. 2007). In this way, hundreds or thousands of defined loci can be sequenced across large numbers of individuals simultaneously.

Improved markers will enable inquiries into numerous aspects of the evolutionary ecology of mycoheterotrophs. In particular, we envision more in-depth investigations of mating systems and gene-flow within and among mycoheterotrophic plant demes, which should provide improved estimates of population and species boundaries. For example, improved likelihood and Bayesian methods for parentage analysis have appeared in recent years (Jones et al. 2010), and could be applied to mycoheterotrophic plants to reveal patterns of pollen movement and gene exchange within and among populations, ESUs and species. Detailed, fine-scale genetic data can also be used to infer population parameters, such as effective population sizes. This should be particularly informative in the case of mycoheterotrophic plants, in which true population sizes are poorly represented through censuses of above-ground flowering individuals in any one year (Shefferson et al. 2001). Bayesian methods have also been developed for assigning individuals to populations, some of which can be used without any a *priori* population definitions (Pritchard et al. 2000). These methods, in part, have been used to define genetically distinct ESU's within *C. striata sensu stricto*, and to investigate specific cases of admixed multilocus genotypes based on multiple nuclear markers (Barrett and Freudenstein 2011). This will also be particularly useful for the elucidation of population boundaries in *C. maculata*, where widely divergent genotypes occur in sympatry, and population boundaries are thus difficult to define using geographic criteria.

Additional markers will also improve our ability to infer the phylogeographic and demographic histories of mycoheterotrophic plant species. We may be able to infer the migration routes that have led to current geographic ranges, along with the timing and coincident climatic conditions under which changes in distributions occurred. These inferences can be approached by combining geographic reconstruction methods such as nested clade analysis (Templeton 1998), coalescent methods that provide insights into demographic histories in the process of reconstructing the likely time course of mutation events (Donnelly and Tavare 1995; Carbone and Kohn 2001), and niche/climate modeling (Peterson 2001; see Beatty and Provan 2011b for an example). A focal issue may be the geographic, demographic, and temporal scenarios under which mycorrhizal specialization has changed.

New markers and analytical tools also open the way to asking more direct questions about adaptation in non-model organisms. For example, one approach seeks to elucidate the genetic architecture underlying adaptation via estimation of heritability and the use of association tests, which search for statistical correlations between variation in particular alleles and variation in traits of interest. Controlled crosses combined with phenotypic measurements in common environments form the basis for the field of quantitative genetics (Falconer and Mackay 1996). For organisms, such as mycoheterotrophic plants, which cannot be easily crossed or cultivated, related methods have been developed for wild organisms (Ritland 2000). The most powerful of these methods start with at least some pedigree information (e.g., mother-offspring relationships), and attempt to reconstruct missing pedigree information (e.g., via paternity assignment methods), then model trait values as a function of allelic segregation (Garant and Kruuk 2005). Another approach relies on the principles of population genetics and molecular evolution, without reference to observable phenotypic variation. DNA sequences themselves can carry signatures of drift, hitchhiking, population bottlenecks, directional selection, balancing selection, and other evolutionary phenomena (Nordborg 1997; Nielsen 2000; Nielsen and Wakeley 2001; Charlesworth 2006). These methods are best viewed as tools for data exploration, i.e., ways to search for candidate loci, leading to the formulation of testable hypotheses concerning the adaptive significance of particular loci (Storz and Wheat 2010).

In the case of mycoheterotrophic plants, we recommend pursuit of both improved molecular markers and field and/or laboratory manipulations to begin to elucidate the processes and selective pressures underlying ecological variation. For example, hand-pollination experiments combined with seed packet methods (Rasmussen and Whigham 1993) and subsequent genomic scans could be used to begin to quantify the contribution of various mechanisms to reproductive isolation among populations with overlapping geographic ranges, i.e., to ascertain whether isolation is prezygotic or postzygotic, and, therefore, whether there may be selection against hybrids.

Lastly, mycoheterotrophic plant research would benefit from the application of advancing methods in functional genomics. Transcriptional profiling can be used to reveal patterns of gene expression under varying conditions, even in non-model organisms. A flood of gene expression data has come in recent years from the use of microarrays. However, microarrays are very expensive to design and construct, and, hence, have only been developed for model organisms. There have been several successful uses of microarrays to study gene expression in non-model organisms that are closely related to model species for which microarrays are available (Travers et al. 2007). Unfortunately, to our knowledge, there are no model organisms that are closely related to a mycoheterotrophic plant species. On the other hand, direct next-generation expressed sequence tag (EST) sequencing methods, such as Illumina Hi-seq, can be applied to any organism at costs much lower than microarray design. These approaches could be used to survey the plant and fungal genes that are up or down-regulated at different stages of the interaction (Seddas et al. 2009). Alternatively, gene expression could be compared across species or ESUs with differing fungal specificity, flowering times, or other traits of interest, to search for patterns of gene expression that may underlie these differences.

6.8 Conclusions

While mycoheterotrophic plants may never become common "lab-rat" model organisms, in our view, they have great value as models of specificity and symbiosis-shaped evolution in the unique circumstance in which it is the plant that attacks the fungus, as opposed to the traditional role of plants as victims of attack by numerous pathogens and herbivores. The value as a study system in evolutionary ecology is magnified by the fact that the specialized plant-fungus interaction is evolving rapidly in some mycoheterotrophic plant lineages. The rapid advance of methods in ecological genetics and genomics that can be applied to wild organisms holds tremendous promise for mycoheterotrophic plant research. To most effectively pursue these opportunities, we suggest that one or a few mycoheterotrophic species should receive increased attention from the research community as mycoheterotrophic plant models. *C. trifida* comes to mind, because it has been grown in tripartite symbiotic microcosms (McKendrick et al. 2000a, 2000b) and because a significant amount of physiological work has been carried out (Zimmer et al. 2008; Cameron et al. 2009). However, we still know relatively little about patterns of genetic variation and fungal associations within this taxon (McKendrick et al. 2000b). While the new methods democratize genomics, we cannot yet apply them to every mycoheterotrophic species due to constraints of funding and feasibility. There are still synergistic gains to be made by focusing on species that are most tractable in terms of crossing and growth (in the field, if necessary), and working as a community to develop tools for these species.

References

Alsos IG, Ehrich D, Thuiller W, Eidesen PB, Tribsch A, Schönswetter P, Lagaye C, Taberlet P, Brochmann C (2012) Genetic consequences of climate change for northern plants. Proc R Soc B 279:2042–2051

Baird NA, Etter PD, Atwood TS, Currey MC, Shiver AL, Lewis ZA, Selker EU, Cresko WA, Johnson EA (2008) Rapid SNP discovery and genetic mapping using sequenced RAD markers. PLoS One 3:e3376

Barrett C, Freudenstein J (2008) Molecular evolution of rbcL in the mycoheterotrophic coralroot orchids (*Corallorhiza* Gagnebin, Orchidaceae). Mol Phylogenet Evol 47:665–679

Barrett CF, Freudenstein JV (2009) Patterns of morphological and plastid DNA variation in the *Corallorhiza striata* species complex (Orchidaceae). Syst Bot 34: 496–504

Barrett CF, Freudenstein JV (2011) An integrative approach to delimiting species in a rare but widespread mycoheterotrophic orchid. Mol Ecol 20:2771–2786

Barrett CF, Freudenstein JV, Taylor DL, Kõljalg U (2010) Rangewide analysis of fungal associations in the fully mycoheterotrophic *Corallorhiza striata* complex (Orchidaceae) reveals extreme specificity on ectomycorrhizal *Tomentella* (Thelephoraceae) across North America. Am J Bot 97:628–643

Baxter SW, Davey JW, Johnston JS, Shelton AM, Heckel DG, Jiggins CD, Blaxter ML (2011) Linkage mapping and comparative genomics using next-generation RAD sequencing of a non-model organism. PLoS One 6:e19315

Beatty GE, Provan J (2010) Refugial persistence and postglacial recolonization of North America by the cold-tolerant herbaceous plant *Orthilia secunda*. Mol Ecol 19:5009–5021

Beatty GE, Provan J (2011a) High clonal diversity in threatened peripheral populations of the yellow bird's

nest (*Hypopitys monotropa*; syn. *Monotropa hypopitys*). Ann Bot 107:663–670
Beatty GE, Provan J (2011b) Phylogeographic analysis of North American populations of the parasitic herbaceous plant *Monotropa hypopitys* L. reveals a complex history of range expansion from multiple late glacial refugia. J Biogeogr 38:1585–1599
Beatty GE, Provan J (2011c) Comparative phylogeography of two related plant species with overlapping ranges in Europe, and the potential effects of climate change on their intraspecific genetic diversity. BMC Evol Biol 11:29
Beatty GE, McEvoy PM, Sweeney O, Provan J (2008) Range-edge effects promote clonal growth in peripheral populations of the one-sided wintergreen *Orthilia secunda*. Divers Distrib 14:546–555
Bidartondo MI, Bruns TD (2001) Extreme specificity in epiparasitic Monotropoideae (Ericaceae): widespread phylogenetic and geographical structure. Mol Ecol 10:2285–2295
Bidartondo MI, Bruns TD (2002) Fine-level mycorrhizal specificity in the Monotropoideae (Ericaceae): specificity for fungal species groups. Mol Ecol 11:557–569
Bidartondo MI, Bruns TD (2005) On the origins of extreme mycorrhizal specificity in the Monotropoideae (Ericaceae): performance trade-offs during seed germination and seedling development. Mol Ecol 14:1549–1560
Björkman E (1960) *Monotropa hypopitys* L. – an epiparasite on tree roots. Physiol Plant 13:308–327
Blouin MS (2003) DNA-based methods for pedigree reconstruction and kinship analysis in natural populations. Trends Ecol Evol 18:503–511
Cameron DD, Preiss K, Gebauer G, Read DJ (2009) The chlorophyll-containing orchid *Corallorhiza trifida* derives little carbon through photosynthesis. New Phytol 183:358–364
Carbone I, Kohn L (2001) A microbial population–species interface: nested cladistic and coalescent inference with multilocus data. Mol Ecol 10:947–964
Catling P, Engel V (1993) Systematics and distibution of *Hexalectris spicata* var. *arizonica* (Orchidaceae). Lindleyana 8:119–125
Charlesworth D (2006) Balancing selection and its effects on sequences in nearby genome regions. PLoS Genet 2:e64
Coleman R (1995) The wild orchids of California. Comstock Publishing, Ithaca, USA
Coleman RA (2002) The wild orchids of Arizona and New Mexico. Cornell University Press, Ithaca, USA
Cullings K (2000) Reassessment of phylogenetic relationships of some members of the Monotropoideae based on partial 28S ribosomal RNA gene sequencing. Can J Bot 78:1–2
Davey JL, Blaxter MW (2010) RADSeq: next-generation population genetics. Brief Funct Genomics 9:416–423
De Leeneer K, De Schrijver J, Clement L, Baetens M, Lefever S, De Keulenaer S, Van Criekinge W, Deforce D, Van Nieuwerburgh F, Bekaert S, Pattyn F, De Wilde B, Coucke P, Vandesompele J, Claes K, Hellemans J (2011) Practical tools to implement massive parallel pyrosequencing of PCR products in next generation molecular diagnostics. PLoS One 6:e25531
Delannoy E, Fujii S, Colas des Francs C, Brundrett M, Small I (2011). Rampant gene loss in the underground orchid *Rhizanthella gardneri* highlights evolutionary constraints on plastid genomes. Mol Biol Evol 28:2077–2086
DePamphilis CW, Palmer JD (1990) Loss of photosynthetic and chlororespiratory genes from the plastid genome of a parasitic flowering plant. Nature 348:337–339
Dizon AE, Lockyer C, Perrin WF, Demaster DP, Sisson J (1992) Rethinking the stock concept: a phylogeographic approach. Conserv Biol 6:24–36
Donnelly P, Tavare S (1995) Coalescents and genealogical structure under neutrality. Annu Rev Genet 29:401–421
Eastwood A (1897) Studies in the herbarium and the field. Proc Calif Acad Sci 1:71–89
Eastwood A (1902) Some new species of California plants. Bull Torrey Bot Club 29:75–82
Elshire RJ, Glaubitz JC, Sun Q, Poland JA, Kawamoto K, Buckler ES, Mitchell SE (2011) A robust, simple genotyping-by-sequencing (GBS) approach for high diversity species. PLoS One 6:e19379
Falconer DS, Mackay TFC (1996) Introduction to quantitative genetics, 4th edn. Longman, Essex, England
Freudenstein JV (1987) A preliminary study of *Corallorhiza maculata* (Orchidaceae) in eastern North America. Contrib Univ Michigan Herb 16:145–153
Freudenstein J V (1992). Systematics of Corallorhiza and the Corallorhizinae (Orchidaceae). PhD thesis. Ithaca, NY, USA: Cornell University
Freudenstein JV (1997) A monograph of *Corallorhiza* (Orchidaceae). Harv Pap Bot 1:5–51
Freudenstein JV, Doyle JJ (1994) Plastid DNA, morphological variation, and the phylogenetic species concept: the *Corallorhiza maculata* (Orchidaceae) complex. Syst Bot 19:273–290
Freudenstein JV, Senyo DM (2008) Relationships and evolution of *matK* in a group of leafless orchids (*Corallorhiza* and Corallorhizinae; Orchidaceae: Epidendroideae). Am J Bot 95:498–505
Garant D, Kruuk LEB (2005) How to use molecular marker data to measure evolutionary parameters in wild populations. Mol Ecol 14:1843–1859
Goldman DH, Freudenstein JV, Kores PJ, Molvray M, Jarrell DC, Whitten WM, Cameron KM, Jansen RK, Chase MW (2001) Phylogenetics of Arethuseae (Orchidaceae) based on plastid *matK* and *rbcL* sequences. Syst Bot 26:670–695
Goldman DH, Coleman RA, Magrath LK, Catling PM (2002) *Hexalectris*. In: The flora of North America (eds) Flora of North America editorial committee, Oxford: University Press. pp. 603–606
Guichoux E, Lagache L, Wagner S, Chaumeil P, Léger P, Lepais O, Lepoittevin C, Malausa T, Revardel E, Salin F, Petit RJ (2011) Current trends in microsatellite genotyping. Mol Ecol Resour 11:591–611

Hafner MS, Page RDM (1995) Molecular phylogenies and host-parasite cospeciation: gophers and lice as a model system. Philos Trans R Soc Lond B Biol Sci 349:77–83

Hampe A, Petit RJ (2005) Conserving biodiversity under climate change: the rear edge matters. Ecol Lett 8:461–467

Hawthorne DJ, Via S (2001) Genetic linkage of ecological specialization and reproductive isolation in pea aphids. Nature 412:904–907

Helyar S, Hemmer-Hansen J, Bekkevold D, Taylor M, Ogden R, Limborg M, Cariani A, Maes G, Diopere E, Carvalho G, Nielsen E (2011) Application of SNPs for population genetics of nonmodel organisms: new opportunities and challenges. Mol Ecol Resour 11:123–136

Hewitt GM (1999) Post-glacial re-colonization of European biota. Biol J Linn Soc 68:87–112

Hohenlohe PA, Bassham S, Etter PD, Stiffler N, Johnson EA, Cresko WA (2010) Population genomics of parallel adaptation in threespine stickleback using sequenced RAD tags. PLoS Genet 6:e1000862

Hopkins SE, Taylor DL (2011) Microsatellite loci development in mycoheterotrophic *Corallorhiza maculata* (Orchidaceae) with amplification in *C. mertensiana*. Am J Bot 98:e253–e255

Huang W, Takebayashi N, Qi Y, Hickerson M (2011) MTML-msBayes: Approximate Bayesian comparative phylogeographic inference from multiple taxa and multiple loci with rate heterogeneity. BMC Bioinformatics 12:1

Jarosz A, Burdon J (1991) Host-pathogen interactions in natural populations of *Linum marginale* and *Melampsora lini*: II. Local and regional variation in patterns of resistance and racial structure. Evolution 45:1618–1627

Jones AG, Small CM, Paczolt KA, Ratterman NL (2010) A practical guide to methods of parentage analysis. Mol Ecol Resour 10:6–30

Kennedy AH, Watson LE (2010) Species delimitations and phylogenetic relationships within the fully mycoheterotrophic *Hexalectris* (Orchidaceae). Syst Bot 35:64–76

Kennedy AH, Taylor DL, Watson LE (2011) Mycorrhizal specificity in the fully mycoheterotrophic *Hexalectris* Raf. (Orchidaceae: Epidendroideae). Mol Ecol 20: 1303–1316

Kipping, J. L. 1971. Pollination studies of native orchids. M. S. Thesis, San Francisco, CA, USA: San Francisco State College.

Klooster MR, Culley TM (2009) Comparative analysis of the reproductive ecology of *Monotropa* and *Monotropsis*: Two mycoheterotrophic genera in the Monotropoideae (Ericaceae). Am J Bot 96:1337

Klooster MR, Culley TM (2010) Population genetic structure of the mycoheterotroph *Monotropa hypopitys* L. (Ericaceae) and differentiation between red and yellow color forms. Int. J. Plant Sci 171:167–174

Kron KA, Judd WS, Stevens PF, Crayn DM, Anderberg AA, Gadek PA, Quinn CJ, Luteyn JL (2002) Phylogenetic classification of Ericaceae: molecular and morphological evidence. Bot Rev 68:335–423

Leake JR (1994) The biology of myco-heterotrophic ("saprophytic") plants. New Phytol 127:171–216

Leake J, McKendrick S, Bidartondo MI, Read DJ (2004) Symbiotic germination and development of the mycoheterotroph *Monotropa hypopitys* in nature and its requirement for locally distributed *Tricholoma* spp. New Phytol 163:405–423

Liggio J, Liggio A (1999) Wild Orchids of Texas. University of Texas Press, Austin, TX, USA

Logacheva MD, Schelkunov MI, Penin AA (2011) Sequencing and analysis of plastid genome in mycoheterotrophic Orchid *Neottia nidus-avis*. Genome Biol Evol 3:1296–1303

Luer CA (1975) The native orchids of the United States and Canada excluding Florida. New York Botanical Garden, New York, USA

McKendrick S, Leake JR, Read DJ (2000a) Symbiotic germination and development of myco-heterotrophic plants in nature: transfer of carbon from ectomycorrhizal *Salix repens* and *Betula pendula* to the orchid *Corallorhiza trifida* through shared hyphal connections. New Phytol 145:539–548

McKendrick S, Leake JR, Taylor DL, Read DJ (2000b) Symbiotic germination and development of mycoheterotrophic plants in nature: Ontogeny of *Corallorhiza trifida* and characterization of its mycorrhizal fungi. New Phytol 145:523–537

Moritz C (1994) Defining "Evolutionarily Significant Units" for conservation. Trends Ecol Evol 9:373–375

Mortazavi A, Williams BA, McCue K, Schaeffer L, Wold B (2008) Mapping and quantifying mammalian transcriptomes by RNA-Seq. Nat Methods 5:621–628

Nadeau NJ, Jiggins CD (2010) A golden age for evolutionary genetics? Genomic studies of adaptation in natural populations. Trends Genet 26:484–492

Nei M, Tajima F (1981) DNA polymorphism detectable by restriction endonucleases. Genetics 97:145

Neyland R (2004) The systematic significance of color variation in *Monotropa hypopithys* (Ericaceae) inferred from large ribosomal subunit (26S) rRNA gene sequences. Madroño 51:275–279

Neyland R (2005) Systematic relationships of *Pityopus californicus* inferred from large ribosomal subunit (26S) rRNA gene sequences. West N Am Naturalist 65:528–535

Nielsen R (2000) Estimation of population parameters and recombination rates from single nucleotide polymorphisms. Genetics 154:931–942

Nielsen R, Wakeley J (2001) Distinguishing migration from isolation: a Markov chain Monte Carlo approach. Genetics 158:885–896

Nordborg M (1997) Structured coalescent processes on different time scales. Genetics 146:1501–1514

Okou DT, Steinberg KM, Middle C, Cutler DJ, Albert TJ, Zwick ME (2007) Microarray-based genomic selection for high-throughput resequencing. Nat Methods 4:907–909

Olson A (1990) Observations on the floral shoots of *Monotropa hypopitys* (Monotropaceae). Rhodora 92:54–56

Olson AR (1993) Patterns of embryo and endosperm formation in *Monotropa hypopitys* (Monotropaceae) from North America and Western Sweden. Am J Bot 80:839–846

Peterson AT (2001) Predicting species' geographic distributions based on ecological niche modeling. Condor 103:599–605

Porreca GJ, Zhang K, Li JB, Xie B, Austin D, Vassallo SL, LeProust EM, Peck BJ, Emig CJ, Dahl F, Gao Y, Church GM, Shendure J (2007) Multiplex amplification of large sets of human exons. Nat Methods 4: 931–936

Pritchard JK, Stephens M, Donnelly P (2000) Inference of population structure using multilocus genotype data. Genetics 155:945–959

Provan J, Bennett K (2008) Phylogeographic insights into cryptic glacial refugia. Trends Ecol Evol 23: 564–571

Provan J, Maggs CA (2012) Unique genetic variation at a species's rear edge is under threat from global climate change. Proc R Soc Lond B 279:39–47

Rasmussen HN, Whigham DF (1993) Seed ecology of dust seeds in situ: a new study technique and its application in terrestrial orchids. Am J Bot 80:1374–1378

Ritland K (2000) Marker-inferred relatedness as a tool for detecting heritability in nature. Mol Ecol 9:1195–1204

Seddas PMA, Arias CM, Arnould C, van Tuinen D, Godfroy O, Benhassou HA, Gouzy J, Morandi D, Dessaint F, Gianinazzi-Pearson V (2009) Symbiosis-related plant genes modulate molecular responses in an arbuscular mycorrhizal fungus during early root interactions. Mol Plant Microbe Interact 22:341–351

Selosse M-A, Bauer R, Moyersoen B (2002a) Basal hymenomycetes belonging to the Sebacinaceae are ectomycorrhizal on temperate deciduous trees. New Phytol 155:183–195

Selosse M-A, Weiss M, Jany J, Tillier A (2002b) Communities and populations of sebacinoid basidiomycetes associated with the achlorophyllous orchid *Neottia nidus-avis* (L.) LCM Rich. and neighbouring tree ectomycorrhizae. Mol Ecol 11:1831–1844

Selosse M-A, Dubois MP, Alvarez N (2009) Do Sebacinales commonly associate with plant roots as endophytes? Mycol Res 113:1062–1069

Shefferson RP, Sandercock BK, Proper J, Beissinger SR (2001) Estimating dormancy and survival of a rare herbaceous perennial using mark-recapture models. Ecology 82:145–156

Shefferson RP, Taylor DL, Weiß M, Garnica S, McCormick MK, Adams S, Gray HM, McFarland JW, Kull T, Tali K, Yukawa T, Kawahara T, Miyoshi K, Lee Y (2007) The evo- lutionary history of mycorrhizal specificity among lady's slipper orchids. Evolution 61:1380–1390

Sosa V (2007) A molecular and morphological phylogenetic study of subtribe Bletiinae (Epidendreae, Orchidaceae). Syst Bot 32:34–42

Stamatakis A (2006) RAxML-VI-HPC: maximum likelihood-based phylogenetic analyses with thousands of taxa and mixed models. Bioinformatics 22:2688–2690

Storz JF, Wheat CW (2010) Integrating evolutionary and functional approaches to infer adaptation at specific loci. Evolution 64:2489–2509

Taberlet P, Fumagalli L, Wust-saucy AG, Cosson JF (1998) Comparative phylogeography and postglacial colonization routes in Europe. Mol Ecol 7:453–464

Taylor, D. L. 1997. The evolution of myco-heterotrophy and specificity in some North American orchids. PhD. thesis, Berkeley, CA, USA: University of California, Berkeley.

Taylor DL, Bruns TD (1999) Population, habitat and genetic correlates of mycorrhizal specialization in the "cheating" orchids *Corallorhiza maculata* and *C. mertensiana*. Mol Ecol 8:1719–1732

Taylor DL, Bruns TD, Szaro T, Hodges SA (2003) Divergence in mycorrhizal specialization within *Hexalectris spicata* (Orchidaceae), a nonphotosynthetic desert orchid. Am J Bot 90:1168–1179

Taylor DL, Bruns TD, Hodges SA (2004) Evidence for mycorrhizal races in a cheating orchid. Proc R Soc Lond B 271:35–43

Templeton AR (1998) Nested clade analyses of phylogeographic data: testing hypotheses about gene flow and population history. Mol Ecol 7:381–397

Thompson JN (1994) The coevolutionary process. University of Chicago Press, Chicago, IL, USA

Thompson JN (2005) The geographic mosaic of coevolution. University of Chicago Press, Chicago, IL, USA

Travers SE, Smith MD, Bai J, Hulbert SH, Leach JE, Schnable PS, Knapp AK, Milliken GA, Fay PA, Saleh A, Garrett KA (2007) Ecological genomics: making the leap from model systems in the lab to native populations in the field. Front Ecol Environ 5:19–24

Wallace GD (1975) Studies of the Monotropoideae (Ericaceae): taxonomy and distribution. Wasmann J Biol 33:1–88

Zimmer K, Meyer C, Gebauer G (2008) The ectomycorrhizal specialist orchid *Corallorhiza trifida* is a partial myco-heterotroph. New Phytol 178:395–400

7 Species Interactions of Mycoheterotrophic Plants: Specialization and its Potential Consequences

Richard J. Waterman, Matthew R. Klooster, Heiko Hentrich, and Martin I. Bidartondo

7.1 Species Interactions of Mycoheterotrophic Plants

Both the origin and maintenance of biodiversity are dependent on interactions between species. All species exist as part of a complex network of interactions, ranging from antagonistic examples such as competition and parasitism, to mutualisms such as those shared between over 90 % of plant species and mycorrhizal fungi (Thompson 2005; Montoya et al. 2006). For example, the diversification of two of Earth's major lineages, the insects and flowering plants, is thought to have been driven by their mutual interactions (Ehrlich and Raven 1964). An even more diverse lineage, the fungi, is notably dominated by biotrophic species that interact with plants and animals. Diverse communities of plants and fungi can be linked physically by common mycorrhizal networks (Smith and Read 2008), and these ecological networks extend to other symbioses as well, with nearly 90 % of flowering plants being pollinated by insects or other animals (Ollerton et al. 2011). Enhancing our understanding of plants engaged in multiple, potentially highly specialized, symbiotic interactions has become the focus of many recent empirical studies, with mycoheterotrophic plants serving as an ideal model system for such analyses.

Mycoheterotrophs are a phylogenetically diverse plant group, defined by their unique interaction with fungi; whereas the vast majority of plants exchange photosynthetically derived carbon resources for fungal-acquired soil nutrients, mycoheterotrophs have evolved to cheat this mutualistic exchange by extracting these resources from fungal associates. These plants therefore offer a fascinating case-study for research into the exploitation of mutualisms, and for determining how a highly specialized life history strategy for obtaining nutrition impacts the wider web of symbiotic interactions invaded by mycoheterotrophs. In this chapter we discuss recent research into the fungal interactions of mycoheterotrophs, and various hypotheses on the impacts of mycoheterotrophy on plant reproduction.

R.J. Waterman (✉)
Department of Animal and Plant Sciences, University of Sheffield, Alfred Denny Building, Western Bank, Sheffield, S10 2TN, UK
e-mail: r.waterman@sheffield.ac.uk

M.R. Klooster
Centre College, Biology Program, 121 Young Hall, Danville, KY 40422, USA

H. Hentrich
Ecosystem Analysis, Institute for Environmental Sciences, University of Koblenz-Landau, Campus Landau, Fortstrasse 7, 76829, Landau in der Pfalz, Germany

M.I. Bidartondo
Imperial College London and Royal Botanic Gardens, Kew, TW9 3DS, UK

V.S.F.T. Merckx (ed.), *Mycoheterotrophy: The Biology of Plants Living on Fungi*,
DOI 10.1007/978-1-4614-5209-6_7,

7.2 The Exploitation of a Mutualism

The two major mycorrhizal symbioses are the arbuscular mycorrhizas involving members of the Glomeromycota and about 80 % of plant species, and the ectomycorrhizas involving some members of the Basidiomycota and Ascomycota and less than 5 % of plants, mainly trees and shrubs. These ancient interactions are generally obligate and mutualistic for both plants and fungi; mycorrhizal plants and fungi exchange carbon for soil minerals. Mycorrhization leads to the development of unique fungal structures among and/or within the cells of plants (e.g., arbuscule, ectomycorrhizal mantle and Hartig net, orchid peloton, monotrope peg).

Mutualisms are arrangements to exchange goods between two distantly related lineages. Mutualists offer their partners goods that are relatively inexpensive for them to produce or obtain, in exchange for goods that are more expensive or unfeasible for them to produce, even at an unmeasurably low profit (Schwartz and Hoeksema 1998; Bronstein 2001). Mutualisms can be subverted by intra-specific cheaters, who are individuals of one of the two interacting lineages that defect from the mutualism by failing to reciprocate. If an entire lineage defects, the mutualism is replaced by a parasitism, or two autonomous lineages emerge. Mutualisms can also be subverted by cheaters from a third lineage that invade a mutualism between two other lineages, or evolve from one of the partners of the mutualism. Thus, cheaters become part of a tripartite symbiosis and depend on the fate of the mutualism they target.

Mycoheterotrophic plants associated with the mycorrhizal fungi of neighboring plants may be considered third-party cheaters (Sachs and Simms 2006). Nonetheless, we should bear in mind that (1) the fitness costs of cheaters are typically lower than those of parasites; (2) the fitness costs (and the fitness) of fungi is a challenging research area (Pringle and Taylor 2002); and (3) it may be difficult to show that no benefit of any kind at no time is accrued by the cheated. For instance, both a "delayed payback" by the initial mycoheterotrophic orchid *Goodyera repens* to the saprobic fungus *Ceratobasidium cornigerum*, and a form of "parental nurture" of mycoheterotrophic gametophytes of *Huperzia hypogeae* by its photosynthetic sporophytes via an arbuscular mycorrhizal *Glomus* sp., have been hypothesized (Cameron et al. 2008; Winther and Friedman 2008). In fact, full mycoheterotrophy could be viewed as an evasion of delayed payback (Bidartondo 2005). The question of whether cheaters are parasites or mutualists has been considered unsettled in both pollination (Maloof and Inouye 2000) and mycorrhizas (Leake 1994). In many cases, the answer will be context-dependent.

Specialized cheating of common mycorrhizal networks has evolved multiple times across several plant lineages, and in both arbuscular and ectomycorrhizal networks (Chap. 5). It is thought that the majority of mycoheterotrophic lineages evolved from photosynthetic ancestors that were mutualistic members of mycorrhizal networks. A potential intermediate step in the evolution of mycoheterotrophy has been proposed for various orchids, where close relatives of mycoheterotrophs have been shown to rely on a combination of mycoheterotrophy and photosynthesis for their carbon needs (Tedersoo et al. 2007). This nutritional mode is termed partial mycoheterotrophy, or mixotrophy, and has been putatively demonstrated in several green forest orchids, which obtain up to 80 % of their carbon via associations with ectomycorrhizal fungi (Gebauer and Meyer 2003; Bidartondo et al. 2004; Julou et al. 2005). This strategy may compensate for low efficiency photosynthesis among green orchids growing in the shaded understory of forests (Julou et al. 2005; Girlanda et al. 2006). These partial mycoheterotrophs are closely related to fully mycoheterotrophic orchids, thus suggesting an evolutionary pathway where low light conditions have selected for a switch from free-living, rhizoctonia-forming fungi to fungal partners involved in mycorrhizal networks. A transition to partial mycoheterotrophy is thought to evolve to compensate for reduced photosynthetic capabilities, allowing for the eventual rise of fully mycoheterotrophic plants. A parallel evolutionary path to mycoheterotrophy has also been demonstrated in the Ericaceae. Understory Pyroleae, related to the fully mycoheterotrophic Monotropoideae (both Ericaceae), obtain between 10 and 67 % of their carbon from fungi, and have associations with networks of ectomycorrhizal fungi (Tedersoo et al. 2007).

Mycoheterotrophs provide numerous case-studies for investigations into the subversion of a mutualism. Mutualisms are widespread in nature, and play a key role in the maintenance of biodiversity, yet in theory should be susceptible to exploitation and eventual breakdown (Bronstein 2001, Chap. 5). The exploitation of mycorrhizal networks can provide insights into how mutualisms remain robust in the face of cheaters, the evolution of parasites and mutualists, and the ecological distribution of cheaters; i.e., are particular mutualisms or lineages particularly susceptible to exploitation, and why?

Although full mycoheterotrophy has evolved independently on at least 40 occasions in diverse plants, including liverworts, monocots, and eudicots (Chap. 5), it appears to evolve repeatedly in certain plant lineages, such as Orchidaceae and Burmanniaceae, perhaps indicating preadaptations to this lifestyle in these families. As mentioned before, being part of a common mycorrhizal network appears to be a prerequisite to the evolution of mycoheterotrophy, even though some orchids can be supported by non-mycorrhizal fungi, e.g., pathogenic rhizoctonia-forming fungi and saprobic *Psathyrella*. The frequency of mycoheterotrophy among orchids, all of which rely on fungi to support germination and early growth, and are therefore mycoheterotrophic for at least the early part of their life cycle, suggests that initial mycoheterotrophy may also make a switch to cheating a more likely occurrence. Orchids must already possess the necessary biochemical and structural adaptations to actively sequester carbon from their fungal associates, reversing the typical plant-to-fungus direction of carbon flow in mycorrhizal symbioses. In addition, certain photosynthetic ancestors of mycoheterotrophic orchids and Ericaceae, both show the ability to form mycorrhizas in which otherwise ectomycorrhizal fungi penetrate the epidermal root cells. This suggests an ability to manipulate their fungal symbionts, and may explain the high mycorrhizal specificity generally shown by mycoheterotrophic plants in these families. Although there are exceptions, mycoheterotrophs are typically small plants that presumably have a relatively low carbon demand, and so incur limited costs to their fungal hosts, and to the trees that act as their ultimate carbon source if the mycoheterotrophs happen to also be epiparasites. Ancestral characteristics related to growth and nutrient acquisition may make some plant groups more likely to evolve mycoheterotrophy than others. It is thought that there is just a single Gymnosperm species that is potentially mycoheterotrophic (Chap. 2); *Parasitaxus usta*, a woody shrub that grows up to 1.8 m, is considered to obtain carbon via a shared mycorrhizal connection with its host tree, whilst obtaining water from the host via a connection similar to that found in holoparasitic plants (Feild and Brodribb 2005). The rarity of mycoheterotrophy among Gymnosperms is, in some ways, surprising—as a group they have a considerable dependence on mycorrhizal fungi, and their seedlings are thought to rely on a partial mycoheterotrophic supply of nutrients from common mycorrhizal networks during establishment (Simard 2012). This may indicate that large woody plants with a higher carbon demand are precluded from becoming cheats due to their negative impact on fungal hosts. Fitness costs to fungi are notoriously challenging to measure, but knowledge of the costs incurred by fungi would provide a number of insights into the evolution of this lifestyle in different plant groups.

7.3 Fungal Hosts

We can refer to fungi as the hosts of mycoheterotrophic plants because the fungi are likely to be older and larger than the plants, and the fungi may belong to older lineages that are being tracked over evolutionary time by the plants (Bidartondo and Bruns 2001; Bidartondo 2005; Merckx et al. 2009). In many cases of mycoheterotrophy, neighboring photosynthetic mycorrhizal plants may be referred to as ultimate hosts, because they represent the ultimate source of carbon for both their mycorrhizal fungi and associated mycoheterotrophic plants (i.e., epiparasitic mycoheterotrophy).

Although the reliance of mycoheterotrophic plants on mycorrhizal fungi has been known for some time (e.g., Björkman 1960), for many years our understanding of the biology of these plants was limited by the inability to identify the plants' fungal hosts. Prior to the "molecular revolution," identification relied on the ability to culture fungi in the laboratory. Although this methodology was sometimes effective, there were several limitations. Culturing methods could be biased towards fast-growing fungi, often precluded many fungi that were un-culturable, and were sometimes based on flawed morphological species concepts. The development of DNA-based molecular tools greatly increased the capacity to identify and sample fungal associates. Fungal-specific primers allowed direct amplification of fungal DNA from plant roots. By applying these methods, the mycorrhizal fungi of many mycoheterotrophs have been identified over the last 20 years (Table 7.1), providing insights into the diversity and identity of the fungi utilized by these plants.

One would expect certain fungal lineages to be more susceptible to exploitation by mycoheterotrophs, or be more attractive as hosts, than others. For example, it might be expected that fungi with large genet size, long lifespan, wide distribution, and a large and reliable source of carbon would be exploited preferentially. Yet among the fungal lineages in which mycorrhizal symbioses have evolved, exploitation seems widespread (Table 7.1). Only two orders of Glomeromycota, and one class of Ascomycota contain mycorrhizal fungi that have not yet been found to associate with mycoheterotrophs, whilst every order of Agaricomycotina includes fungi that are exploited (Hynson and Bruns 2010). Nevertheless, there are surprising exceptions of fungi that have not yet been found to associate with mycoheterotrophs. The widespread *Hymenoscyphus ericae* complex includes the dominant mycorrhizal associates of most plants with ericoid mycorrhizas. The lack of exploiters of this fungus is striking, considering that it is thought to form mycorrhizal networks between understory and tree vegetation (Grelet et al. 2009), and the Ericaceae with which it associates are closely related to the mycoheterotrophic Monotropoideae.

All orchids are mycoheterotrophic during development, and the majority associate with fungi from the form genus Rhizoctonia (a polyphyletic group, consisting of fungi from the Ceratobasidaceae, Sebacinales, and *Tulasnella*). However, where orchids have become mycoheterotrophic throughout their life they have nearly always switched to ectomycorrhizal fungal partners. For example, *Cephalanthera* orchids associate with ectomycorrhizal fungi in the Thelephoraceae family (Taylor and Bruns 1997; Abadie et al. 2006; Bidartondo and Read 2008), and similarly, *Corallorhiza* associate with fungi in the Russulaceae (Table 7.1; Taylor and Bruns 1997, 1999). Surprisingly, there are no examples of achlorophyllous orchids that associate with non-ectomycorrhizal rhizoctonias. The fully mycoheterotrophic orchids *Rhizanthella gardneri* and *Chamaegastrodia sikokiana*, do associate with Ceratobasidiaceae (Yagame et al. 2008; Bougoure et al. 2009), but in these cases the fungal partner is also ectomycorrhizal, thought to be from a distinct clade of the family, separate from the saprobic/pathogenic Ceratobasidaceae that many green orchids associate with. Similarly, although some mycoheterotrophic orchids associate with *Sebacina* (McKendrick et al. 2002; Selosse et al. 2002; Taylor et al. 2003), this is an ectomycorrhizal clade of Sebacinales, phylogenetically distinct from those used by green orchids (Weiss et al. 2004). It may be that as free-living fungi, unconnected to mycorrhizal networks, most rhizoctonia-forming fungi do not possess sufficient carbon supplies to sustain mycoheterotrophic plants throughout their life; or that as the sole victim of exploitation, with no epiparasitised tree as the carbon source, selection to avoid exploitation is stronger among these fungi. Selection pressure to obtain alternative sources of carbon will be highest in understory habitats where ectomycorrhizal fungi are more common, although given their cosmopolitan distributions, it would be surprising if saprobic/pathogenic rhizoctonias were not also present in these environments. Interestingly, there are rare examples of non-photosynthetic orchids that exploit fungi and are not part of a mycorrhizal network—this ability to recruit non-mycorrhizal fungi and channel

Table 7.1 Selected studies reporting the mycorrhizal fungi of initial and fully mycoheterotrophic plants in nature using DNA sequence data generated without fungal cultivation

Taxon	Samples	Fungi	Trophic group	Reference
Aneuraceae				
Aneura mirabilis	32 thalli	*Tulasnella*	Ectomycorrhizal	Bidartondo et al. (2003)
Burmanniaceae				
Apteria aphylla	6 adults	*Glomus* Group A	Arbuscular mycorrhizal	Courty et al. (2011); Merckx et al. (2012)
Burmannia hexaptera	3 adults	*Glomus* group A	Arbuscular mycorrhizal	Franke et al. (2006); Merckx and Bidartondo (2008)
Campylosiphon congestus	3 adults	*Glomus* Group A, Acaulosporaceae	Arbuscular mycorrhizal	Franke et al. (2006); Merckx et al. (2012)
Dictyostega orobanchoides	6 adults	*Glomus* Group A	Arbuscular mycorrhizal	Merckx et al. (2010, 2012)
Gymnosiphon capitatus	2 adults	*Glomus* Group A, Acaulosporaceae	Arbuscular mycorrhizal	Merckx et al. (2012)
Gymnosiphon divaricatus	1 adult	*Glomus* Group A	Arbuscular mycorrhizal	Merckx et al. (2012)
Gymnosiphon longistylus	1 adult	*Glomus* Group A	Arbuscular mycorrhizal	Merckx et al. (2012)
Gymnosiphon minutus	1 adult	*Glomus* Group A	Arbuscular mycorrhizal	Leake (2005)
Gymnosiphon sp.	4 adults	*Glomus* Group A	Arbuscular mycorrhizal	Courty et al. (2011)
Hexapterella gentianoides	1 adult	*Glomus* Group A	Arbuscular mycorrhizal	Merckx et al. (2012)
Corsiaceae				
Arachnitis uniflora	8 adults	*Glomus* group A	Arbuscular mycorrhizal	Bidartondo et al. (2002)
Ericaceae				
Allotropa virgata	18 adults	*Tricholoma magnivelare*	Ectomycorrhizal	Bidartondo and Bruns (2001)
Hypopithys monotropa	46 adults, seedlings	*Tricholoma*	Ectomycorrhizal	Bidartondo and Bruns (2001); Leake et al. (2004)
Hemitomes congestum	7 adults	*Hydnellum*	Ectomycorrhizal	Bidartondo and Bruns (2001)
Monotropa uniflora	56 adults, seedlings	Russulaceae	Ectomycorrhizal	Cullings et al. (1996); Bidartondo and Bruns (2001, 2005); Young et al. (2002)
Monotropastrum humile	106 adults	Russulaceae	Ectomycorrhizal	Bidartondo and Bruns (2001); Yokoyama et al. (2005); Matsuda et al. (2011)
Monotropastrum humile var. *Glaberrimum*	2 adults	Thelephoraceae	Ectomycorrhizal	Yokoyama et al. (2005)
Monotropsis odorata	2 adults	*Hydnellum*	Ectomycorrhizal	Bidartondo and Bruns (2001)
Pityopus californicus	12 adults, seedlings	*Tricholoma*	Ectomycorrhizal	Bidartondo and Bruns (2001, 2005)
Pleuricospora fimbriolata	42 adults	*Gautieria monticola*	Ectomycorrhizal	Bidartondo and Bruns (2001)
Pterospora andromedea	234 adults, seedlings	*Rhizopogon salebrosus*, *R. Arctostaphyli*, *R. ellenae*	Ectomycorrhizal	Cullings et al. (1996); Bidartondo and Bruns (2001, 2005); Dowie et al. (2011, 2012); Hazard et al. (2012)

(continued)

Table 7.1 (continued)

Taxon	Samples	Fungi	Trophic group	Reference
Pyrola aphylla	12 adults	various Asco- and Basidiomycota	Ectomycorrhizal, Saprothrophic?	Hynson and Bruns (2009)
Sarcodes sanguinea	150 adults, seedlings	*Rhizopogon ellenae, R. subpurpurascens*	Ectomycorrhizal	Kretzer et al. (2000); Bidartondo and Bruns (2001, 2005)
Gentianaceae				
Exochaenium oliganthum	1 adult	*Glomus* Group A	Arbuscular mycorrhizal	Franke et al. (2006)
Voyria aphylla	12 adults	*Glomus* Group A, Gigasporaceae	Arbuscular mycorrhizal	Merckx et al. (2010); Courty et al. (2011)
Voyria aurantiaca	2 adults	*Glomus* group A	Arbuscular mycorrhizal	Bidartondo et al. (2002)
Voyria caerulea	1 adult	*Glomus* group A	Arbuscular mycorrhizal	Bidartondo et al. (2002)
Voyria corymbosa	1 adult	*Glomus* group A	Arbuscular mycorrhizal	Bidartondo et al. (2002)
Voyria rosea	2 adults	*Glomus* group A	Arbuscular mycorrhizal	Bidartondo et al. (2002)
Voyria tenuifolia	2 adults	*Glomus* group A, *Gigaspora*	Arbuscular mycorrhizal	Bidartondo et al. (2002)
Voyriella parviflora	3 adults	*Glomus* group A	Arbuscular mycorrhizal	Bidartondo et al. (2002)
Lycopodiaceae				
Huperzia hypogeae	2 gametophytes	*Glomus* group A	Arbuscular mycorrhizal	Winther and Friedman (2008)
Ophioglossaceae				
Botrychium lanceolatum	8 gametophytes	*Glomus Group A*	Arbuscular mycorrhizal	Winther and Friedman (2007)
Botrychium crenulatum	8 gametophytes	*Glomus Group A*	Arbuscular mycorrhizal	Winther and Friedman (2007)
Orchidaceae				
Aphyllorchis caudata	13 adults	Russulaceae, Thelephoraceae and others	Ectomycorrhizal	Roy et al. (2009a)
Aphyllorchis montana	39 adults	Russulaceae, Thelephoraceae and others	Ectomycorrhizal	Roy et al. (2009a)
Cephalanthera austinae	26 adults	Thelephoraceae	Ectomycorrhizal	Taylor and Bruns (1997)
Cephalanthera damasonium	seedlings	Thelephoraceae	Ectomycorrhizal	Bidartondo and Read (2008)
Cephalanthera damasonium (albino)	2 adults	Thelephoraceae and others	Ectomycorrhizal	Julou et al. (2005)
Cephalanthera exigua	9 adults	Thelephoraceae and others	Ectomycorrhizal	Roy et al. (2009a)
Cephalanthera longifolia (albino)	2 adults	Thelephoraceae and others	Ectomycorrhizal	Abadie et al. (2006)
Cephalanthera longifolia	seedlings	Thelephoraceae	Ectomycorrhizal	Bidartondo and Read (2008)
Chamaegastrodia shikokiana	4 adults	Ceratobasidiaceae	Ectomycorrhizal	Yagame et al. (2008)
Corallorhiza bentleyi	xx adults	*Tomentella*	Ectomycorrhizal	Barrett and Freudenstein (2008)

Corallorhiza maculata	224 adults	Russulaceae	Ectomycorrhizal	Taylor and Bruns (1997, 1999); Taylor et al. (2004)
Corallorhiza mertensiana	27 adults	Russulaceae	Ectomycorrhizal	Taylor and Bruns (1999)
Corallorhiza striata	115 adults	Thelephoraceae	Ectomycorrhizal	Taylor (1997); Barrett et al. (2010)
Corallorhiza trifida	4 adults, seedlings	Thelephoraceae	Ectomycorrhizal	McKendrick et al. (2000)
Cymbidium macrorhizon	6 adults	Russulaceae	Ectomycorrhizal	Motomura et al. (2010)
Dipodium hamiltonianum	4 adults	Russulaceae	Ectomycorrhizal	Dearnaley and Le Brocque (2006)
Dipodium variegatum	3 adults	Russulaceae and others	Ectomycorrhizal	Bougoure and Dearnaley (2005)
Epipactis atrorubens	seedlings	Thelephoraceae and others	Ectomycorrhizal	Bidartondo and Read (2008)
Epipactis microphylla (albino)	2 adults	Tuberaceae and others	Ectomycorrhizal	Selosse et al. (2004)
Epipogium aphyllum	34 adults	*Inocybe* and others	Ectomycorrhizal	Roy et al. (2009b)
Epipogium roseum	13 adults	Coprinaceae	Saprotrophic	Yamato et al. (2005)
Erythrorchis cassythoides	7 adults	*Gymnopus, Russula* and others	Saprotrophic and ectomycorrhizal	Dearnaley (2006)
Eulophia zollingeri	12 adults	*Psathyrella candolleana*	Saprotrophic	Ogura-Tsujita and Yukawa (2008)
Gastrodia confusa	44 adults	*Mycena, Marasmiellus*-like	Saprotrophic	Ogura-Tsujita et al. (2009)
Gastrodia sesamoides	6 adults	*Marasmiaceae*	Saprotrophic	Dearnaley and Bougoure (2010)
Gastrodia similis	15 adults	*Resinicium*	Saprotrophic	Martos et al. (2009)
Hexalectris arizonica	33 adults	Sebacinaceae, Thelephoraceae and others	Ectomycorrhizal	Taylor et al. (2003); Kennedy et al. (2011)
Hexalectris brevicaulis	7 adults	Russulaceae, Sebacinaceae, Ceratobasidiaceae	Ectomycorrhizal	Kennedy et al. (2011)
Hexalectris colemanii	9 adults	Sebacinaceae	Ectomycorrhizal	Kennedy et al. (2011)
Hexalectris grandiflora	10 adults	Russulaceae, Ceratobasidiaceae, and others	Ectomycorrhizal	Kennedy et al. (2011)
Hexalectris nitida	9 adults	Sebacinaceae and others	Ectomycorrhizal	Kennedy et al. (2011)
Hexalectris parviflora	3 adults	Sebacinaceae, Ceratobasidaceae	Ectomycorrhizal	Kennedy et al. (2011)
Hexalectris revoluta	6 adults	Sebacinaceae and others	Ectomycorrhizal	Taylor et al. (2003); Kennedy et al. (2011)
Hexalectris spicata	56 adults	Sebacinaceae and others	Ectomycorrhizal	Kennedy et al. (2011)
Hexalectris spicata var. spicata	18 adults	*Sebacina, Thanatephorus ochraceus*	Ectomycorrhizal, pathogenic?	Taylor et al. (2003)
Hexalectris spicata var. unknown	1 adult	*Thanatephorus ochraceus*	Pathogenic?	Taylor et al. (2003)
Hexalectris warnockii	7 adults	Thelephoraceae	Ectomycorrhizal	Kennedy et al. (2011)

(continued)

Table 7.1 (continued)

Taxon	Samples	Fungi	Trophic group	Reference
Limodorum arbortivum	76 adults	Russulaceae	Ectomycorrhizal	Girlanda et al. (2006)
Limodorum trabutianum	1 adult	Russulaceae	Ectomycorrhizal	Girlanda et al. (2006)
Neottia nidus-avis	69 adults, seedlings	*Sebacina*	Ectomycorrhizal	McKendrick et al. (2002); Selosse et al. (2002)
Rhizanthella gardneri	12 adults	Ceratobasidiaceae	Ectomycorrhizal	Bougoure et al. (2010)
Rhizanthella slateri	1 adult	Ceratobasidiaceae	Ectomycorrhizal	Bougoure et al. (2010)
Wullschlaegelia aphylla	16 adults	*Mycena* and others	Saprotrophic	Martos et al. (2009)
Petrosaviaceae				
Petrosavia sakuraii	17 adults	*Glomus* Group A	Arbuscular mycorrhizal	Yamato et al. (2011a, b)
Psilotaceae				
Psilotum nudum	16 gametophytes	*Glomus Group A*	Arbuscular mycorrhizal	Winther and Friedman (2009)
Thismiaceae				
Afrothismia foertheriana	4 adults	*Glomus group A*	Arbuscular mycorrhizal	Franke et al. (2006); Merckx and Bidartondo (2008)
Afrothismia gesnerioides	2 adults	*Glomus group A*	Arbuscular mycorrhizal	Franke et al. (2006); Merckx and Bidartondo (2008)
Afrothismia hydra	7 adults	*Glomus group A*	Arbuscular mycorrhizal	Franke et al. (2006); Merckx and Bidartondo (2008)
Afrothismia korupensis	3 adults	*Glomus group A*	Arbuscular mycorrhizal	Franke et al. (2006); Merckx and Bidartondo (2008)
Afrothismia saingei	1 adult	*Glomus* Group A	Arbuscular mycorrhizal	Franke et al. (2006)
Afrothismia winkleri	5 adults	*Glomus group A*	Arbuscular mycorrhizal	Franke et al. (2006); Merckx and Bidartondo (2008)
Afrothismia sp.	1 adult	*Glomus* Group A	Arbuscular mycorrhizal	Franke et al. (2006)
Thismia rodwayi	2 adults	*Glomus* Group A	Arbuscular mycorrhizal	Merckx et al. (2012)
Triuridaceae				
Kupea martinetugei	2 adults	*Glomus* group A	Arbuscular mycorrhizal	Franke et al. (2006); Merckx and Bidartondo (2008)
Sciaphila japonica	7 adults	*Glomus* Group A	Arbuscular mycorrhizal	Yamato et al. (2011a, b)
Sciaphila ledermanii	4 adult	*Glomus* group A, Acaulosporaceae, Gigasporaceae	Arbuscular mycorrhizal	Franke et al. (2006); Merckx and Bidartondo (2008); Merckx et al. (2012)
Sciaphila tosaensis	6 adult	*Glomus* group A	Arbuscular mycorrhizal	Yamato (2001); Yamato et al. (2011a, b)

them into novel mycorrhizal-like symbioses appears so far unique to orchids. Martos et al. (2009) demonstrated that *Gastrodia similis* has a specific association with wood-decaying *Resinicium*, and *Wullschlaegelia aphylla* associates with leaf-decaying *Mycena*. *Gastrodia confusa* has also been found to be specialized towards species of *Mycena* (Ogura-Tsujita et al. 2009). This could suggest that these fungi provide a more consistent supply of carbon than saprobic rhizoctonias in these environments.

The genus *Epipogium* provides an interesting example of fungal partner choice among mycoheterotrophic orchids, and how it is influenced by the environment. *Epipogium aphyllum* occurs predominantly in temperate areas, and has been found to depend on *Inocybe* fungi (Roy et al. 2009b) that are ectomycorrhizal with woody plants. The closely related *Epipogium roseum* occurs in tropical forests and associates with saprobic Coprinaceae (Yamato et al. 2005; Yagame et al. 2007). Other orchids associated with saprobic fungi also tend to be found in tropical forests (Umata 1995), although tropical mycoheterotrophic orchids have also been found that associate with ectomycorrhizal fungi (Roy et al. 2009a). It may be that the hot, humid conditions found in tropical forests mean that saprobic fungi in these environments are a better source of carbon than those found in temperate forests, where decomposition rates are lower. It is notable that *E. roseum* has a shorter life cycle, and produces smaller roots and fruits than *E. aphyllum* (Roy et al. 2009b), perhaps as a consequence of saprobic fungi providing less carbon than ectomycorrhizal fungi. Further comparisons of phylogenetically related mycoheterotrophs that differ in environment and/or fungal host ecology will be informative.

7.4 The Causes and Consequences of Fungal Specialization

Specificity is a multifaceted and relative concept with far-reaching ecological and evolutionary implications. For instance, in mycorrhizal symbioses, narrow receptivity implies low ecological redundancy whereby a particular fungus is essential for the establishment, survival, and/or diversification of a plant. If mycoheterotrophic plants are biotrophic parasites, we should expect them to be specialized relative to (1) mutualistic mycorrhizal plants and (2) the fungi potentially available in mycoheterotroph habitats (Bidartondo 2005). This is because continuous intimate contact between host and symbiont throughout life generally correlates with high specificity in parasites (Price 1980; Thompson 2005). Indeed, the ongoing molecular ecology revolution has revealed several examples of exceptionally high specificity by mycoheterotrophic plants towards narrow clades of ecto- and arbuscular mycorrhizal fungi relative to generalist autotrophic mycorrhizal plants, or towards narrow clades of saprobic fungi, within what are typically highly diverse fungal communities in nature. These novel findings have in turn led to numerous new ideas and interpretations (Bruns et al. 2002; Gardes 2002; Hibbett 2002; Taylor et al. 2002; Sanders 2003; Taylor 2004; Bidartondo 2005; Leake 2005; Selosse et al. 2006; Smith and Read 2008; Waterman and Bidartondo 2008; Merckx et al. 2009; Hynson and Bruns 2010).

Notably, molecular ecology allows researchers to test the first of Koch's four postulates for demonstrating a plant-fungal link in mycoheterotrophic mycorrhizas; researchers can identify fungi using DNA sequences generated from field mycorrhizas without the need for fungal isolation (Table 7.1). Some studies have supported these molecular fungal identifications, to the phylum or family level, using light microscopy (Taylor and Bruns 1997; Ogura-Tsujita and Yukawa 2008) or electron microscopy of mycorrhizas (Winther and Friedman 2008; Roy et al. 2009b). However, due to the rarity of some mycoheterotrophic plants many studies have only examined a few plants from a few populations and some studies have only used fungal DNA sequence identification of fungi cultured from field-collected mycorrhizas (Yamato et al. 2005; Yagame et al. 2008). Due to the unculturability or slow growth of most mycorrhizal fungi under non-mycorrhizal conditions, there have been only two examples of testing all three subsequent postulates—isolation of fungi in pure culture, re-mycorrhization and re-isolation—in mycoheterotrophic systems. These are the orchid *Epipogium roseum* with the litter-decomposer *Coprinus disseminatus* (Yagame et al. 2007) and the underground orchid *Rhizanthella gardneri* with the ectomycorrhizal/saprobic *Thanatephorus gardneri* (Bougoure et al. 2009). In the future, more studies will use multi-locus fungal identification from field mycorrhizas and will reach asymptotes for mycorrhizal fungus accumulation curves at multiple populations.

Obligate manipulation of the host may select for extreme specificity in parasites and host manipulation may be explained by the need of a parasite to complete its life cycle on one host individual (Thompson 2005). Mycoheterotrophic plants appear to be highly manipulative of their symbionts' growth (see Chap. 4) while avoiding recognition and/or suppressing defenses. The evolution and maintenance of the unknown biochemical and structural mechanisms that allow mycoheterotrophic plants to actively sequester fungal carbon may also generate selective pressures toward specialization.

Also in keeping with the general characteristics of parasitism, mycoheterotrophy has led to phylogenetic tracking of several fungi (e.g., *Tricholoma, Russula, Glomus*), strict mycorrhizal loyalty even when different mycoheterotrophic plant lineages co-occur, and the formation of geographic mosaics with regards to mycorrhizal specificity (Bidartondo 2005). Nevertheless, as shown by Jackson (2004), the phylogenetic co-divergence pattern of monotrope mycoheterotrophs and their mycorrhizal fungi is radically different from that of plants and their pathogenic fungi. The difficulty lies in explaining the odd combination of outlandish host jumps with narrowly conservative host shifting observed in the monotropes, despite the plants having ample ecological opportunities for host jumping within forest ecosystems (Bidartondo and Bruns 2002).

It should be noted that not all mycoheterotrophs are specialized towards fungi. For example, two species of *Aphyllorchis*, growing in tropical Dipterocarp forests in Asia were shown to associate with a wide range of ectomycorrhizal fungi—individual orchids associated with up to nine species of fungi (Roy et al. 2009a). The authors speculated that the higher photosynthetic capability of trees in the tropics may make the costs of mycoheterotrophs lower, and so make functional co-adaptation, or co-evolutionary arms-races less likely. The challenges of quantifying costs to host fungi make it difficult to test this hypothesis. Mycoheterotrophs that exploit arbuscular mycorrhizal networks also show differing levels of specificity. Mycoheterotrophs found to associate with multiple distantly related arbuscular mycorrhizal fungi are known in the tropics (Bidartondo et al. 2002; Franke et al. 2006; Merckx et al. 2010; Courty et al. 2011), but there are also examples of extreme specialist cheaters of arbuscular mycorrhizal networks in tropical forests (Bidartondo et al. 2002; Merckx and Bidartondo 2008). Some members of the Ericaceae also show low specificity—*Pyrola aphylla* associates with a diverse range of ectomycorrhizal fungi (Hynson and Bruns 2009). It is notable that a partially mycoheterotrophic ancestor of *Pyrola* is also a generalist (Hynson and Bruns 2009; Toftegaard et al. 2010). The degree of specialization among the photosynthetic ancestors of mycoheterotrophs is bound to influence the breadth of possible fungal hosts once mycoheterotrophy has evolved. It is possible that the high specialization found in mycoheterotrophic orchids is simply an ancestral trait—many photosynthetic orchids show equal, or even greater specialization towards mycorrhizal fungi (McCormick et al. 2004).

If mycorrhization within a short time frame after seed germination is necessary for survival, then germination must be a critical life history checkpoint for all mycoheterotrophic plants. Some orchids may imbibe water and break their seed coat in the absence of fungi, but further development is arrested unless mycorrhization occurs (Smith and Read 2008). The seeds of monotrope mycoheterotrophs are able to recognize specific molecules released by fungi as germination cues, as demonstrated by in vitro experiments with *Sarcodes sanguinea* and *Pterospora andromedea* seeds exposed across a permeable membrane to hyphae of *Rhizopogon* section *Amylopogon* fungi (Bruns and Read 2000).

Data on germination of mycoheterotrophic plants in nature is currently scarce; it stems from the application of the laborious seed packet burial technique (Rasmussen and Whigham 1993) or the fortuitous unearthing of seedlings. In some monotropes there is evidence of specificity to the same fungi colonizing adults, with occasional "mistakes" when seeds are buried in locations where fungi are naïve (Bidartondo and Bruns 2005). Furthermore, the seeds of some partial mycoheterotrophic *Cephalanthera* orchids are known to undergo a narrow specificity checkpoint during their initial mycoheterotrophic development relative to their generalist mycorrhizal behavior as adults (Bidartondo and Read 2008).

A useful conceptual counterpart to mycoheterotrophic germination, and one that has been the focus of extensive research, is direct plant–plant parasitism. There is in vitro data available for some phases of seed germination and early seedling development from direct plant parasitic Scrophulariaceae (i.e., *Striga*, *Orobanche*). The seeds of many direct plant parasites are microsperms with longevities in soil of several years. Low concentrations of small labile diffusible compounds known as xenognosins serve as host plant recognition factors that probably activate developmental programs in the parasitic plant such as breaking seed dormancy, starting a viability clock that lasts for a few days without host plant attachment, and signaling haustorium development (Yoder 1999; Keyes et al. 2001; Yoder 2001). Surprisingly, some of these same signals are now known to be targeted at germinating Glomeromycota spores and only secondarily coopted by direct plant parasites to germinate their seeds (Akiyama et al. 2005). The first two programs are of interest from the perspective of mycoheterotrophic plant seeds interacting with potential host fungi. The last may also be of interest for some mycoheterotrophs; the root hairs of the mycoheterotrophic *Voyria aphylla* (Gentianaceae) grow towards the roots of photosynthetic plants prior to the formation of hyphal connections (Imhof 1999) much like the haustorial hairs of the parasitic *Triphysaria pusilla* (Scrophulariaceae) grow towards host roots prior to penetrating between epidermal cells (Yoder 1999).

In the context of mycoheterotrophy, specific elicitors of seed germination would provide a ready co-evolutionary arena (in addition to any potential arms-races post-colonization). Seeds produced by outcrossing should have a high frequency of mismatches between the plant and fungus and hence low colonization frequencies. Ultimately, if hybrids have lower fitness at germination than specifically coevolved local genotypes, coevolution could precipitate reproductive isolation and diversification. It will be fascinating to test these hypotheses in mycorrhizal symbioses.

Being so reliant on fungi for nutrition, the choice of host will have major consequences for the potential distribution of mycoheterotrophs, especially those dependent on fungi for germination. High specialization will place constraints on the plant distribution, limiting it to be at least less than that of the fungus. This may explain the extreme rarity of many of these plants. Due to the conditions within which this lifestyle was initially selected, most mycoheterotrophs are found in dark forest floor habitats, where plants relying on photosynthesis would be under a competitive disadvantage. However, choice of fungal partner will still limit which of these habitats the plant can occur in. For example, *Corallorhiza trifida* was shown to germinate in *Betula* and *Salix* forests, but not within pine forests, probably due to the lack of a host fungus in this environment (McKendrick et al. 2000). However, specialization towards a single fungus does not preclude a wide distribution—the orchid *Eulophia zollingeri* has a wide geographic range despite being specialized towards a single fungal species, because its fungal host also has an extremely wide distribution (Ogura-Tsujita and Yukawa 2008) and the same is true for Monotropoideae—diverse clades with wide ranges depend on diverse widespread fungi (*Monotropa hypopithys/Tricholoma*, *Monotropa uniflora/Russula*) and more restricted plants depend on less widespread fungi (*Pleuricospora/Gautieria*, *Monotropsis/Hydnellum*, *Pterospora-Sarcodes/Rhizopogon* sect. *Amylopogon*) (Bidartondo and Bruns 2005). The nested structure of most interaction networks (Bascompte and Jordano 2007), whereby specialists interact with generalists, may limit the negative consequences of specialization.

One of the key findings from the simultaneous increase in molecular phylogenetic and mycoheterotrophy research over the last two decades has been the realization that this nutritional strategy has evolved independently on so many occasions. However, despite this progress, there has been very little work performed to elucidate the subsequent consequences of mycoheterotrophy on the diversification of those lineages. Any change that so profoundly alters the life history of a plant is likely to have some impact on the subsequent evolution of that lineage, yet so far we have been largely unable to investigate these impacts. A study of the mycorrhizal partners of *Afrothismia* in Cameroon shows that they appear to have diversified on to a preexisting diversity of fungi within a narrow clade of arbuscular mycorrhizal *Glomus* fungi (Merckx and Bidartondo 2008). The diversity of mycoheterotrophic lineages may therefore be restricted to equaling or remaining less than the diversity of their associated fungi, unless they are able to shift to an alternate fungal host lineage.

Ecological specialization has historically been thought of as an evolutionary dead end (Cope 1896; Gould 1970; Futuyma and Moreno 1988). It is thought that the numerous adaptations required to specialize on a particular resource can lead lineages to become increasingly committed to that niche, making reversals or shifts to generalization very unlikely (Futuyma and Moreno 1988). Generalists meanwhile, have greater potential to adapt to new environments, providing more opportunities for speciation and a reduced risk of extinction (Zayed et al. 2005). Such directionality can clearly be seen in the evolution of mycoheterotrophy in lineages such as Neottieae, in which mycoheterotrophy is thought to have evolved repeatedly, but reversals to autotrophy are unknown. The "mutational freefall" in which the metabolic pathways of photosynthesis are rapidly lost following the evolution of mycoheterotrophy, makes any reversion back to autotrophy highly unlikely. The mutation of plastid genes essential for photosynthesis into non-functional pseudogenes has been demonstrated among several mycoheterophic lineages (Cameron 2004; Cameron and Molina 2006; Barrett and Freudenstein 2008). The subsequent dependence on and specialization towards particular fungal hosts is therefore likely to be a conserved trait within mycoheterotrophic lineages, and this specialization may cause similar constraints on the subsequent evolution of these clades.

The impacts of this specialized interaction on processes such as speciation, adaptation, and specialization towards other organisms remains unknown. The high degree of specialization towards particular fungal partners certainly does not prevent diversifying selection from acting on those species. Indeed, some mycoheterotrophic lineages have diversified to occupy a wide range of habitats following the loss of photosynthetic capabilities e.g., *Neottia*. There is anecdotal evidence of ongoing morphological and molecular diversification—e.g., *Hypopitys monotropa* shows great morphological diversity and has a record 85 taxonomic synonyms, indicating its specialization towards *Tricholoma* has not constrained any subsequent evolution. As previously mentioned, sympatric genotypes of mycoheterotrophs have been shown to consistently associate with distinct fungi—in these cases the diversity of a particular fungal clade may limit the diversity of mycoheterotrophic plants able to occur at a given site. It may be worth investigating whether this partitioning of fungi between co-occurring mycoheterotrophs is a mechanism to avoid competing for nutrition, as has been proposed for specialized photosynthetic orchids in diverse communities (Waterman et al. 2011).

Although mycoheterotrophy is certainly not a barrier to further diversification, the evolution of this life history trait is still likely to have had an impact on speciation and extinction, and a number of specific hypotheses can be raised. The reliance on clonal reproduction, and specialization towards a narrow resource may limit mycoheterotrophic plants' ability to adapt, and thus lower speciation rates. Alternatively, the strongly subdivided populations associated with mycoheterotrophy may promote speciation (Chap. 6). Mycoheterotrophy may also affect extinction rates—loss of genetic diversity through increased clonal reproduction, and extreme reliance on a narrow range of fungal partners may render mycoheterotrophic lineages particularly vulnerable to extinction. The simplest way to investigate the effects of nutritional mode on diversification rates is to compare the diversity of mycoheterotrophic lineages with their autotrophic sister clades. In the Vanilleae tribe, mycoheterotrophic lineages are clearly less diverse than their autotrophic sister clades (Cameron 2009), suggesting reduced speciation rates or increased extinction. In contrast, mycoheterotrophic Thismiaceae (minus *Afrothismia*) appears to be more diverse than its autotrophic sister clade, *Tacca* (Chap. 2). A global analysis of sister clades would be useful in uncovering major trends, and identifying how they differ between angiosperm families, but at present it is difficult to confidently assign sister clades to the vast majority of mycoheterotrophic lineages. Mycoheterotrophic plants present particular challenges to the construction of molecular phylogenetic trees, not least of which is the scarcity of study material (Merckx and Freudenstein 2010). However, as new data on mycoheterotrophs and their relatives gradually becomes available we should be provided with greater insights into the consequences of mycoheterotrophy for diversification. Near-complete phylogenies can be subjected to more sophisticated tests to detect changes in speciation or extinction rates associated with shifts in nutritional mode from ancestral autotrophy to mixotrophy to full mycoheterotrophy (e.g., BiSSE analysis, Maddison et al. 2007).

7.5 Non-fungal Interactions

As a group of plants defined by their nutritional strategy, it is understandable that most research into mycoheterotrophs has focused on their nutritional physiology, and the nature of their fungal interactions. The wider interaction networks of mycoheterotrophs have received little attention, despite offering an excellent model to test how the inherent constraints of a highly specialized relationship with fungi might impact the evolution of other species interactions, such as those with herbivores (see Box 7.1) and pollinators. Their limitation to environments where fungal hosts are available, in dark understory habitats, and possible nutrient limitations, has led some to propose that a syndrome of characteristics may evolve with mycoheterotrophy and lead to convergence in traits related to reproduction. However, this could manifest itself in various ways. It is notable that many mycoheterotrophs produce a vast number of tiny seeds, although it is uncertain if this is a cause or consequence of reliance on fungi. As this convergent trait is also found among many parasitic plants, it could be a reproductive strategy aimed at increasing the likelihood of at least a few offspring locating a suitable host. The production of vast quantities of seeds requires large amounts of pollen to be transferred between plants, and this has the potential to lead to specialized pollination systems, seen most notably in the Orchidaceae. Certainly, there are mycoheterotrophic orchids with specialized pollination systems—*Cryptostylis hunteriana* is pollinated specifically by an Ichneumon wasp (*Lissopimpla excelsa*) via pseudo-copulation (Nicholls 1938). It is notable in this example that the wasp has a wide distribution—as with fungal interactions, a common or widely distributed partner may promote specialization.

Another theory is that limited carbon supply, patchy distributions, and a restriction to habitats with few pollinators, may lead to a reliance

Box 7.1 Cryptic Coloration in *Monotropsis odorata*

Camouflage (cryptic) coloration that allows an organism to blend in with its surroundings can increase individual fitness by reducing attacks from visually guided predators or by increasing concealment from possible prey. While this is a well-studied and accepted strategy in the animal kingdom, until recently cryptic coloration has only been proposed to exist in plants. Such an absence of empirical evidence is not surprising, given that the coloration of the vast majority of plants is constrained to consist of green pigmentation necessary for photosynthetic activities. However, non-photosynthetic plants, such as

(continued)

Box 7.1 (continued)

Fig. 7.1 (**a**, **b**) The reproductive stems of *Monotropsis odorata*, with (**a**) bracts intact, and (**b**) bracts experimentally removed

mycoheterotrophs, do not require such green pigmentation and are therefore open to a world of possibility in coloration, much of which may be adaptive.

Monotropsis odorata (Ericaceae) is a mycoheterotroph that has been proposed to exhibit cryptic coloration, as it is small in stature, notoriously difficult to locate in the wild, and seemingly reliant on fragrance cues for both pollinator and seed dispersal agents (Klooster and Culley 2009). Standing only 3.5–6 cm tall, *M. odorata* is covered with dried, sterile bracts that resemble dead leaf litter under casual observation (see Fig. 7.1a). In the first empirical investigation of plant cryptic coloration, Klooster et al. (2009) set out to determine (1) if manipulating these bracts can impact herbivory rates and (2) if the bracts, as dead vegetative material, actually resemble leaf litter in the visual spectrum. A large population of *M. odorata* was studied in Tennessee prior to reproductive maturity during a 2-year study and, each year, a control group of unmanipulated plants was established (see Fig. 7.1a) and an experimental group of plants had its bracts removed (see Fig. 7.1b). Throughout the blooming period, the control and experimental plants were monitored to assess herbivory rates. Overall, 27 % and 20 % higher herbivory rates were observed in the experimental group consisting of plants without bracts compared to the control group of unmanipulated plants across 2 years, respectively. These significant findings supported the hypothesis that bracts play a role in herbivore avoidance for *M. odorata*. Additional measurements of the reflectance spectra of *M. odorata* bracts, flower petals, and stems were compared to reflectance measurements taken of ambient leaf litter collected from the base of the plants. Such analyses demonstrated that bracts are indistinguishable from leaf litter in the visual spectrum and fleshy flower petals and stems strongly contrast with leaf litter. Therefore, plants with bracts intact resemble leaf litter in the visual spectrum, whereas plants without bracts are much more visually conspicuous, supporting the proposed hypothesis of cryptic coloration.

In conclusion, mycoheterotrophic plants, such as *M. odorata*, can play a valuable role in testing hypotheses that may otherwise be difficult to assess in most plant systems. This study of herbivore interactions with *M. odorata* shows that mycoheterotrophic plants acquire unique adaptations, convergent with those of other heterotrophs, by eliminating aspects of photosynthesis from their life history. Further empirical investigations into the ecological interactions of mycoheterotrophs with members of their community will likely elucidate more novel life history traits.

on autogamy among mycoheterotrophs (Dressler 1981; Leake 1994), although detailed studies indicate that autogamy may be less common among these plants than sometimes thought (see Box 7.2). It has also been hypothesized that mycoheterotrophs will converge on floral characters to attract pollinators more common in understory forests. This could potentially lead to a mycoheterotroph floral syndrome, consisting of small white flowers, with a scent attractive to fungus gnats, as seen in *Neottia cordata* (Ackerman and Mesler 1979), and tight synchrony of blooming period among co-occurring population members. However, a review of the anecdotal evidence available on reproduction among mycoheterotrophs shows a diversity of breeding systems and pollinators (Table 7.2). Although there are no obvious patterns in the available data, there is only one study that looks at the diversity of pollination systems among related plants within a monophyletic lineage (Klooster and Culley 2009), and this is where any reciprocal evolution of nutritional and reproductive strategies may be most apparent.

Box 7.2 Is Autogamy an Adaptation Towards Mycoheterotrophy? *Voyria* as a Case-Study

Several investigations of mycoheterotrophic species have revealed that self-pollination seems to be a common characteristic of these plants (Oehler 1927; Bakshi 1959; Takahashi et al. 1993; Lehnebach et al. 2005; Klooster and Culley 2009). In this context, it has often been hypothesized that autogamy represents an adaptation towards the mycoheterotrophic mode of life. Leake (1994) argued that mycoheterotrophic plants invest large amounts of carbohydrates into their reproductive organs so that they probably cannot afford failure to produce seeds. Since they typically bloom in moist shaded places near the forest floor, where flowers are sometimes covered with leaf litter, their habitat may be unfavorable for common pollinating insects. In addition, single plants or plant patches are often widely separated, and as they depend on the nutrient supply of the fungus, simultaneous flowering cannot always be attained. It appears, therefore, that autonomous self-pollination may be a logical consequence for the plants to assure seed production. Nonetheless, most mycoheterotrophic plants exhibit floral characteristics, raising doubts about reproduction purely by self-pollination. Many flowers develop special morphological structures like long appendages, trap-like cavities, broadened thecae, hairs, etc. Some are conspicuously colored and/or fragrant and even possess glands for nectar production. All these observations do not fit in the image of a typical self-pollinated plant. Why should an organism spend its precious carbohydrate reserves on large floral tissues, pigments, volatiles, and sugars if its final aim is self-pollination?

An interesting plant group in this respect is the genus *Voyria* (Gentianaceae). The genus comprises 19 species and has a vast distribution in tropical and subtropical America with its center in the Guianas and one species occurring in western tropical Africa (Maas and Ruyters 1986; Albert and Struwe 1997). Most taxa possess brightly colored flowers, which emit scent and offer nectar (Fig. 7.2). Consequently, they are considered to be cross-pollinated (Maas and Ruyters 1986). Nevertheless, floral visitors have only been observed once (*Brachycera* sp. at *V. tenella*; Imhof et al. 1994), and there was insufficient evidence to prove that these insects were also pollinators of the plant. The sole other publication dealing with the reproductive biology of this genus postulated that *Voyria rosea* and *Voyriella parviflora* (a close relative of *Voyria*) are strictly autogamous (Oehler 1927). Therefore, in a recent study, Hentrich et al. (2010) studied the reproductive biology of three *Voyria* species—*V. caerulea*, *V. clavata*, and *V. rosea*—to document the reproductive traits of the plants and determine the importance of floral visitors for their propagation. For this purpose, in the early to mid-rainy

(continued)

Box 7.2 (continued)

Fig. 7.2 Floral diversity in the genus *Voyria*, illustrating the brightly colored flowers and long corolla tubes indicative of specialization towards long-tongued insect pollinators. Above ground organs of *V. caerulea*, *V. clavata*, *V. rosea*, *V. tenuiflora*, *V. corymbosa* (from *left* to *right*)

season of the years 2004–2006, they studied the plant's floral morphology, characterized UV-reflections of the flowers, and determined the ratio of pollen to ovules, since this is thought to be a good indicator of breeding systems (Cruden 1977). Further, the composition of floral nectar and floral scent was analysed, flowers were observed to identify potential pollinators, and finally bagging experiments were carried out on *V. rosea* to test for autonomous self-pollination.

Germinating pollen grains were observed in the anthers of all species, forming large interwoven clumps that were transferred by the pollinators as a unit. In *V. caerulea* and *V. rosea*, stamens and stigmas were positioned close together, enabling the plants to self-pollinate. In addition, a pollinator exclusion experiment clearly demonstrated that *V. rosea* produced seeds by autonomous self-pollination. In contrast, male and female organs were widely separated in *V. clavata*, making self-pollination very unlikely in this species. Pollen-to-ovule ratios of all three species were low (33–78), indicating efficient pollen transfer, either by selfing or by outcrossing. Observed flower visitors were butterflies (*V. caerulea*, *V. rosea*) and/or long-tongued bees of the genus *Euglossa* (*V. clavata*, *V. rosea*). However, the level of pollinator specialization differed between species and this may be the key factor influencing the plant's breeding systems and levels of autogamy. The floral characteristics of *V. caerulea* and *V. rosea* allow a large group of long-tongued insects to gather nectar from the flowers, indicating a generalist pollination strategy. Nevertheless, these species were visited only rarely and irregularly, and may therefore follow a mixed selfing-outcrossing strategy to ensure seed production when pollen transfer by visitors failed. In contrast, *V. clavata* showed a high degree of specialization towards a small species group of euglossine-pollinators, which were frequent visitors. The

(continued)

Box 7.2 (continued)

close plant–pollinator interaction increases the probability of outcrossings and might enable this species to completely rely on pollinators for reproduction.

This study highlighted the risks of making generalizations for breeding systems in mycoheterotrophic plants. Breeding systems differ significantly within the genus *Voyria*, and pollinators also exist in the seemingly hostile environment of forest under-stories. This is also likely for other genera of mycoheterotrophic plants which have been demonstrated to be autogamous in earlier studies; extended visitor observations might reveal that they do not exclusively rely on autonomous self-pollination for reproduction. Of course, selfing entails several advantages. The nutrients provided by the fungus can be directly used for the build-up of progeny instead of investing them in showy floral organs and costly floral resources for attracting visitors. Moreover, genetic similarity to the parental generation can be an advantage when offspring grow in a similar environment, which is often the case in mycoheterotrophic plants (Solbrig 1976; Catling 1990; Lloyd 1992). On the other hand, relying on selfing can also lead to serious disadvantages. It can result in inbreeding depression, and the adaptive capability to react to changing environmental conditions is limited due to a lowered genetic variability. This might additionally restrain the plants from colonizing new habitats (Ornduff 1969; Lloyd 1992). Mycoheterotrophic species that are not xenogamous might therefore follow a mixed selfing-outcrossing strategy. The conspicuous flowers of many "autogamous" mycoheterotrophic plants, which are often provided with nectar glands and sometimes emit scent, support this theory.

In a review of mycoheterotroph biology, Bidartondo (2005) hypothesized that mycoheterotrophic plants specialized on fungi will rarely be specialized towards pollinators due to the evolutionary instability inherent on specializing on two interactions (Fig. 7.3). Because mycoheterotrophic plants already engage in highly specialized symbiotic interactions with fungi in one aspect of their life history, Bidartondo argued it would be an evolutionarily unstable strategy to engage in additional compulsory associations. Consequently, mycoheterotrophs are likely to evolve reproductive traits free from highly specialized symbiotic associations, with characteristics including a generalist pollination syndrome, high occurrence of autogamous self-pollination, and resource allocation away from metabolically expensive reproductive structures, such as large showy flowers, instead dedicating resources to seed production.

To empirically address this gap in our understanding of mycoheterotrophic plant reproduction, Klooster and Culley (2009) assessed the reproductive strategies of related mycoheterotrophic plants. The Monotropoideae (Ericaceae) was targeted for this investigation, as it has been an important study system throughout the scientific history of plant–fungal mycorrhizal interactions and is useful as a model system for enhancing our understanding of mycoheterotrophic plant biology. Specifically, genera within this subfamily possess many distinct traits, with broad variation in geographic distribution, degree of mycorrhizal specificity, and reproductive characteristics; including size and color of reproductive stems, flower number per stem, fragrance production, seed shape, and fruit type. Given this broad variation in traits, the Monotropoideae present an opportunity to assess reproductive strategies and infer limitations in the evolution of reproductive traits of closely related mycoheterotrophic plants, using a comparative framework.

Three consecutive years of field observations and empirical manipulations were performed upon four populations of *Monotropa uniflora* L., seven of *Hypopitys monotropa* Crantz (both red and yellow color forms), and two of *Monotropsis odorata* Schwein. ex Elliot located in the eastern

Table 7.2 Selected studies reporting the breeding system and pollinators of fully mycoheterotrophic plants

Taxon	Family	Breeding system	Additional information	Reference
	Burmanniaceae	Possible xenogamy	No visitor observations were made, but flower morphology strongly indicates xenogamy	Engler and Prantl (1889)
		Possible xenogamy	No visitor observations were made, but entomophily is very likely	Wettstein (1911)
		Possible xenogamy	Strongly colored flowers with prominent wings may indicate cross-pollination by animals	Maas et al. (1986)
Euburmannieae	Burmanniaceae	Assumed tendency towards autogamy		Ernst and Bernard (1912)
Apteria aphylla	Burmanniaceae	Assumed tendency towards autogamy		Ernst and Bernard (1912)
		Facultative autogamy	Pollen germinates within anthers, pollen tubes of self-pollen penetrate stigma; possible geitonogamy or visitor-mediated autogamy by flower mites (*Frankliniella* spp.)	Warming (1901); Uphof (1929); Maas et al. (1986)
Burmannia spp.	Burmanniaceae	Possible xenogamy	Functioning of the lower lip of the stigmatic lobes avoid autogamy	Malme (1896)
		Assumed tendency towards autogamy		Ernst and Bernard (1912)
Burmannia candida	Burmanniaceae	Facultative autogamy	Pollen germinates within anthers, pollen tubes of self-pollen penetrate stigma	Ernst and Bernard (1912)
Burmannia championii	Burmanniaceae	Facultative autogamy	Pollen germinates within anthers, pollen tubes of self-pollen penetrate stigma	Ernst and Bernard (1912); Schoch (1920); Rübsamen (1980)
Burmannia larseniana	Burmanniaceae	Facultative autogamy	Flower throat is blocked by three stigmatic branches preventing cross-pollination	Zhang and Saunders (1999)
Burmannia lutescens	Burmanniaceae	Facultative autogamy	Pollen germinates within anthers, pollen tubes of self-pollen penetrate stigma	Ernst and Bernard (1912); Schoch (1920); Rübsamen (1980)
		Possible xenogamy?	Flowers visited by Culicidae (*Armingeres* sp., *Culex* sp.)	Kato (1996); Momose et al. (1998)
Burmannia stuebelii	Burmanniaceae	Facultative autogamy	Pollen germinates within anthers, pollen tubes of self-pollen penetrate stigma	Spitmann (1975); Zhang and Saunders (2000)
Burmannia wallichii	Burmanniaceae	Facultative autogamy	No visitors observed, corolla lobes cover anthers and prevent pollen transfer by visitors. nevertheless—outcrossing cannot be precluded	Zhang and Saunders (2000)
Cymbocarpa spp.	Burmanniaceae	Facultative autogamy	Pollen germinates within anthers, pollen tubes of self-pollen penetrate stigma	Maas et al. (1986)

Dictyostega spp.	Burmanniaceae	Facultative autogamy	Pollen germinates within anthers, pollen tubes of self-pollen penetrate stigma	Miers (1841)
		Assumed tendency towards autogamy		Ernst and Bernard (1912)
Dictyostega orobanchoides	Burmanniaceae	Facultative autogamy	Pollen germinates within anthers, pollen tubes of self-pollen penetrate stigma	Miers (1841); Warming (1901)
Gymnosiphon spp.	Burmanniaceae	Facultative autogamy	Pollen germinates within anthers, pollen tubes of self-pollen penetrate stigma	Maas et al. (1986)
Hexapterella spp.	Burmanniaceae	Facultative autogamy	Pollen germinates within anthers, pollen tubes of self-pollen penetrate stigma	Rübsamen (1980)
Miersiella umbellata	Burmanniaceae	Facultative autogamy	Pollen germinates within anthers, pollen tubes of self-pollen penetrate stigma	Miers (1841); Warming (1901)
Arachnitis spp.	Corsiaceae	Possible xenogamy	Pollination by fungus gnats suggested	Vogel (1978)
Allotropa virgata	Ericaceae	Facultative xenogamy?	Visits by bumblebees (*Psithyrus insularis*, *Bombus mixtus*)	Wallace (1975, 1977)
		Possible autogamy	Seed set always abundant	Wallace (1975)
Hemitomes congestum	Ericaceae	Facultative xenogamy?	Visits by bumblebees (*Bombus caliginosus*, *B. mixtus*, *B. sitkensis*, *B. vosnesenski*)	Wallace (1975, 1977)
		Assumed autogamy		Wallace (1975)
Hypopithys monotropa	Ericaceae	Facultative autogamy	Strong variation between moderate and high fruit set in bagging experiments; visitors: *Bombus* spp., Syrphidae, Vespidae	Wallace (1975); Klooster and Culley (2009)
Monotropa uniflora	Ericaceae	Facultative xenogamy?	Visits by bumblebees (*Bombus caliginosus*)	Wallace (1975, 1977)
		Facultative autogamy	Low levels of fruit set in bagging experiments; visitors: *Bombus* spp., Syrphidae, Halicitidae	Klooster and Culley (2009)
Monotropastrum humile	Ericaceae	Facultative xenogamy?	Visits by bumblebees (*Bombus ardens ardens*)	Tanaka (1978); Ushimaru and Imamura (2002)
Monotropsis odorata	Ericaceae	Obligate xenogamy	Visitors: *Bombus* spp., Apriocrita, *Erynnis* spp., Tachinidae, *Epargyreus* spp.	Klooster and Culley (2009)
Pityopus californicus	Ericaceae	Facultative xenogamy?	Visits by bumblebees (*Bombus caliginosus*)	Wallace (1975, 1977)
Pterospora andromedea	Ericaceae	Facultative autogamy?	Anthers may dehisce in bud stage and pollen is released. Pollen accumulates on stigma and pollen tubes germinate. Observation of unidentified insect visitors	Bakshi (1959)
		Facultative xenogamy?	Pollination by bumblebees	Wallace (1977); M. Klooster pers. obs.

(continued)

Table 7.2 (continued)

Taxon	Family	Breeding system	Additional information	Reference
Sarcodes sanguinea	Ericaceae	Facultative xenogamy?	Visits by hummingbirds and bumblebees (*Bombus caliginosus*, *B. melanopygos*)	Wallace (1975, 1977)
Exacum tenue	Gentianaceae	Apomixis		Oehler (1927)
Exacum spp.	Gentianaceae	Possible xenogamy	Brightly colored enantiostylous flowers suggest pollination by bees	
Exochaenium oliganthum	Gentianaceae	Cleistogamy and possible xenogamy	Underground cleistogamous and aerial chasmogamous flowers are reported with heterostyly in the aerial ones	Raynal-Roques (1967)
Voyria spp.	Gentianaceae	Possible xenogamy	Brightly colored flowers that emit a pleasant floral scent and produce nectar indicate cross-pollination	Maas and Ruyters (1986)
Voyria caerulea	Gentianaceae	Possible autogamy	Germinating pollen on stigma surface = self-pollen?	Oehler (1927)
		Facultative autogamy	Pollen germinates within anthers, visiting butterfly observed	Hentrich et al. (2010)
Voyria clavata	Gentianaceae	Obligate xenogamy	Visits by Euglossini (Euglossa imperialis agg.)	Hentrich et al. (2010)
Voyria rosea	Gentianaceae	Strict autogamy	Pollen germinates within anthers, pollen tubes of self-pollen penetrate stigma	Oehler (1927)
		Facultative autogamy	Visitors observed (Hesperiidae, Heliconiidae, *Euglossa despecta*, *E. imperialis* agg.), one visitor was demonstrated to transfer pollen (Hesperiidae) = pollinator	Hentrich et al. (2010)
Voyria tenella	Gentianaceae	Facultative xenogamy?	Visits of Brachycera indicate possible pollination by insects	Imhof et al. (1994)
Voyriella parviflora	Gentianaceae	Strict autogamy	Pollen germinates within anthers, pollen tubes of self-pollen penetrate stigma	Oehler (1927)
Petrosavia sakuraii	Petrosaviaceae	Insect-mediated facultative autogamy	Visits by ants (*Paratrechina flavipes*) strongly increased fruitset, cross-pollination might be possible by visiting bees (*Lasioglossum* sp.)	Takahashi et al. (1993)
Parasitaxus usta	Podocarpaceae	Obligate xenogamy	Wind-pollinated	
Epirixanthes elongata	Polygalaceae	Facultative autogamy?		Wirz (1910)
Afrothismia sp.	Thismiaceae	Possible xenogamy	Observation of visiting drosophilid fly	Franke (2004)
Afrothismia pachyantha	Thismiaceae	Possible xenogamy	Observation of dipterans	Cheek and Williams (1999)
Afrothismia winkleri	Thismiaceae	Possible xenogamy	Observation of small dipterans	Engler (1905)
Thismia spp.	Thismiaceae	Assumed xenogamy	Colorful flowers, tentacle-like tepal tips, and the presence of scent tissue indicate possible pollination by small Diptera/flies	Vogel (1962); Stone (1980); Maas et al. (1986)

Thismia americana	Thismiaceae	Facultative autogamy	Pollen germinates within anthers, pollen tubes of self-pollen penetrate stigma	Pfeiffer (1918)
Thismia clandestina	Thismiaceae	Apomixis		Ernst and Bernard (1912)
Thismia clavarioides	Thismiaceae	Possible xenogamy	Trap-like structures indicate cross-pollination	Thiele and Jordan (2002)
Thismia fungiformis	Thismiaceae	Assumed xenogamy	Assumed deceit pollination by fungus gnats	Vogel (1978)
Thismia hyalina	Thismiaceae	Cleistogamy		Miers (1866)
Thismia javanica	Thismiaceae	Apomixis		Ernst and Bernard (1912)
Thismia winkleri	Thismiaceae	Assumed autogamy		Engler and Prantl (1889)
Tiputinia foetida	Thismiaceae	Assumed xenogamy	Floral morphology and odor indicate sapromyiophily	Woodward et al. (2007)
Oxygyne yamashitae	Thismiaceae	Possible xenogamy?	Observation of flower visitors (ants, mites) that did not transfer pollen	Yahara and Tsukaya (2008)
	Triuridaceae	Assumed xenogamy	Species either with unisexual flowers in protog ynous inflorescences or dioecious plants	Maas-van de Kamer (1995)
		Possible xenogamy	Filamentous osmophores indicate sapromyiophily	Rudall (2003)
Lacandonia schismatica	Triuridaceae	Cleistogamy		Márquez-Guzmán et al. (1993)
Sciaphila secundiflora	Triuridaceae	Possible xenogamy?	Flowers visited by Culicidae (*Armingeres* sp., *Culex* sp.) and Calliphoridae	Kato (1996); Momose et al. (1998)
Corallorhiza bentleyi	Orchidaceae	Cleistogamy		Freudenstein (1999)
Corallorhiza ekmanii	Orchidaceae	Cleistogamy		Freudenstein (1997)
Corallorhiza maculata	Orchidaceae	Facultative autogamy	Facultative self-pollination by stipe rotation occurs in 5–50 % of flowers, although some insect pollination has been observed	Freudenstein (1997)
Corallorhiza odonto-rhiza var. *odontorhiza*	Orchidaceae	Cleistogamy		Catling (1983)
Corallorhiza odonto-rhiza var. *pringlei*	Orchidaceae	Assumed xenogamy		Catling (1983)
Corallorhiza striata	Orchidaceae	Xenogamy	Pollinated by a parasitic wasp in the genus *Coccygominus*	Freudenstein (1997)
Corallorhiza trifida	Orchidaceae	Facultative autogamy	Some insect visitors noted, but self-pollination is common	Catling (1983)
Epipogium roseum	Orchidaceae	Autogamous	Fertilization occurs prior to flowers opening	Jones (1985)
Erythrorchis cassythoides	Orchidaceae	Xenogamy	Flowers are said to be pollinated by small bees attracted by perfume	Jones (1988)
Gastrodia cunninghamii	Orchidaceae	Facultative autogamy	Possible pollination or mediated self-pollination by visiting Aphidae	Lehnebach et al. (2005)

(continued)

Table 7.2 (continued)

Taxon	Family	Breeding system	Additional information	Reference
Gastrodia elata	Orchidaceae	Xenogamy	Pollinated by female halactid bees	Kato et al. (2006)
Gastrodia exilis	Orchidaceae	Assumed xenogamy	Around 25 % of flowers self-pollinate. Thought to be pollinated by minute insects, possibly thrips	Pedersen et al. (2004)
Gastrodia sesamoides	Orchidaceae	Xenogamy	Pollinated by a small native bee in the genus *Exoneura*	Jones (1985)
Hexalectris nitida	Orchidaceae	Cleistogamy	Only occasionally displays open flowers	Liggio and Liggio (1999)
Hexalectris spicata var. *arizonica*	Orchidaceae	Autogamous	Self-pollination occurs when portions of pollinia crumble away and come in contact with the stigma	Catling and Engels (1993)
Neottia cordata	Orchidaceae	Xenogamy	Pollinated primarily by fungus gnats, but also some Hymenoptera	Ackerman and Mesler (1979)
Neottia listeroides	Orchidaceae	Xenogamy	Pollinated by ants	Wang et al. (2008)
Neottia ovata	Orchidaceae	Xenogamy	Pollinated primarily by ichneumonid wasps, but also sawflies and beetles	Nilsson (1981)
Pseudovanilla foliata	Orchidaceae	Xenogamy	Flowers are said to be pollinated by small bees attracted by perfume	Jones (1988)
Rhizanthella gardneri	Orchidaceae	Assumed xenogamy	Observation of insect pollination	George (1980)
Stereosandra javanica	Orchidaceae	Assumed autogamy	Ovaries observed as already swollen when flowers are opening	Seidenfaden and Wood (1992)

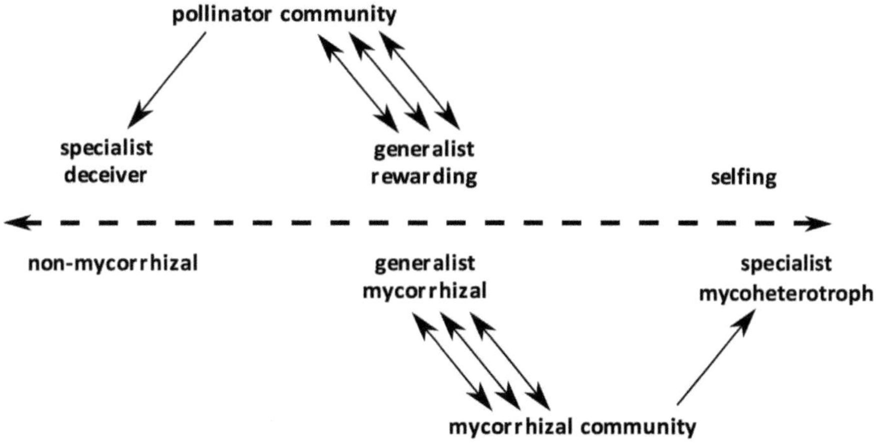

Fig. 7.3 A potential trade-off between above- and below-ground symbioses shown as a single symbiotic continuum (*dashed*). Specialization for both symbioses may limit ecological opportunities and be evolutionarily unstable, leading to a trade-off between specialization and generalization above- and below-ground. Most plant species are generalist mutualists towards both pollinators and mycorrhizal fungi (*center*). Some plant species are specialized towards pollinators and they are non-mycorrhizal (*extreme left*). Other plant species cheat their mycorrhizal fungi by failing to provide photosynthates and they are selfers (*extreme right*). Mutualists and cheater plants interact with subsets of the available community of symbionts. Figure modified from Waterman and Bidartondo (2008)

United States and included investigations of flowering phenology, pollination ecology, breeding system, and reproductive effort and output. Results of these analyses revealed that, in addition to differing strongly in the seasonal timing and duration of reproductive phenological traits, taxa exhibited unique breeding systems. Specifically, pollinator exclusion tents and hand-pollination treatments, used to assess levels of self-compatibility and autogamous self-pollination, demonstrated that red and yellow color forms within *Hypopitys monotropa* were both self-compatible but differed substantially in levels of autogamy, with the yellow color form highly autogamous compared to the facultatively xenogamous red color form. *Monotropa uniflora* was moderately self-compatible but showed very little sign of autogamous fruit set. In sharp contrast, *Monotropsis odorata* showed negligible fruit set in the self-compatibility treatment and a corresponding lack of autogamy. While these findings uncover useful results, important to understanding reproduction in these fascinating plants, much more variation in breeding system traits was observed than what might be expected under Bidartondo's theory, which proposed convergence of taxa upon reproductive self-compatibility and autogamy.

In contrast to the variation in breeding systems documented in this study, pollinator observations revealed that all taxa specialize on *Bombus* spp. as their primary pollen dispersal agents (Fig. 7.4). *Bombus* spp. were anecdotally reported by Wallace (1975, 1977) as floral visitors across many taxa in the Monotropoideae (Table 7.1), and observations by Klooster and Culley demonstrated that *Bombus* spp. visited populations of *Hypopitys, Monotropa,* and *Monotropsis* with high frequency and fidelity, both within and across seasons. *Monotropa uniflora* and the red color form of *H. monotropa* relied upon *Bombus* spp. for movement of pollen among flowers on the same plant (i.e., geitonogamous self-pollination), and among plants within a population. The yellow color form of *H. monotropa* received few visitations by *Bombus* spp., although the presence of a mixed breeding system with high instances of self-pollination accounted for consistent fruit set among plants of this color form. *Monotropsis odorata* also obtained a large number of visits from *Bombus* spp. pollinators both within and across years of study. Interestingly, in contrast to

Fig. 7.4 Pollination of *Monotropa uniflora* by *Bombus* spp.

Monotropa spp., which produce anthers that dehisce along a single slit spilling pollen throughout the inside of the flower, anthers of *M. odorata* possess two pores at the apex, leaving pollen encased within the anther sacs after pore dehiscence. Consequently, the release of pollen from anthers of *M. odorata* and subsequent pollen dispersal required the high frequency vibration of the wing muscles of *Bombus* floral visitors, through an evolved mechanism known as "buzz pollination" (e.g., Buchmann 1983). The resultant low diversity of pollen dispersal agents among taxa in this study is indicative of pollinator specialization, which was not expected under Bidartondo's theory, which proposes low pollinator reliance by mycoheterotrophic plants. Despite the lack of diversity in pollinators, these monotropoid plants do appear to have specialized on a common, abundant, and reliable pollen dispersal agent across their respective ranges. In this way, the already restricted ranges of these mycoheterotrophic plants are not further limited by their symbiosis with *Bombus* pollinators, offering some evidence that specialization in multiple symbiotic interactions may evolve in mycoheterotrophs.

While it is certainly feasible that further specialization in mycoheterotrophic plants might be the path to an evolutionary "dead end," as argued by Bidartondo (2005), this may be possible if symbiotic associates do not exert redundant selective pressures that further restrict an already constrained suite of life history traits. For instance, specialization on an abundant and widespread *Bombus* pollinator would not restrict Monotropoid mycoheterotrophs from locating and associating with their mycorrhizal fungal hosts, as the ranges of these pollinator and fungal symbionts neatly overlap. Additional thorough investigations into this and other similar study systems will help to further elucidate any discernible trends in the evolutionary ecology of mycoheterotroph reproductive strategies. Long-term evolutionary instability is not a barrier to the evolution of a trait, provided it brings short-term benefits, which specialization often does. The question of whether specialization towards fungi and pollinators are in conflict therefore requires not only case-studies on the presence or absence of species specialized towards both interactions, but a knowledge of the evolutionary persistence of any such strategy. Using molecular phylogenetics it may be possible to compare branch lengths, and test tree topologies against randomizations, in order to predict whether lineages specialized in multiple interactions are less able to persist in the long term. Such analyses of diversification in mycoheterotrophic lineages could ultimately provide insights into the conflict between evolutionary forces that drive the selection of the trait, and those that determine the subsequent fate of that lineage; be it extinction, further diversification or long-term persistence in a co-evolutionary arms race.

Despite our present lack of resolution in understanding the long-term impacts of mycoheterotrophy on the evolution of reproductive strategies, this life history trait does likely affect various seasonal aspects of plant reproductive timing, output, and success (Bidartondo 2005; Selosse et al. 2006). For example, Luoma (1987) demonstrated that reproduction in mycoheterotrophs is naturally variable, with observations that *Sarcodes sanguinea* plants reproduced in no more than two out of five consecutive seasons of monitoring. This finding was further supported by Klooster and Culley (2009), who identified significant variation in seasonal reproductive output across populations of *Monotropa* spp. and *Monotropsis odorata* with an average of 87 % of plants exhibiting at least 1 year of dormancy in the 3 years of observations. This and other like anecdotes of inconsistent reproductive output in mycoheterotrophic plants (e.g., Leake 1994) suggest the possibility that the necessary resources for reproduction provided by fungal associates may be somewhat limited and are likely seasonally variable. Also, temporal variation in the reproductive effort of a plant species may result from association with unique mycorrhizal taxa at the population level and could lead to speciation if this temporal variation functions as a persistent mechanism for reproductive isolation (Taylor and Bruns 1999; Bidartondo and Bruns 2002; Taylor et al. 2003). Plants for which successful fertilization takes place must further rely upon this web of interactions for resources necessary for seed and fruit production. Specialization upon different mycorrhizal fungi and differences in the dynamics of resource allocation by particular mycorrhizas might help explain the observed differences in the timing and duration of reproductive phenological traits at both the intra- and inter-specific levels.

It has also been shown that variation in abiotic factors may have profound impacts upon mycoheterotroph reproduction. For instance, removal of the forest canopy and alteration of litter composition can severely impact mycoheterotroph reproductive effort (Luoma 1987; Moola and Vasseur 2004; M. Klooster pers. observ.), with observed declines in the reproductive output of mycoheterotrophs possibly corresponding to unfavorable shifts in vital abiotic factors affecting the composition of mycorrhizal communities and/or the vitality of obligate fungal associates. Also, Klooster and Culley (2009) found a significant decline in the reproductive output of *Monotropa uniflora* and *H. monotropa* strongly associated with an exceptionally dry reproductive season, with lower than average levels of precipitation. Although correlations between specific ecological factors responsible for successful and reliable reproduction in mycoheterotrophic plants have not yet been empirically assessed, these observations highlight particular ecosystem dynamics that likely influence monotropoid plant reproduction.

Although the floral morphology of mycoheterotrophic plants has frequently been described in great detail (Chap. 2), there is often little information available about the reproductive strategies of mycoheterotrophic plants. Recent studies have made use of available case-studies to test hypotheses about the effects of mycoheterotrophic nutrition on the interaction network of these plants. However, further studies identifying the fungal and pollinator networks of entire mycoheterotrophic lineages could be a productive future avenue of research to test ideas about the effects of cheating on mutualistic networks and the consequences of specialization in multiple life history traits.

References

Abadie JC, Puttsepp U, Gebauer G, Faccio A, Bonfante P, Selosse MA (2006) *Cephalanthera longifolia* (Neottieae, Orchidaceae) is mixotrophic: a comparative study between green and nonphotosynthetic individuals. Can J Bot 84:1462–1477

Ackerman JD, Mesler MR (1979) Pollination biology of *Listera cordata* (Orchidaceae). Am J Bot 66:820–824

Akiyama K, Matsuzaki K, Hayashi H (2005) Plant sesquiterpenes induce hyphal branching in arbuscular mycorrhizal fungi. Nature 435:824–827

Albert VA, Struwe L (1997) Phylogeny and classification of *Voyria* (saprophytic Gentianaceae). Brittonia 49:466–479

Bakshi TS (1959) Ecology and morphology of *Pterospora andromeda*. Bot Gaz 120:203–217

Barrett CF, Freudenstein JV (2008) Molecular evolution of *rbc*L in the mycoheterotrophic coralroot orchids (*Corallorhiza* Gagnebin, Orchidaceae). Mol Phylogenet Evol 47:665–679

Barrett CF, Freudenstein JV, Taylor DL, Koljalg U (2010) Rangewide analysis of fungal association in the fully mycoheterotrophic *Corallorhiza striata* complex (Orchidaceae) reveals extreme specificity on ectomycorrhizal *Tomentella* (Thelephoraceae) across North America. Am J Bot 97:628–643

Bascompte J, Jordano P (2007) Plant-animal mutualistic networks: the architecture of biodiversity. Annu Rev Ecol Evol Syst 38:567–593

Bidartondo MI (2005) The evolutionary ecology of mycoheterotrophy. New Phytol 167:335–352

Bidartondo MI, Bruns TD (2001) Extreme specificity in epiparasitic Monotropoideae (Ericaceae): widespread phylogenetic and geographical structure. Mol Ecol 10:2285–2295

Bidartondo MI, Bruns TD (2002) Fine-level mycorrhizal specificity in the Monotropoideae (Ericaceae): specificity for fungal species groups. Mol Ecol 11:557–569

Bidartondo MI, Bruns TD (2005) On the origins of extreme mycorrhizal specificity in the Monotropoideae (Ericaceae): performance trade-offs during seed germination and seedling development. Mol Ecol 14:1549–1560

Bidartondo MI, Bruns TD, Weiss M, Sergio C, Read DJ (2003) Specialized cheating of the ectomycorrhizal symbiosis by an epiparasitic liverwort. Proc R Soc Lond B 270:835–842

Bidartondo MI, Burghardt B, Gebauer G, Bruns TD, Read DJ (2004) Changing partners in the dark: isotopic and molecular evidence of ectomycorrhizal liaisons between forest orchids and trees. Proc R Soc Lond B 271:1799–1806

Bidartondo MI, Read DJ (2008) Fungal specificity bottlenecks during orchid germination and development. Mol Ecol 17:3707–3716

Bidartondo MI, Redecker D, Hijri I, Wiemken A, Bruns TD, Dominguez L, Sersic A, Leake JR, Read DJ (2002) Epiparasitic plants specialized on arbuscular mycorrhizal fungi. Nature 419:389–392

Björkman E (1960) *Monotropa hypopitys* L.—an epiparasite on tree roots. Physiol Plant 13:308–327

Bougoure J, Ludwig M, Brundrett M, Grierson P (2009) Identity and specificity of the fungi forming mycorrhizas with the rare mycoheterotrophic orchid *Rhizanthella gardneri*. Mycol Res 113:1097–1106

Bougoure JJ, Brundrett MC, Grierson PF (2010) Carbon and nitrogen supply to the underground orchid, *Rhizanthella gardneri*. New Phytol 186:947–956

Bougoure JJ, Dearnaley J (2005) The fungal endophytes of *Dipodium variegatum* (Orchidaceae). Austral Mycol 24:15–19

Bronstein JL (2001) The exploitation of mutualisms. Ecol Lett 4:277–287

Bruns TD, Bidartondo MI, Taylor DL (2002) Host specificity in ectomycorrhizal communities: what do the exceptions tell us? Integr Comp Biol 42:352–359

Bruns TD, Read DJ (2000) In vitro germination of nonphotosynthetic, myco-heterotrophic plants stimulated by fungi isolated from the adult plants. New Phytol 148:335–342

Buchmann SL (1983) Buzz pollination in angiosperms. In: Jones CE, Little RJ (eds) Handbook of experimental pollination biology. S. & E. Scientific and Academic Editions, New York, pp 73–113

Cameron DD, Johnson I, Read DJ, Leake JR (2008) Giving and receiving: measuring the carbon cost of mycorrhizas in the green orchid, *Goodyera repens*. New Phytol 180:176–184

Cameron KM (2004) Utility of plastid *psaB* gene sequences for investigating intrafamilial relationships within Orchidaceae. Mol Phylogenet Evol 31:1157–1180

Cameron KM (2009) On the value of nuclear and mitochondrial gene sequences for reconstructing the phylogeny of vanilloid orchids (Vanilloideae, Orchidaceae). Ann Bot 104:377–385

Cameron KM, Molina MC (2006) Photosystem II gene sequences of *psbB* and *psbC* clarify the phylogenetic position of *Vanilla* (Vanilloideae, Orchidaceae). Cladistics 22:239–248

Catling PM (1983) Autogamy in eastern Canadian Orchidaceae: a review of current knowledge and some new observations. Nat Can 110:37–53

Catling PM (1990) Auto-pollination in the Orchidaceae. In: Arditti J (ed) Orchid biology, reviews and perspectives V. Timber Press, Portand, pp 121–158

Catling PM, Engels VS (1993) Systematics and distribution of *Hexalectris spicata* var. *arizonica* (Orchidaceae). Lindleyana 8:119–125

Cheek M, Williams SA (1999) A review of African saprophytic flowering plants. In: Timberlake J, Kativu S (eds) African plants: biodiversity, taxonomy & uses. Proceedings of the 1997 AETFAT congress, Harare, Zimbabwe. Royal Botanic Gardens, Kew, pp 39–49

Cope ED (1896) The primary factors of organic evolution. Open Court Publishing, Chicago

Courty PE, Walder F, Boller T, Ineichen K, Wiemken A, Rousteau A, Selosse MA (2011) Carbon and nitrogen metabolism in mycorrhizal networks and mycoheterotrophic plants of tropical forests: a stable isotope analysis. Plant Physiol 156:952–961

Cruden RW (1977) Pollen-ovule ratios: a conservative indicator of breeding systems in flowering plants. Evolution 31:32–46

Cullings KW, Szaro TM, Bruns TD (1996) Evolution of extreme specialization within a lineage of ectomycorrhizal epiparasites. Nature 379:63–66

Dearnaley J (2006) The fungal endophytes of *Erythrorchis cassythoides*—is this orchid saprophytic or parasitic? Austral Mycol 25:51–57

Dearnaley JDW, Bougoure JJ (2010) Isotopic and molecular evidence for saprotrophic Marasmiaceae mycobionts in rhizomes of Gastrodia sesamoides. Fungal Ecol 3:288–294

Dearnaley JDW, Le Brocque AF (2006) Molecular identifcation of the primary root fungal endophytes of *Dipodium hamiltonianum* (Orchidaceae). Aust J Bot 54:487–491

Dowie NJ, Hemenway JJ, Miller SL (2012) Identity, genetic lineages and putative hybrids of an obligate mycobiont associated with the mycoheterotrophic plant *Pterospora andromedea* in the south-central Rocky Mountains. Fungal Ecol 5:137–146

Dowie NJ, Hemenway JJ, Trowbridge SM, Miller SL (2011) Mycobiont overlap between two mycoheterotrophic genera of Monotropoideae (*Pterospora andromedea* and *Sarcodes sanguinea*) found in the Greater Yellowstone Ecosystem. Symbiosis 54:29–36

Dressler RL (1981) The orchids—natural history and classification. Harvard University Press, Cambridge, Massachusets and London

Ehrlich PR, Raven PH (1964) Butterflies and plants: a study in coevolution. Evolution 18:586–608

Engler A (1905) *Thismia winkleri* Engl., eine neue afrikanische Burmanniacee. Bot Jahrb Syst 38:89–91

Engler A, Prantl K (1889) Die natürlichen Pflanzenfamilien. Verlag von Wilhem Engelmann, Leipzig

Ernst A, Bernard C (1912) Beiträge zur Kenntnis der Saprophyten Javas. IX. Entwicklungsgeschichte des Embryosacks und des Embryos von Burmannia candida Engl. und B. championii Thw. Ann Jard Bot Buitenzorg 25:161–184

Feild TS, Brodribb TJ (2005) A unique mode of parasitism in the conifer coral tree *Parasitaxus ustus* (Podocarpaceae). Plant Cell Environ 28:1316–1325

Franke T (2004) *Afrothismia saingei* (Burmanniaceae, Thismieae), a new myco-heterotrophic plant from Cameroon. Syst Geogr Pl 74:287–291

Franke T, Beenken L, Doring M, Kocyan A, Agerer R (2006) Arbuscular mycorrhizal fungi of the *Glomus*-group A lineage (Glomerales; Glomeromycota) detected in myco- heterotrophic plants from tropical Africa. Mycol Prog 5:24–31

Freudenstein JV (1997) A monograph of *Corallorhiza* (Orchidaceae). Harvard Pap Bot 10:5–51

Freudenstein JV (1999) A new species of *Corallorhiza* (Orchidaceae) from West Virginia, U.S.A. Novon 9:511–513

Futuyma DJ, Moreno G (1988) The evolution of ecological specialization. Annu Rev Ecol Syst 19:207–233

Gardes M (2002) An orchid-fungus marriage—physical promiscuity, conflict and cheating. New Phytol 154:4–7

Gebauer G, Meyer M (2003) 15N and 13C natural abundance of autotrophic and mycoheterotrophic orchids provides insight into nitrogen and carbon gain from fungal association. New Phytol 160:209–223

George AS (1980) *Rhizanthella gardneri* R.S. Rogers the underground orchid of Western Australia. Am Orchid Soc Bull 49:631–646

Girlanda M, Selosse MA, Cafasso D, Brilli F, Delfine S, Fabbian R, Ghignone S, Pinelli P, Segreto R, Loreto F, Cozzolino S, Perotto S (2006) Inefficient photosynthesis in the Mediterranean orchid *Limodorum abortivum* is mirrored by specific association to ectomycorrhizal Russulaceae. Mol Ecol 15:491–504

Gould SJ (1970) Dollo on Dollo's law: irreversibility and the status of evolutionary laws. J Hist Biol 3:189–212

Grelet GA, Johnson D, Paterson E, Anderson IC, Alexander IJ (2009) Reciprocal carbon and nitrogen transfer between an ericaceous dwarf shrub and fungi isolated from *Piceirhiza bicolorata* ectomycorrhizas. New Phytol 182:359–366

Hazard C, Lilleskov EA, Horton TR (2012) Is rarity of pinedrops (*Pterospora andromedea*) in eastern North America linked to rarity of its unique fungal symbiont? Mycorrhiza 22:393–402

Hentrich H, Kaiser R, Gottsberger G (2010) The reproductive biology of *Voyria* (Gentianaceae) species in French Guiana. Taxon 59:867–880

Hibbett DS (2002) Plant-fungal interactions: when good relationships go bad. Nature 419:345–346

Hynson NA, Bruns TD (2009) Evidence of a mycoheterotroph in the plant family Ericaceae that lacks mycorrhizal specificity. Proc R Soc Lond B 276: 4053–4059

Hynson NA, Bruns TD (2010) Fungal hosts for mycoheterotrophic plants: a nonexclusive, but highly selective club. New Phytol 185:598–601

Imhof S (1999) Root morphology, anatomy and mycotrophy of the achlorophyllous *Voyria aphylla* (Jacq.) Pers. (Gentianaceae). Mycorrhiza 9:33–39

Imhof S, Weber HC, Gómez-Pignataro LD (1994) Ein Beitrag zur Biologie von *Voyria tenella* Hook. und *Voyria truncata* (Standley) Standley & Steyermark (Gentianaceae). Beiträge zur Biologie der Pflanzen 68:113–123

Jackson AP (2004) A reconciliation analysis of host switching in plant-fungal symbioses. Evolution 58:1909–1923

Jones DL (1985) The self-pollination of *Epipogium roseum* (D. Don) Lindley. Orchadian 8:91–92

Jones DL (1988) Native orchids of Australia. Reed Books, New South Wales

Julou T, Burghardt B, Gebauer G, Berveiller D, Damesin C, Selosse MA (2005) Mixotrophy in orchids: insights from a comparative study of green individuals and nonphotosynthetic individuals of *Cephalanthera damasonium*. New Phytol 166:639–653

Kato M (1996) Plant-pollinator interactions in the understory of a lowland mixed dipterocarp forest in Sarawak. Am J Bot 83:732–743

Kato M, Tsuji K, Kawakita A (2006) Pollinator and stem- and corm-boring insects associated with mycoheterotrophic orchid *Gastrodia elata*. Ann Entomol Soc Am 99:851–858

Kennedy AH, Taylor DL, Watson LE (2011) Mycorrhizal specificity in the fully mycoheterotrophic *Hexalectris* Raf. (Orchidaceae: Epidendroideae). Mol Ecol 20:1303–1316

Keyes WJ, Taylor JV, Apkarian RP, Lynn DG (2001) Dancing together. Social controls in parasitic plant development. Plant Physiol 127:1508–1512

Klooster MR, Clark DL, Culley TM (2009) Cryptic bracts facilitate herbivore avoidance in the mycoheterotrophic plant *Monotropsis odorata* (Ericaceae). Am J Bot 96:2197–2205

Klooster MR, Culley TM (2009) Comparative analysis of the reproductive ecology of *Monotropa* and *Monotropsis*: two mycoheterotrophic genera in the Monotropoideae (Ericaceae). Am J Bot 96: 1337–1347

Kretzer AM, Bidartondo MI, Grubisha LC, Spatafora JW, Szaro TM, Bruns TD (2000) Regional specialization of *Sarcodes sanguinea* (Ericaceae) on a single fungal symbiont from the *Rhizopogon ellenae* (Rhizopogonaceae) species complex. Am J Bot 87: 1778–1782

Leake JR (1994) The biology of myco-heterotrophic (saprophytic) plants. New Phytol 127:171–216

Leake JR (2005) Plants parasitic on fungi: unearthing the fungi in myco-heterotrophs and debunking the 'saprophytic' plant myth. Mycologist 19:113–122

Leake JR, McKendrick SL, Bidartondo M, Read DJ (2004) Symbiotic germination and development of the myco-heterotroph *Monotropa hypopitys* in nature and its requirement for locally distributed *Tricholoma* spp. New Phytol 163:405–423

Lehnebach CA, Robertson AW, Hedderley D (2005) Pollination studies of four New Zealand terrestrial orchids and the implication for their conservation. New Zealand J Bot 43:467–477

Liggio J, Liggio AO (1999) Wild orchids of Texas. University of Texas Press, Austin

Lloyd DG (1992) Self-fertilization and cross-fertilization in Plants. 2. The selection of self-fertilization. Int J Plant Sci 153:370–380

Luoma DL (1987) Synecology of the Monotropoideae within Limpy Rock Research Natural Area, Umpqua National Forest, Oregon. MSc thesis. Corvallis, Oregon State University

Maas-van de Kamer H (1995) Triuridiflorae—Gardener's delight? In: Rudall PJ, Cribb PJ, Cutler DF, Humphries CJ (eds) Monocotyledones: systematics. Royal Botanic Gardens, Kew, pp 287–301

Maas PJM, Maas-van de Kamer H, van Benthem J, Snelders H, Rübsamen T (1986) Burmanniaceae. The New York Botanical Garden, Bronx, New York

Maas PJM, Ruyters P (1986) *Voyria* and *Voyriela* (saprophytic Genrianaceae). Fl Neotrop 41:1–93

Maddison WP, Midford PE, Otto SP (2007) Estimating a binary character's effect on speciation and extinction. Syst Biol 56:701–710

Malme GOA (1896) Die Burmannien der ersten Regnel'schen Expedition. Bih Kongl Svenska Vetensk Akad Handl 22:1–32

Maloof JE, Inouye DW (2000) Are nectar Robbers cheaters or mutualists? Ecology 81:2651–2661

Márquez-Guzmán J, Vázquez-Santana S, Engleman EM, Martínez-Mena A, Martínez E (1993) Pollen development and fertilization in *Lacandonia schismatica* (Lacandoniaceae). Ann Missouri Bot Gard 80:891–897

Martos F, Dulormne M, Pailler T, Bonfante P, Faccio A, Fournel J, Dubois MP, Selosse MA (2009) Independent recruitment of saprotrophic fungi as mycorrhizal partners by tropical achlorophyllous orchids. New Phytol 184:668–681

Matsuda Y, Okochi S, Katayama T, Yamada A, Ito S (2011) Mycorrhizal fungi associated with *Monotropastrum humile* (Ericaceae) in central Japan. Mycorrhiza 21:569–576

McCormick MK, Whigham DF, O'Neill J (2004) Mycorrhizal diversity in photosynthetic terrestrial orchids. New Phytol 163:425–438

McKendrick SL, Leake JR, Read DJ (2000) Symbiotic germination and development of myco-heterotrophic plants in nature: transfer of carbon from ectomycorrhizal *Salix repens* and *Betula pendula* to the orchid *Corallorhiza trifida* through shared hyphal connections. New Phytol 145:539–548

McKendrick SL, Leake JR, Taylor DL, Read DJ (2002) Symbiotic germination and development of the mycoheterotrophic orchid *Neottia nidus-avis* in nature and its requirement for locally distributed *Sebacina* spp. New Phytol 154:233–247

Merckx V, Bidartondo MI (2008) Breakdown and delayed cospeciation in the arbuscular mycorrhizal mutualism. Proc R Soc Lond B 275:1029–1035

Merckx V, Bidartondo MI, Hynson NA (2009) Mycoheterotrophy: when fungi host plants. Ann Bot 104:1255–1261

Merckx V, Freudenstein JV (2010) Evolution of mycoheterotrophy in plants: a phylogenetic perspective. New Phytol 185:605–609

Merckx V, Stockel M, Fleischmann A, Bruns TD, Gebauer G (2010) 15N and 13C natural abundance of two mycoheterotrophic and a putative partially mycoheterotrophic species associated with arbuscular mycorrhizal fungi. New Phytol 188:590–596

Merckx VSFT, Janssens SB, Hynson NA, Specht CD, Bruns TD, Smets EF (2012) Mycoheterotrophic interactions are not limited to a narrow phylogenetic range of arbuscular mycorrhizal fungi. Mol Ecol 21:1524–1532

Miers J (1841) On some new Brasilian plants allied to the natural order Burmanniaceae. Trans Linn Soc Lond 18:535–556

Miers J (1866) On *Myostoma*, a new genus of the Burmanniaceae. Trans Linn Soc Lond 25:461–476

Momose K, Yumoto T, Nagamitsu T, Kato M, Nagamasu H, Sakai S, Harrison RD, Itioka T, Hamid AA, Inoue T (1998) Pollination biology in a lowland dipterocarp forest in Sarawak, Malaysia. I. Characteristics of the plant-pollinator community in a lowland dipterocarp forest. Am J Bot 85:1477–1501

Montoya JM, Pimm SL, Sole RV (2006) Ecological networks and their fragility. Nature 442:259–264

Moola FM, Vasseur L (2004) Recovery of late-seral vascular plants in a chronosequence of post-clearcut forest stands in coastal Nova Scotia. Can Plant Ecol 172:183–197

Motomura H, Selosse MA, Martos F, Kagawa A, Yukawa T (2010) Mycoheterotrophy evolved from mixotrophic ancestors: evidence in *Cymbidium* (Orchidaceae). Ann Bot 106:573–581

Nicholls WH (1938) A new species of the genus *Cryptostylis* R.Br. Vic Nat 54:182–183

Nilsson LA (1981) The pollination ecology of *Listera ovata* (Orchidaceae). Nord J Bot 1:461–480

Oehler E (1927) Entwicklungsgeschichtlich-zytologische Untersuchungen an einigen saprophytischen Gentianaceen. Planta 3:671–733

Ogura-Tsujita Y, Gebauer G, Hashimoto T, Umata H, Yukawa T (2009) Evidence for novel and specialized mycorrhizal parasitism: the orchid *Gastrodia confusa* gains carbon from saprotrophic *Mycena*. Proc R Soc Lond B 276:761–767

Ogura-Tsujita Y, Yukawa T (2008) *Epipactis helleborine* shows strong mycorrhizal preference towards ectomycorrhizal fungi with contrasting geographic distributions in Japan. Mycorrhiza 18:331–338

Ollerton J, Winfree R, Tarrant S (2011) How many flowering plants are pollinated by animals? Oikos 120:321–326

Ornduff R (1969) Reproductive biology in relation to systematics. Taxon 18:121–133
Pedersen HA, Watthana S, Suddee S, Sasurat S (2004) Breeding system, post-pollination growth, and seed dispersal in *Gastrodia exilis* (Orchidaceae). Nat Hist Bull Siam Soc 52:9–26
Pfeiffer NE (1918) The sporangia of *Thismia americana*. Bot Gaz 66:354–363
Price PW (1980) Evolutionary biology of parasites. Princeton University Press, Princeton
Pringle A, Taylor JW (2002) The fitness of filamentous fungi. Trends Microbiol 10:474–481
Rasmussen HN, Whigham DF (1993) Seed ecology of dust seeds *in-situ*=a new study technique and its application in terrestrial orchids. Am J Bot 80: 1374–1378
Raynal-Roques A (1967) Sur un sebaea Africain saprophyte (Gentianaceae). Adansonia 7:207–219
Roy M, Watthana S, Stier A, Richard F, Vessabutr S, Selosse MA (2009a) Two mycoheterotrophic orchids from Thailand tropical dipterocarpacean forests associate with a broad diversity of ectomycorrhizal fungi. BMC Biol 7:17
Roy M, Yagame T, Yamato M, Iwase K, Heinz C, Faccio A, Bonfante P, Selosse MA (2009b) Ectomycorrhizal *Inocybe* species associate with the mycoheterotrophic orchid *Epipogium aphyllum* but not its asexual propagules. Ann Bot 104:595–610
Rübsamen T (1980) Beiträge zur Mikromorphologie der Testa und zur Embryologie amerikanischer Burmanniaceen und Triuridaceen. Diplomarbeit. Ruhr-Universität, Bochum
Rudall PJ (2003) Monocot pseudanthia revisited: floral structure of the mycoheterotrophic family Triuridaceae. Int J Plant Sci 164:S307–S320
Sachs JL, Simms EL (2006) Pathways to mutualism breakdown. Trends Ecol Evol 21:585–592
Sanders IR (2003) Preference, specificity and cheating in the arbuscular mycorrhizal symbiosis. Trends Plant Sci 8:143–145
Schoch M (1920) Entwicklungsgeschichtlich-cytologische Untersuchungen über die Pollenbildung und Bestäubung bei einigen Burmannia-Arten. Dissertation. Universität Zürich
Schwartz MW, Hoeksema JD (1998) Specialization and resource trade: biological markets as a model of mutualisms. Ecology 79:1029–1038
Seidenfaden G, Wood JJ (1992) The orchids of peninsular Malaysia and Singapore. Olsen and Olsen, Fredensborg
Selosse MA, Faccio A, Scappaticci G, Bonfante P (2004) Chlorophyllous and achlorophyllous specimens of *Epipactis microphylla* (Neottieae, Orchidaceae) are associated with ectomycorrhizal septomycetes, including truffles. Microb Ecol 47:416–426
Selosse MA, Richard F, He XH, Simard SW (2006) Mycorrhizal networks: des liaisons dangereuses? Trends Ecol Evol 21:621–628
Selosse MA, Weiss M, Jany JL, Tillier A (2002) Communities and populations of sebacinoid basidiomycetes associated with the achlorophyllous orchid *Neottia nidus-avis* (L.) LCM Rich. and neighbouring tree ectomycorrhizae. Mol Ecol 11:1831–1844
Simard SW (2012) Mycorrhizal networks and seedling establishment in Douglas-fir forests. In: Southworth D (ed) Biocomplexity of plant-fungal interactions. Wiley-Blackwell, Oxford, pp 85–107
Smith SE, Read DJ (2008) Mycorrhizal symbiosis. Academic, London
Solbrig O (1976) On the relative advantages of cross- and self-fertilization. Ann Missouri Bot Gard 63:262–276
Spitmann A (1975) Entwicklungsgeschichtliche Untersuchungen an *Burmannia stuebelii* Hieron. et Schltr. Unveröffentlichte Staatsexamensarbeit. Universität Bochum
Stone BC (1980) Rediscovery of *Thismia clavigera* (Becc.) F. v. M. (Burmanniaceae). Blumea 26:419–425
Takahashi H, Nishio E, Hayashi H (1993) Pollination biology of the saprophytic species *Petrosavia sakuraii* (Makino) van Steenis in Central Japan. J Plant Res 106:213–217
Tanaka H (1978) Pollination biology of *Monotropastrum globosum*. Jap J Bot 53:201–202
Taylor DL (1997) The evolution of myco-heterotrophy and specificity in some North American orchids. PhD Thesis. University of California, Berkeley
Taylor DL (2004) Myco-heterotroph-fungus marriages—is fidelity over-rated? New Phytol 163:217–221
Taylor DL, Bruns TD (1997) Independent, specialized invasions of ectomycorrhizal mutualism by two nonphotosynthetic orchids. Proc Natl Acad Sci USA 94:4510–4515
Taylor DL, Bruns TD (1999) Population, habitat and genetic correlates of mycorrhizal specialization in the 'cheating' orchids *Corallorhiza maculata* and *C. mertensiana*. Mol Ecol 8:1719–1732
Taylor DL, Bruns TD, Hodges SA (2004) Evidence for mycorrhizal races in a cheating orchid. Proc R Soc Lond B 271:35–43
Taylor DL, Bruns TD, Leake JR, Read DJ (2002) Mycorrhizal specificity and function in myco-heterotrophic plants. In: van der Hijden MG, Sanders IR (eds) Mycorrhizal ecology. Springer, Berlin, pp 375–414
Taylor DL, Bruns TD, Szaro TM, Hodges SA (2003) Divergence in mycorrhizal specialization within *Hexalectris spicata* (Orchidaceae), a nonphotosynthetic desert orchid. Am J Bot 90:1168–1179
Tedersoo L, Pellet P, Koljalg U, Selosse MA (2007) Parallel evolutionary paths to mycoheterotrophy in understorey Ericaceae and Orchidaceae: ecological evidence for mixotrophy in Pyroleae. Oecologia 151:206–217
Thiele KR, Jordan P (2002) *Thismia clavarioides* (Thismiaceae), a new species of fairy lantern from New South Wales. Telopea 9:765–771
Thompson JN (2005) The geographic mosaic of coevolution. The University of Chicago Press, Chicago
Toftegaard T, Iason GR, Alexander IJ, Rosendahl S, Taylor AFS (2010) The threatened plant intermediate wintergreen (*Pyrola media*) associates with a wide

range of biotrophic fungi in native Scottish pine woods. Biodivers Conserv 19:3963–3971

Umata H (1995) Seed germination of *Galeola altissima*, an achlorophyllous orchid, with aphyllophorales fungi. Mycoscience 36:369–372

Uphof JCT (1929) Beiträge zur Kenntnis der Burmaniacee *Apteria aphylla* (Nutt.). Barnh Österr Bot Z 78:71–80

Ushimaru A, Imamura A (2002) Large variation in flower size of the myco-heterotrophic plant, *Monotropastrum globosum*: effect of floral display on female reproductive success. Plant Spec Biol 17:147–153

Vogel S (1962) Duftdrüsen im Dienste der Bestäubung. Akad Wiss Abh Math Naturwiss Kl 10:601–763

Vogel S (1978) Pilzmücken als Pilzmimeten. Flora 167:329–398

Wallace GD (1975) Studies of the Monotropoideae (Ericaceae): taxonomy and distribution. Wasmann J Biol 33:1–88

Wallace GD (1977) Studies of the Monotropideae (Ericaceae). Floral nectaries: anatomy and function in pollination ecology. Am J Bot 64:199–206

Wang CQ, Luo YB, Tai YD, An DJ, Kou Y (2008) Ants pollinate *Neottia listeroides* (Orchidaceae) in Sichuan. China J Syst Evol 46:836–846

Warming E (1901) Sur quelques Burmanniacées resueillies au Brésil par le Dr. A. Glaziou. Oversigt over det kong. danske Videnskabernes Selskabs Forhandlingar 6:173–188

Waterman RJ, Bidartondo MI (2008) Deception above, deception below: linking pollination and mycorrhizal biology of orchids. J Exp Bot 59:1085–1096

Waterman RJ, Bidartondo MI, Stofberg J, Combs JK, Gebauer G, Savolainen V, Barraclough TG, Pauw A (2011) The effects of above- and belowground mutualisms on orchid speciation and coexistence. Am Nat 177:E54–E68

Weiss M, Selosse MA, Rexer KH, Urban A, Oberwinkler F (2004) Sebacinales: a hitherto overlooked cosm of heterobasidiomycetes with a broad mycorrhizal potential. Mycol Res 108:1003–1010

Wettstein R (1911) Handbuch der Systematischen Botanik. Dueticke, Wien

Winther JL, Friedman WE (2007) Arbuscular mycorrhizal symbionts in *Botrychium* (Ophioglossaceae). Am J Bot 94:1248–1255

Winther JL, Friedman WE (2008) Arbuscular mycorrhizal associations in Lycopodiaceae. New Phytol 177:790–801

Winther JL, Friedman WE (2009) Phylogenetic affinity of arbuscular mycorrhizal symbionts in *Psilotum nudum*. J Plant Res 122:485–496

Wirz H (1910) Beiträge zur Entwicklungsgeschichte von *Sciaphila* spec. und von *Epirrhizanthes elongata* Bl. Dissertation. Universität Zürich

Woodward CL, Berry PE, Maas-van de Kamer H, Swing K (2007) *Tiputinia foetida*, a new mycoheterotrophic genus of Thismiaceae from Amazonian Ecuador, and a likely case of deceit pollination. Taxon 56:157–162

Yagame T, Yamato M, Mii M, Suzuki A, Iwase K (2007) Developmental processes of achlorophyllous orchid. Epipogium roseum: from seed germination to flowering under symbiotic cultivation with mycorrhizal fungus. J Plant Res 120:229–236

Yagame T, Yamato M, Suzuki A, Iwase K (2008) Ceratobasidiaceae mycorrhizal fungi isolated from nonphotosynthetic orchid *Chamaegastrodia sikokiana*. Mycorrhiza 18:97–101

Yahara T, Tsukaya H (2008) *Oxygyne yamashitae*, a new species of Thismiaceae from Yaku Island, Japan. Acta Phytotax Geobot 59:97–104

Yamato M (2001) Identification of a mycorrhizal fungus in the roots of achlorophyllous *Sciaphila tosaensis* Makino (Triuridaceae). Mycorrhiza 11:83–88

Yamato M, Yagame T, Iwase K (2011a) Arbuscular mycorrhizal fungi in roots of non-photosynthetic plants, *Sciaphila japonica* and *Sciaphila tosaensis* (Triuridaceae). Mycoscience 52:217–223

Yamato M, Yagame T, Shimomura N, Iwase K, Takahashi H, Ogura-Tsujita Y, Yukawa T (2011b) Specific arbuscular mycorrhizal fungi associated with non-photosynthetic *Petrosavia sakuraii* (Petrosaviaceae). Mycorrhiza 21:631–639

Yamato M, Yagame T, Suzuki A, Iwase K (2005) Isolation and identification of mycorrhizal fungi associating with an achlorophyllous plant, *Epipogium roseum* (Orchidaceae). Mycoscience 46:73–77

Yoder JI (1999) Parasitic plant responses to host plant signals: a model for subterranean plant-plant interactions. Curr Opin Plant Biol 2:65–70

Yoder JI (2001) Host-plant recognition by parasitic Scrophulariaceae. Curr Opin Plant Biol 4: 359–365

Yokoyama J, Fukuda T, Tsukaya H (2005) Molecular identification of the mycorrhizal fungi of the epiparasitic plant *Monotropastrum humile* var. *glaberrimum* (Ericaceae). J Plant Res 118:53–56

Young BW, Massicotte HB, Tackaberry LE, Baldwin QF, Egger KN (2002) *Monotropa uniflora*: morphological and molecular assessment of mycorrhizae retrieved from sites in the Sub-Boreal Spruce biogeoclimatic zone in central British Columbia. Mycorrhiza 12:75–82

Zayed A, Packer L, Grixti JC, Ruz L, Owen RE, Toro H (2005) Increased genetic differentiation in a specialist versus a generalist bee: implications for conservation. Conserv Genet 6:1017–1026

Zhang DX, Saunders RMK (1999) *Burmannia larseniana* (Burmanniaceae): a new species from Thailand. Nord J Bot 19:241–244

Zhang DX, Saunders RMK (2000) Reproductive biology of a mycoheterotrophic species, *Burmannia wallichii* (Burmanniaceae). Bot J Linn Soc 132:359–367

8 The Physiological Ecology of Mycoheterotrophy

Nicole A. Hynson, Thomas P. Madsen, Marc-André Selosse, Iris K.U. Adam, Yuki Ogura-Tsujita, Melanie Roy, and Gerhard Gebauer

8.1 Introduction

Mycoheterotrophic (MH) plants have been centerpieces for the study of the physiological aspects of the mycorrhizal symbiosis for over 150 years. A.B. Frank who in 1885 coined the term "mycorrhiza" conducted research on the anatomy and physiology of the fully MH species *Hypopitys monotropa* (syn. *Monotropa hypopitys*) (Frank 1885). Due to the charismatic, often highly derived, appearance of many MH plants, it is no surprise that they have captured the interest of researchers throughout the centuries. These plants continue to draw attention from various realms of science because they consistently require that our fundamental understanding of many ecological, evolutionary, and physiological theories be expanded upon. Both the ecology and evolution of mycoheterotrophy are tightly coupled to plant–fungal interactions or ecophysiology, and this subject is the focus of this chapter. From an ecological perspective, MH plants that associate with mycorrhizal fungi represent the best-known examples of mycorrhizal networks, where unrelated plants transfer elemental compounds via shared fungal symbionts (Chap. 1; Simard and Durall 2004; Selosse et al. 2006). From an evolutionary perspective, fully MH plants that associate with mycorrhizal fungi are the primary example of one extreme in the mycorrhizal continuum, ranging from plants giving carbon (C), to plants receiving C from their fungal symbionts (Chap. 1).

Within the plant kingdom the MH strategy has arisen numerous times throughout evolutionary history and involves not only mycorrhizal fungi, but saprotrophic (SAP) fungi as well. The physiology of mycoheterotrophs and their mycorrhizal fungi represent the only clear example of a complete reversal in the normal flow of nutrients in the mycorrhizal symbiosis, where instead of

N.A. Hynson (✉)
Department of Botany, University of Hawaii at Manoa, Honolulu, HI 96822, USA
e-mail: nhynson@hawaii.edu

T.P. Madsen
Department of Integrative Biology, University of California, Berkeley, Berkeley, CA 94720, USA

M.-A. Selosse
Centre d'Ecologie Fonctionnelle et Evolutive (CEFE-CNRS), Montpellier, France

I.K.U. Adam • G. Gebauer
Laboratory of Isotope Biogeochemistry, Bayreuth Center of Ecology and Environmental Research (BayCEER), University of Bayreuth, 95440, Bayreuth, Germany

Y. Ogura-Tsujita
Department of Chemical and Biological Sciences, Japan Women's University, 2-8-1 Mejirodai, Tokyo, 112-8681, Japan

M. Roy
Centre d'Ecologie Fonctionnelle et Evolutive (CEFE-CNRS), Montpellier, France

Laboratoire Evolution et Diversité Biologique (UMR5174 EDB), Université de Toulouse, Toulouse, France

V.S.F.T. Merckx (ed.), *Mycoheterotrophy: The Biology of Plants Living on Fungi*, DOI 10.1007/978-1-4614-5209-6_8, © Springer Science+Business Media New York 2013 plant-derived C being traded for nutrients acquired by fungi (Kiers et al. 2011; Selosse and Rousset 2011), both C and other nutrients have a net unidirectional flow from fungus to plant. Furthermore, fully MH plants that depend on SAP fungi are the only known examples of plants whose primary C source is derived solely from complex dead organic substrates. Their existence reopens the important, but debated question of reentry of organic C into plants via mycorrhizal interactions (Baldrian 2009; Selosse et al. 2010). Sometimes it is these exceptions to conventional biological systems that provide the greatest insights into ecosystem function as a whole, and we approach here the ecophysiology of mycoheterotrophy with this broader context in mind.

The purpose of this chapter is to provide a practical and theoretical framework for the study of MH plants' ecophysiology. We will highlight the findings from recent studies that have provided inroads into unraveling the curious functioning of MH plants. Of all the chapters in this book, this chapter will most likely be the quickest to become outdated, because at the time of writing, the field of MH plant ecophysiology is in rapid development. Techniques for high-throughput sample analysis especially for elemental and isotope studies are gaining momentum, as are new techniques for studying source–sink relationships between plants, fungi, and the environment in situ. However, with the application of new methodologies, it is important to remember that within the specific field of MH research, the questions being asked have remained fundamentally the same for over 200 years (Rayner 1927 and references therein). These questions include, who are the players in MH associations? What are the plant and fungal trade-offs for MH interactions? What factors, environmental or otherwise, select for mycoheterotrophy? And more specifically, what forms of C and other nutrients are transferred from fungi to mycoheterotrophs? To date, the majority of research on mycoheterotrophy has addressed the first question and we now have a substantial amount of information on the functional and phylogenetic diversity of fungi that associate with mycoheterotrophs (see Chaps. 5, 6, and 7 for more detail).

A frequent pattern among most MH species is specificity for particular fungal hosts. However, these fungi have a wide phylogenetic breadth within the kingdom Fungi. This indicates that rather than phylogenetic conservatism, it is either physiological or evolutionary pressures that select the fungal hosts for mycoheterotrophs (Hynson and Bruns 2010). To date, the most in-depth research has been focused on fully MH orchids and other taxa that are dependent on mycorrhizal fungi, particularly ectomycorrhizal (EM) fungi, to meet their nutrient demands. Although some fully MH species subsist on compounds derived from SAP fungi, they appear to occur only among the orchids (Selosse et al. 2010). However, the vast majority of fully MH species associate with arbuscular mycorrhizal (AM) fungi in the tropics, but due to a bias of field research to temperate regions, much of what we understand about the ecophysiology of mycoheterotrophy comes from temperate species.

What are the potential ecophysiological determinants of mycoheterotrophy? The physiological dependency of mycoheterotrophs on fungi starts with the earliest stages of seedling development when there are chemical cues between seed and fungus that trigger germination (Bruns and Read 2000). However, germination also seems to be the tightest bottleneck for MH plant survival (Bidartondo and Read 2008; Eriksson and Kainulainen 2011; Tesitelova et al. 2012; Chap. 5). This is because the seeds of many MH species will only germinate in the presence of fungi that are closely related to those associated with adult plants, but upon reaching maturity, no plants have been found with the "wrong" fungal host. However, we cannot rule out the possibility that there are rare instances of individuals surviving with alternative hosts, and that this may lead to permanent host switching, explaining jumps from one fungal partner to another.

At the early developmental stages, securing a source of C and other nutrients is of paramount importance to the plant. It has been argued that for C and other nutrient transfer to occur from fungus to MH plant, the plant must create a concentration gradient that has a draw-down or sink effect on the plant-fungal network (Finlay and

Read 1986). This gradient could be created by the rapid transformation of compounds received from the fungus by the mycoheterotroph into forms that are unavailable for fungal use, or by storage apart from the plant–fungal interface. This has been somewhat demonstrated in the mutualistic interactions of mycorrhizal fungi with autotrophic host plants where plant carbohydrates derived from photosynthesis are converted by fungi into trehalose and polyols that are largely unavailable for plant uptake (Smith and Read 2008). The primary example of fungal assimilated C compounds being converted to plant carbohydrates by a mycoheterotroph is from a study by Smith (1966). In her laboratory experiments two rhizoctonia fungi first colonized the MH seedlings of the orchid *Dactylorhiza purpurella*, then fungi were fed ^{14}C labeled sucrose, which they transformed into trehalose, a portion of which was transferred to the orchid seedlings which transformed it into glucose, fructose, and sucrose (Box 8.1). Interestingly, invertases are absent from at least some EM fungi (Martin and Selosse 2008; Parrent et al. 2009) and AM partners (Tisserant et al. 2012), so that sucrose may well be unavailable for fungal use. Similar to

Box 8.1 Radioisotope and Stable Isotope Labeling

The use of naturally rare radioactive isotopes (e.g., ^{14}C, ^{32}P, or ^{35}S) has a long history in biochemistry and biology. Many physiological matter pathways of living organisms were elucidated using radioisotopes as tracers. For example, Melvin Calvin and colleagues discovered in the late 1940s and 1950s the photosynthetic C reduction cycle of plants and mapped the complete route that C travels during photosynthesis, starting from its absorption as CO_2 to its conversion into sugars and other organic compounds by using ^{14}C as a tracer. In 1961 Calvin was awarded the Nobel Prize in Chemistry for his discovery. Nuclear weapon tests in the 1950s and 1960s were responsible for a doubling of the natural ^{14}C concentration in the atmosphere (Levin and Kromer 1997). This increase in ^{14}C concentration in the atmosphere has been used in ecological studies as a "natural" tracer to investigate turnover rates of organic C pools and the age of soil organic C stores (e.g., Trumbore 2000). Later on rare stable isotopes (e.g., ^{13}C, ^{15}N, or ^{18}O) were added to the suite of tracers in field and laboratory experiments on matter fluxes and transformations between ecosystem pools. Isotopes, i.e., atoms belonging to the same element and possessing equal numbers of protons and electrons, but different numbers of neutrons, are considered as ideally suited tracers because of their almost identical chemical and physical properties. In a typical tracer or isotope labeling experiment a known and usually small amount of a naturally rare isotope in a defined chemical form (e.g., $^{13}CO_2$) is added to a source pool (e.g., atmosphere of a growth chamber containing $^{12}CO_2$) and then recovered transformed in a sink pool (e.g., tissue of a photosynthetic plant) after a known amount of time. The total amount of an element (e.g., ^{13}C tracer and ^{12}C from the growth chamber atmosphere) that moved from the labeled source pool (A) to the sink pool (B) can be calculated knowing the atom% excess of pool A (I_A), and the mass (P_B) and atom% excess of pool B (I_B) at the end of the experiment (from Stark 2000):

$$M_{AB} = (P_B \times I_B) / I_A$$

Dividing of M_{AB} by the time of the experiment gives the flux rate from A to B.

Isotope labeling experiments are based on the following assumptions:

- Added isotope and natural isotope behave identically.
- The source pool is uniformly labeled, and this labeling remains constant over the duration of the experiment.
- Addition of the tracer does not change flux rates.
- No labeling gets lost from the sink pool.

(continued)

Box 8.1 (continued)

These assumptions are not always entirely valid. For example, due to fractionation effects isotopes behave very similarly, but not entirely equal (see Box 8.2). However, isotope fractionations can be ignored in labeling experiments as long as isotope enrichments well above natural isotope abundance are used.

Radioisotopes in source and sink pools can easily be quantified using scintillation counting (e.g., Schimel 1993). In addition autoradiography is a long known and elegant technique to visualize the radioisotope distribution in a sink pool (e.g., Schimel 1993). The use of stable isotopes in tracer studies requires more sophisticated analytical techniques (e.g., isotope ratio mass spectrometry, see Box 8.2); however, they bear no health or environmental hazards and therefore can easily be used in tracer experiments on field level. Furthermore, stable isotopes provide access to trace elements for which no natural radioisotopes exist, e.g., nitrogen and oxygen. For further details on the theory and application of radioisotopes and stable isotopes as tracers we refer readers to Schimel (1993).

autotrophic plant-to-plant C transfer via shared mycorrhizal fungi, in MH plants, the demand or sink strength could be life stage dependent, as well as seasonally and environmentally variable (Lerat et al. 2002; Simard and Durall 2004).

The recent revelation of cryptic or partial mycoheterotrophy in green plants has highlighted the variation in plants' dependency on fungi to meet their C demands. Partial mycoheterotrophs are green plants that appear to be fully autotrophic, but meet some portion of their C demands via fungi in a mixotrophic nutrition that will be explored in detail in this chapter. Since these findings, partial mycoheterotrophy has been proposed as a potential evolutionary pathway to full mycoheterotrophy. However, the underlying determinants, geographical and phylogenetic extent of partial mycoheterotrophy are unknown and currently an area of active research. Ecophysiological methods such as radioactive and stable isotope probing, measuring plant assimilatory and respiratory responses to environmental gradients such as light availability, and natural abundance stable isotope analysis are all critical tools for the study of mycoheterotrophy and their applications are discussed in detail in this chapter. Methodological limitations and considerations for the study of physiological ecology of mycoheterotrophy will also be outlined.

The final section of this chapter will address areas for future research and draw attention to the gaps in our current knowledge of MH ecophysiology. This field is ripe for the undertaking by new methods and researchers. In the years to come, many of the current limitations to studying mycoheterotrophs, and plant–fungal interactions in general, will become obsolete as we continue to develop new quantitative and noninvasive methods to study these systems in situ or perhaps even ex situ (Yagame et al. 2012). What is critical now is that robust theoretical frameworks be established a priori that will engage researchers from across fields and provide a sound foundation for the interpretation of forthcoming data on MH plants. In future research, MH plants will continue to be model systems for the study of the ecophysiology of plant–fungal interactions. This is due to their complete dependency on often a sole fungus to meet their nutrient demands, the fact that they are phylogenetically and geographically widespread, and many represent a profound modification of the most common and abundant mutualism on earth, the mycorrhizal symbiosis.

8.2 From Mutualism to Parasitism: Deconstructing the Continuum of Plant–Fungal Interactions

8.2.1 Mycorrhizal Networks

Since the definitive experiments of Erik Björkman in the 1960s where the transfer of a ^{14}C label applied to the phloem of pines was traced to surrounding individuals of the fully MH species *Hypopitys monotropa*, researchers continue to test for the presence and extent of mycorrhizal networks in nature. In its simplest form, a mycorrhizal network consists of two plant individuals of the same species connected via a shared mycorrhizal fungus. Due to the diffusivity of the mycorrhizal symbiosis and the large size of fungal individuals in some species (some genetic individuals can cover several m^2; Douhan et al. 2011) it is almost certain that these connections exist in nature, but to what extent they actually link different plants and act as conduits for C and mineral nutrient exchange between plants is the subject of much debate.

Fully MH plants that associate with mycorrhizal fungi provide the best examples of C transfer from unrelated plants via shared fungi. Because this tripartite network of autotrophic host plants, mycorrhizal fungus, and mycoheterotroph involves only unidirectional C and mineral nutrient flow to the mycoheterotroph, it is often referred to as an epiparasitism rather than a mutualism. However, the impact of mycoheterotrophy on partners' fitness (especially on the autotrophic host plant) is currently unknown. The evidence for mycorrhizal networks where there is transfer of C or other nutrients among plants that engage in a mutualism with their mycorrhizal fungi is less clear-cut. The first quantitative ex situ laboratory studies to show significant transfer of C between autotrophic plants via shared AM or EM fungi were conducted by Francis and Read in 1984 and Finlay and Read in 1986. Since then, numerous field manipulation and laboratory experiments have taken place to test the significance of mycorrhizal networks in plant establishment, survival, and below-ground resource sharing. These studies have had mixed results, leading some researchers to question the overall importance of mycorrhizal networks (Fitter et al. 1998; Robinson and Fitter 1999; Wu et al. 2001; Pfeffer et al. 2004). However, there is mounting evidence that (1) mycorrhizal networks are common in nature (Simard and Durall 2004; Selosse et al. 2006), (2) there is the potential for bidirectional C, nitrogen (N), and phosphorus (P) movement between plants (Lerat et al. 2002; Teste et al. 2009), and (3) depending on environmental factors such as light availability the sink strength of "receiver" plants in mycorrhizal networks can increase (Finlay and Read 1986; Simard et al. 1997). All these factors could have profound effects on interplant competition, plant and fungal diversity, and community dynamics (Simard and Durall 2004; Selosse et al. 2006). Recent field studies of mycorrhizal networks have focused on their role in seedling establishment (Nara 2006; Teste et al. 2009), survival (McGuire 2007), and growth (Booth 2004). These studies provided strong support that mycorrhizal networks can be critically important in early forest succession stages and tree recruitment. However, they suffer from similar limitations such as the difficulties in assessing the physical presence of fungal connections between plants, measuring long-term net C flow from donor to receiver plants, and the contributions of mycorrhizal networks to plant fitness over temporal and life stage gradients. Future efforts in the study of mycorrhizal networks should be focused in these areas as well as gaining a better understanding of the environmental plasticity of mycorrhizal networks; if they are controlled by plants and/or fungi; if the payoffs of networking are stronger selective forces than the benefits of competition; and finally, whether plants receiving benefits from networking are true "cheaters," providing no reciprocity, or if they somehow compensate for what they receive.

It is becoming undoubtedly clear that the mycorrhizal symbiosis is far less static than it was historically thought to be. Furthermore, where a particular plant or fungus falls along the continuum of mutualistic to parasitic in relation to its symbiotic partner(s) appears to be potentially life-stage, environmentally, and community driven. Fully MH plants may provide one exception to this plasticity due to their absolute dependency on fungi. Thus, quantifying complete C budgets for mycoheterotrophs over the course of their lifecycles, as well as the fitness costs to their fungi (and perhaps autotrophic host plants) will provide much needed constraints for modeling plant–fungal interactions.

8.2.2 Determination of Full Mycoheterotrophy

8.2.2.1 Evidence Based on Radioisotope and Stable Isotope Labeling

In 1881 it was hypothesized that *Hypopitys monotropa* shares a symbiotic fungus with neighboring forest trees and is nourished by these trees through a common mycelial network (Kamienski 1881). However, at that time this hypothesis was not widely accepted. It took almost 80 years until Kamienski's hypothesis was for the first time experimentally confirmed based on radioisotope labeling experiments. It was Björkman (1960) who demonstrated in field experiments that ^{14}C-labeled glucose and ^{32}P-labeled phosphate injected into the phloem of spruce and pine trees were translocated within 5 days to adjacent *Hypopitys monotropa* plants. Other neighboring understory plants, like *Vaccinium myrtillus*, *V. vitis-idaea*, and *Calluna vulgaris*, remained unlabeled (for details on isotope labeling see Box 8.1). Björkman (1960) furthermore demonstrated that a trenching of *H. monotropa* plants from adjacent tree roots by metal sheets severely reduced their development. He concluded that these observations confirm the existence of hyphal connections between *H. monotropa* and neighboring trees and indicate a selective C and P transfer from the trees to *H. monotropa* plants through shared fungal hyphae.

It took another 40 years until further substantial evidence on a selective C transfer from trees to an MH plant through linked fungal mycelia was successfully documented—again using radiocarbon as a tracer. In microcosm experiments McKendrick et al. (2000) fed shoots of *Betula pendula* and *Salix repens* plants growing in association with the leafless orchid *Corallorhiza trifida* with $^{14}CO_2$ and traced the movement of the isotope by a combination of digital autoradiography and scintillation counting. Direct C transfer assimilated by both of the autotrophs to *Corallorhiza* plants occurred only in those cases where plants had already been connected to a shared mycorrhizal fungus. *C. trifida* seedlings introduced to the microcosms as controls immediately before isotope labeling and thus lacking these hyphal connections failed to assimilate significant C amounts. McKendrick et al. (2000) furthermore documented that *C. trifida* plants linked to *B. pendula* and *S. repens* through mycorrhizal hyphae gained 6–14% in biomass during the 25–28 weeks period of the microcosm experiment, while *C. trifida* plants growing in microcosms with *Pinus sylvestris* failed to develop hyphal links and lost 13% of their weight over the same period. Nearly at the same time, fungal ribosomal DNA also provided evidence that the same fungal individuals occurred in the roots of surrounding trees and of the MH orchids *Cephalanthera austinae* (Taylor and Bruns 1997) and *Neottia nidus-avis* (Selosse et al. 2002), supporting a link to trees by individual fungal mycelia. In another microcosm labeling experiment it was shown that a ^{14}C label provided as CO_2 to *Betula pendula* seedlings was transferred to the non-photosynthetic liverwort *Aneura* (*Cryptothallus*) *mirabilis* through a shared mycorrhizal fungus (Bidartondo et al. 2003). In this case a *Tulasnella* sp. was identified to form simultaneously an EM association with trees and a connection with *Aneura mirabilis*.

Also using the microcosm approach, but stable isotope labels (^{13}C and ^{15}N) instead of radioactive isotopes, Bougoure et al. (2010) investigated the tripartite matter exchange between the fully subterranean orchid *Rhizanthella gardneri*, a mycorrhizal fungus from the genus *Ceratobasidium* and the photosynthetic shrub *Melaleuca scalena*. They demonstrated that up to 5% of the C applied as $^{13}CO_2$ to the autotrophic shrub was transferred to *R. gardneri*. *Rhizanthella gardneri* also gained 6% of the C and 22% of the N fed as [^{13}C-^{15}N] glycine to the soil through the mycorrhizal fungus. The non-stoichiometric C and N transfer from the glycine source through the fungus to *R. gardneri* is explained by fungal glycine transformation and respiratory ^{13}C-loss (see also Taylor et al. 2004).

Isotope labeling experiments are not only used to trace matter fluxes between ecosystem compartments, but also to document a lack of matter fluxes or reduced fluxes, for example due to missing or reduced metabolic activity. Cameron et al. (2009) provide an example for this kind of tracer application. They compared the potential for CO_2 assimilation by the MH orchid *Neottia nidus-avis* to that of the leafless but chlorophyll containing orchid *Corallorhiza trifida*. CO_2 assimilation of these two orchid species was further compared to the leafy and green (chlorophyllous) orchid *Cephalanthera damasonium* and to *Fagus sylvatica* seedlings using ^{13}C isotope tracers in the field. The $^{13}CO_2$ assimilation rates decreased in the order *Fagus*>*Cephalanthera*>>*Corrallorhiza* >*Neottia*. These results indicated that the photosynthetic capacity of the *Corallorhiza trifida* individuals on the day of this experiment was closer to the fully MH *Neottia nidus-avis* than to the autotrophic *Fagus sylvatica* or the apparently partially MH (PMH) *Cephalanthera damasonium* (for further details on this tracer experiment see Sect. 8.5.1).

8.2.2.2 Evidence Based on Natural Abundance ^{13}C and ^{15}N

Independent investigations by Gebauer and Meyer (2003) and Trudell et al. (2003) discovered a considerable enrichment of heavy C and N isotopes in the tissues of fully MH orchids and monotropoids (Ericaceae) in comparison to surrounding autotrophic plants (for further details on the stable isotope natural abundance approach see Box 8.2). This enrichment in ^{13}C in the investigated fully MH plants was explained by these species tapping into alternative C sources to atmospheric CO_2 utilized by autotrophic plants in photosynthesis. Fully MH plants enrichment in ^{15}N was thought to be due to these plants receiving compounds enriched in ^{15}N compared

Box 8.2 Stable Isotope Natural Abundance

Most elements of biological interest are composed of two or more stable and/or nonstable (radioactive) isotopes. Isotopes of an element have by definition identical numbers of protons and electrons. However, they are distinguished by the number of neutrons in their nucleus and therefore have different atomic mass units. Identical numbers of protons and electrons are responsible for the mostly equal chemical and physical properties of isotopes. Nonetheless, the difference in the number of neutrons causes slightly different symmetries of atom nuclei and the electron sheath of isotopes. These differences in atomic properties cause thermodynamic isotope effects, i.e., the equilibrium constants of isotopes are slightly different, and kinetic isotope effects, i.e., the speed constants of isotopes in (bio)chemical reactions are slightly different. Due to these isotope effects, the isotopic composition of ecosystem compartments changes in predictable ways as elements cycle through the biosphere. Geochemists exploit changes in the isotopic composition of various biospherical pools to understand principles of the global element cycles. During the last two decades biologists have also become increasingly aware of the potential information that can be gained from the analysis of (mostly stable) isotope natural abundance variation. The use of stable isotope natural abundance analysis provides a nondisruptive method for studying ecosystem fluxes that complements the long established use of stable and radioactive isotopes in tracer experiments (see Box 8.1). However, the adoption of stable isotope analysis has been relatively slow to the field of biology due to: (1) Limited access to isotope ratio mass spectrometers (IRMS) which require specific knowledge of sophisticated analytical techniques to operate and are quite expensive. Also, many biologists do not have sufficient backgrounds in analytical chemistry to successfully run this equipment or interpret the resulting data. (2) The framework for advances in many fields of biology, and specifically in ecophysiology, are based on manipulative experiments, i.e., mostly short-term manipulations of organisms or environmental conditions and analysis of responses to these manipulations. Exploiting information from stable isotope natural abundances, however, provides new perspectives based on an essentially different conceptual framework.

(continued)

Box 8.2 (continued)

Gaining information of biological relevance from variations in stable isotope natural abundance does not necessarily require experimental manipulation. The "experiments" are all in situ processes in nature. Stable isotope natural abundance of various ecosystem compartments integrates these processes and associated matter fluxes (sources) over time. A typical example of process information based on stable isotope natural abundance is the isotopic distinction between plants following the C3 and C4 pathways of photosynthesis. Due to the different isotope fractionations by the two enzymes involved in primary CO_2 fixation: Rubisco in C3 plants and PEP carboxylase in C4 plants, C3 and C4 plants have distinct stable carbon isotope signatures. Conversely, typical source information is the origin of compounds in a product pool, e.g., the total N pool of a legume acquired from soil-borne N compounds taken up by the roots versus N gained through symbiotic fixation of isotopically distinguished atmospheric N_2 in root nodules. Thus, analysis of stable isotope natural abundance in various ecosystem compartments provides process and source information, and care must be taken to avoid a mix up of these two kinds of information.

The two elements of major interest in this chapter, carbon (C) and nitrogen (N), are both composed of two stable isotopes: ^{12}C and ^{13}C, or ^{14}N and ^{15}N, respectively. In addition, C also has a natural very rare radioactive isotope: ^{14}C. In analogy to most other elements of biological interest, the light stable isotopes of C and N (^{12}C and ^{14}N) are much more abundant than the heavy ones (^{13}C and ^{15}N). While ^{12}C contributes about 98.89 atom% to the terrestrial C pool, ^{13}C contributes only about 1.11 atom%. The relative abundance of ^{15}N is even lower. This isotope contributes only about 0.37 atom% to the terrestrial N pool, thus leaving 99.63 atom% to the isotope ^{14}N. The variation in relative abundance of heavy isotopes between ecosystem compartments driven by isotope fractionation due to thermodynamic or kinetic isotope effects is rather low. For C and N the natural variation in relative abundance of the respective stable isotopes in nature ranges at a maximum by 0.1 atom% or 0.02 atom%, respectively. Highly precise detection of such small variations in stable isotope natural abundance requires sophisticated analytical techniques. Specifically designed IRMS measuring the frequency ratios of stable isotope pairs of the respective elements in relation to defined reference gases (in our case CO_2 or N_2, respectively) coupled online to different types of gas sample preparation devices like elemental analyzers, pyrolysis units, and gas chromatographs are at present most commonly used to fulfill these requirements. Due to the fact that the analysis of stable isotope natural abundances is based on abundance ratios, in a sample of unknown isotope abundances, its isotope ratios are measured and then related to stable isotope abundance ratios of a defined standard. The so-called δ notation is commonly used to express isotopic compositions:

$$\delta x = (R_{sample} / R_{standard} - 1) \times 1,000[‰]$$

where x in our case is either ^{13}C or ^{15}N and R is the corresponding ratio $^{13}C/^{12}C$ or $^{15}N/^{14}N$, respectively. International standard reference material for C stable isotope abundances is PDB (limestone from a fossil belemnite of the PeeDee strata in South Carolina) with a $^{13}C/^{12}C$ ratio of 0.0112372 and for N stable isotope abundances is N_2 gas in the atmosphere with a $^{15}N/^{14}N$ ratio of 0.0036765. By subtracting 1 from the $R_{sample}/R_{standard}$ ratio and subsequent multiplying by 1,000 the δ value provides information about the deviation in the respective heavy isotope abundance between sample and standard in ‰. For example, a δ value of +1,000‰ is equivalent to a twice as high heavy isotope abundance in a sample than in the respective standard.

For further details on principles of the stable isotope natural abundance technique and on the exponentially increasing spectrum of applications in plant ecophysiology and

(continued)

Box 8.2 (continued)

ecology in general we refer readers to reviews by Dawson et al. (2002) and Fry (2006). Three aspects out of the huge range of stable isotope natural abundance applications that are of specific importance for the study of mycoheterotrophy are: (1) the principles of how environmental factors affect carbon isotope signatures of plants with C3 photosynthesis (Farquhar et al. 1989), (2) the concept of an isotopic shift along food chains (DeNiro and Epstein 1978, 1981), and (3) the finding of a significant enrichment in heavy C (Gleixner et al. 1993) and N isotopes (Gebauer and Dietrich 1993) in fungal fruit bodies in comparison to plant tissues.

to surrounding autotrophic plants that share the same mycorrhizal fungi. Owing to MH plants' obligate association with various functional groups of fungi (see Chap. 7), and based on the findings from earlier isotope labeling studies, fungi were proposed as the most likely alternative C and N source of fully MH plants. As further support for this, many fungi were already known to be enriched in the heavy isotopes ^{13}C and ^{15}N in comparison to autotrophic plants from the same habitat due to their specific physiology and access to C and N sources also enriched in heavy isotopes (see review by Mayor et al. 2009). Following the food chain concept (Fry 2006) the C and N isotope signatures of MH plants should be similar to, or even more enriched in heavy isotopes than in their fungal source (Trudell et al. 2003). Moreover, the relative enrichment in heavy isotopes in fungi is not a uniform feature, but is specific to different functional and taxonomic fungal groups. For example, fungi forming EM associations tend to be more enriched in heavy C and N isotopes than the majority of SAP fungi (Kohzu et al. 1999; Taylor et al.; 2003; Mayor et al. 2009). In contrast, AM fungi tend to be more depleted in ^{15}N than EM fungi, and are apparently not enriched in ^{13}C compared to their autotrophic host plants (Courty et al. 2011 and references therein). The relative ^{13}C enrichment in EM fungi is related to the gain of ^{13}C-enriched carbohydrates from photosynthetic plant partners (Gleixner et al. 1993). Among EM fungi, species associated with overstory trees are more enriched in ^{13}C than species associated with understory trees (Högberg et al. 1999), and species capable of decomposing recalcitrant soil organic compounds are more enriched in ^{15}N than species with a preference for inorganic N compounds (Gebauer and Taylor 1999). For SAP fungi their respective C and N source (wood, leaf litter, humus, etc.) determines their isotope signature (Gebauer and Taylor 1999; Kohzu et al. 1999). Thus, according to the isotope food chain concept, the differing patterns in the isotope signatures of the various functional groups of fungi is expected to be mirrored by MH plants associated with them. In the following section we provide a comparative overview of our current knowledge on C and N isotope natural abundance in fully MH plants associated with EM, wood- and litter-decomposer SAP, and AM fungi and discuss their ecophysiological implications.

8.3 Isotopic Patterns in Full Mycoheterotrophs Based on Fungal Host

8.3.1 Fully Mycoheterotrophic Plants Associated with Ectomycorrhizal Fungi

C and N stable isotope natural abundance data of MH plants associated with ectomycorrhizal fungi (EM-MH plants) are already available for a considerable number of species collected in a broad range of habitats of wide geographic distribution (Fig. 8.1). These data include field collections from Europe (Gebauer and Meyer 2003; Bidartondo et al. 2004; Julou et al. 2005; Abadie et al. 2006; Tedersoo et al. 2007; Zimmer et al. 2007, 2008; Liebel et al. 2010; Liebel and Gebauer 2011), North America (Trudell et al. 2003; Zimmer et al. 2007; Hynson et al. 2009b),

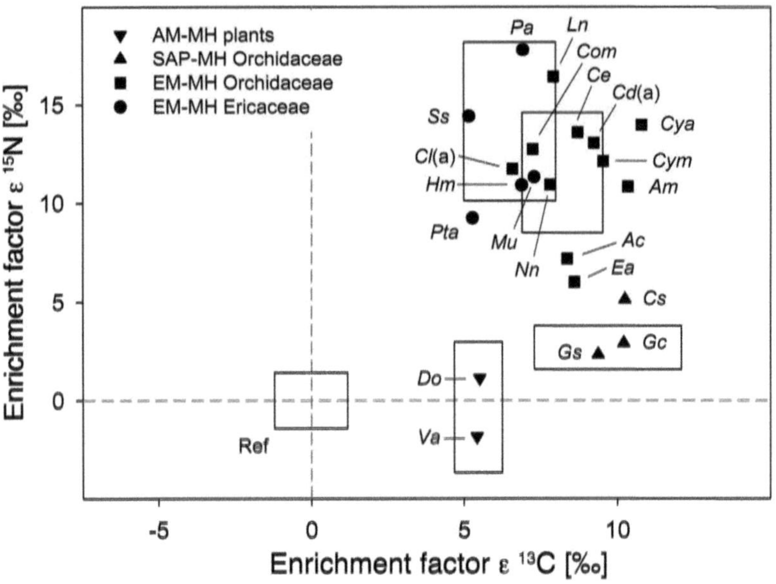

Fig. 8.1 Mean enrichment factors (ε, see Box 8.3) for ^{13}C and ^{15}N of fully mycoheterotrophic (MH) plants associated with fungi forming ectomycorrhizas (EM-MH Orchidaceae and EM-MH Ericaceae), with saprotrophic wood-decomposer fungi (SAP-MH Orchidaceae) and with fungi forming arbuscular mycorrhizas (AM-MH plants). The *boxes* represent one SD of the mean ε values for the four significantly distinguished groups of MH plants and for the photosynthetic reference plants (Ref, $n = 659$) collected together with each of the mycoheterotrophs. Abbreviations of the respective species and numbers of replicates (n) as following: *Ac* = *Aphyllorchis caudata* ($n = 3$); *Am* = *A. montana* ($n = 4$); *Cd(a)* = *Cephalanthera damasonium albino* ($n = 10$); *Ce* = *C. exigua* ($n = 5$); *Cl(a)* = *C. longifolia albino* ($n = 9$); *Com* = *Corallorhiza maculata* ($n = 10$); *Cs* = *Cyrtosia septentrionalis* ($n = 1$); *Cya* = *Cymbidium aberrans* ($n = 3$); *Cym* = *C. macrorhizon* ($n = 6$); *Do* = *Dictyostega orobanchoides* ($n = 8$); *Ea* = *Epipogium aphyllum* ($n = 5$); *Gc* = *Gastrodia confusa* ($n = 5$); *Gs* = *G. similis* ($n = 10$); *Hm* = *Hypopitys monotropa* ($n = 23$); *Ln* = *Lecanorchis nigricans* ($n = 3$); *Mu* = *Monotropa uniflora* ($n = 8$); *Nn* = *Neottia nidus-avis* ($n = 36$); *Pa* = *Pyrola aphylla* ($n = 39$); *Pta* = *Pterospora andromedea* ($n = 10$); *Ss* = *Sarcodes sanguinea* ($n = 15$); *Va* = *Voyria aphylla* ($n = 8$). Data compiled from Gebauer and Meyer (2003), Bidartondo et al. (2004), Julou et al. (2005), Abadie et al. (2006), Tedersoo et al. (2007), Zimmer et al. (2007, 2008), Hynson et al. (2009b), Martos et al. (2009), Ogura-Tsujita et al. (2009), Roy et al. (2009a), Liebel et al. (2010), Merckx et al. (2010), Motomura et al. (2010) and Liebel and Gebauer (2011)

and Asia (Ogura-Tsujita et al. 2009; Roy et al. 2009a; Motomura et al. 2010) covering habitats from boreal coniferous forests, deciduous or mixed temperate forests, evergreen forests in Mediterranean climates to tropical forests. All of these habitats are forests at least partially composed of tree species known to form EM—an essential prerequisite for a tripartite matter flux between trees, EM fungi, and EM-MH plants. The C and N stable isotope natural abundance data of EM-MH plants currently available from the literature mostly also contains stable isotope natural abundance data of accompanying autotrophic plants. Using these autotrophic plants as references for C gains independent of fungi, site-independent differences between autotrophic "reference" and MH "target" species can be calculated. These calculations allow for comparisons of isotope enrichment between species across broad geographic regions (enrichment factors ε, see Box 8.3). Figure 8.1 compiles the entire currently available C and N stable isotope natural abundance data for EM-MH plants suited to calculate enrichment factors. The data set is composed of 11 orchid species belonging to five different tribes within the subfamily Epidendroideae and five species of Ericaceae belonging to three different tribes within the subfamily Monotropoideae. Among this data set are EM-MH species that have served as model organisms to elucidate matter fluxes between trees, EM fungi, and mycoheterotrophs (like

Box 8.3 Normalizing Stable Isotope Natural Abundance Data: The ε Approach

As detailed in Box 8.2, the isotopic composition of an ecosystem compartment (e.g., the leaf of a chlorophyllous, photosynthetically active plant) is driven by the isotope signature of its C source (CO_2 in the atmosphere) and by the isotope fractionation during the process of assimilation, for example CO_2 uptake through the stomata and enzymatic reduction in photosynthesis. Thus, with known C isotope signature of CO_2 in the atmosphere and known isotope fractionation during CO_2 assimilation the C isotope signature of a green plant leaf following the C3 pathway of photosynthesis can be predicted as following:

$$\delta^{13}C_{leaf} = \delta^{13}C_{air} - a - (b - a)c_i / c_a$$

with *a* as the fractionation caused by the slower gas phase diffusion of $^{13}CO_2$ relative to $^{12}CO_2$ (4.4‰), *b* as the fractionation caused by discrimination of the enzyme Rubisco against $^{13}CO_2$ (27‰), c_a as the atmospheric CO_2 concentration and c_i as the leaf internal CO_2 concentration (Farquhar et al. 1989).

While *a*, *b*, and c_a might be considered as fairly constant, neither $\delta^{13}C_{air}$ nor c_i is constant in nature. For example, CO_2 assimilated by the leaf of a plant from a temperate forest understory is composed of CO_2 from the free atmosphere (current $\delta^{13}C \approx -8$‰) and CO_2 from soil respiration ($\delta^{13}C \approx -25$‰) with nonconstant mixing ratios. The leaf internal CO_2 concentration c_i depends on all of those environmental parameters that affect stomata regulation and CO_2 assimilation (light climate, atmospheric water vapor pressure deficit, soil water status, leaf temperature, nutrient supply, etc.) and can thus be highly variable over time and in space.

To make realistic comparisons of stable isotope values between autotrophic and mycoheterotrophic (MH) plants from different habitats it is essential to have either the variables mentioned above remain constant (which is almost impossible under field conditions) or to normalize δ values for site-specific environmental stochasticity. For the study of MH plants this normalization is done by analyzing the isotope signatures of target plants (e.g., putative MH species) together with reference plants (a selection of chlorophyllous C3 plants of different life forms and taxonomic groups that are associated with different types of mycorrhizas and thus representing the spectrum of plants living together with the target plants) growing in close spatial proximity and thus experiencing over time the same microenvironmental conditions as the target plants (for further details of the sampling concept see Gebauer and Meyer 2003). This approach allows calculating (1) the variability of isotope signatures of reference plants and (2) differences between reference and target plants, and (3) testing of these differences for their statistical significance. Following this approach, the conventional δ values (see Box 8.2) are converted into differences. Stable isotope nomenclature uses the enrichment factor ε to describe the difference in relative stable isotope abundance of a product and its substrate (Högberg 1997). To calculate ε the mean δ value per element of the reference plants from a single sampling plot is subtracted from the individual δ values of each plant collected at the same plot (irrespective of whether target or reference plant) as follows:

$$\varepsilon_{Sx} = \delta_{Sx} - \delta_{Refx} [\text{‰}]$$

with *S* as a single value of a sample from a reference or target plant, *x* as a sampling plot within a certain study site, and Ref as the mean value of all reference plants (Preiss and Gebauer 2008). Only in cases from the literature where target and reference plants of a study site were not collected plot-wise do we use a simplified site-wise instead of plot-wise ε value calculation. This simplified approach is accompanied by a loss in resolution (Preiss and Gebauer 2008). Irrespective of these approaches the mean ε value of all reference plants at a certain study site is always 0‰, with an associated standard deviation that represents the small variance in the differing physiologies of autotrophic plants. Thus, converting site-dependent δ values into site-independent ε values opens the option of meta-analyses over sites and the presentation of multiple datasets within a single graph.

Hypopitys monotropa: Björkman 1960) or to identify the specificity towards certain EM fungi (like *Corallorhiza maculata* and *Neottia nidus-avis*: Taylor and Bruns 1997; McKendrick et al. 2002; Selosse et al. 2002). Also among this data set are, however, also species that were just recently identified as EM-MH species based on isotope abundance analyses (like *Pyrola aphylla*, the only known fully EM-MH species among the tribe Pyroleae: Zimmer et al. 2007; Hynson et al. 2009b), vegetative albino forms of usually chlorophyllous orchid species (*Cephalanthera damasonium* and *C. longifolia*: Julou et al. 2005; Abadie et al. 2006; see Sects. 8.4.1 and 8.4.3 for further information on albino phenotypes), and MH species with taxonomically close relatives associated with functional types of fungi other than EM fungi (like the relatives of *Epipogium aphyllum*: Roy et al. 2009b; Liebel and Gebauer 2011). Furthermore, the data set contains species strictly specialized on narrow EM fungal strains (like *Corallorhiza maculata, Neottia nidus-avis,* and *Hypopitys monotropa*: Taylor and Bruns 1997; McKendrick et al. 2002; Bidartondo and Bruns 2005) and species associated with a fairly constant set of EM fungi (like *Epipogium aphyllum* and the albino variants of *Cephalanthera damasonium* and *C. longifolia*: Julou et al. 2005; Abadie et al. 2006; Roy et al. 2009b; Liebel and Gebauer 2011) or even with multiple EM fungi (like *Pyrola aphylla, Aphyllorchis caudata,* and *A. montana*: Hynson and Bruns 2009; Roy et al. 2009a).

Irrespective of their wide geographical distribution, phylogenetic breadth within the plant kingdom, their occurrence in different forest types, and their varying degree of EM fungal specificity, what all of these EM-MH species have in common is that they are significantly enriched in the heavy C and N isotopes compared to neighboring reference plants. However, the enrichment factors for both C and N are not constant between MH species, presumably due to variations in the isotope signature of their respective fungal associates. The mean enrichment ranges from 5.3 ± 1.2(SD)‰ (*Sarcodes sanguinea*, Ericaceae) to 10.8 ± 0.6(SD)‰ (*Cymbidium aberrans*, Orchidaceae) for ^{13}C and from 6.0 ± 2.1‰ (*Epipogium aphyllum*, Orchidaceae) to 17.8 ± 2.7‰ (*Pyrola aphylla*, Ericaceae) for ^{15}N. Thus, the variation between EM-MH species in heavy isotope enrichment is greater for N than for C and probably reflects the broader variation in fungal N isotope abundance than fungal—and specifically EM fungal—C isotope abundance (Mayor et al. 2009). Current data furthermore indicate a difference in the heavy isotope enrichment of EM-MH species belonging to the Orchidaceae and to the Ericaceae (Fig. 8.1). The investigated 11 EM-MH species belonging to the Orchidaceae are significantly (Mann–Whitney *U*-test: $U = 1731.0$, $p < 0.001$) more enriched in ^{13}C (mean $\varepsilon^{13}C = 8.2 \pm 1.3$‰, $n = 94$) than the five EM-MH species belonging to the Ericaceae (mean $\varepsilon^{13}C = 6.5 \pm 1.5$‰, $n = 95$), and the EM-MH Orchidaceae (mean $\varepsilon^{15}N = 11.6 \pm 3.1$‰, $n = 94$) are significantly ($U = 2941.5$, $p < 0.001$) less enriched in ^{15}N than the EM-MH Ericaceae (mean $\varepsilon^{15}N = 14.2 \pm 4.0$‰, $n = 95$). The significant difference in C and N stable isotope enrichment between the investigated EM-MH representatives of the two plant families becomes obvious from this meta-analysis thanks to a continuously increasing number of data points available from the literature. Though yet to be confirmed, these differences between orchids and ericaceous MH plants may also relate to the evolution of different mechanisms for MH nutrition in these two families.

8.3.2 Fully Mycoheterotrophic Plants Associated with Saprotrophic Fungi

Besides the tripartite associations with trees and EM fungi, some fully MH orchids associate with free-living wood-decomposer or litter decaying fungi. Interestingly, all of these fully MH plants are orchids and belong mostly to the subfamily Epidendroideae, similar to many EM-MH Orchidaceae. Such saprotrophic-MH (SAP-MH) orchids are typically found in the litter-rich forest floor or beside decomposed woody materials such as decayed tree trunks, stumps, logs, and pruned branches. In 1911, Kusano reported on an association of the MH orchid *Gastrodia elata* with the normally plant pathogenic wood-decomposing fungus *Armillaria*, and Hamada (1939) also found this fungus associating with the MH orchid *Cyrtosia sepentrionalis*. Subsequently, various *Armillaria* species have been isolated and identified from these orchids (Terashita and Chuman 1987, 1989; Cha and Igarashi 1995, 1996; Matsushita et al. 1996; Terashima et al. 1998; Ota et al. 2000; Kikuchi et al. 2008a, b; Sekizaki et al. 2008). In the world's largest MH climbing orchid, *Erythrorchis ochobiensis*, an association with the SAP *Erythromyces crocicreas* (Basidiomycota) was reported by Hamada and Nakamura (1963) and, later, a wide range of wood-decomposing fungi, such as *Lentinula edodes* and *Pleurotus ostreatus*, were shown to induce seed germination and plantlet formation in symbiotic culture (Umata 1995, 1997, 1998a, b, 1999). The MH orchid *Epipogium roseum*, the sister species of the EM-MH orchid *Epipogium aphyllum*, has been shown to associate with fungi from the SAP family Psathyrellaceae (Basidiomycota), including *Coprinellus disseminatus* (Yamato et al. 2005; Yagame et al. 2008a). Notable is that *Epipogium roseum* (Yagame et al. 2007) can be cultivated for its whole lifecycle with wood-decomposer fungi under laboratory conditions. Such instances of in vitro cultivation are also known for other SAP-MH orchids (Burgeff 1932, 1936; Umata et al. 2007), and *Gastrodia elata* is now routinely cultivated symbiotically with *Armillaria* spp. for commercial use of its tubers (Chou 1974; Sung et al. 1995; Xu and Guo 2000). In addition to these reports, several other SAP decomposer fungal genera such as *Fomes*, *Marasmius*, *Xerotus*, *Campanella*, and *Gymnopus* have been reported to be associated with MH *Gastrodia* species, its related MH genus *Didymoplexis*, and some species of the MH genera *Galeola* and *Erythrorchis* (Burgeff 1932; Campbell 1962, 1964; Dearnaley 2006; Dearnaley and Bougoure 2010). Although information about fungal specificity of SAP-MH orchids is still rather scarce, recent reports showed a wide range of specificities, from a strict association between *Eulophia zollingeri* and the *Psathyrella candolleana* species group (Ogura-Tsujita and Yukawa 2008), to an association with a constant set of fungal partners, like *Gastrodia confusa* and representatives of the Mycenaceae (Ogura-Tsujita et al. 2009), to a broad spectrum of associated fungi in the case of *Wullschlaegelia aphylla* or *Gastrodia similis* and multiple Basidiomycetes related to the genera *Mycena*, *Marasmius*, *Psathyrella*, and *Resinicium* (Martos et al. 2009). Apparently different lineages within Psathyrellaceae and Mycenaceae have been repetitively targeted by independent MH orchid lineages (Selosse et al. 2010).

Associations between decomposer SAP fungi and fully MH plants as well as laboratory cultivation experiments provide a hint that fully MH plants also can cover their C and N demand through associations with these fungi in nature. As with EM-MH plants, stable isotope natural abundance analysis is a powerful tool to elucidate the nutritional sources of MH plants associated with decomposer fungi. The still very small stable isotope natural abundance data set available for SAP-MH orchids and accompanying reference plants indicates a significant enrichment in ^{13}C and ^{15}N due to C and N gain from the fungal source, but with a different pattern than found for EM-MH plants (Fig. 8.1). The so far investigated SAP-MH orchids (*Gastrodia confusa* from a wet-temperate bamboo forest in Japan and associated with several species of wood-decomposer *Mycena* fungi (Ogura-Tsujita et al. 2009), *Gastrodia similis* from tropical rainforests and secondary forests in La Réunion and associated with the wood-decomposer *Resinicium* (Martos et al. 2009), and one individual of *Cyrtosia septentrionalis* collected in a warm-temperate evergreen broadleaf forest in Japan (Motomura et al. 2010)) indicate a ^{13}C enrichment (mean $\varepsilon^{13}C = 9.7 \pm 2.4‰$, $n = 16$) that is significantly higher than EM-MH (vs. EM-MH orchids $U = 456, p = 0.012$; vs. EM-MH Ericaceae $U = 212$, $p < 0.001$) and a ^{15}N enrichment (mean $\varepsilon^{15}N = 2.7 \pm 1.1‰$, $n = 16$) that is significantly lower than found for EM-MH orchids ($U = 2$, $p < 0.001$) and EM-MH Ericaceae ($U = 0$, $p < 0.001$). Though not suited for the ε approach due to the absence of true reference plant data, the C and N isotope signature of one other MH orchid species associated with wood-decomposer fungi (*Gastrodia sesamoides* from open woodland in Queensland, Australia; (Dearnaley and Bougoure 2010) also points towards a considerably high ^{13}C enrichment and a ^{15}N enrichment lower than found for EM-MH plants. The most likely reason for the different pattern of C and N isotope natural abundance between SAP-MH plants and EM-MH plants is a difference in the isotope signature of the C and N sources utilized by their respective fungal hosts. These few investigated SAP-MH plants are consistently associated with wood-decomposer fungi, and wood is known to be enriched in ^{13}C and depleted in ^{15}N (Gebauer and Schulze 1991; Gebauer and Dietrich 1993; Gebauer and Taylor 1999). Presently, stable isotope natural abundance data are available only for one SAP-MH orchid associated with litter decaying fungi (*Wullschlaegelia aphylla* from a tropical rainforest in Guadeloupe; Martos et al. 2009). These data point towards lower ^{13}C enrichment in SAP-MH orchids associated with litter decaying fungi than in SAP-MH orchids with wood-decomposer fungi. Undoubtedly more data on stable isotope signatures of SAP-MH orchids (especially associated to litter decaying fungi) and accompanying reference plants are required to confirm these preliminary conclusions. Also, the geographic distribution of SAP-MH orchids and their apparent preference for wood-decomposers among the SAP fungi require further investigation. At present we only can conclude that SAP-MH orchids obviously prefer humid climate conditions under which wood or litter decomposition rates are high.

8.3.3 Fully Mycoheterotrophic Plants Associated with Arbuscular Mycorrhizal Fungi

The AM symbiosis is one of the oldest plant symbioses on earth, estimated to be at least 400 million years old (Smith and Read 2008). It is therefore not surprising that this lineage of fungi has been infiltrated by the cheating strategy of fully MH plants on many more independent occasions than EM or SAP fungi. The majority of fully MH plants grow in tropical forests dominated by photosynthetic plants that associate with AM fungi. For this reason it was assumed in analogy to the EM-MH plants that among these fully MH plants in AM dominated forests an association with AM fungi occurs. Based on fungal DNA analysis from the mycorrhizal roots of several fully MH plants, five *Voyria*, and one *Voyriella* species (Gentianaceae) from tropical forests in French Guiana and *Arachnitis uniflora* (Corsiaceae) from three subantarctic forest sites in Argentina, and neighboring green plants, Bidartondo et al. (2002) showed that these plants indeed associate with AM fungi and display a fungal host specificity similar to many EM-MH plants. This finding suggested that AM fungi mediate a C transfer between autotrophic AM plants and AM-MH plants. Recent stable isotope natural abundance data confirm this suggestion (Fig. 8.1). Though phylogenetically distant, the fully MH plants *Voyria aphylla* (Gentianaceae) and *Dictyostega orobanchoides* (Burmanniaceae) also collected in a tropical forest in French Guiana are both significantly enriched in the heavy isotope ^{13}C in comparison to neighboring photosynthetic plants (Merckx et al. 2010). The relative ^{13}C enrichment of these AM-MH plants (mean $\varepsilon^{13}C = 5.5 \pm 0.8$ ‰, $n = 16$) is, however, significantly lower than in EM-MH orchids ($U=40.5$, $p<0.001$), EM-MH Ericaceae ($U=393.5$, $p=0.002$), and SAP-MH orchids ($U=4$, $p<0.001$). Similar trends were found by Courty et al. (2011) for *V. aphylla* and *V. tenella* (Gentianaceae) as well as *Aptera aphylla* and *Gymnosiphon sphaerocarpus* (Burmanniaceae) from a Caribbean island, La Guadeloupe. In addition, the latter study recovered spores from soil to investigate the isotope signature of AM fungal propagules, which proved similar to that of leaves of canopy trees. Thus, while the relative ^{13}C enrichment of EM-MH plants compared to surrounding autotrophic plants is primarily related to the ^{13}C enrichment of their fungal associates, the enrichment of AM-MH plants compared to surrounding understory autotrophic plants is likely unrelated to differences in AM fungal C acquisition and rather due to differences in photosynthetic rates between canopy trees and understory plants. Because of low photosynthetic rates and more humid conditions supporting longer and more numerous stomatal openings, understory plants are more depleted in ^{13}C than canopy trees (see Gebauer and Schulze 1991). Noteworthy is the finding of no significant ^{15}N enrichment in the AM-MH plants by both Merckx et al. (2010) (mean $\varepsilon^{15}N = -0.4 \pm 3.3$ ‰, $n = 16$; see Fig. 8.1) and Courty et al. (2011) who, additionally, reported that the ^{15}N signature was similar in AM fungi. Courty et al. (2011) also showed that, compared to their accompanying reference plants, the investigated AM-MH plants have similar total N concentrations as compared to green plants and fungi, whereas higher total N concentrations are common among EM-MH plants (Gebauer and Meyer 2003; Julou et al. 2005; Liebel et al. 2010; Stöckel et al. 2011; see also Sect. 8.6.1). Lack of differentiation in ^{15}N natural abundance and total N concentrations between AM-MH plants and reference plants suggests utilization by all of these plants of similar N sources, presumably inorganic N compounds obtained through their AM fungal partners. Taken together, these observations, which deserve replicates from other sites and taxa, suggest that the matter exchanged between fungus and host plant differs in EM-MH vs. AM-MH plants, a situation not fully unexpected owing to the ancient divergence between AM Glomeromycota and EM-forming taxa of Asco- and Basidiomycota.

8.4 Partial Mycoheterotrophy

8.4.1 Evolution of Partial and Full Mycoheterotrophy

PMH plant lineages add to a long list of taxa where heterotrophic abilities evolved from autotrophic ancestors. In the broadest sense, such a strategy is a kind of mixotrophy. The word mixotrophy has been used as a synonym for partial mycoheterotrophy, but is a more encompassing term than partial mycoheterotrophy, because mixotrophic strategies and mechanisms are very diverse. For example, many independent phyla of planktonic algae are mixotrophic, either by uptake of dissolved organic matter (Kamjunke and Tittel 2009) or by phagotrophy on unicellular preys (Jones 2000). Uptake of C or phagotrophy is a plesiomorphic condition in these algae, since ancestors of plastid-bearing taxa are considered to have been heterotrophs (phagotrophy likely being the process through which plastids were indeed acquired). In contrast, mixotrophy in land plants is secondarily evolved, i.e., represents subversion from full autotrophy, and its ecological relevance is yet to be estimated in terrestrial ecosystems. Here we use the phrase "partial mycoheterotrophy" whenever mixotrophy is achieved in a green plant by partial use of C from a mycorrhizal fungus.

In land plants, mixotrophic strategies encompasses partial mycoheterotrophy, but also the use of C from prey (Adamec 1997) and of the host sap in hemiparasitic plants (Schulze et al. 1991; Press and Graves 1995; Těšitel et al. 2010). The latter are green plants that obtain some nourishment, especially mineral nutrients and sometimes C, by parasitizing other plants: C is obtained from connection to the xylem sap, or even to the phloem sap. For example, mistletoes (Loranthaceae) derive up to 63% of their C from their host (Schulze et al. 1991; Bannister and Strong 2001); *Olax phyllanthi* (Olacaceae) derives 19–30% (Tennakoon and Pate 1996); *Rhinanthus alectorolophus* (Orobanchaceae) derives up to 50% (Těšitel et al. 2011). In the latter case, as is true for some PMH species (see Sect. 8.4.3), shading enhanced the contribution of host-derived C; moreover, achlorophyllous variants of *Striga hermonthica* (Orobanchaceae) can survive (Press and Graves 1995).

When considering mycoheterotrophy in a phylogenetic framework, some fully MH species are nested within PMH lineages, e.g., in the tribe Neottieae (Fig. 8.2a; Abadie et al. 2006; Selosse and Roy 2009) and in the genus *Cymbidium* (Motomura et al. 2010); and the same may have happened in the genus *Platanthera* (Yagame et al. 2012). Pyroloids (= tribe Pyroleae) are closely related to the MH Monotropoideae (Monotropeae and Pterosporeae; Kron et al. 2002), suggesting a similar scenario (Tedersoo et al. 2007); however, the subtribe relationships in Monotropoideae deserve new analyses and a basal position of Pyroleae remains uncertain. Interestingly, in pyroloids, there are two well-supported clades (Freudenstein 1999; Kron et al. 2002), *Pyrola* + *Orthilia* on the one hand and *Moneses* + *Chimaphila* on the other; the second clade tends to encompass less frequent reports of

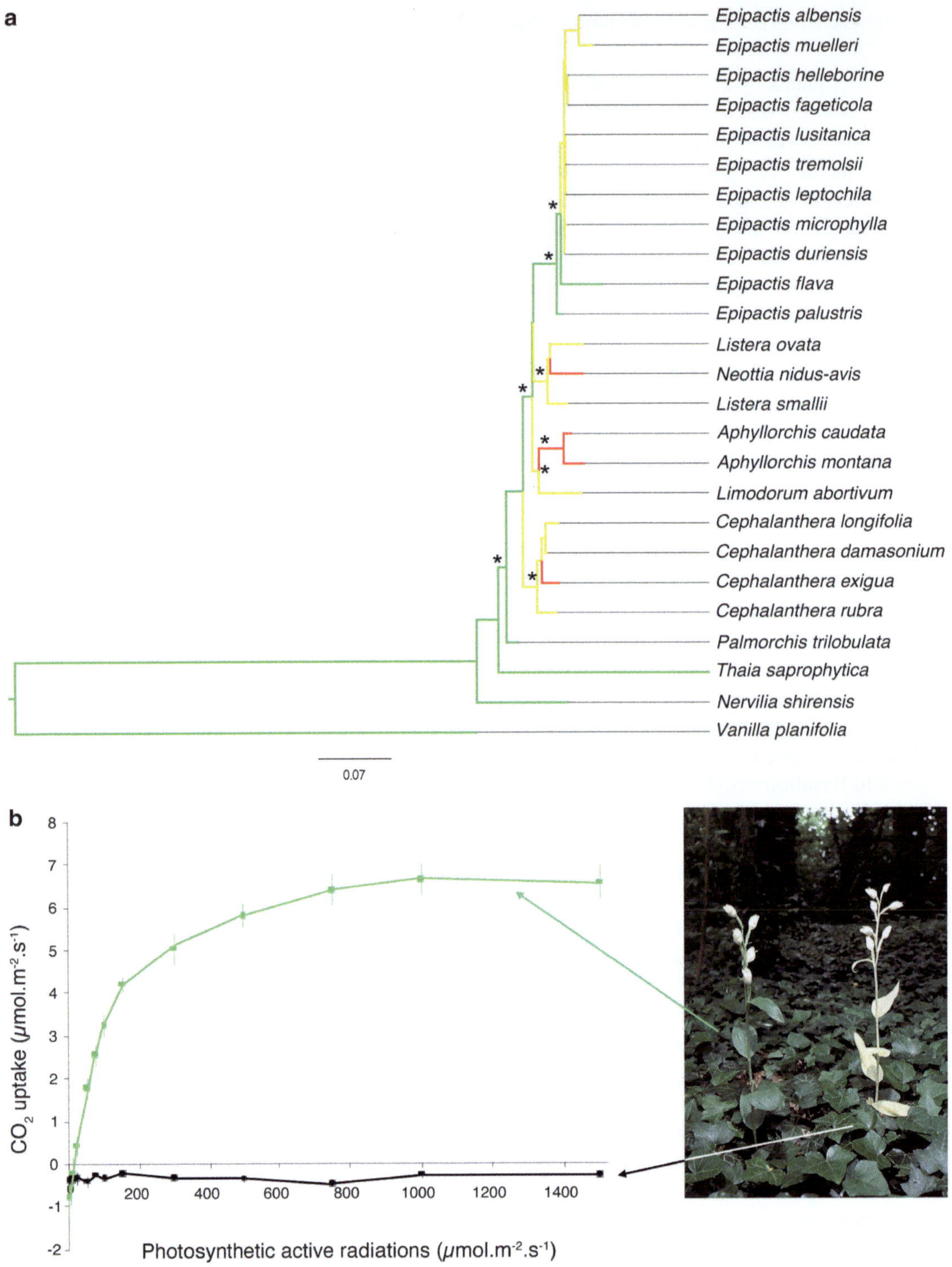

Fig. 8.2 (**a**) A phylogeny of Neottieae (based on ITS + rbcL phylogenies, inferred by Maximum Likelihood and 1,000 bootstrap repetitions, using a GTR model, from Roy et al. 2009a) with reconstruction inferred ancestral trophic status, assuming the most parsimonious scenario for its evolution. *Green*, autotrophic nutrition; *yellow*, partially mycoheterotrophic (PMH) nutrition; *red*, mycoheterotrophic nutrition. *Stars* indicate node supported by >80% bootstrap support and (**a**) indicates green species for which albino, i.e., fully achlorophyllous variants are reported. (**b**) CO_2 exchanges at various light levels in *Cephalanthera damasonium*, a green species where rare fully non-chlorophyllous albinos exist. Each *curve* is the mean of five individuals (M. Roy and M.-A. Selosse, unpublished data; see Julou et al. 2005 for site, material and methods). On the *right* a typical green *C. damasonium* individual together with a non-chlorophyllous variant

Table 8.1 A convergent scenario for the evolution of heterotrophy in plants through mixotrophic steps that exploit a living source for mineral nutrients

Nutrition	Evolution to parasitism in plants:	Evolution to mycoheterotrophy	Positively selected for
#1 Autotrophy	Free-living, autotrophic non-mycorrhizal plant.		
#2 Autotrophy	Hemiparasitic xylem tapping plants of its host mainly for mineral nutrition.	Mineral nutrition by mycorrhizal fungi shared with surrounding plants that also contribute to fungal nutrition.	improved and less costly mineral nutrition.
#3 Mixotrophy (low to high heterotrophy)[a]	Hemiparasitic xylem tapping plant tapping xylem accessing both minerals and carbon from its host.	Mixotrophic plant deriving mineral nutrients and some carbon from shared mycorrhizal fungi, but still photosynthetic.	improved carbon nutrition, improved tolerance to low light.
#4 Mixotrophy (high heterotrophy)[a]	Mixotrophic hemiparasitic plant with reduced photosynthetic abilities with improved growth and reproduction when associated with it's host plant.	Partially mycoheterotrophic plant with reduced photosynthetic abilities.	improved carbon nutrition in low light.
#5 Full heterotrophy	Heterotrophic, nongreen plant, obtaining all its mineral nutrient and carbon supply by tapping the xylem and phloem of its host plant.	Mycoheterotrophic, nongreen plant obtaining all its mineral and carbon supply from fungi.	Complete carbon nutrition, even in absence of light.

[a]Possibility of survival for albino (=achlorophyllous) variants

partial MH than the first, which harbors the fully MH *P. aphylla* (Zimmer et al. 2007; Hynson et al. 2009b) and PMH *P. japonica* (Matsuda et al. 2012).

The evolution to full mycoheterotrophy through partial mycoheterotrophy is thus reminiscent of the evolution of holoparasitic plants (= full heterotrophs) from hemiparasitic ancestors, as reported in the Orobanchaceae (Bennett and Mathews 2006) and Convolvulaceae (McNeal et al. 2007). A common scenario for evolution of plant heterotrophy can be suggested (Table 8.1; Cameron and Leake 2007; Selosse and Roy 2009), where biological interactions which only formerly selected for mineral nutrition allow (1) some indirect mixotrophy, and then (2) select for emergences of full heterotrophy, directly targeting C. Basically, this is an exaptation from mineral to C nutrition by "baiting the feeding hand." This scenario awaits confirmation from other taxa such as AM-MH plants that associate with one of the oldest extant mycorrhizal lineages of fungi on earth, the Glomeromycota (Selosse and Roy 2009).

In terms of selective mechanisms, evolutionary transition to mycoheterotrophy can be fuelled by the fact that some PMH plants, at least in some environments, grow in young forests whose initially open canopies allows photosynthesis but tend to close due to ongoing succession. Thus, they undergo increasing shading that selects for more light-independent C supply (Selosse and Roy 2009); indeed, this may explain the convergent evolution to full MH observed within the Neottieae (Fig. 8.2a; Roy et al. 2009a). In this tribe, full MH is never shown to be evolutionarily reversible, maybe due to loss of photosynthetic genes both in the plastid and nucleus. However, there are possible reversions from partial mycoheterotrophy at adulthood to autotrophy at adulthood. For example, the *Epipactis palustris-gigantea* clade (Fig. 8.2a) is autotrophic at adulthood, as shown by survival upon transplantation

(Sadovsky 1965), ^{13}C abundances indicative of autotrophy (Fig. 8.3), and association with rhizoctonia fungi that are common associates of initially MH orchids (Bidartondo et al. 2004; Zimmer et al. 2007; Illyés et al. 2009). Similarly, some green *Listera* species, despite close taxonomic relatedness to *Neottia nidus-avis* and Asian MH species (Roy et al. 2009a), are likely autotrophic as adults based on their ^{13}C abundances (Fig. 8.3) and associations with rhizoctonias (Bidartondo et al. 2004). Although Neottieae phylogenies remain poorly resolved, these clades are unlikely to be basal (Pridgeon et al. 2008; Roy et al. 2009a), making reversion from partial mycoheterotrophy to autotrophy at adulthood the most parsimonious scenario, while multiple shifts to partial mycoherotrophy remain a possible, but less parsimonious, alternative. Reversion to autotrophy at adulthood is not unexpected, since photosynthetic abilities remain in partial mycoheterotrophs.

Despite their successful survival, albinos observed in some orchid species show reduced fitness (Salmia 1986, 1989b; Roy et al. 2012), reduced demography as compared to green individuals (Abadie et al. 2006; Tranchida-Lombardo et al. 2010), and more or less impaired vegetative traits (Salmia 1989b; Julou et al. 2005). Their lower basal metabolic rates (Julou et al. 2005; see CO_2 evolution in the dark on Fig. 8.2b) suggest C limitation. They often display higher rates of mycorrhizal colonization (Salmia 1989a; Selosse et al. 2004; Abadie et al. 2006), a fact that could compensate for absence of photosynthesis, but it is unknown if more extensive colonization is linked to greater carbon gains. In a study of two *C. damasonium* populations with albino and green individuals (Roy et al. 2012), albinos comparatively displayed (1) more frequent shoot drying at time of fruiting, possibly due to stomatal dysfunctions, (2) higher frequency of dormancy, and (3) fewer seeds, with lower germination capacity. This results in a 500–1,000× fitness reduction as compared to green individuals. Among other factors, two observed features were proposed to cause a C limitation and fitness reduction in these albinos: they displayed higher pathogen and herbivore load, and a sharp reduction of mycorrhizal colonization at time of fruiting, which would likely be compensated by photosynthesis in green individuals, but may be critical for the survival of albinos (Roy et al. 2012). Albinos likely represent unique snapshots of failed transitions from partial mycoheterotrophy to full mycoheterotrophy, and the analysis by Roy et al. (2012) suggests that successful transitions at least require degeneration of leaves and stomata, optimization of the temporal pattern of fungal colonization and shoot sprouting, and new defences against pathogens and herbivores. In Neottieae, albinos suggest that the transition from partial to full mycoheterotrophy cannot be sudden, and that additional traits are required to become successfully MH. Moreover, their absence in PMH Ericaceae suggests that the transition to full mycoheterotrophy can occur in lineages devoid of albinos, so that they are not a necessary step toward full mycoheterotrophy. Albinos are ecological equivalents to mutants in genetics, i.e., their dysfunctions may suggest what makes mycoheterotrophy successful. Although their determinism remains unknown, they offer fascinating models for comparing the physiology of mixo- and autotrophs within very similar genetic backgrounds. The options for physiological investigations on the transition from autotrophy to full mycoheterotrophy become even wider when including MH species possessing variegated leaves. In addition to the frequent green and rare albino forms of some orchid species, individuals with a continuous range between these extremes have episodically been found (Renner 1938; Salmia 1989b; Stöckel et al. 2011).

8.4.2 Initial Mycoheterotrophy-Autotrophy

In contrast with full mycoheterotrophs, many plants are MH during and after seed/spore germination, but eventually develop into autotrophic individuals. While the duration of mycoheterotrophy in such species is relatively short, it remains an obligate and critical part of their life cycle. Most orchids (aside from a few hundred species

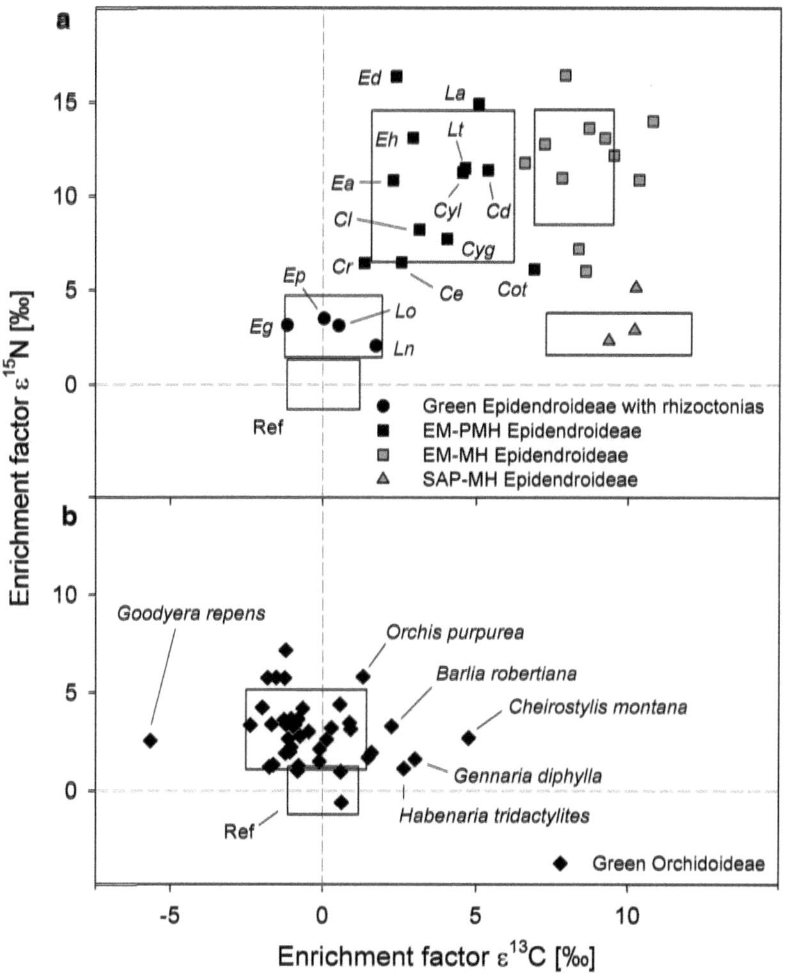

Fig. 8.3 Mean enrichment factors (ε, see Box 8.3) for ^{13}C and ^{15}N of (**a**) 16 green orchid species (*black symbols*) and 14 fully mycoheterotrophic (MH) orchid species (*grey symbols*, for further details see Fig. 8.1) belonging to the subfamily Epidendroideae and associated either with fungi from the polyphyletic rhizoctonia group or, fungi simultaneously forming ectomycorrhizas with surrounding trees (EM-PMH and EM-MH) or with saprotrophic wood-decomposer fungi (SAP-MH) and of autotrophic reference plants (Ref, *n*=765) collected together with each of the respective orchids and (**b**) of 44 green orchid species belonging to the subfamily Orchidoideae and of autotrophic reference plants (Ref, *n*=863) collected together with each of the respective orchids. The *boxes* represent one SD of the mean ε values for the different groups of green and MH orchids and for the autotrophic reference plants. Abbreviations of the green orchid species belonging to the Epidendroideae and numbers of replicates (*n*) as following: *Cd*=*Cephalanthera damasonium* (*n*=21); *Ce*=*C. erecta* (*n*=3); *Cyl*=*C. longifolia* (*n*=19); *Cr*=*C. rubra* (*n*=7); *Cot*=*Corallorhiza trifida* (*n*=9); *Cyg*=*Cymbidium goeringii* (*n*=7); *Cl*=*C. lancifolium* (*n*=6); *Ea*=*Epipactis atrorubens* (*n*=11); *Ed*=*E. distans* (*n*=4); *Eg*=*E. gigantea* (*n*=5); *Eh*=*E. helleborine* (*n*=18); *Ep*=*E. palustris* (*n*=4); *La*=*Limodorum abortivum* (*n*=10); *Lt*=*L. trabutianum* (*n*=5); *Ln*=*Liparis nervosa* (*n*=3); *Lo*=*Listera ovata* (*n*=25). The data set of the green orchids belonging to the Orchidoideae includes the following species and replicates (*n*): *Aceras anthropophorum* (*n*=10), *Anacamptis laxiflora* (*n*=5), *Barlia metlesicsiana* (*n*=5), *B. robertiana* (*n*=5), *Cheirostylis montana* (*n*=2), *Dactylorhiza majalis* (*n*=4), *D. sambucina* (*n*=11), *Gennaria diphylla* (*n*=10), *Goodyera oblongifolia* (*n*=18), *G. repens* (*n*=5), *G. schlechtendaliana* (*n*=1), *Gymnadenia conopsea* (*n*=8), *Habenaria tridactylites* (*n*=5), *Ludisia discolour* (*n*=5), *Neotinea maculata* (*n*=5), *Ophrys fuciflora* (*n*=9), *O. insectifera* (*n*=12), *O. apifera* (*n*=5), *O. incubacea* (*n*=5), *O. sicula* (*n*=5), *O. sphegodes* (*n*=5), *Orchis brancifortii* (*n*=5), *O. canariensis* (*n*=5), *O. ichnusae* (*n*=5), *O. laxiflora* (*n*=5), *O. longicornu* (*n*=5), *O. mascula* (*n*=18), *O. morio* (*n*=5), *O. papilionacea* (*n*=5), *O. pauciflora* (*n*=5), *O. provincialis* (*n*=5), *O. purpurea* (*n*=10), *O. tridentata* (*n*=5), *O. ustulata* (*n*=5), *Platanthera bifolia* (*n*=7), *P. chlorantha* (*n*=4), *P. leucostachys* (*n*=13), *Serapias cordigera* (*n*=5), *S. lingua* (*n*=5), *S. nurrica* (*n*=5), *S. parviflora* (*n*=10), *S. vomeracea* (*n*=10), *Spiranthes spiralis* (*n*=5), and *Zeuxine agyokuana* (*n*=1). Data compiled from Gebauer and Meyer (2003), Bidartondo et al. (2004), Julou et al. (2005), Abadie et al. (2006), Tedersoo et al. (2007), Zimmer et al. (2007, 2008), Hynson et al. (2009a), Roy et al. (2009a), Liebel et al. (2010), Motomura et al. (2010) and Girlanda et al. (2011)

that remain fully or PMH as adults) appear to be initially mycoheterotrophic-autotrophic (IMH), although direct evidence is available for only a small number of species. With over 20,000 species in the Orchidaceae, the number of IMH species is likely far greater than all fully and PMH species combined. Initial mycoheterotrophy-autotrophy is not limited to the orchid family; however, additional taxa within the Ericaceae (Pyroleae), Lycopodiaceae, and several fern families appear to be IMH, as well (see Chap. 2).

MH seedlings and sporelings of IMH plants are typically subterranean and non-photosynthetic; however, despite their cryptic nature, such germlings belonging to the Orchidaceae (Salisbury 1804), Pyroleae (Irmisch 1855), Ophioglossaceae (Hofmeister 1857), and Lycopodiaceae (Mettenius 1856) had already been observed by the nineteenth century. Like fully MH plants, they were often erroneously described as "saprophytic" (Leake 2005). The understanding of IMH orchid seedlings advanced in the late nineteenth to early twentieth centuries, with discoveries by Noël Bernard (Bernard 1899) that fungal symbionts are necessary for the heterotrophic growth of orchid seedlings, Beau (1920) that growth of symbiotic orchid seedlings occurs only when the fungi have access to a C source, and Bernard (1908) and Knudson (1922) that orchid seedlings exhibit asymbiotic growth on media enriched with simple sugars. More recently, studies involving labeled and naturally abundant isotopes have indicated that adult plants of photosynthetic orchid spp., as has generally been assumed, are usually autotrophic (Gebauer and Meyer 2003; Cameron et al. 2006, 2008; Hynson et al. 2009a; Liebel et al. 2010; Girlanda et al. 2011). Autotrophy at adulthood is indicated by ^{13}C natural abundance for the majority of green species from the subfamily Orchidoideae investigated thus far (Fig. 8.3b) and for species from the subfamily Epidendroideae that are solely associated with fungi of the polyphyletic rhizoctonia group (Fig. 8.3a). While isotopic evidence for initial mycoheterotrophy-autotrophy is not yet available for most putatively IMH taxa, this lifestyle can be inferred by characteristic dust seed morphology (in angiosperms), subterranean and non-photosynthetic germling development, consistent association of germlings with fungi, absence of hemiparasitic interactions, and photosynthetic, readily cultivated adults.

8.4.2.1 Initial Mycoheterotrophy-Autotrophy in the Orchidaceae

Like fully MH angiosperms, initially mycoheterotrophic-autotrophic orchids (IMHOs) have highly reduced "dust seeds" which contain an embryo but no endosperm (see Chap. 5). Germination occurs when the embryo enlarges enough to rupture the testa, with the seedling at this stage known as a protocorm. Seeds of IMHOs may be stimulated to germinate by host or non-host fungi, or may germinate in the absence of fungi all together (Downie 1959; Hadley 1970; Warcup 1973; Rasmussen 1995); however, further protocorm growth is usually dependent on mycoheterotrophy. Protocorms of most terrestrial IMHOs are subterranean, non-photosynthetic, and unambiguously MH. While protocorms of many epiphytic and some terrestrial species are superficial and green, they usually cannot progress beyond germination without a fungal host (or exogenous supply of sugar). A small number of terrestrial species (e.g., in the genera *Disa* (Section *Disa*, *Bletilla*, and *Sobralia*) have been observed to germinate and develop to the leafy stage without host fungi (Burgeff 1959). However, it has been suggested that seedling development in some of these taxa occurs more rapidly with fungal symbionts than without.

Shoot production in symbiotically cultured seedlings frequently occurs within several months of germination (Warcup 1973; Muir 1989; reviewed in Rasmussen 1995), though the first leaf may not emerge until the second growing season in some temperate spp. (e.g., Zettler et al. 2001; Sharma et al. 2003); production of the first root usually occurs concurrently or shortly thereafter. Early field reports suggesting IMHO seedlings remain underground and MH for several years (e.g., Curtis 1943) were based on conjectural interpretation of seedling "growth segments"; as MH periods of similar duration have not been observed in symbiotic cultural studies, it would appear that such claims are exaggerated. Whether seedlings become fully autotrophic upon emergence of the first leaf or remain PMH for a period of time thereafter is not known. However, asymbiotic seedlings propagated following Knudson's (1922) protocol are commonly removed from sugar-enriched media when they have well-developed shoots and roots, apparently transitioning readily from partial/full heterotrophy to full autotrophy.

With very few exceptions (e.g., McCormick et al. 2004), host fungi of IMHO seedlings belong to *Ceratobasidium sensu lato* (incl. *Thanatephorus*), *Tulasnella*, and the Sebacinales (Dearnaley et al. 2012). Before Warcup and Talbot (1967) identified the perfect states of host fungi in culture, the identity of such fungi was hidden behind the veil of the morphologically defined, asexual genus *Rhizoctonia*, now known to be polyphyletic. Identification of host fungi has at times been further confounded by the propensity of IMHO seeds to germinate—and even for protocorms to undergo limited MH development—with non-host fungi, and of adult plants to allow peloton-formation of such fungi in their roots. These fungi may include rhizoctonia strains capable of hosting seedlings of other orchid species (Harvais and Hadley 1967), as well as Basidio- and Ascomycota not known to host seedlings of any orchid species (Currah et al. 1997; Vujanovic et al. 2000). Consequently, true host fungi of IMHOs are most appropriately identified as those supporting seedling development to the first leaf stage (e.g., Warcup 1973; Zettler and Hofer 1998).

Host specificity of IMHO seedlings is variable; while some IMHO species are highly host-specific (e.g., Wright et al. 2010; Phillips et al. 2011), compatibility with multiple strains of *Ceratobasidium* and/or *Tulasnella* is not uncommon (Hadley 1970). However, compatibility with *Tulasnella* and Sebacinales has only been observed in *Microtis* spp. (Warcup 1981; Milligan and Williams 1988; Bonnardeaux et al. 2007). *Ceratobasidium* and *Tulasnella*, by far the most common hosts of IMHOs, are thought to be predominantly SAP and/or pathogenic (Rasmussen 2002), though some lineages can be EM (Bidartondo et al. 2003; Yagame et al. 2012), and the ecology of these genera deserves further study. Sebacinales are known as hosts of IMHO seedlings within the Caladeniinae and occasionally the Prasophyllinae and Acianthinae (three closely related and predominantly Australian subtribes within the Diuridae). Sebacinales have been implicated in a wide variety of mycorrhizal and non-mycorrhizal (endophytic) interactions with plants (Selosse et al. 2009; Weiss et al. 2011), though notably, almost all such hosts of IMHOs belong to Sebacinales clade B, a group that is endophytic in many plants (Weiss et al. 2004; Selosse et al. 2009).

In symbiotic culture, growth of IMHO seedlings commonly occurs when C is available to host fungi in the form of starch or cellulose, and there is field evidence that organic matter may enhance germination in the presence of host fungi (McCormick et al. 2012). The extent to which C from living plants—acquired by orchid host fungi as pathogens or endophytes—may contribute to IMHO seedling development in nature is unknown. Unlike *Ceratobasidium* and *Tulasnella* spp., Sebacinales are often difficult to culture, suggesting relatively poor SAP capability. Nevertheless, they may consume dead plant tissues (Zuccaro et al. 2011) and are capable of supporting IMHO seedling development in vitro via utilization of starch (Warcup 1981; Ramsay et al. 1986; Bonnardeaux et al. 2007; Wright et al. 2009). Seedlings of *Microtis* spp., which are sometimes compatible with EM sebacinoid hosts (Warcup 1988), are the only known examples of EM-hosted IMHOs. However, given the recent discovery of EM *Ceratobasidium* and *Tulasnella* strains hosting fully and PMH orchids (Mursidawati 2004; Bougoure et al. 2009, 2010; Yagame et al. 2008b, 2012) and a fully MH liverwort (Bidartondo et al. 2003), respectively, it is possible that IMHO seedlings are hosted by EM fungi in additional genera.

8.4.2.2 Isotopic Evidence for Initial Mycoheterotrophy-Autotrophy in Orchidaceae

Cameron et al. (2006, 2008) demonstrated reciprocal transport of labeled C between adult, photosynthetic plants of *Goodyera repens* and their *Ceratobasidium* symbiont, with more than five times as much C transferred from plant to fungus as in the opposite direction. These findings suggest the direction of net C flow may reverse as seedlings transition from mycoheterotrophy to autotrophy. Although Cameron et al. did not confirm that the fungal symbiont associated with adult plants is capable of hosting MH development of conspecific seedlings, this has frequently been found in other studies of *Goodyera* spp. (Rasmussen 1995; McCormick et al. 2004). It should be noted, however, that the placement within the subfamily Orchidoideae of taxa capable of reciprocal C transfer is not sufficient to support that the trait is ancestral to the orchid family; isotopic data from taxa in the Apostasioideae, Vanilloideae, and Cypripedioideae are needed in order to determine the evolutionary polarity of this trait. Further, the repeated acquisition of EM hosts by partially and fully MH orchids, as stated above, suggests an innate ability to exploit C of fungi with which they had no apparent preexisting mutualism.

Natural abundance ^{13}C and ^{15}N data indicate that adult IMHOs may, as expected, exhibit ^{13}C and ^{15}N abundances equivalent to autotrophic reference plants (Fig. 8.3). More often, however, they are significantly depleted in ^{13}C and/or enriched in ^{15}N (Fig. 8.3). Hynson et al. (2009a) suggest that because adult individuals of *Goodyera* spp. exhibit such ^{13}C depletion (Fig. 8.3b), it may result from the net C transfer from plant to fungus, as documented in *G. repens* by Cameron et al. (2006, 2008). Given that ^{13}C depletion has been found in other spp. within the Orchideae (Liebel et al. 2010), it is possible that reciprocal C transfer occurs in additional taxa.

Why C transfer from orchids to fungi might result in ^{13}C depletion relative to autotrophic reference plants (which also participate in such transfer) is not known. It is possible that physiological differences in C transfer pathways between orchids and SAP *Ceratobasidium/Tulasnella* fungi and non-orchid reference plants and EM fungi result in different patterns of ^{13}C natural abundance in these groups of plants. The mechanism by which adults of some IMHOs are enriched in ^{15}N, typically to a level intermediate between autotrophic and MH reference plants (Fig. 8.3), is also unknown. Liebel et al. (2010) suggest such species may continue to obtain N via a pathway analogous to that in fully MH plants, resulting in ^{15}N enrichment, but that concurrent ^{13}C enrichment of plant tissue may be counteracted by simultaneous transfer of ^{13}C from plant to fungus. This suggestion is supported by the finding of significantly increased total N concentrations in many IMHOs in comparison to accompanying reference plants (Gebauer and Meyer 2003; Abadie et al. 2006; Liebel et al. 2010; see also Sect. 8.6.1).

While it has been assumed that cultivable, *Ceratobasidium/Tulasnella*-hosted orchid species are invariably IMH, Liebel et al. (2010) and Girlanda et al. (2011) found levels of natural ^{13}C enrichment consistent with partial mycoheterotrophy in adult plants of several such species (Fig. 8.3b). While these taxa represent a minority of the orchids investigated thus far, the discovery of their trophic status suggests the total number of IMHO species may be significantly smaller than previously expected. Further, the known cultivability of some of the study taxa suggests that partial mycoheterotrophy in the adults is facultative rather than obligate, in which case initial mycoheterotrophy-autotrophy and initial mycoheterotrophy-partial mycoheterotrophy may not represent mutually exclusive categories. Additionally, the association of these orchids with *Ceratobasidium* and *Tulasnella* suggest that partial mycoheterotrophy in adult orchids may not always be accompanied by a switch to EM hosts, though the ability to form mycorrhizal networks among the strains identified by Liebel et al. (2010) and Girlanda et al. (2011) is not yet known.

Among putatively IMHOs, natural abundance stable isotope analyses have largely been limited to species that are (1) hosted by *Ceratobasidium* and/or *Tulasnella*, (2) members of the Orchidoideae and Epidendroideae subfamilies, and (3) terrestrial. Given the utility of these analyses in confirming the trophic status of photosynthetic adults, they deserve to be more widely applied to species that are hosted by Sebacinales members of the Apostasioideae, Vanilloideae, and Cypripedioideae subfamilies; and/or epiphytic species. However, a potentially confounding factor for examining partial mycoheterotrophy among epiphytic orchids is that many of these species are either obligate or facultative CAM (crassulacean acid metabolism) plants (Neales and Hew 1975; Motomura et al. 2008; Silvera et al. 2010). Plants that rely on the CAM photosynthetic pathway are enriched in ^{13}C compared to those that utilize C3 photosynthesis. Thus, based on their C stable isotope profiles some epiphytic CAM orchids may, like PMH orchids, be enriched in ^{13}C even though they do not rely on fungi to meet their C demands. While analyses of IMHO seedlings have not yet been published, it would seem that these seedlings are likely to exhibit enrichment in ^{13}C and ^{15}N similar to fully MH plants. Enrichment in ^{13}C and ^{15}N compared to surrounding autotrophs has been observed in some adult orchids associated with *Ceratobasidium* and *Tulasnella* (Liebel et al. 2010; Girlanda et al. 2011), the most common hosts of IMHO seedlings.

8.4.2.3 Initial Mycoheterotrophy-Autotrophy in Pyroleae

The Pyroleae have dust seeds that, like most terrestrial orchids, germinate underground and develop into MH seedlings. While such subterranean, non-photosynthetic seedlings, made up of root-like organs, have been observed in culture (Lihnell 1942) and in the field (Irmisch 1855; Velenovsky 1892), the duration of initial mycoheterotrophy remains unknown. The few data available indicate that individual seedlings of two species of Pyroleae, *Pyrola chlorantha* and *Orthilia secunda,* are hosted by fungi in Sebacinales clade B (*sensu* Weiss et al. 2004), and a suite of EM fungi (N.A. Hynson unpublished; Smith and Read 2008). When investigating the fungal hosts to seedlings of *P. asarifolia* in Japan, Hashimoto et al. (2012) found a higher degree of host specificity for fungi only in Sebacinales clade B. However, seedlings of *Pyrola chlorantha* and *Orthilia secunda* have also been found to associate with ectomycorrhizal Sebacinales fungi from clade A (N.A. Hynson unpublished). It is surprising that single seedlings associate with single fungal hosts, and that some appear to be rather specific to non-EM fungi given that adult Pyroleae commonly associate with a diversity of EM fungi (Tedersoo et al. 2007; Zimmer et al. 2007; Vincenot et al. 2008; Hynson and Bruns 2009; Toftegaard et al. 2010).

With the exception of the fully MH *Pyrola aphylla* (Zimmer et al. 2007; Hynson et al. 2009b), many adult Pyroleae are leafy and primarily dependent on photosynthesis. Partial mycoheterotrophy in adult plants is frequently inconsistent between conspecific populations, with some individuals significantly more enriched in ^{13}C than autotrophic reference plants and others not (see Sect. 8.4.3; Tedersoo et al. 2007; Zimmer et al. 2007; Hynson et al. 2009b). Nevertheless, the populations that appear to be primarily dependent on photosynthetic C gains upon reaching adulthood, together with the survival of adult plants in cultivation (e.g., Hunt and Hope-Simpson 1990), suggests that some adult Pyroleae may be facultatively, if not consistently, autotrophic. Clearly, this group of plants deserves further investigation into the trophic status of adult individuals.

8.4.2.4 Initial Mycoheterotrophy-Autotrophy in Other Taxa

A number of taxa in the Lycopodiaceae, Psilotaceae, Ophioglossaceae, Schizaeaceae, and Gleicheniaceae have unambiguously MH gametophytes and preemergent sporophytes (Boullard 1979). With few exceptions, adult sporophytes are consistently photosynthetic, and many can be cultivated. It appears that many of these taxa may be IMH, although the trophic status of adult sporophytes under field conditions has yet to be investigated (see Sect. 8.4.3).

8.4.3 Initial Mycoheterotrophy-Partial Mycoheterotrophy

Contrasting with the previous scenario where adult plants are fully autotrophic, several species that turn green at adulthood after MH seedling development were discovered to remain PMH, i.e., maintain a C flow from the fungus to the plant over their whole lifespan. Although orchids were instrumental in the emergence of the concept, the

Table 8.2 Mean proportional C and N gains from fungi via mycoheterotrophy (Cd [%] and Ndf [%] ± 1 SD) by green orchids from the subfamily Epidendroideae and green pyroloids (Ericaceae) based on the data of all so far published literature (see Figs. 8.3 and 8.5) and on the linear two-source isotopic mixing model (Box 8.4) with fully EM-MH orchids or Ericaceae species used as the upper end point of the model representing 100% C and N gains via mycoheterotrophy, respectively

Species	Association with EM fungi	N_{df} (%)	C_{df} (%)	*n*
Orchidaceae				
Cephalanthera damasonium	+	*98 ± 21*	65 ± 24	21
C. erecta	?	56 ± 17	30 ± 6	3
C. longifolia	+	71 ± 21	38 ± 22	19
C. rubra	+	56 ± 8	16 ± 14	7
Corallorhiza trifida	+	53 ± 10	84 ± 12	9
Cymbidium goerginii	+	67 ± 33	49 ± 32	7
C. lancifolium	+	*97 ± 9*	55 ± 26	6
Epipactis atrorubens	+	*94 ± 30*	28 ± 19	11
E. distans	+	142 ± 32	*29 ± 23*	4
E. helleborine	+	*113 ± 38*	36 ± 33	18
Limodorum abortivum	+	129 ± 27	62 ± 6	10
L. trabutianum	+	*99 ± 30*	56 ± 13	5
Epipactis gigantea	–	27 ± 6	−14 ± 8	5
E. palustris	–	30 ± 3	*0 ± 7*	4
Liparis nervosa	–	18 ± 10	*21 ± 21*	3
Listera ovata	–	27 ± 16	*6 ± 20*	25
Ericaceae				
Chimaphila umbellata	+	61 ± 19	*0 ± 20*	36
Orthilia secunda	+	45 ± 16	33 ± 20	17
Pyrola chlorantha	+	67 ± 26	20 ± 18	17
P. minor	+	33 ± 13	*−4 ± 12*	9
P. picta	+	73 ± 13	*6 ± 23*	54
P. rotundifolia	+	68 ± 5	55 ± 7	6

Note: N_{df} data in *italics* are not significantly distinguished from 100%. C_{df} data in *italics* are not significantly distinguished from 0%

phenomenon is now suspected, and partly demonstrated, to be more widespread (Tedersoo et al. 2007; Zimmer et al. 2007; Cameron and Bolin 2010; reviewed in Selosse and Roy 2009).

8.4.3.1 Discovery of Partial Mycoheterotrophy in Adult Orchids

The suspicion of partial mycoheterotrophy came from two lines of observation in species of the Neottieae orchid tribe: unique C and N isotope natural abundances compared to autotrophic reference plants, and existence of achlorophyllous, albino individuals in otherwise green species. Gebauer and Meyer (2003) discovered unexpected isotope abundances in some forest orchids, with ^{13}C and ^{15}N abundances intermediate between those of autotrophic reference plants and full mycoheterotrophs from the same site. This was confirmed for additional European, North American and Asian species by several studies (Fig. 8.3a; Bidartondo et al. 2004; Julou et al. 2005; Abadie et al. 2006; Tedersoo et al. 2007; Zimmer et al. 2007; Liebel et al. 2010; Motomura et al. 2010). Stable isotope analyses from putative PMH plants have also been used to calculate these plants degree of heterotrophy (Table 8.2, Box 8.4).

Independent observations that some fully achlorophyllous individuals (= albinos), with colors ranging from white to pinkish due to anthocyanins (Fig. 8.4a, b), indicated partial

Box 8.4 The Linear Two-Source Mixing Model Approach

Stable isotope natural abundance data not only contain qualitative information about the C and N origin from different sources in the various types of MH plants, but are also a tool to estimate the proportions of C and N gained from different sources by PMH plants. Stable isotope natural abundance data and mixing model approaches have already frequently been used in a broad range of ecological field investigations to partition the origin of different nutrient sources utilized by various kinds of organisms (see Fry 2006). In our specific case we have to distinguish between two kinds of sources for C or N. For C the sources are quite obvious. PMH plants can gain C from photosynthesis and organic C compounds from their fungal hosts. For N the nature of the sources is less clear, because N uptake through the roots of terrestrial plants is usually mediated by mycorrhizal fungi irrespective of whether these plants are autotrophs or full or partial mycoheterotrophs. Nonetheless, it is obvious that there must be a distinction between N compounds gained by autotrophic and MH plants through their fungal associates. Otherwise, autotrophic and fully MH plants would not be distinguished by their N isotope signatures. Though the nature of N compounds utilized by autotrophic and MH plants is not fully understood, we can use their different N isotope signatures to estimate the contribution of either of these N fractions to cover the N demand of PMH plants. The fact that we have to consider only two kinds of sources provides the opportunity to use the most simple type of mixing model approaches, namely the linear two-source mixing model. This model requires as endpoints information about the isotopic composition of the two respective sources potentially utilized by a PMH plant and assumes that these sources are mixing in a linear manner in the tissue of the target plant. The mixing model approach in our case furthermore assumes (1) that the isotopic composition of the C source from photosynthesis and the isotopic composition of the N gained by autotrophic plants as one of the two endpoints are *represented* by the isotopic composition of fully autotrophic reference plants living in close spatial proximity and under identical micro climate conditions as the PMH target plant (=0% C or N gain of MH origin) and (2) that the isotopic composition of C and N gained from MH nutrition through the fungal source as the other endpoint is *represented* by fully MH orchids or Ericaceae, respectively (=100% C or N gain of mycoheterotrophic origin). According to this model the C and N isotope signature of PMH plants should range between the two endpoints and the proportional C and N gain from the source utilized by mycoheterotrophs can be calculated according to

$$\%x_{df} = \frac{(\delta x_{PMH} - -\delta x_{Ref})}{\varepsilon_{MH}} 100$$

with $\%x_{df}$ as percentage of C or N in the tissue of a PMH plant derived through MH nutrition from fungi, δx_{PMH} as the individual $\delta^{13}C$ or $\delta^{15}N$ value of a PMH plant, δx_{Ref} as the mean $\delta^{13}C$ or $\delta^{15}N$ value of autotrophic reference plants from a certain study site, and ε_{MH} as the mean enrichment factor of fully MH plants relative to obligate autotrophic reference plants from the same site (Gebauer and Meyer 2003).

mycoheterotrophy in some orchids. The surviving of such individuals in species that are normally green, especially from the Neottieae tribe, hinted at partial heterotrophy in these species. Albinos occur especially frequently in the genera *Epipactis* (Beau 1920; Renner 1938; Salmia 1986, 1989a, b; Selosse et al. 2004) and *Cephalanthera* (Renner 1938; Julou et al. 2005; Abadie et al. 2006; Stöckel et al. 2011). In some populations, the phenotype remains stable for green individuals and nearby albinos over many years (Renner 1938; Tranchida-Lombardo et al.

Fig. 8.4 (**a**) Albino *Cephalanthera rubra* (courtesy of J.-P. Amardeihl); see also Fig. 8.2b for albino *C. damasonium*. (**b**) Variegated individual of *Epipactis helleborine* (photo M.-A. Selosse). (**c**) Correlation between relative enrichments in ^{13}C (ε, see Box 8.3) and mean light availability for two PMH and one fully mycoheterotrophic (MH) orchid species and for autotrophic reference plants. Regression lines (±95% confidence intervals) represent the range of isotope signatures of autotrophic plants including the initially mycoheterotrophic (IMH) orchid *Cypripedium calceolus* (*green*), of the two PMH orchids *Cephalanthera damasonium* (*light red*) and *C. rubra* (*dark red*) and of the MH orchid *Neottia nidus-avis* (*blue*). *Arrows* indicate the variable proportions of C derived from fungi (C_{df}) or photosynthesis (C_{dp}), respectively (reproduced with permission from Preiss et al. 2010)

2010), up to 14 years for albinos (Abadie et al. 2006), while in others, variegated individuals or reversion to green shoots can occur (Salmia 1986; Stöckel et al. 2011; Fig. 8.4a, b). Although they tend to perform less well than green individuals (as stated above, see Sect. 8.4.1), some albinos form flowers and fruits (Salmia 1986, 1989a, b; Julou et al. 2005; Tranchida-Lombardo et al. 2010). Albinos were suggested to depend on their mycorrhizal fungi for C nutrition (Selosse et al. 2004): Beau (1920), having observed albino *Epipactis* and *Cephalanthera* spp. nearly one century ago, wrote that "the exact complementation of the photosynthetic function by the symbiosis permits us to understand how green orchids can exceptionally grow and flower in more or less etiolating conditions" (see also Renner 1938). Indeed, most of these species tend to inhabit shaded forest sites. Albinos' mycoheterotrophy is now further corroborated by the demonstration of their low chlorophyll content and lack of CO_2 absorption in the light (Fig. 8.4b; Julou et al. 2005); congruently, they display ^{13}C enrichment similar to that of fully MH plants (Fig. 8.1; Julou et al. 2005; Abadie et al. 2006). This supported the likelihood of partial mycoheterotrophy in green conspecifics (Selosse et al. 2004; Julou et al. 2005). Accordingly, survival of albinos is also reported from green parasitic plants such as *Striga hermonthica* (Press et al. 1991) that use other plants' sap to support part of their C needs (see Sect. 8.4.1 and Table 8.1).

Moreover, some green orchids were found to have high ^{13}C abundance, which correlates with the potential for mycoheterotrophy via their EM fungal partners (Bidartondo et al. 2004; see also Dearnaley et al. 2012, for review). These orchid species also have a trend to low or no colonization by rhizoctonias. However, many orchids from more or less open environments associate with rhizoctonias mainly but occasionally display EM fungi, e.g. *Cypripedium* (Shefferson et al. 2005, 2007), *Gymnadenia* (Stark et al. 2009), or *Orchis* (Liebel et al. 2010; Lievens et al. 2010; Girlanda et al. 2011). These associates were likely hidden in studies based on fungal cultivation, because EM fungi do not grow easily in culture, and may even be discarded as "molecular scraps" in some molecular studies (Selosse et al. 2010), but, as suggested by Girlanda et al. (2011), they could allow a C flow to the plant, at least in some light environments.

PMH Neottieae display various levels of specificity to non-rhizoctonia fungi that are known to form EM associations with forest trees: most *Epipactis* species show a preference for Pezizomycetes related to truffles, sometimes with additional fungi (Bidartondo et al. 2004; Selosse et al. 2004; Ouanphanivanh et al. 2008; Ogura-Tsujita and Yukawa 2008; Shefferson et al. 2008; Liebel et al. 2010); *Cephalanthera* species display a large fungal spectrum including Cortinariaceae, Hymenogastraceae and Thelephoraceae (Bidartondo et al. 2004; Julou et al. 2005; Abadie et al. 2006; Matsuda et al. 2008; Yamato and Iwase 2008) whereas *Russula* are specific associates of *Limodorum* species (Girlanda et al. 2006; Liebel et al. 2010). This is much reminiscent of the EM fungal associates in many fully MH orchids (see Sect. 8.3.1). A similar feature occurs in Japanese *Cymbidium* species (Motomura et al. 2010; Ogura-Tsujita et al. 2012). The MH *C. macrorhizon* and *C. aberrans* exclusively associate with EM taxa (Russulaceae, Thelephoraceae, and Sebacinaceae), the green *C. lancifolium* and *C. goeringii*, which have ^{13}C abundances indicative of partial mycoheterotrophy (Fig. 8.3a), associate simultaneously with EM taxa and Tulasnellaceae. However, there is evidence that Ceratobasidiaceae species alone can support the ex situ growth of *C. goeringii*, although the heterotrophy level of these individuals and ecology of these fungi remain unknown Wu et al. 2010). The pattern is further complicated by the fact that some orchids associated with potentially EM taxa have not demonstrated partial mycoheterotrophy (Waterman et al. 2011), and that a few rhizoctonia lineages within the Sebacinales (Selosse et al. 2004), the Tulasnellaceae (Bidartondo et al. 2003) or the Ceratobasidiaceae (Bougoure et al. 2009) have EM abilities. Indeed, EM *Ceratobasidium* associate with the PMH *Plantanthera minor* and autotrophic trees such as *Pinus densiflora* (Yagame et al. 2008b, 2012). Surrounding trees are thus likely to be the ultimate C source of all known PMH orchids, and the availability of EM fungi could be a limitation for the geographic distribution of partial mycoheterotrophy (Liebel et al. 2010). The reason why non-EM rhizoctonia, which support MH seedling development in green orchids, only support minor MH gains in adult orchids (Liebel et al. 2010; Girlanda et al. 2011) remains unclear: one interpretation is that they may not be capable of transferring a sufficient amount of C to support adult orchid individualss (Taylor and Bruns 1997; Martos et al. 2009), but there is no direct evidence for this.

Experiments carried out by Sadovsky (1965), at a time European protection laws did not forbid destructive manipulation of native orchids, and in repeated attempts to transplant various orchids, he listed some species that did not survive the process. The list interestingly mixes full mycoheterotrophs (such as *Neottia nidus-avis*) and PMH species (*Corallorhiza trifida*, *Cephalanthera*, *Limodorum*, and *Epipactis* spp.), suggesting that in these cases, disconnecting the fungus from its resources (=nearby mycorrhizal tree roots) entailed plant death.

Within the Epidendroideae (Fig. 8.3a), partial mycoheterotrophy is not limited to the Neottieae. It also occurs in the genus *Cymbidium* (*tribe* Cymbidieae) as stated above, and in the interesting case of *Corallorhiza trifida* (tribe Calypsoeae). Despite being leafless and having only chlorophyllous stems and seed capsules *Corallorhiza trifida*, in contrast to all other species of its genus (Freudenstein and Doyle 1994), is not fully MH upon reaching adulthood (Zimmer et al. 2008; see also Sect. 8.5.1). PMH species with apparently low heterotrophy levels were found among Mediterranean and Macaronesian orchids from the tribe Orchideae (such as *Barlia robertiana*, *Gennaria diphylla*, *Habenaria tridactylites*, and *Orchis purpurea*; Fig. 8.3b; Liebel et al. 2010; Girlanda et al. 2011). Interestingly, *G. diphylla* is associated with EM fungi (Liebel et al. 2010), and it is possible that some Tulasnellaceae and Ceratobasidiaceae found in association with other orchid taxa (Liebel et al. 2010; Girlanda et al. 2011) are EM as well.

The situation is more complicated in Orchidoideae (Fig. 8.3b), where some green species show ^{13}C depletion compared to surrounding autotrophs (Hynson et al. 2009a). This is especially pronounced for *Goodyera repens* (Fig. 8.3b), which is demonstrated to be autotrophic and to transfer C to its fungal associate (Cameron et al. 2008). Finding a relevant autotrophic baseline in terms of ^{13}C abundance is thus difficult for Orchidoideae phylogenetically related to *Goodyera* spp.: using the surrounding green plants that are enriched in ^{13}C relative to autotrophic *Goodyera* spp. may underestimate the heterotrophy level of adult PMH plants in some Orchidoideae. Nevertheless, even using such a baseline, partial mycoheterotrophy was successfully demonstrated in the Japanese *Platanthera minor* (Yagame et al. 2012), and in *Cheirostylis montana* from Thailand (Roy et al. 2009a; Fig. 8.3b), suggesting the need for more investigations to find PMH Orchidoideae within North America and Europe. Thus, many, if not all, orchid taxa appear predisposed to partial mycoheterotrophy (as well as to full mycoheterotrophy), perhaps because of their required MH germination.

We currently have no data on the impact of adult PMH plants on their associated fungi (exact load of the C loss, existence of any reward such as nutrients and/or protection). If one assumes, as suggested by C flow, that the plant is parasitic on the fungus, then some conflicts may occur and fungi may undergo selection to avoid this parasitism. In initial mycoheterotrophy-autotrophy, the interaction maybe positively selected on the fungal side, since investing C in a seedling may later allow it to recover C from the adult roots (see below) and thus compensate the C flow to germinating seedlings. Thus, one may speculate that initial mycoheterotrophy-autotrophy systems are more evolutionarily stable, which is corroborated by this strategy being more ancient in orchids than initial mycoheterotrophy-partial mycoheterotrophy. However, one might also speculate that initial mycoheterotrophy-autotrophy is more evolutionarily stable because the plant—by being MH only at the seedling stage—limits the duration and magnitude of the parasitism, thereby reducing the selective pressure on host fungi to evolve resistance.

8.4.3.2 Partial Mycoheterotrophy in Adult Pyroloids

Until recently, orchids have been the primary models for investigations of partial mycoheterotrophy, but this nutritional mode is more widespread, and some Ericaceae were essential in demonstrating that this strategy had evolved convergently in other plant lineages. Pyroloids, for instance, are small shade-tolerant perennial shrubs, often with impressive underground rhizome networks, and grow in temperate, alpine and boreal forests. They are candidates for partial mycoheterotrophy for three reasons: (1) they have a close, although not fully resolved, phylogenetic relationship with the MH Monotropeae and Pterosporeae (all members of the Monotropoideae are MH; Kron et al. 2002; Tedersoo et al. 2007; Matsuda et al. 2012). As stated in Sect. 8.3.1, one species of Pyroleae, *Pyrola aphylla*, is even fully MH (Fig. 8.1; Zimmer et al. 2007; Hynson et al. 2009b), although its recognition as a separate species, or simply a leafless variant of the green *P. picta*, remains unclear (Camp 1940; Haber 1987; Freudenstein 1999). (2) As in orchids, seeds are very small, with few C reserves (Eriksson and Kainulainen 2011), and pyroloids undergo an MH subterranean germination (Christoph 1921; Lihnell 1942) where various fungi have been shown to provide C nutrition (Smith and Read 2008). (3) As in many other basal Ericaceae (Selosse et al. 2007), pyroloids form EM-like arbutoid associations: the fungus often forms a sheath around the root and a Hartig net, but in addition, hyphal coils penetrate root cells (Khan 1972; Robertson and Robertson 1985; Vincenot et al. 2008; Hynson and Bruns 2009). The fungal taxa involved are also capable of forming typical EM associations with surrounding tree roots (Robertson and Robertson 1985; Tedersoo et al. 2007; Zimmer et al. 2007; Vincenot et al. 2008; Hynson and Bruns 2009; Toftegaard et al. 2010; Matsuda et al. 2012), thus offering a link to autotrophic trees.

Investigations on stable isotope abundances of leafy Pyroleae species have provided variable results. An investigation in two boreal Estonian forests revealed that some pyroloids had higher ^{13}C and ^{15}N abundances and leaf N concentrations than surrounding autotrophs, including other plants from Ericaceae (Fig. 8.5; Tedersoo et al. 2007). Based on a linear mixing model using *Hypopitys monotropa* as the fully MH end-member (Box 8.4), 10–68% of C was of fungal origin in *Orthilia secunda*, *Pyrola chlorantha*, *P. rotundifolia*, and *Chimaphila umbellata*. However, the latter species did not differ from autotrophs in one of the investigated sites, and the levels of conspecific heterotrophy did not correlate among sites, indicating a complex regulation of this parameter among species. In a second set of investigations in more southern sites from Germany and California, (Fig. 8.5; Zimmer et al. 2007; Hynson et al. 2009b), ^{15}N abundance was intermediate between autotrophic and MH plants for *O. secunda*, *C. umbellata*, *P. chlorantha*, *P. minor*, and *P. picta*; however, based on ^{13}C abundance, only *O. secunda* showed significant C gain via mycoheterotrophy at a single low irradiance site. Given the broad range of irradiances in investigated sites, light availability is unlikely to be the only driver of the heterotrophy level in the study by Zimmer et al. (2007). Conversely, Matsuda et al. (2012) demonstrated that the Japanese *P. japonica* derived 50% of it's C from fungi, and that individuals growing in most shaded forest microsites and in darker months (due to canopy closure) tended to display higher ^{15}N and ^{13}C abundances, and higher leaf N concentrations. This was linked to a more specific association to *Russula* spp. in these conditions, a feature speculated to allow a better supply of fungal C. In a recent study by Hynson et al. (2012), the first truly experimental study on PMH in pyroloids, light and access to mycorrhizal networks, was manipulated in the field for two species of Pyroleae, *P. picta* and *C. umbellata*. The C stable isotope values from these species' leaf soluble sugars were then used to track changes in their relative dependency on MH C gains over the course of a growing season. The major findings of this study were that (1) *P. picta* and

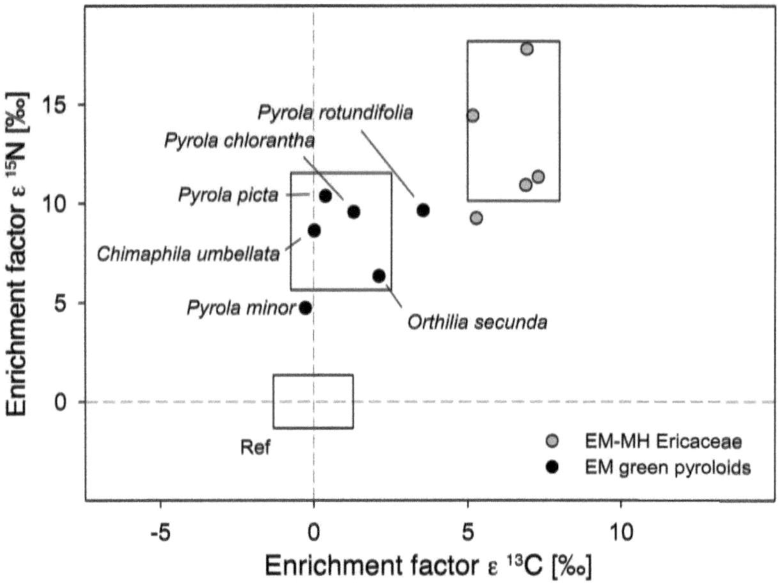

Fig. 8.5 Mean enrichment factors (ε, see Box 8.3) for ^{13}C and ^{15}N of six green species belonging to the tribe Pyroleae within the subfamily Monotropoideae of the Ericaceae (*black symbols*) and five fully mycoheterotrophic (MH) species belonging to three different tribes within the subfamily Monotropoideae of the Ericaceae (*grey symbols*, for further details see Fig. 8.1) and of autotrophic reference plants (Ref, $n=392$) collected together with each of the respective target species. The *boxes* represent one SD of the mean ε values for the different groups of green and fully MH Monotropoideae and for the autotrophic reference plants. Number of replicates for the six target species as following: *Chimaphila umbellata* ($n=36$), *Orthilia secunda* ($n=38$), *Pyrola chlorantha* ($n=17$), *P. minor* ($n=9$), *P. picta* ($n=54$), *P. rotundifolia* ($n=6$). Data compiled from Tedersoo et al. (2007), Zimmer et al. (2007), Hynson et al. (2009b) and Liebel et al. (2009)

C. umbellata respond differently to a decrease in light availability: due to an increase in ^{13}C abundance over the course of the experiment *P. picta* showed signs of partial mycoheterotrophy, whereas *C. umbellata* did not. (2) Assaying the leaf soluble sugars from the two target species rather than bulk tissue revealed subtle changes along the autotrophy-mycoheterotrophy continuum in these species that would have otherwise gone undetected. And (3), the relative dependency of *P. picta* compared to *C. umbellata* on MH C gains does not appear to be solely driven by light availability.

In the cases where light levels do not drive the heterotrophy level, it was suggested that the C gains were not solely based on C needs, but could arise as a "side-product" of the N and P nutrition (Selosse and Roy 2009; see stage #3 in Table 8.1). Some fungal C could move along with organic forms of N, which is the case in mixotrophic algae where C gains arise as a consequence of mineral nutrients from prey (see Sect. 8.4.1). Indeed, a specific strategy to obtain fungal organic N in pyroloids may also explain their unusual ^{15}N abundance (Fig. 8.5), as well as their elimination after anthropogenic N deposition (Allen et al. 2007). A study of two species (*O. secunda* and *P. asarifolia*) in Canada confirmed high ^{15}N abundance and leaf N concentrations (Kranabetter and MacKenzie 2010), but the responses of these parameters to an edaphic gradient of N availability strongly suggested difference in N metabolism, not only as compared to other plants, but also between these two species. Much remains to be studied on N and C acquisition and metabolism in pyroloids. Interestingly, no evidence for lysis or degradation of intracellular hyphae is reported in living root cells of pyroloids (Tedersoo et al. 2007; Vincenot et al. 2008), adding to the idea, against previous postulates, that hyphal lysis may not be necessary for C transfer.

Furthermore, a ^{14}C labeling experiment was claimed to show transfer from autotrophs to pyroloids, in dual pot cultures of *Larix kaempferi* seedlings and *Pyrola incarnata* from Japan (Kunishi et al. 2004; Hashimoto et al. 2005), but this experiment and its controls remain unpublished. Carbon gains from fungi may explain other ecophysiological features of pyroloids, such as the few C reserves in winter (the fungal C may contribute to development of new organs in spring for *P. incarnata*; Isogai et al. 2003), the low capacity for adjusting vegetative growth and leaf area after shading (in *P. rotundifolia* at least; Hunt and Hope-Simpson 1990), or the sensitivity to forest disturbances that affect mycorrhizal networks (logging, burning; Zimmer et al. 2007). However, at least some green pyroloids survive, at least temporarily, outplanting to glasshouse conditions (e.g., Hunt and Hope-Simpson 1990). In contrast with PMH orchids, the use of fungal C in pyroloids may be more facultative, allowing disconnections from mycorrhizal network in some conditions; If one assumes that populations display variable level of dependency on mycorrhizal networks, this may also resolve the discrepancies between studies reporting variable levels of heterotrophy (Tedersoo et al. 2007; Zimmer et al. 2007; Hynson et al. 2009b).

8.4.3.3 Partial Mycoheterotrophy in Plants Associated with Arbuscular Mycorrhizal Fungi

The two lineages investigated above, orchids and Pyroleae, present two unusual features among MH lineages: (1) they associate with Asco- or Basidiomycota (EM or SAP) and (2) are primarily from temperate ecosystems. Yet, the majority of fully MH species are from tropical regions, and associate with the obligatorily biotrophic AM fungi (Leake 1994, 2004). Is there some evidence of partial mycoheterotrophy in these or other families mycorrhizal with Glomeromycota? Indeed, competition for light is high in wet tropical forests, and may select for partial mycoheterotrophy. In these conditions, some families contain both green and MH species, such as Burmanniaceae (Merckx et al. 2006, 2010), Gentianaceae (Struwe et al. 2002; Cameron and Bolin 2010), Polygalaceae (Imhof 2008), Iridaceae (Reeves et al. 2001), Pandanales (Rudall and Bateman 2006), and Petrosaviaceae (Cameron et al. 2003). Also, some green species have reduced leaves that suggest that photosynthesis is not their sole C source (Cameron and Bolin 2010).

There are unique challenges to assessing partial mycoheterotrophy in AM plants. For instance, it is difficult to sample the tiny, submillimetric spores of individual AM fungi (that do not form conspicuous fruitbodies) for stable isotope analyses, although this has sometimes been done (Allen and Allen 1990; Nakano et al. 1999; Courty et al. 2011). Also, ^{15}N abundances and N concentrations are similar in AM-MH plants and nearby green plants (Merckx et al. 2010; Courty et al. 2011; see Sect. 8.3.3): this suggests similar N nutrition and forbids any specific detection of mixotrophy on these parameters only—at least in Burmanniaceae and Gentianaceae. Finally, Courty et al. (2011) found few or no difference between canopy leaves, Glomeromycota spores, and AM-MH plants, so that specific ^{13}C enrichment is not generally expected for AM-partial mycoheterotrophy. Only in dense forests, light-starved understory plants may display some differences with AM-partial mycoheterotrophy (see Sect. 8.3.3). In more open conditions (field or canopy-open forest), similar ^{13}C enrichment in potential AM-PMH and surrounding autotrophic plants can be expected. Moreover, in some non-forest ecosystems, the occurrence of C4 plants as partners of AM fungi may further complicate the pattern, as these have lower ^{13}C depletion than the C3 plants.

Evidence for partial mycoheterotrophy among AM-MH plants, although expected (e.g., Selosse and Roy 2009), thus remains unclear. Investigating two Gentianaceae species with reduced leaves from a Virginia hardwood forest, Cameron and Bolin (2010) found no ^{13}C enrichment, but higher leaf N concentrations in *Bartonia virginica* and higher ^{15}N abundance in *Obolaria virginica* as compared to nearby autotrophs. Given the statements above, it is difficult to conclude whether this reflects some heterotrophy or some physiological particularity of N nutrition. Investigating the potential candidate for AM-partial mycoheterotrophy

Burmannia capitata from French Guiana forest gaps, Merckx et al. (2010) showed no enrichment in ^{13}C as compared to nearby herbaceous plants. Moreover, this species and three other from the same genus were successfully grown from seeds in isolated pots under controlled conditions (i.e., no mycorrhizal network). Thus, partial mycoheterotrophy is lacking, or at least facultative in these cases.

Green relatives of fully AM-MH plants could provide promising targets for future mechanistic studies on mycoheterotrophy. If partial mycoheterotrophy is found among AM plants, these plants target the most widespread mycorrhizal symbiosis, and could prove ideal for further studies of the ecophysiology of MH plants. Field and laboratory studies of the tripartite symbiosis between autotrophic host plant, mycorrhizal fungus and mycoheterotroph that have been difficult with EM mycoheterotrophs (but see Yagame et al. 2012) may receive new attention among AM mycoheterotrophs. Although many AM fungi cannot be grown axenically in culture, the putative autotrophic C source for AM mycoheterotrophs are relatively smaller than the trees that host EM fungi, and are possibly annuals which can be grown in combination with their AM mutualists. Therefore, these systems lend themselves to both in situ and laboratory experiments that follow the fate of C as it is passed from the autotrophic host, onto its associated fungus and finally the mycoheterotroph. Owing to these factors and, the plasticity of specificity among AM mycoheterotrophs, the physiological pathways and biochemical signaling between mycoheterotrophs and fungi may finally gain traction in the field of MH research.

8.5 Challenges in the Study of Partial Mycoheterotrophy

8.5.1 Gas Exchange Measurements and Photosynthesis in Partially Mycoheterotrophic Plants

Investigations of photosynthesis and in situ gas exchange support partial heterotrophy for some green orchids. CO_2 exchanges in *C. damasonium* revealed that, as expected from their phenotype, albinos were fully heterotrophic and did not respond to light, while green individuals exhibited a normal photosynthetic response to light (Fig. 8.4b; Julou et al. 2005). After in situ ^{13}C labeling, *C. damasonium* showed reduced CO_2 assimilation compared to surrounding green plants, while *C. trifida* showed nearly no assimilation, close to the level of the fully MH control (Cameron et al. 2009). This was in agreement with chlorophyll content and fluorescence values (reflecting proper assembly of pigments into photosystems): these parameters suggested subnormal and very reduced potential for photosynthesis in *C. damasonium* and *C. trifida* respectively (Fig. 8.6; Cameron et al. 2009). In *C. trifida*, the chlorophyll content is only 1% of that of *C. damasonium* (Zimmer et al. 2008). However, Montfort and Küsters (1940), measuring CO_2 exchange in inflorescences and maturing infrutescences, found that CO_2 assimilation was 2.2 times higher than respiration—whether differences in technology or plant tissue origin explains this remains unclear. Intrinsic photosynthetic limitations also exist in *Limodorum abortivum*, where photosynthetic abilities do not compensate respiration even in full light (Girlanda et al. 2006). In some *C. damasonium* populations, there is evidence that low light conditions force the plant to survive near its compensation point, where C losses through respiration are equal to C gains from photosynthesis (Julou et al. 2005). Based on variegated albinos, *C. damasonium* individuals with a mosaic of chlorophyllous and achlorophyllous tissues, Stöckel et al. (2011) found a positive correlation between leaf chlorophyll concentrations and degree of mycoheterotrophy, as shown by ^{13}C abundance. However, the recent finding that some meadow Mediterranean orchids, living in places devoid of canopy, are PMH (Girlanda et al. 2011) demonstrates that light limitation is not the sole driver of this nutritional strategy. Thus, both intrinsic and environmental factors determine partial mycoheterotrophy, depending on the species and site. Partial mycoheterotrophy may also allow buffering against herbivory or shading damage: in a defoliation and shading experiment involving the autotrophic

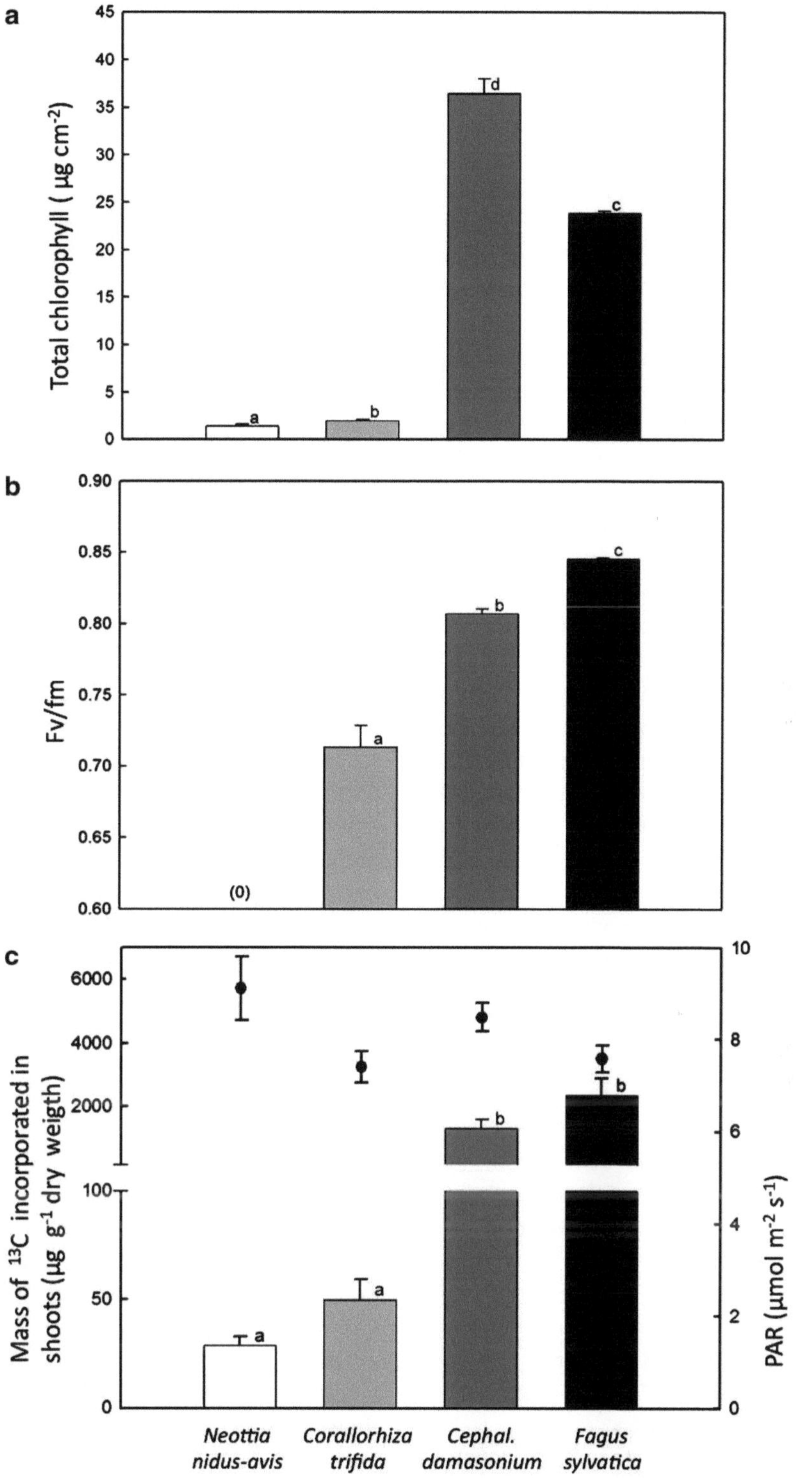

Fig. 8.6 A comparison of photosynthetic abilities between *Neottia nidus-avis* (mycoheterotroph) and *Corallorhiza trifida* (leafless partial mycoheterotroph) and the leaves of *Cephalanthera damasonium* (partial mycoheterotroph) and *Fagus sylvatica* (autotroph), reproduced with permission from Cameron et al. (2009). (**a**) Total chlorophyll content as a function of surface area of the organ (stems for *Corallorhiza trifida*, which is leafless, and leaves in all other cases). (**b**) Maximum quantum yield (Fv/Fm). (**c**) Total amount of ^{13}C present in plant shoots after a 4-h exposure to a $^{13}CO_2$ atmosphere; mean photosynthetically active radiation (PAR) is given above each bar (values on the *right*). In each panel, bars with differing letters are significantly different

Cypripedium calceolus and the PMH *C. longifolia*, Shefferson et al. (2006) demonstrated that both treatments led to significant declines in vegetative and reproductive vigor of *C. calceolus*, but had more limited impacts in *C. longifolia.*

Measurements of photosynthetic activity and stable isotope natural abundance are both powerful tools for investigating partial mycoheterotrophy. However, the information gained from these techniques differs. Gas exchange measurements provide snapshot information about photosynthetic activity, while stable isotope natural abundance data integrate the sources of C gain over the entire life history of a plant or plant organ (Liebel et al. 2010). This difference may explain the diverging results for *C. trifida*, where isotopic abundances suggest partial mycoheterotrophy (Zimmer et al. 2008) whereas CO_2 fixation was detected (Montfort and Küsters 1940) or not (Cameron et al. 2009). However, one advantage of stable isotope analyses is that they can be used in a quantitative manner way to compare levels of heterotrophy: the percent of C derived from the fungi can be estimated from a linear mixing model (Box 8.4, Table 8.2, but see 8.5.2 for considerations when applying linear mixing models to the study of PMH). Not unexpectedly, this reveals a continuum from autotrophy to full mycoheterotrophy (as also suggested in Fig. 8.3). Moreover, a given species shows a variable heterotrophy level from one site to another: for the well-studied *C. damasonium*, this level ranges for green individuals from 33% in an open, sunny pine forest to 85% in a dark beech forest (Gebauer and Meyer 2003; Bidartondo et al. 2004; Gebauer 2005).

8.5.2 Considerations for the Application of Linear Isotope Mixing Models to Mycoheterotrophic Food Webs

With the revelation of partial mycoheterotrophy, two primary challenges for researchers have been (1) determining the degree of dependence on fungal C gains in these plants and (2) elucidating the factors that select for partial mycoheterotrophy in nature. Attempts to quantify the percent fungal C and N gains in PMH plants have been made through the application of an isotope mixing model (Box 8.4; Gebauer and Meyer 2003; Bidartondo et al. 2004; Tedersoo et al. 2007; Zimmer et al. 2007). The two end-members of this linear mixing model are the ^{13}C and ^{15}N isotope abundances of fully MH plants that theoretically receive all of their C and mineral nutrients from fungi, and the ^{13}C and ^{15}N isotope abundances of autotrophic plants, which are theoretically not receiving any C from their associated mycorrhizal fungi, but meeting basically all of their N demands via their mycorrhizal fungi (Gebauer and Meyer 2003). However, while this model generates estimates of the potential dependency on fungal derived C, it also involves numerous assumptions about the isotopic behavior of MH plants, many of which are not yet fully understood. For instance, it is unknown if the relationship between the ^{13}C isotope signatures of partial mycoheterotrophs and autotrophs across environmental gradients is linear. Most PMH orchids and some pyroloids have been found living in low-light environments (<20 μmol photons m^{-2} s^{-1}; Gebauer and Meyer 2003; Bidartondo et al. 2004; Julou et al. 2005; Abadie et al. 2006; Zimmer et al. 2007; Liebel et al. 2010). Under low light conditions, fully autotrophic plants in the understory would theoretically become more depleted in ^{13}C. This depletion is owed to a combination of biochemical and leaf-level processes such as a decrease in photosynthetic rate, along with an increase in C_i/C_a (the ratio of the CO_2 concentration inside the leaf to the atmosphere outside the leaf) which would lead to greater discrimination against $^{13}CO_2$ by the primary caboxylating enzyme for C3 photosynthesis, Rubisco (Farquhar et al. 1989). In contrast, PMH often have significant increases in δ13C in low light conditions (Zimmer et al. 2007; Liebel et al. 2010; Preiss et al. 2010; Hynson et al. 2012; Fig. 8.4b). Thus the assumption that the $\delta^{13}C$ of target species and the autotrophic end-members of the mixing model are linearly related may be invalid, especially if they are from different light environments or have differing rates of photosynthesis (Fig. 8.2b). In fact, PMH species are likely have lower photosynthesis rates than autotrophic plants, resulting in better equilibration of $^{13}CO_2$ concentration between environmental air and stomatal chamber, therefore entailing increased ^{13}C discrimination (= lower ^{13}C abundance). They may also use respired CO_2 in relatively larger amounts as compared to plants with higher CO_2 need; and this respiratory CO_2 is more depleted in ^{13}C than atmospheric CO_2. Thus the photosynthetic contributions to PMH's δ13C may be more depleted in ^{13}C than surrounding autotrophic plants. The linear relationship between leaf chlorophyll concentration (= photosynthetic capacity) and $\delta^{13}C$ observed in the variegated leaves of *Cephalanthera damasonium* (Stöckel et al. 2011) suggests that these mechanisms may have limited effects. However, it should be kept in mind that CO_2 equilibrium effects and respiratory effects may increase ^{13}C depletion in leaves, thus potentially reducing the apparent heterotrophy level of some PMH plants.

Furthermore, the factors that lead to interspecific differences in the $\delta^{13}C$ and $\delta^{15}N$ of fully MH plants are unclear (Fig. 8.1; Kranabetter and MacKenzie 2010). Therefore, the species of fully MH plants used as end-members in the mixing model can greatly affect the estimates of percent C or N gains via mycoheterotrophy. For example, in the case of the MH *Pyrola aphylla* (Ericaceae), if the percent C and N gains via fungi are calculated (Hynson et al. 2009b) using the mixing model approach with an MH end-member based on the mean relative ^{13}C enrichment of seven fully MH plants associated with EM fungi (Table 8.2, Preiss and Gebauer 2008) the estimated percent C derived from fungi would be 96±12%. This indicates that *P. aphylla* essentially gains all of its C from a source that is similar to other fully MH plants associated with EM fungi. In contrast, if the same mixing model is used to calculate percent N gain via mycoheterotrophy, *P. aphylla* would gain over 100% (149±18%) of its N from the source(s) utilized by the MH end-members. Thus, the MH species used as end-members in this scenario do not fully represent the extent of variability in ^{15}N signatures of mycoheterotrophs. However, from the meta-analysis of MH plant enrichment factors (Fig. 8.1), we have found significant differences in C and N stable isotope enrichment between EM-MH Orchidaceae and Ericaceae. In order to improve the resolution of the mixing model approach, we recommend that for future quantifications of mycoheterotrophy among orchidaceous and ericaceous species the MH end-member of the mixing model should be chosen based on the test species' respective plant family.

Other potential issues with using a mixing model approach to quantify the degree of mycoheterotrophy are the types of compounds that are being analyzed for their stable isotope abundances. To date, all mixing model calculations of partial mycoheterotrophy except a single study by Hynson et al. (2012), have been based on structural (bulk) leaf or stem tissue isotope signatures. Because these values represent a time-integrated measurement of C and other nutrient gains over the lifespan of a leaf, they may mask short-term fluctuations of mycoheterotrophy with organ or plant developmental stage, or with season. This factor may be particularly important for long-lived evergreen species such as Pyroleae species or orchids that have long belowground dormancy periods. For example, in the latter case, when the dormancy period ends and aboveground shoots are formed, the C for building these tissues will be derived from a heterotrophic source—either fungal or the mobilization of stored C such as starch. Both these scenarios would lead to ^{13}C enrichment in newly formed orchid leaves, potentially leading to erroneous conclusions regarding the MH status of the plants at time of observation (although the conclusion is exact regarding the C origin). The only study to date to have targeted compounds other than bulk C and N is by Hynson et al. (2012) where the leaf soluble sugars of two Pyroleae species were used to track changes in relative levels of mycoheterotrophy throughout a single growing season. Though the identities of the transfer compounds from fungi to MH plants remain unknown, sugars are reasonable candidates. Sugars are the major transport materials in biological systems, including the mycorrhizal symbiosis, due to their high energy and chemical stability (Smith and Read 2008). The results of the work by Hynson et al. (2012) revealed that indeed sugars were a sensitive assay for shifts of C assimilation over short time periods, whereas bulk C gave an average value for the global origin of the C in the investigated plants' biomass.

8.5.3 Cryptic Mycoheterotrophy

One additional gap remaining in the current literature are labeling experiments directly demonstrating the C transfer from plant to (1) fungus and (2) PMH orchid: thus, the fact that nearby trees are the ultimate C source remains to be formally demonstrated. There are two indirect additional sources of evidence for the use of mycorrhizal fungi as trophic source: ^{15}N abundance and N concentrations in plant tissues (N content, best expressed as mmol N gdw^{-1}, also expressed as C/N ratios in some papers) are high in PMH orchids as compared to surrounding autotrophs (Gebauer and Meyer 2003; Abadie et al. 2006; Tedersoo et al. 2007; Selosse and Roy 2009; Liebel et al. 2010). However, these values are lower than in fully MH plants, supporting a mixotrophic strategy (see also Sect. 8.6.1). Stöckel et al. (2011) demonstrated a linear relationship between MH C gain and leaf N concentration in *Cephalanthera damasonium* populations with albinos displaying various level of variegation with green tissues. The same studies also demonstrate that some orchids considered as autotrophic based on their ^{13}C abundance exhibit higher (although less extreme) ^{15}N abundance (Fig. 8.3) and N concentrations than other autotrophs, again raising the possibility of a cryptic partial mycoheterotrophy.

8.6 Future Directions and Unanswered Questions

8.6.1 Nitrogen

Soil N availability is for many plants in natural terrestrial habitats a growth-limiting factor (e.g., Marschner 1995). Therefore, plants repeatedly developed strategies to gain access to N sources alternative to soil-born mineral N. For example, legumes through their symbiosis with N_2-fixing bacteria use atmospheric N and carnivorous plants use their preys as alternative N sources. The N availability to plants is mirrored by the N concentration in their tissues (Gebauer et al. 1988). Legumes have N concentrations in their tissues that are up to twice as high as in non-legumes living in the same environment. A major part of N, specifically in photosynthetic plant tissues, is invested in the enzyme Rubisco (Feller et al. 2008), which is directly involved in photosynthetic C fixation. Thus, the capacity for leaf photosynthesis and plant productivity increases with increasing leaf total N concentration (Schulze et al. 2005).

As already mentioned above (see Sects. 8.3.3 and 8.5.3) many MH and PMH plants, specifically those associated with EM fungi, but also orchids associated with rhizoctonias, have surprisingly high total N concentrations in their tissues (Gebauer and Meyer 2003; Abadie et al. 2006). Liebel et al. (2010) investigated orchids and autotrophic reference plant species in the Mediterranean and Macaronesian regions and found that representatives of the tribe Neottieae (subfamily Epidendroideae) mostly associated with EM fungi were PMH or even MH had total N concentrations of 2.9 ± 0.6(SD) mmol N gdw^{-1} ($n=42$). Conversely, species belonging to the subfamily Orchidoideae that were mostly associated with rhizoctonias did not show signs of mycoheterotrophy based on their bulk stable isotope signatures, but had tissue total N concentrations of 1.7 ± 0.4(SD) mmol N gdw^{-1} ($n=150$). Both groups of orchids' total N concentrations are significantly distinguished from each other and significantly above those found for non-orchid reference plants from the same environments: 1.4 ± 0.5(SD) mmol N gdw^{-1} ($n=513$). The finding of increased tissue N concentrations in orchids associated with rhizoctonias supports the idea of a cryptic C gain from the fungus through organic N compounds (see Sect. 8.5.3). Moreover, the high total N concentrations in PMH and MH orchids raise the question as to the function of these high N concentrations in their tissues. Use of increased photosynthetic activity as in "normal" autotrophic plants seems to be unlikely, because PMH plants mostly live under conditions under which light availability limits their photosynthetic activity (Preiss et al. 2010) and fully MH plants have given up photosynthetic activity altogether. Furthermore, in variegated individuals of *Cephalanthera damasonium*, Stöckel et al. (2011) documented a linear increase in leaf total N concentration with decreasing chlorophyll concentration and thus, decreasing photosynthetic capacity. At the moment we only can speculate about the reasons why MH and PMH plants accumulate unusually high N concentrations in their tissues. One possible reason may be related to the fact that fungal tissues and specifically those of EM fungi also have total N concentrations far above those known for the majority of autotrophic plants (Gebauer and Dietrich 1993; Gebauer and Taylor 1999) and MH plants may have an N metabolism more similar to their fungal hosts than other plants. Future studies that focus specifically on the N assimilation strategies among MH plants are greatly needed. These studies may in turn elucidate modes of C acquisition among MH plants because the two nutrients are inherently linked in their organic forms.

8.6.2 Liverworts, Lycopods, and Ferns

Gametophytes of most species in the liverwort family Aneuraceae are photosynthetic, readily grown in asymbiotic culture, and lack a morphologically unambiguous MH stage. However, in the field, most species maintain specialized relationships with tulasnelloid fungi, exhibiting patterns of fungal peloton formation and digestion that appear cytologically identical to those in the confamilial MH liverwort *Aneura mirabilis* and in IMH orchid seedlings (Ligrone et al. 1993). As in other liverworts, the sporophytes of the Aneuraceae are minute, ephemeral, dependent on the gametophytes, and uncolonized by mycorrhizal fungi. While it appears likely that gametophytes of many species in the Aneuraceae exhibit facultative partial mycoheterotrophy, their trophic status has yet to be investigated via isotope analyses.

Most members of the Lycopodiaceae, all members of the Ophioglossaceae and Psilotaceae, some members of the Schizaeaceae, and one member of the Gleicheniaceae have subterranean, non-photosynthetic, and unambiguously MH gametophytes and preemergent sporophytes. Both gametophytes and sporophytes in the Lycopodiaceae (Winther and Friedman 2008), Ophioglossaeae (Winther and Friedman 2007), and Psilotaceae (Winther and Friedman 2009) have specialized associations with Glomeromycota. Adult sporophytes of all species are typically photosynthetic, although albino and/or non-emergent forms have been observed in *Lycopodium clavatum* (Bruce and Beitel 1979), *Psilotum nudum* (Whittier 1988), and *Botrychium mormo* (Johnson-Groh 1998; Johnson-Groh and Lee 2002). In *B. paradoxum*, sporophyte leaves, while green, are reduced to small, bladeless, sporangia-bearing stipes (Wagner and Wagner 1981). Adult sporophytes of Lycopodiaceae, Ophioglossaceae, and Psilotaceae are commonly cultivated in the absence of AM host plants, although some *Botrychium* spp. are noted to be uncultivable (Donald Farrar, pers. comm.). In contrast, adult sporophytes of the Gleicheniaceae are difficult to cultivate, and those of the Schizaeaceae are rarely, if ever, cultivated.

Basic life history data, such as the duration of initial mycoheterotrophy, are unknown for most fern and lycopod taxa. The identities of autotrophic plants supporting MH development of ferns and lycopods via AM fungal networks are also largely unknown. Adult sporophytes are apparently autotrophic or PMH, depending on the taxa, but this has yet to be investigated via isotope analyses. Whether surficial, green gametophytes possessed by some taxa in the Lycopodiaceae, Schizaeaceae, and Gleicheniaceae are autotrophic or PMH is also uncertain. While gametophytes belonging to some species in the Lycopodiaceae (Whittier and Renzaglia 2005) and Gleicheniaceae (Stokey 1950) have been cultured asymbiotically on sugar-free media (indicating they are at least facultatively autotrophic), relatively few such taxa have been studied and the trophic status of gametophytes under field conditions remains unknown. While this represents an interesting avenue for further research, collecting isotope data from minute—and in some cases, filamentous—gametophytes may prove difficult.

8.6.3 Fungi

Because the study of PMH and MH plants reveals constraints that we currently are unable to explain, future studies should also focus on the physiology of soil mycelia in regards to carbon and nutrient cycling. For example, apparently only in wet climates can SAP fungi support fully MH orchids. Although, as mentioned above (see Sect. 8.3.2), this may be linked to longer periods of C assimilation (growing seasons) and higher efficiency of fungal exoenzymes, which allow the access to larger C amounts, we must acknowledge these ideas are speculative, and face the reality that current knowledge of fungal ecophysiology may limit our understanding of these processes. Similarly, although SAP, parasitic and endophytic rhizoctonias support MH seedling development in most orchid species and some pyroloids, they are never used by fully MH orchids. Here again, we currently can speculate that the maximal C provided by rhizoctonias is not sufficient (Taylor and Bruns 1997; Hashimoto et al. 2012), but we lack direct evidence for this. In the future, simple designs that are tractable in the lab, or data from the growing number of sequenced fungal genomes, that reveal metabolic abilities (Martin and Selosse 2008) may help understanding these observations through the knowledge gained on the physiology of fungal partners.

8.6.4 An Ecological View

Up to now, as exemplified in this chapter, our approach to the study of partial and full mycoherotrophy is centered on the side of plants receiving C, as well as on the description of the fungal associates. We now need to know more about the overlooked components of the association. First, what is the impact on the fungus? Although widespread opinion holds mycoheterotrophy as a parasitism on the fungus, and EM-MH/AM-MH nutrition as "epiparasitism," on the autotrophic host plant the final evidence for this would be to show that the fungal or autotrophic host plant fitness, i.e., growth and/or sporulation, is impaired. Again here, tractable ex situ designs, which may be the most plausible on plants using SAP fungi, may help to resolve the costs to fungi associating with MH plants. It may be discovered that there is a variable level of impact of mycoheterotrophy on fungal fitness, or even some compensation by MH plants to their fungi: This compentation could possibly in the form of nutrients, growth factors (as early suggested by Björkman 1960) or shelter provided to the fungi in the roots of MH plants.

Furthermore, future studies should address the impact of PMH and MH plants on surrounding green plant(s) that provide C to these plants, in every case where a mycorrhizal fungus is providing C. From these ideas comes the question of the ecological relevance of MH nutrition. Parasitic plants are well known to impact ecosystem productivity (Westwood et al. 2010) and hemiparasitic plants, by influencing dominance of plants, have important impact on diversity of plant communities (Watson 2009). It is fascinating to observe how common some MH plants are in some tropical and temperate forests while completely absent from forests that appear similar in both their above and belowground biotic and abiotic conditions. Future studies, will hopefully reveal the conditions that select for the presence of MH plants in nature, and discern the impacts that they have on ecosystem dynamics.

References

Abadie J-C, Püttsepp Ü, Gebauer G, Faccio A, Bonfante P, Selosse M-A (2006) *Cephalanthera longifolia* (Neottieae, Orchidaceae) is mixotrophic: a comparative study between green and non-photosynthetic individuals. Can J Bot 84:1462–1477

Adamec L (1997) Mineral nutrition of carnivorous plants: a review. Bot Rev 63:273–299

Allen EB, Allen MF (1990) The mediation of competition by mycorrhizae in successional and patchy environments.

In: Grace JR, Tilman D (eds) Perspectives on plant competition. Academic, New York, pp 367–389

Allen EB, Temple PJ, Bytnerowicz A, Arbaugh MJ, Sirulnik AG, Rao LE (2007) Patterns of understory diversity in mixed coniferous forests of southern California impacted by air pollution. ScientificWorldJournal 7, http://www.thescientificworld.com

Baldrian P (2009) Ectomycorrhizal fungi and their enzymes in soils: is there enough evidence for their role as facultative soil saprotrophs? Oecologia 161:657–660

Bannister P, Strong GL (2001) Carbon and nitrogen isotope ratios, nitrogen content and heterotrophy in New Zealand mistletoes. Oecologia 126:10–20

Beau C (1920) Sur le role trophique des endophytes d'orchid6es. Comp Rend Acad Sci 171:675–677

Bennett JR, Mathews S (2006) Phylogeny of the parasitic plant family Orobanchaceae inferred from phytochrome A. Am J Bot 93:1039–1051

Bernard N (1899) Sur la germination du *Neottia nidus-avis*. C R Hebd Seances Acad Sci 128:1253–1255

Bernard N (1908) La culture des orchidées dans ses rapports avec la symbiose. Société Royale d'Agriculture et de Botanique de Gand, Gand

Bidartondo MI, Bruns TD (2005) On the origins of extreme mycorrhizal specificity in the Monotropoideae (Ericaceae): performance trade-offs during seed germination and seedling development. Mol Ecol 14:1549–1560

Bidartondo MI, Read DJ (2008) Fungal specificity bottlenecks during orchid germination and development. Mol Ecol 17:3707–3716

Bidartondo MI, Redecker D, Hijri I, Wiemken A, Bruns TD, Dominguez L, Sérsic A, Leake JR, Read DJ (2002) Epiparasitic plants specialized on arbuscular mycorrhizal fungi. Nature 419:389–392

Bidartondo MI, Bruns TD, Weiß M, Sérgio C, Read DJ (2003) Specialized cheating of the ectomycorrhizal symbiosis by a epiparasitic liverwort. Proc Biol Sci 270:835–842

Bidartondo MI, Burghardt B, Gebauer G, Bruns TD, Read DJ (2004) Changing partners in the dark: Isotopic and molecular evidence of ectomycorrhizal liaisons between forest orchids and trees. Proc Biol Sci 271:1799–1806

Björkman E (1960) *Monotropa hypopitys* L.—an epiparasite on tree roots. Physiol Plant 13:308–327

Bonnardeaux Y, Brundrett M, Batty A, Dixon K, Koch J, Sivasithamparam K (2007) Diversity of mycorrhizal fungi of terrestrial orchids: compatibility webs, brief encounters, lasting relationships and alien invasions. Mycol Res 111:51–61

Booth MG (2004) Mycorrhizal networks mediate overstorey-understorey competition in a temperate forest. Ecol Lett 7:538–546

Bougoure J, Ludwig M, Brundrett M, Grierson P (2009) Identity and specificity of the fungi forming mycorrhizas with the rare mycoheterotrophic orchid *Rhizanthella gardneri*. Mycol Res 113:1097–1106

Bougoure JJ, Brundrett MC, Grierson PF (2010) Carbon and nitrogen supply to the underground orchid, *Rhizanthella gardneri*. New Phytol 186:947–956

Boullard B (1979) Considérations sur la symbiose fongique chez les Ptéridophytes. Syllogeus 19:1–58

Bruce JG, Beitel JM (1979) A community of *Lycopodium* gametophytes in Michigan. Am Fern J 69:33–41

Bruns TD, Read DJ (2000) In vitro germination of non-photosynthetic, myco-heterotrophic plants stimulated by fungi isolated from the adult plants. New Phytol 148:335–342

Burgeff H (1932) Saprophytismus und Symbiose. Studien an tropischen Orchideen. Gustav Fischer, Jena, Germany

Burgeff H (1936) Samenkeimung der Orchideen. Gustav Fischer, Jena, Germany

Burgeff H (1959) Mycorrhiza of orchids. In: Withner CL (ed) The orchids. Ronald Press, New York, pp 361–395

Cameron DD, Bolin JF (2010) Isotopic evidence of partial mycoheterotrophy in the Gentianaceae: *Bartonia virginica* and *Obolaria virginica* as case studies. Am J Bot 97:1272–1277

Cameron DD, Leake JR (2007) A different kind of parasitic plant: a brief history of mycoheterotrophy and epi-parasitism. Haustorium 50:4–6

Cameron KM, Chase MW, Rudall PJ (2003) Recircumscription of the monocotyledonous family Petrosaviaceae to include *Japanolirion*. Brittonia 55:214–225

Cameron DD, Leake JR, Read DJ (2006) Mutualistic mycorrhiza in orchids: evidence from the plant-fungus carbon and nitrogen transfers in the green-leaved terrestrial orchid *Goodyera repens*. New Phytol 171:405–416

Cameron DD, Johnson I, Read DJ, Leake JR (2008) Giving and receiving: measuring the carbon cost of mycorrhizas in the green orchid, *Goodyera repens*. New Phytol 180:176–184

Cameron DD, Preiss K, Gebauer G, Read DJ (2009) The chlorophyll-containing orchid *Corallorhiza trifida* derives little carbon through photosynthesis. New Phytol 183:358–364

Camp WH (1940) Aphyllous forms in pyrola. Bull Torrey Bot Club 67:453–465

Campbell EO (1962) The mycorrhiza of *Gastrodia cunninghamii* Hook. F. Trans R Soc New Zealand 1:289–296

Campbell EO (1964) The fungal association in a colony of *Gastrodia sesamoides*. R Br Trans R Soc New Zealand 2:237–246

Cha JY, Igarashi T (1995) *Armillaria* species associated with *Gastrodia elata* in Japan. Eur J Forest Pathol 25:319–326

Cha JY, Igarashi T (1996) *Armillaria jezoensis*, a new symbiont of *Galeola septentrionalis* (Orchidaceae) in Hokkaido. Mycoscience 37:21–24

Chou H (1974) In propagation of *Gastrodia elata* Bl. Acta Bot Sin 16:288–290

Christoph H (1921) Untersuchungen über mykotrophen Verhältnisse der "Ericales" und die Keimung von Pirolaceen. Beihefte Botan Centralblatt 38:115–157

Courty P-E, Walder F, Boller T, Ineichen K, Wiemken A, Selosse M-A (2011) C and N metabolism in mycorrhizal networks and mycoheterotrophic plants of tropical forests: a stable isotope analysis. Plant Physiol 156:952–961

Currah RS, Zelmer CD, Hambleton S, Richardson KA (1997) Fungi from orchid mycorrhizas. In: Arditti J, Pridgeon AM (eds) Orchid biology: reviews and perspectives, VII. Kluwer Academic Publishers, Dordrecht, The Netherlands, pp 117–170

Curtis JT (1943) Germination and seedling development in five species of *Cypripedium*. Am J Bot 30:199–206

Dawson TE, Mambelli S, Plamboeck AH, Templer PH, Tu KP (2002) Stable isotopes in plant ecology. Annu Rev Ecol Syst 33:507–559

Dearnaley JDW (2006) The fungal endophytes of *Erythrorchis cassythoides*—is this orchid saprophytic or parasitic? Aust Mycol 25:51–57

Dearnaley JDW, Bougoure JJ (2010) Isotopic and molecular evidence for saprotrophic Marasmiaceae mycobionts in rhizomes of *Gastrodia sesamoides*. Fungal Ecol 3:288–294

Dearnaley JDW, Martos F, Selosse M-A (2012) Orchid mycorrhizas: molecular ecology, physiology, evolution and conservation aspects. In: Hock B (ed) Fungal associations, The mycota IX, 2nd edn. Springer-Verlag, Berlin

DeNiro MJ, Epstein S (1978) Influence of diet on the distribution of carbon isotopes in animals. Geochim Cosmochim Acta 42:495–506

DeNiro MJ, Epstein S (1981) Influence of diet on the distribution of nitrogen isotopes in animals. Geochim Cosmochim Acta 45:341–351

Douhan GW, Vincenot L, Selosse M-A (2011) Population genetics of ectomycorrhizal fungi: from current knowledge to emerging directions. Fungal Biol 115:569–597

Downie DG (1959) *Rhisoctonia solani* and orchid seed. Trans Bot Soc Edinb 37:279–285

Eriksson O, Kainulainen K (2011) The evolutionary ecology of dust seeds. Perspect Plant Ecol Evol Syst 13:73–87

Farquhar GD, Ehleringer JR, Hubick KT (1989) Carbon isotope discrimination and photosynthesis. Annu Rev Plant Physiol Plant Mol Biol 40:503–537

Feller U, Anders I, Mae T (2008) Rubiscolytics: fate of Rubisco after its enzymatic function in a cell is terminated. J Exp Bot 59:1615–1624

Finlay RD, Read DJ (1986) The structure and function of the vegetative mycelium of ectomycorrhizal plants: 1. Translocation of C-14-labeled carbon between plants interconnected by a common mycelium. New Phytol 103:143–156

Fitter AH, Graves JD, Watkins NK, Robinson D, Scrimgeour C (1998) Carbon transfer between plants and its control in networks of arbuscular mycorrhizas. Funct Ecol 12:406–412

Francis R, Read DJ (1984) Direct transfer of carbon between plants connected by vesicular arbuscular mycorrhizal mycelium. Nature 307:53–56

Frank AB (1885) Neue Mittheilungen über die Mycorrhiza der Bäume und der *Monotropa hypopitys*. Ber Dtsch Bot Ges 3:27–33

Freudenstein JV (1999) Relationships and character transformation in Pyroloideae (Ericaceae) based on ITS sequences, morphology, and development. Syst Bot 24:398–408

Freudenstein JV, Doyle JJ (1994) Character transformation and relationships in *Corallorhiza* (Orchidaceae: Epidendroideae). I. Plastid DNA. Am J Bot 81:1449–1457

Fry B (2006) Stable isotope ecology. Springer, New York, 308pp

Gebauer G (2005) Partnertausch im dunklen Wald—Stabile Isotope geben neue Einblicke in das Ernährungsverhalten von Orchideen. In: Bayer (ed) Auf Spurensuche in der Natur: Stabile Isotope in der ökologischen Forschung. Rundgespräche der Kommission für Ökologie Bd. 30. Akademie der Wissenschaften. Verlag Dr. Friedrich Pfeil, München, Germany, pp 55–67

Gebauer G, Dietrich P (1993) Nitrogen isotope ratios in different compartments of a mixed stand of spruce, larch and beech trees and of understorey vegetation including fungi. Isotopenpraxis 29:35–44

Gebauer G, Meyer M (2003) ^{15}N and ^{13}C natural abundance of autotrophic and mycoheterotrophic orchids provides insight into nitrogen and carbon gain from fungal association. New Phytol 160:209–223

Gebauer G, Schulze E-D (1991) Carbon and nitrogen isotope ratios in different compartments of a healthy and a declining *Picea abies* forest in the Fichtelgebirge, NE Bavaria. Oecologia 87:198–207

Gebauer G, Taylor AFS (1999) ^{15}N natural abundance in fruit bodies of different functional groups of fungi in relation to substrate utilization. New Phytol 142:93–101

Gebauer G, Rehder H, Wollenweber B (1988) Nitrate, nitrate reduction and organic nitrogen in plants from different ecological and taxonomic groups of Central Europe. Oecologia 75:371–385

Girlanda M, Selosse M-A, Cafasso D, Brilli F, Delfine S, Fabbian R, Ghignone S, Pinelli P, Segreto R, Loreto F, Cozzolino S, Perotto S (2006) Inefficient photosynthesis in the Mediterranean orchid *Limodorum abortivum* is mirrored by specific association to ectomycorrhizal Russulaceae. Mol Ecol 15:491–504

Girlanda M, Segreto R, Cafasso D, Liebel HT, Rodda M, Ercole E, Cozzolino S, Gebauer G, Perotto S (2011) Photosynthetic Mediterranean meadow orchids feature partial mycoheterotrophy and specific mycorrhizal associations. Am J Bot 98:1148–1163

Gleixner G, Danier H-J, Werner RA, Schmidt H-L (1993) Correlations between the ^{13}C content of primary and secondary plant products in different cell compartments and that in decomposing fungi. Plant Physiol 102:1287–1290

Haber E (1987) Variability distribution and systematics of pyrola-picta sensu-lato Ericaceae in Western North America. Syst Bot 12:324–335

Hadley G (1970) Non-specificity of symbiotic infection in orchid mycorrhiza. New Phytol 69:1015–1023

Hamada M (1939) Studien über die Mykorrhiza von *Galeola septentrionalis* Reichb. F.—Ein neuer Fall der Mycorrhizabildung durch intraradicale Rhizomorpha. Jpn J Bot 10:151–211

Hamada M, Nakamura S (1963) Wurzelsymbiose von *Galeola altissima* Reichb. F., einer chlorophyllfreien Orchidee, mit dem holzzerstörenden Pilz *Hymenochaete crocicreas* Berk et Br. Science Reports of the Tohoku University, Fourth Series (Biology), vol 29, pp 227–238

Harvais G, Hadley G (1967) The relation between host and endophyte in orchid mycorrhiza. New Phytol 66:205–215

Hashimoto Y, Kunishi A, Hasegawa S (2005) Interspecific C transfers from *Larix kaempferi* Carr. to *Pyrola incarnata* Fischer by way of mycorrhizal fungi. Inoculum (Supplement to Mycologia) 56:23–24

Hashimoto Y, Fukukawa S, Kunishi A, Suga H, Richard F, Sauve M, Selosse M-A (2012) Mycoheterotrophic germination of *Pyrola asarifolia* dust seeds reveals convergences with germination in orchids. New Phytol 195(3):620–630

Hofmeister W (1857) Beiträge zur Entwicklungsgeschichte der Gefässkryptogamen. II. Abh Math Phys Cl Königl Sächs Ges Wiss 3:601–682, Taf. 1–13

Högberg P (1997) Tansley review no. 95. ^{15}N natural abundance in soil-plant systems. New Phytol 137:179–203

Högberg P, Plamboeck AH, Taylor AFS, Fransson PMA (1999) Natural ^{13}C abundance reveals trophic status of fungi and host-origin of carbon in mycorrhizal fungi in mixed forests. Proc Natl Acad Sci U S A 96:8534–8539

Hunt R, Hope-Simpson JF (1990) Growth of *Pyrola rotundifolia* ssp. *maritima* in relation to shade. New Phytol 114:129–137

Hynson NA, Bruns TD (2009) Evidence of a myco-heterotroph in the plant family Ericaceae that lacks mycorrhizal specificity. Proc Biol Sci 276:4053–4059

Hynson NA, Bruns TD (2010) Fungal hosts for mycoheterotrophic plants: a nonexclusive, but highly selective club. New Phytol 185:598–601

Hynson NA, Preiss K, Gebauer G (2009a) Is it better to give than receive? A stable isotope perspective to orchid-fungal carbon transport in the green orchid species *Goodyera repens* and *Goodyera oblongifolia*. New Phytol 182:8–11

Hynson NA, Preiss K, Gebauer G, Bruns TD (2009b) Isotopic evidence of full and partial myco-heterotrophy in the plant tribe Pyroleae (Ericaceae). New Phytol 182:719–726

Hynson NA, Mambelli S, Amend AS, Dawson TE (2012) Measuring carbon gains from fungal networks in understory plants from the tribe Pyroleae (Ericaceae): a field manipulation and stable isotope approach. Oecologia 169:307–317. doi:10.1007/s00442-011-2198-3

Illyés Z, Halász K, Rudnóy S, Ouanphanivanh N, Garay T, Bratek Z (2009) Changes in the diversity of the mycorrhizal fungi of orchids as a function of the water supply of the habitat. J Appl Bot Food Qual 83:28–36

Imhof S (2008) Specialized mycorrhizal colonization pattern in achlorophyllous *Epirixanthes* spp. (Polygalaceae). Plant Biol 9:786–792

Irmisch T (1855) Bemerkungen über einige Pflanzen der deutschen Flora. Flora 13:625–638

Isogai N, Yamamura Y, Mariko S, Nakano T (2003) Seasonal pattern of photosynthetic production in a subalpine evergreen herb, *Pyrola incarnata*. J Plant Res 116:199–206

Johnson-Groh CL (1998) Population demographics, underground ecology and phenology of *Botrychium mormo*. In: Berlin N, Miller P, Borovansky J, Seal US, Byers O (eds) Population and habitat viability assessment (PHVA) for the goblin fern (*Botrychium mormo*), Final Report. Conservation Biology Specialist Group, Apple Valley, Minnesota, pp 103–108

Johnson-Groh CL, Lee JM (2002) Phenology and demography of two species of *Botrychium* (Ophioglossaceae). Am J Bot 89:1624–1633

Jones RI (2000) Mixotrophy in planktonic protists: an overview. Freshwater Biol 45:219–226

Julou T, Burghardt B, Gebauer G, Berveiller D, Damesin C, Selosse M-A (2005) Mixotrophy in orchids: insights from a comparative study of green individuals and non-photosynthetic mutants of *Cephalanthera damasonium*. New Phytol 166:639–653

Kamienski F (1881) Die Vegetationsorgane der *Monotropa hypopitys* L. Bot Zeit 29:458–461

Kamjunke N, Tittel J (2009) Mixotrophic algae constrain the loss of organic carbon by exudation. J Phycol 45:807–811

Khan AG (1972) Mycorrhizae in the Pakistan Ericales Pak. J Bot 4:183–194

Kiers ET, Duhamel M, Beesetty Y, Mensah JA, Franken O, Verbruggen E, Felbaum CR, Kowalchuk GA, Hart MM, Bago A, Palmer TM, West SA, Vandenkoornhuyse P, Jansa J, Bücking H (2011) Reciprocal rewards stabilize cooperation in the mycorrhizal symbiosis. Science 333:880–882

Kikuchi G, Higuchi M, Yoshimura H, Morota T, Suzuki A (2008a) In vitro symbiosis between *Gastrodia elata* Blume (Orchidaceae) and *Armillaria* Kummer (Tricholomataceae) species isolated from the orchid tuber. J Jpn Bot 83:77–87

Kikuchi G, Higuchi M, Morota T, Nagasawa E, Suzuki A (2008b) Fungal symbiont and cultivation test of *Gastrodia elata* Blume (Orchidaceae). J Jpn Bot 83:88–95

Knudson L (1922) Nonsymbiotic germination of orchid seeds. Bot Gaz 73:1–25

Kohzu A, Yoshioka T, Ando T, Takahashi M, Koba K, Wada E (1999) Natural ^{13}C and ^{15}N abundance of field-collected fungi and their ecological implications. New Phytol 144:323–330

Kranabetter JM, MacKenzie WH (2010) Contrasts among mycorrhizal plant guilds in foliar nitrogen concentration and d^{15}N along productivity gradients of a boreal forest. Ecosystems 13:108–117

Kron KA, Judd WS, Stevens PF, Crayn DM, Anderberg AA, Gadek PA, Quinn CJ, Luteyn JL (2002) Phylogenic classification of Ericaceae: molecular and morphological evidence. Bot Rev 68:335–423

Kunishi A, Hasegawa S, Hashimoto Y (2004) Effects of mycorrhiza on *Pyrola incarnata* growing in dark forest floor. In: Proceedings of the 51st annual meeting of the Ecological Society of Japan (JES51)

Kusano S (1911) *Gastrodia elata* and its symbiotic association with *Armillaria mellea*. J Coll Agric Imp Univ Tokyo 4:1–65

Leake JR (1994) Tansley review No. 69. The biology of myco-heterotrophic ('saprophytic') plants. New Phytol 127:171–216

Leake JR (2004) Myco-heterotroph/epiparasitic plant interactions with ectomycorrhizal and arbuscular mycorrhizal fungi. Curr Opin Plant Biol 7:422–428

Leake JR (2005) Plants parasitic on fungi: unearthing the fungi in myco-heterotrophs and debunking the "saprophytic" plant myth. Mycologist 19:113–122

Lerat S, Gauci R, Catford JG, Vierheilig H, Piché Y, Lapointe L (2002) ^{14}C transfer between the spring ephemeral *Erythronium americanum* and sugar maple saplings *via* arbuscular mycorrhizal fungi in natural stands. Oecologia 132:181–187

Levin I, Kromer B (1997) Twenty years of atmospheric $^{14}CO_2$ observations at Schauinsland Station, Germany. Radiocarbon 39:205–218

Liebel HT, Gebauer G (2011) Stable isotope signatures confirm carbon and nitrogen gain through ectomycorrhizas in the ghost orchid *Epipogium aphyllum* Swartz. Plant Biol 13:270–275

Liebel HT, Preiss K, Gebauer G (2009) Parsiell mycoheterotrofi i norske vintergrønnarter – relevans for vernetiltak av truede vintergrønnarter (partial myco-heterotrophy in Norwegian wintergreen species—relevance for endangered species protection). Blyttia J Norw Bot Soc 67:138–143

Liebel HT, Bidartondo MI, Preiss K, Segreto R, Stöckel M, Rodda M, Gebauer G (2010) C and N isotope signatures reveal constraints to nutritional modes in orchids of the Mediterranean and Macaronesia. Am J Bot 97:903–912

Lievens B, van Kerckhove S, Justé A, Cammue BPA, Honnay O, Jacquemyn H (2010) From extensive clone libraries to comprehensive DNA arrays for the efficient and simultaneous detection and identification of orchid mycorrhizal fungi. J Microbiol Methods 80:76–85

Ligrone R, Pocok K, Duckett JG (1993) A comparative ultrastructural study of endophytic basidiomycetes in the parasitic achlorophyllous hepatic *Cryptothallus mirabilis* and the closely allied photosynthetic species *Aneura pinguis* (Metzgeriales). Can J Bot 71:666–679

Lihnell D (1942) Keimungsversuche mit Pyrola-Samen. Symb Bot Upsal 6:1–37

Marschner H (1995) Mineral nutrition of higher plants, 2nd edn. Academic, London, UK

Martin F, Selosse M-A (2008) The *Laccaria* genome: a symbiont blueprint decoded (Tansley review—introduction to a special issue). New Phytol 180:296–310

Martos F, Dulormne M, Pailler T, Bonfante P, Faccio A, Fournel J, Dubios M-P, Selosse M-A (2009) Independent recruitment of saprotrophic fungi as mycorrhizal partners by tropical achlorophyllous orchids. New Phytol 184:668–681

Matsuda Y, Amiya A, Ito S-I (2008) Colonization patterns of mycorrhizal fungi associated with two rare orchids, *Cephalanthera falcata* and *C. erecta*.. Ecol Res 24:1023–1031

Matsuda Y, Shimizu S, Mori M, Ito S-I, Selosse M-A (2012) Seasonal and spatial changes of mycorrhizal associations and heterotrophy levels in mixotrophic *Pyrola japonica* growing under different light environments. In revision for Am. J, Bot

Matsushita N, Fukuda K, Nagasawa E, Terashita T, Suzuki K (1996) *Armillaria* species in Japan identified by isozyme patterns with special reference to the biological species of the Northern hemisphere. J For Res 1:155–160

Mayor JR, Schuur EAG, Henkel TW (2009) Elucidating the nutritional dynamics of fungi using stable isotopes. Ecol Lett 12:171–183

McCormick MK, Whigham DF, O'Neill J (2004) Mycorrhizal diversity in photosynthetic terrestrial orchids. New Phytol 163:425–438

McCormick MK, Taylor DL, Juhaszova K, Burnett RK, Whigham DF, O'Neill JP (2012) Limitations on orchid recruitment: not a simple picture. Mol Ecol 21:1511–1523

McGuire KL (2007) Common ectomycorrhizal networks may maintain monodominance in a tropical rain forest. Ecology 88:567–574

McKendrick SL, Leake JR, Read DJ (2000) Symbiotic germination and development of myco-heterotrophic plants in nature: transfer of carbon from ectomycorrhizal *Salix repens* and *Betula pendula* to the orchid *Corallorhiza trifida* through shared hyphal connections. New Phytol 145:539–548

McKendrick SL, Leake JR, Taylor DL, Read DJ (2002) Symbiotic germination and development of the myco-heterotrophic orchid *Neottia nidus-avis* in nature and its requirement for locally distributed *Sebacina* spp. New Phytol 154:233–247

McNeal JR, Arumugunathan K, Kueh JV, Boore JL, dePamphilis CW (2007) Systematics and plastid genome evolution of the cryptically photosynthetic parasitic plant genus *Cuscuta* (Convolvulaceae). BMC Biol 5:55

Merckx V, Schols P, Van De Kamer HM, Maas HM, Huysmans S, Smets E (2006) Phylogeny and evolution of Burmaniaceae (Dioscoreales) based on nuclear and mitochondrial data. Am J Bot 93:1684–1698

Merckx V, Stöckel M, Fleischmann A, Bruns TD, Gebauer G (2010) ^{15}N and ^{13}C natural abundance of two mycoheterotrophic and a putative partially mycoheterotrophic species associated with arbuscular mycorrhizal fungi. New Phytol 188:590–596

Mettenius G (1856) Filices Horti Botanici Lipsiensis. L. Voss, Leipzig

Milligan MJ, Williams PG (1988) The mycorrhizal relationship of multinucleate rhizoctonias from

non-orchids with *Microtis* (Orchidaceae). New Phytol 108:205–209
Montfort C, Küsters E (1940) Saprophytismus und Photosynthese. I. Biochemische und physiologische Studien an Humus-Orchideen. Bot Arch 40: 571–633
Motomura H, Yukawa T, Ueno O, Kagawa A (2008) The occurrence of Crassulacean acid metabolism in *Cymbidium* (Orchidaceae) and its ecological and evolutionary implications. J Plant Res 121:163–177
Motomura H, Selosse M-A, Martos F, Kagawa A, Yukawa T (2010) Mycoheterotrophy evolved from mixotrophic ancestors: evidence in *Cymbidium* (Orchidaceae). Ann Bot 106:573–581
Muir HJ (1989) Germination and mycorrhizal fungus compatibility in European orchids. In: Pritchard HW (ed) Modern methods in orchid conservation: the role of physiology, ecology and management. Cambridge University Press, England, pp 39–56
Mursidawati S (2004) Mycorrhizal association, propagation and conservation of the myco-heterotrophic orchid *Rhizanthella gardneri*. MSc thesis, The University of Western Australia
Nakano A, Takahashi K, Kimura M (1999) The carbon origin of arbuscular mycorrhizal fungi estimated from δ13C values of individual spores. Mycorrhiza 9:41–47
Nara K (2006) Ectomycorrhizal networks and seedling establishment during early primary succession. New Phytol 169:169–178
Neales TF, Hew CS (1975) Two types of carbon fixation in tropical orchids. Planta 123:303–306
Ogura-Tsujita Y, Yukawa T (2008) High mycorrhizal specificity in a widespread mycoheterotrophic plant, *Eulophia zollingeri* (Orchidaceae). Am J Bot 95:93–97
Ogura-Tsujita Y, Gebauer G, Hashimoto T, Umata H, Yukawa T (2009) Evidence for novel and specialized mycorrhizal parasitism: the orchid *Gastrodia confusa* gains carbon from saprotrophic *Mycena*. Proc Biol Sci 276:761–767
Ogura-Tsujita Y, Yokoyama J, Miyoshi K, Yukawa T (2012) Shifts in mycorrhizal fungi during the evolution of autotrophy to mycoheterotrophy in *Cymbidium* (Orchidaceae). Am J Bot 99(7):1158–1176
Ota Y, Intini M, Hattori T (2000) Genetic characterization of heterothallic and non-heterothallic *Armillaria mellea* sensu stricto. Mycol Res 104:1046–1054
Ouanphanivanh N, Merényi Z, Orczán ÁK, Bratek Z, Szigeti Z, Illyés Z (2008) Could orchids indicate truffle habitats? Mycorrhizal association between orchids and truffles. Acta Biol Szeg 52:229–232
Parrent JL, James TY, Vasaitis R, Taylor AFS (2009) Friend or foe? Evolutionary history of glycoside hydrolase family 32 genes encoding for sucrolytic activity in fungi and its implications for plant-fungal symbioses. BMC Evol Biol 9:148
Pfeffer PE, Douds DD, Bucking H, Schwartz DP, Shachar-Hill Y (2004) The fungus does not transfer carbon to or between roots in an arbuscular mycorrhizal symbiosis. New Phytol 163:617–627
Phillips RD, Barrett MD, Dixon KW, Hopper SD (2011) Do mycorrhizal symbioses cause rarity in orchids? J Ecol 99:858–869
Preiss K, Gebauer G (2008) A methodological approach to improve estimates of nutrient gains by partially myco-heterotrophic plants. Isotopes Environ Health Stud 44:393–401
Preiss K, Adam IKU, Gebauer G (2010) Irradiance governs exploitation of fungi: fine-tuning of carbon gain by two partially myco-heterotrophic orchids. Proc Biol Sci 277:1333–1336
Press MC, Graves JD (1995) Parasitic plants. Chapmann & Hall, London
Press MC, Smith JD, Stewart GR (1991) Carbon acquisition and assimilation relations in parasitic plants. Funct Ecol 5:278–283
Pridgeon A, Cribb PJ, Chase MM (2008) Genera orchidacearum: vol 4: Epidendroidae. Oxford University Press, New York
Ramsay RR, Dixon KW, Sivasithamparam K (1986) Patterns of infection and endophytes associated with Western Australian orchids. Lindleyana 1:203–214
Rasmussen HN (1995) Terrestrial orchids from seed to mycotrophic plant. Cambridge University Press, Cambridge, UK
Rasmussen HN (2002) Recent developments in the study of orchid mycorrhiza. Plant Soil 244:149–163
Rayner MC (1927) Mycorrhiza: an account of non-pathogenic infection by fungi in vascular plants and bryophyes. New phytologist reprint no. 15. Wheldon & Wesley, London, UK
Reeves P, Chase MW, Goldblatt P, Rudall P, Fay MF, Cox AV, Lejeune B, Souza-Chies T (2001) Molecular systematics of Iridaceae: evidence from four plastid DNA regions. Am J Bot 88:2074–2087
Renner O (1938) Über blasse, saprophytische *Cephalanthera alba* und *Epipactis latifolia*. Flora 132:225–233
Robertson DC, Robertson JA (1985) Ultrastructural aspects of *Pyrola* mycorrhizae. Can J Bot 63:1089–1098
Robinson D, Fitter A (1999) The magnitude and control of carbon transfer between plants linked by a common mycorrhizal network. J Exp Bot 50:9–13
Roy M, Watthana S, Stier A, Richard F, Vessabutr S, Selosse M-A (2009a) Two mycoheterotrophic orchids from Thailand tropical dipterocarpacean forests associate with a broad diversity of ectomycorrhizal fungi. BMC Biol 7:51
Roy M, Yagame T, Yamato M, Iwase K, Heinz C, Faccio A, Bonfante P, Selosse M-A (2009b) Ectomycorrhizal *Incybe* species associate with the mycoheterotrophic orchid *Epipogium aphyllum* but not its asexual propagules. Ann Bot 104:595–610
Roy M, Gonneau C, Rocheteau A, Berveiller D, Thomas J-C, Damesin C, Selosse M-A (in press) Why do mixotrophic plants stay green? A comparison between

green and achlorophyllous orchid individuals *in situ*. Ecol Monographs
Rudall PJ, Bateman R (2006) Morphological phylogenetic analysis of Pandanales: testing contrasting hypotheses of floral evolution. Syst Bot 31:223–238
Sadovsky O (1965) Orchideen im eigenen Garten. Bayerischer Landwirtschaftsverlag, Munich, Germany
Salisbury RA (1804) On the germination of the seeds of Orchidaceae. Trans Linn Soc Lond 7:29–32
Salmia A (1986) Chlorophyll-free form of *Epipactis helleborine* (Orchidaceae) in SE Finland. Ann Bot Fenn 23:49–57
Salmia A (1989a) Features of endomycorrhizal infection of chlorophyll-free and green forms of *Epipactis helleborine* (Orchidaceae). Ann Bot Fenn 26:15–26
Salmia A (1989b) General morphology and anatomy of chlorophyll-free and green forms of *Epipactis helleborine* (Orchidaceae). Ann Bot Fenn 26:95–105
Schimel DS (1993) Theory and application of tracers. Academic, San Diego, USA
Schulze E-D, Lange OL, Ziegler H, Gebauer G (1991) Carbon and nitrogen isotope ratios of mistletoes growing on nitrogen fixing and non-nitrogen fixing hosts and on CAM plants in the Namib desert confirm partial heterotrophy. Oecologia 88:457–462
Schulze E-D, Beck E, Müller-Hohenstein K (2005) Plant ecology. Springer, Berlin
Sekizaki H, Kuninaga S, Yamamoto M, Asazu SN, Sawa S, Kojoma M, Yokosawa R, Yoshida N (2008) Identification of *Armillaria nabsnona* in *Gastrodia* tubers. Biol Pharm Bull 31:1410–1414
Selosse M-A, Rousset F (2011) The plant-fungal marketplace. Science 333:828–829
Selosse M-A, Roy M (2009) Green plants eating fungi: facts and questions about mixotrophy. Trends Plant Sci 14:64–70
Selosse M-A, Weiss M, Jany JL, Tillier A (2002) Communities and populations of sebacinoid basidiomycetes associated with the achlorophyllous orchid *Neottia nidus-avis* (L.) L.C.M. Rich. and neighbouring treeectomycorrhizae. Mol Ecol 11:1831–1844
Selosse M-A, Faccio A, Scappaticci G, Bonfante P (2004) Chlorophyllous and achlorophyllous specimens of *Epipactis microphylla* (Neottieae, Orchidaceae) are associated with ectomycorrhizal septomycetes, including truffles. Microb Ecol 47:416–426
Selosse M-A, Richard F, He X, Simard SW (2006) Mycorrhizal networks: des liaisons dangereuses? Trends Ecol Evol 22:621–628
Selosse M-A, Setaro S, Glatard F, Richard F, Urcelay C, Weiß M (2007) Sebacinales are common mycorrhizal associates of Ericaceae. New Phytol 174:864–878
Selosse M-A, Dubois M-P, Alvarez N (2009) Do Sebacinales commonly associate with plant roots as endophytes? Mycol Res 113:1062–1069
Selosse M-A, Martos F, Perry BA, Padamsee M, Roy M, Pailler T (2010) Saprotrophic fungal symbionts in tropical achlorophyllous orchids: finding treasures among the 'molecular scraps'? Plant Signal Behav 5:1–5
Selosse M-A, Boullard B, Richardson D (2011) Noel Bernard (1874–1911): orchids to symbiosis in a dozen years, one century ago. Symbiosis 54:61–68
Sharma J, Zettler LW, Van Sambeek JW, Ellersieck MR, Starbuck CJ (2003) Symbiotic seed germination and mycorrhizae of federally threatened *Platanthera praeclara* (Orchidaceae). Am Midl Nat 149:104–120
Shefferson RP, Weiß M, Kull T, Taylor DL (2005) High specificity generally characterizes mycorrhizal association in rare lady's slipper orchids, genus *Cypripedium*. Mol Ecol 14:613–626
Shefferson RP, Kull T, Tali K (2006) Demographic response to shading and defoliation in two woodland orchids. Folia Geobot 41:95–106
Shefferson RP, Taylor DL, Weiß M, Garnica S, McCormick MK, Adams S, Gray HM, McFarland JW, Kull T, Tali K, Yukawa T, Kawahara T, Miyoshi K, Lee Y-I (2007) The evolutionary history of mycorrhizal specificity among lady's slipper orchids. Evolution 61:1380–1390
Shefferson RP, Kull T, Tali K (2008) Mycorrhizal interactions of orchids colonizing Estonian mine tailings hills. Am J Bot 95:156–164
Silvera K, Santiago LS, Cushman JC, Winter K (2010) The incidence of crassulacean acid metabolism in Orchidaceae derived from carbon isotope ratios: a checklist of the flora of Panama and Costa Rica. Bot J Linn Soc 163:194–222
Simard SW, Durall DM (2004) Mycorrhizal networks: a review of their extent, function, and importance. Can J Bot 82:1140–1165
Simard SW, Perry DA, Jones MD, Myrold DD, Durall DM, Molina R (1997) Net transfer of carbon between ectomycorrhizal tree species in the field. Nature 388:579–582
Smith SE (1966) Physiology and ecology of orchid mycorrhizal fungi with reference to seedling nutrition. New Phytol 65:488–499
Smith SE, Read DJ (2008) Mycorrhizal symbiosis, 3rd edn. Academic, London, UK
Stark JM (2000) Nutrient transformations. In: Sala OE, Jackson RB, Mooney HA, Howarth RW (eds) Methods in ecosystem science. Springer, New York, pp 215–234
Stark C, Babik W, Durka W (2009) Fungi from the roots of the common terrestrial orchid *Gymnadenia conopsea*. Mycol Res 113:952–959
Stöckel M, Meyer C, Gebauer G (2011) The degree of mycoheterotrophic carbon gain in green, variegated and vegetative albino individuals of *Cephalanthera damasonium* is related to leaf chlorophyll concentrations. New Phytol 189:790–796
Stokey AG (1950) The gametophyte of the Gleicheniaceae. Bull Torrey Bot Club 77:323–339
Struwe L, Kadereit J, Klackenberg J, Nilsson S, Thiv M, von Hagen KB, Albert VA (2002) Systematics, character evolution, and biogeography of Gentianaceae, including a new tribal and subtribal classification. In: Struwe L, Albert VA (eds) Gentianaceae—systematics and natural history. Cambridge University Press, Cambridge, pp 21–309

Sung JM, Jung BS, Yang KJ, Lee HK, Harrington TC (1995) Production of *Gastrodia elata* tuber using *Armillaria* spp. Korean J Mycol 23:61–70

Taylor DL, Bruns TD (1997) Independent, specialized invasions of ectomycorrhizal mutualism by two nonphotosynthetic orchids. Proc Natl Acad Sci U S A 94:4510–4515

Taylor AFS, Fransson PM, Högberg P, Högberg MN, Plamboeck AH (2003) Species level patterns in ^{13}C and ^{15}N abundance of ectomycorrhizal and saprotrophic fungal sporocarps. New Phytol 159:757–774

Taylor AFS, Gebauer G, Read DJ (2004) Uptake of nitrogen and carbon from double-labelled (^{15}N and ^{13}C) glycine by mycorrhizal pine seedlings. New Phytol 164:383–388

Tedersoo L, Pellet P, Kõljalg U, Selosse M-A (2007) Parallel evolutionary paths to mycohetereotrophy in understorey Ericaceae and Orchidaceae: ecological evidence for mixotrophy in Pyroleae. Oecologia 151:206–217

Tennakoon KU, Pate JS (1996) Heterotrophic gain of carbon from hosts by the xylem-tapping root hemiparasite *Olax phyllanthi* (Olacaceae). Oecologia 105:369–376

Terashima K, Kawashima Y, Cha JY, Miura K (1998) Identification of *Armillaria* species from Hokkaido by analysis of the intergenic spacer (IGS) region of ribosomal DNA using PCR-RFLP. Mycoscience 39:179–183

Terashita T, Chuman S (1987) Fungi inhabiting wild orchids in Japan. IV. *Armillariella tabescens*, a new symbiont of *Galeola septentrionalis*. Trans Mycol Soc Jpn 28:145–154

Terashita T, Chuman S (1989) Armillarias, isolated from the wild orchid, *Galeola septentrionalis*. In: Proceedings of the 7th IUFRO international conference on root and butt rots of forest trees, pp 364–370

Těšitel J, Plavcová L, Cameron DD (2010) Interactions between hemiparasitic plants and their hosts. The importance of organic carbon transfer. Plant Signal Behav 5:1072–1076

Těšitel J, Lepš J, Vráblová M, Cameron DD (2011) The role of heterotrophic carbon acquisition by the hemiparasitic plant *Rhinanthus alectorolophus* in seedling establishment in natural communities: a physiological perspective. New Phytol 192:188–199

Tesitelova T, Těšitel J, Jersáková J, Říhová G, Selosse M-A (2012) Symbiotic germination capability of four *Epipactis* species (Orchidaceae) is broader than expected from adult ecology. Am J Bot 99(6):1020–1032

Teste FP, Simard SW, Durall DM, Guy RD, Jones MD, Schoonmaker AL (2009) Access to mycorrhizal networks and roots of trees: importance for seedling survival and resource transfer. Ecology 90:2808–2822

Tisserant E, Kohler A, Dozolme-Seddas P, Balestrini R, Benabdellah K, Colard A, Croll D, Da Silva C, Gomez SK, Koul R, Ferrol N, Fiorilli V, Formey D, Franken P, Helber N, Hijri M, Lanfranco L, Lindquist E, Liu Y, Malbreil M, Morin E, Poulain J, Shapiro H, van Tuinen D, Waschke A, Azcón-Aguilar C, Bécard G, Bonfante P, Harrison MJ, Küster H, Lammers P, Paszkowski U, Requena N, Rensing SA, Roux C, Sanders IR, Shachar-Hill Y, Tuskan G, Young JP, Gianinazzi-Pearson V, Martin F (2012) The transcriptome of the arbuscular mycorrhizal fungus *Glomus intraradices* (DAOM 197198) reveals functional tradeoffs in an obligate symbiont. New Phytol 193:755–769

Toftegaard T, Iason GR, Alexander IJ, Rosendahl S, Taylor AFS (2010) The threatened plant intermediate wintergreen (*Pyrola media*) associates with a wide range of biotrophic fungi in native Scottish pine woods. Biodivers Conserv 19:3963–3971

Tranchida-Lombardo V, Roy M, Bugot E, Santoro G, Püttsepp U, Selosse M-A, Cozzolino S (2010) Spatial repartition and genetic relationship of green and albino individuals in mixed populations of *Cephalanthera* orchids. Plant Biol 12:659–667

Trudell SA, Rygiewicz PT, Edmonds RL (2003) Nitrogen and carbon stable isotope abundances support the myco-heterotrophic nature and host-specificity of certain achlorophyllous plants. New Phytol 160: 391–401

Trumbore SE (2000) Age of soil organic matter and soil respiration: radiocarbon constraints on belowground C dynamics. Ecol Appl 10:399–411

Umata H (1995) Seed germination of *Galeola altissima*, an achlorophyllous orchid, with aphyllophorales fungi. Mycoscience 36:369–372

Umata H (1997) Formation of endomycorrhizas by an achlorophyllous orchid, *Erythrorchis ochobiensis*, and *Auricularia polytricha*. Mycoscience 38:335–339

Umata H (1998a) In vitro symbiotic association of an achlorophyllous orchid, *Erythrorchis ochobiensis*, with orchid and non-orchid fungi. Memoirs of the Faculty of Agriculture, Kagoshima University, vol 34, pp 97–107

Umata H (1998b) A new biological function of shiitake mushroom, *Lentinula edodes*, in a myco-heterotrophic orchid, *Erythrorchis ochobiensis*. Mycoscience 39: 85–88

Umata H (1999) Germination and growth of *Erythrorchis ochobiensis* (Orchidaceae) accelerated by monokaryons and dikaryons of *Lenzites betulinus* and *Trametes hirsuta*. Mycoscience 40:367–371

Umata H, Kaneko M, Miyagi T, Nakahira Y (2007) The application and utilization of fungi for the propagation of the endangered achlorophyllous plant, *Erythrorchis ochobiensis* (Hayata) Garay (Orchidaceae) in natural situations. Research Bulletin of the Kagoshima University Forests 35:31–48

Velenovsky J (1892) Über die Biologie und Morphologie der Gattung *Moneses*. Rozpravy Královské české společnosti nauk, řada, matematicko-přírodovědná 11:147–159

Vincenot L, Tedersoo L, Richard F, Horcine H, Kõljalg U, Selosse M-A (2008) Fungal associates of *Pyrola rotundifolia*, a mixotrophic Ericaceae, from two Estonian boreal forests. Mycorrhiza 19:15–25

Vujanovic V, St-Arnaut M, Barab D, Thibeault G (2000) Viability testing of orchid seed and the promotion of colouration and germination. Ann Bot 86:79–86

Wagner WH Jr, Wagner FS (1981) New species of moonworts, *Botrychium* subg. *Botrychium* (Ophioglossaceae), from North America. Am Fern J 71:20–30

Warcup JH (1973) Symbiotic germination of some Australian terrestrial orchids. New Phytol 72:387–392

Warcup JH (1981) The mycorrhizal relationships of Australian orchids. New Phytol 87:371–381

Warcup JH (1988) Mycorrhizal associations of isolates of *Sebacina vermifera*. New Phytol 110:227–231

Warcup JH, Talbot PHB (1967) Perfect states of rhizoctonias associated with orchids. New Phytol 66:631–641

Waterman RJ, Bidartondo MI, Stofberg J, Combs JK, Gebauer G, Savolainen V, Barraclough TG, Pauw A (2011) The effects of above- and belowground mutualism in orchid speciation and coexistence. Am Nat 177:E54–E68

Watson DM (2009) Parasitic plants as facilitators: more dryad than dracula? J Ecol 97:1151–1159

Weiss M, Selosse M-A, Rexer K, Urban A, Oberwinkler F (2004) Sebacinales: a hitherto overlooked cosm of heterobasidiomycetes with a broad mycorrhizal potential. Mycol Res 108:1003–1010

Weiss M, Sýkorová Z, Garnica S, Riess K, Martos F, Krause C, Oberwinkler F, Bauer R, Redecker D (2011) Sebacinales everywhere: previously overlooked ubiquitous fungal endophytes. PLoS One 6:e16793

Westwood JH, Yoder JI, Timko MP, dePamphilis CW (2010) The evolution of parasitism in plants. Trends Plant Sci 15:227–235

Whittier P (1988) Dark-grown psilotum. Am Fern J 78:109–116

Whittier DP, Renzaglia KS (2005) The young gametophyte of *Lycopodiella lateralis* and the role of the intermediate shaft in development of *Lycopodiella* gametophytes. Am Fern J 95:153–159

Winther JL, Friedman WE (2007) Arbuscular mycorrhizal symbionts in *Botrychium* (Ophioglossaceae). Am J Bot 94:1248–1255

Winther JL, Friedman WE (2008) Arbuscular mycorrhizal associations in Lycopodiaceae. New Phytol 177:790–801

Winther JL, Friedman WE (2009) Phylogenetic affinity of arbuscular mycorrhizal symbionts in *Psilotum nudum*. J Plant Res 122:485–496

Wright M, Cross R, Dixon K, Huynh T, Lawrie A, Nesbitt L, Pritchard A, Swarts N, Thomson R (2009) Propagation and reintroduction of *Caladenia*. Aust J Bot 57:373–387

Wright MM, Cross R, Cousens RD, May TW, McLean CB (2010) Taxonomic and functional characterisation of fungi from the *Sebacina vermifera* complex from common and rare orchids in the genus *Caladenia*. Mycorrhiza 20:375–390

Wu BY, Nara K, Hogetsu T (2001) Can C-14-labeled photosynthetic products move between *Pinus densiflora* seedlings linked by ectomycorrhizal mycelia? New Phytol 149:137–146

Wu J, Ma H, Lü M, Han S, Zhu Y, Jin H, Liang J, Liu L, Xu J (2010) *Rhizoctonia* fungi enhance the growth of the endangered orchid *Cymbidium goeringii*. Botany 88:20–29

Xu J, Guo S (2000) Retrospect on the research of the cultivation of *Gastrodia elata* Bl, a rare traditional Chinese medicine. Chin Med J 113:686–692

Yagame T, Yamato M, Mii M, Suzuki A, Iwase K (2007) Developmental processes of achlorophyllous orchid, *Epipogium roseum*: from seed germination to flowering under symbiotic cultivation with mycorrhizal fungus. J Plant Res 120:229–236

Yagame T, Fukiharu T, Yamato M, Suzuki A, Iwase K (2008a) Identification of a mycorrhizal fungus in *Epipogium roseum* (Orchidaceae) from morphological characteristics of basidiomata. Mycoscience 49:147–151

Yagame T, Yamato M, Suzuki A, Iwase K (2008b) Ceratobasidiaceae mycorrhizal fungi isolated from nonphotosynthetic orchid *Chamaegastrodia sikokiana*. Mycorrhiza 18:97–101

Yagame T, Orihara T, Selosse M-A, Yamato M, Iwase K (2012) Mixotrophy of *Platanthera minor*, an orchid associated with ectomycorrhiza-forming Ceratobasidiaceae fungi. New Phytol 193:178–187

Yamato M, Iwase K (2008) Introduction of asymbiotically propagated seedlings of *Cephalanthera falcata* (Orchidaceae) into natural habitat and investigation of colonized mycorrhizal fungi. Ecol Res 23:329–337

Yamato M, Yagame T, Suzuki A, Iwase K (2005) Isolation and identification of mycorrhizal fungi associating with an achlorophyllous plant, *Epipogium roseum* (Orchidaceae). Mycoscience 46:73–77

Zettler LW, Hofer CJ (1998) Propagation of the little clubspur orchid (*Platanthera clavellata*) by symbiotic seed germination and its ecological implications. Environ Exp Bot 39:189–195

Zettler LW, Stewart SL, Bowles ML, Jacobs KA (2001) Mycorrhizal fungi and cold-assisted symbiotic germination of federally threatened eastern prairie fringed orchid, *Platanthera leucophaea* (Nuttall) Lindley. Am Midl Nat 145:168–175

Zimmer K, Hynson NA, Gebauer G, Allen EB, Allen MF, Read DJ (2007) Wide geographical and ecological distribution of nitrogen and carbon gains from fungi in pyroloids and monotropoids (Ericaceae) and in orchids. New Phytol 175:166–175

Zimmer K, Meyer C, Gebauer G (2008) The ectomycorrhizal specialist orchid *Corallorhiza trifida* is a partial myco-heterotroph. New Phytol 178:395–400

Zuccaro A, Lahrmann U, Güldener U, Langen G, Pfiffi S, Biedenkopf D, Wong P, Samans B, Grimm C, Basiewicz M, Murat C, Martin F, Kogel K-H (2011) Endophytic life strategies decoded by genome and transcriptome analyses of the mutualistic root symbiont *Piriformospora indica*. PLoS Pathog 7:e1002290

Glossary

Achlorophyllous Lacking chlorophyll.

Allopatry Having geographically nonoverlapping distributions.

Arbuscule A finely branched structure produced by arbuscular mycorrhizal fungi inside the root cells of their host plants.

Arbuscular mycorrhizas Widespread type of mycorrhizal interactions involving fungi of the phylum Glomeromycota and diverse plants. The fungal hyphae grow into the inner cortex of the plant root and form finely branched exchange structures called "arbuscules."

Ascomycota Phylum of the kingdom Fungi characterized by a microscopic sexual reproduction structure called ascus.

Autotrophy Nutritional strategy in which organic compounds are synthesized from inorganic sources, for example by photosynthesis.

Basidiomycota Phylum of the kingdom Fungi of which the sexual reproduction is accomplished in club-shaped cells called "basidia," which bear external spores.

Chloroplast Cell organelle found in plants and algae in which chlorophylls are contained. Photosynthesis occurs here.

Chlorophyll The green pigment of plant cells, which is the receptor of light energy in photosynthesis. Also found in algae and photosynthetic bacteria.

Coevolution Evolutionary process in which the genetic composition of one species (or group) changes in response to a genetic change in another.

Community All organisms occurring and interacting with each other in a common environment.

Convergent evolution Independent evolutionary development of similar structures in organisms that are not directly related.

Cospeciation Synchronous speciation in two or more interacting populations.

Disjunct distribution Distribution with large geographic gaps.

Ectomycorrhizas Symbiosis between various plants and fungi belonging to the Ascomycota or Basidiomycota. The fungal hyphae surround the root tip and grow between the epidermal cells of the plant root, but never enter the cell lumen.

Endemic Restricted to a certain area.

Endosperm Tissue in the seeds of most flowering plants that contains stored food. It is digested by the growing sporophyte.

Epiparasitism Interaction in which one parasitic species is parasitized by another. In the context of mycoheterotrophy epiparasitism defines the tripartite symbiosis between a parasitic lineage (mycoheterotroph), an intermediate host lineage (mycorrhizal fungus), and an ultimate host lineage (mycorrhizal autotroph).

Fruit body Reproductive structure of many fungi.

Full mycoheterotroph A plant that solely depends on fungal carbon during its entire life cycle.

Gametophyte The haploid (n), gamete-producing generation.

Glomeromycota Phylum of the kingdom Fungi of which the majority of taxa form arbuscular mycorrhizas with the roots or thalli of land plants.

V.S.F.T. Merckx (ed.), *Mycoheterotrophy: The Biology of Plants Living on Fungi*,
DOI 10.1007/978-1-4614-5209-6, © Springer Science+Business Media New York 2013

Haustorium The modified root of parasitic angiosperms capable of penetrating and absorbing materials from host tissues.

Heterotrophy Nutritional strategy in which organic compounds are obtained by feeding on organic materials.

Host An organism on which a parasite lives.

Hybridization The formation of offspring between two different species or varieties.

Hyphae Tubular filaments of a fungus.

Initial mycoheterotroph A plant that is fully dependent on associated fungi for its carbon supply during the early stages of its development.

Mixotrophy Nutritional strategy in which a mix of different sources of energy and carbon are used.

Mutualism A symbiosis between two lineages of organisms based on a balanced reciprocal exchange of resources.

Mycelium The mass of hyphae forming the body of a fungus.

Mycoheterotroph A plant capable of mycoheterotrophy during at least one stage of its development.

Mycoheterotrophy Nutritional strategy in which a plant obtains carbon from fungi.

Mycorrhiza Symbiosis between a plant and a fungus. This interaction is typically mutualistic, obligate, and based on an exchange of photosynthates for soil minerals.

Obligate symbiosis When symbiosis is necessary for completion of the life cycle of one or both interacting lineages.

Organic Referring to compounds formed by living organisms.

Parasite An organism that lives on or in an organism of a different species and derives nutrients from it through parasitism.

Parasitism Interaction between organisms of different species where one organism, the parasite, benefits at the expense of the other, the host.

Partial mycoheterotroph A plant that combines autotrophy and mycoheterotrophy to obtain carbon during at least one stage of its life cycle.

Partial mycoheterotrophy Nutritional strategy in which a plant obtains carbon simultaneously through autotrophy and mycoheterotrophy.

Partner choice Occurs when preference is due to organisms being capable of assessing the quality of potential partners.

Pathogen An organism that causes a disease.

Photosynthesis The synthesis of carbohydrates from carbon dioxide and water in the presence of chlorophyll by using light energy.

Phylogeny Evolutionary relationships among organisms.

Phylogenetic conservatism When host shifts or jumps are prevented and lead to phylogenetic tracking of the host lineage by the symbiont lineage.

Preference Disproportionate association between one lineage and a subset of its compatible symbiotic lineages in the absence of external constraints.

Saprotrophy Nutritional strategy in which organic compounds are synthesized from dead organic matter.

Specificity Limitation in the number of other lineages with which a particular lineage interacts.

Sporophyte The diploid ($2n$), spore-producing generation.

Symbiosis Intimate long-term interaction between two different lineages, or symbionts.

Symbiotic continuum The ecological or evolutionary space between exclusive parasitism of one symbiont and exclusive parasitism of the other symbiont, with reciprocal parasitism (i.e., mutualism) at the center.

Sympatry Having geographically overlapping distributions.

Index

V.S.F.T. Merckx (ed.), *Mycoheterotrophy: The Biology of Plants Living on Fungi*,
DOI 10.1007/978-1-4614-5209-6, © Springer Science+Business Media New York 2013

R

S

T

MIX
Papier aus verantwortungsvollen Quellen
Paper from responsible sources
FSC® C105338

If you have any concerns about our products,
you can contact us on
ProductSafety@springernature.com

In case Publisher is established outside the EU,
the EU authorized representative is:
Springer Nature Customer Service Center GmbH
Europaplatz 3, 69115 Heidelberg, Germany

Printed by Libri Plureos GmbH
in Hamburg, Germany